D1091064

Resource Allocation and Productivity in
National and International Agricultural Research

Resource Allocation and Productivity in National and International Agricultural Research

Thomas M. Arndt
Dana G. Dalrymple
and
Vernon W. Ruttan
Editors

UNIVERSITY OF MINNESOTA PRESS, MINNEAPOLIS

Published in Canada by Burns & MacEachern
Limited, Don Mills, Ontario

Library of Congress Catalog Card Number 76-44064

ISBN 0-8166-0805-9

Preface

Within the last decade significant steps have been taken to narrow the agricultural technology gap among countries. New international agricultural research centers have been established, and a number of developing countries have made substantial progress in strengthening their national agricultural research capacities. It is estimated that world expenditures on agricultural research (in constant 1971 U.S. dollars) have risen from approximately $1.3 billion in 1959 to $3.8 billion in 1974. In the less developed countries of Latin America, Africa, and Asia, the estimated increase in research expenditures for the same period was from $141 million to $957 million.

The creation of two international agricultural research institutes in the early 1960s was an important factor in this surge of research investment. The establishment of the International Rice Research Institute (IRRI) in the Philippines in 1960 and the International Maize and Wheat Improvement Center (CIMMYT) in Mexico in 1966 signaled a wider contemporary refocusing on research as an essential instrument of agricultural productivity change in the less developed countries (LDC's).

The ensuing story is well known. CIMMYT and IRRI developed short-strawed, fertilizer-responsive, high-yielding wheat and rice varieties which were rapidly adapted and adopted in parts of Asia, Africa, and Latin America and which produced an upsurge in grain production popularly known as the "green revolution." By 1975, nine international research institutes and two other international programs were either in operation or in the process of being established in the LDC's (see Table 11-1).

The view that effective research capacity in developing countries is a primary means to raising agricultural productivity is now widely shared. The World Bank, the United States Agency for International Development (USAID), and other donors have begun to give agricultural research a higher priority for assistance. Investment in agricultural research in developing countries has grown rapidly in recent years. As the agricultural research system has continued to expand, research productivity and research resource allocation have become important issues for development planners and science managers.

In order to examine these issues, an international conference was held at Airlie House, Virginia, in January 1975. The conference had two main objectives: to examine recent evidence on the returns to investment in national and international agricultural research systems; and to explore the relevance of social and economic factors for the organization and management of national and international research systems. Technical issues related to the measurement of research productivity, the planning of research programs, and the management of research systems were also discussed.

The conference drew together over fifty natural scientists, social scientists, and administrators from national and international research agencies, some of whom wore more than one hat. The format of the conference was somewhat unusual. Individual papers were not read but rather circulated before the meeting and orally summarized at the conference by preselected discussants, several of whose commentaries proved so stimulating and original that they are included in this volume (chapters 26 and 27). Most of the conference, therefore, was devoted to what turned out to be a rich and vigorous discussion.

Following the conference, contributors were asked to review and revise their papers in the light of the discussions. The revised papers were then edited and returned to the authors for their further review, a painstaking process which was completed in November 1975. As they appear in this book, the twenty-eight papers have been organized into six sections. They are preceded by an introduction (by Thomas M. Arndt and Vernon W. Ruttan) which highlights the main issues discussed at the conference.

The first section of this book is devoted to a series of studies of the *productivity of national research systems* in both developed and less developed countries. There are two papers on the organization and productivity of research systems in developed countries: the first (by Yujiro Hayami and Masakatsu Akino) describes the national-prefectural system in Japan; the second (by Willis L. Peterson and Joseph C. Fitzharris) discusses the federal-state system in the United States. There are also two papers on returns to investment in agricultural research in developing countries, one drawing upon Colombian experience (by Reed Hertford, Jorge Ardila, Andrés Rocha, and Carlos Trujillo) and one focusing upon India (by A. S. Kahlon, P. N. Saxena, H. K. Bal, and Dayanath Jha). The final paper in this section (by Reed Hertford and

Andrew Schmitz), prepared after the conference, reviews some of the theoretical foundations of, and empirical considerations involved in, the evaluation of returns to research.

The second section includes three papers on the *productivity of international research systems*. The first paper (by Dana G. Dalrymple) documents the impact of the CIMMYT and IRRI varietal development programs on wheat and rice production in Asia. A second paper (by Robert E. Evenson) reviews the history of cycles in research productivity and in international diffusion patterns for three commodities: sugarcane, wheat, and rice. A third paper (also by Evenson) presents measures of the rates of return to the resources invested in research at IRRI and CIMMYT. In a fourth paper (by Yoav Kislev) an attempt is made to develop a theoretical model of applied research consistent with the historical experience reviewed in the earlier papers in the first two sections.

The third section is devoted to the *organization and development of the international agricultural research institute system*. The first chapter (by J. G. Crawford) traces the development of a system for organizing, funding, and managing the new institutes, spearheaded by the Ford and Rockefeller foundations and now guided by the Consultative Group on International Agricultural Research. The programs that have been developed to achieve closer articulation of the research programs of the international and national systems are outlined by the director of IRRI (Nyle C. Brady) and the director of CIMMYT (Haldore Hanson). The problem of establishing effective working relationships between the international and national agricultural research systems is then reviewed (by Sterling Wortman). The final paper in this section (by Burton E. Swanson) presents a comparison of the impact of the IRRI and CIMMYT training programs on the career patterns and effectiveness of young scientists who have studied at these two institutes.

The fourth section is devoted to issues bearing directly on the *organization and management of agricultural research systems*. The first paper (by Albert H. Moseman) outlines the evolution of coordinated national research projects for improving food crop production. Two papers focus on the problem of reorganizing and reforming national research systems: in the United Kingdom (by Tilo L. V. Ulbricht) and in Brazil (by José Pastore and Eliseu R. A. Alves). Next, the contribution of private sector international agricultural research is described (by S. M. Sehgal), based on the experience of Pioneer Hi-Bred International. A systems approach to research resource allocation is then outlined and evaluated (by Per Pinstrup-Andersen and David Franklin). The final chapter in this section (by C. Richard Shumway) presents a review and evaluation of the literature on formal models and methods for allocating resources in research.

The fifth section examines the role of *economic and social factors in re-*

search resource allocations. The first paper (by Martin E. Abel and Delane E. Welsch) outlines a theoretical model for exploring the relative effects of environmental constraints and commodity mix on research resource allocation. The second paper (by John W. Mellor) is devoted to an empirical evaluation of the effects of efforts to relate research resource allocation to alternative goals such as labor absorption. In a third paper (by J. P. Ramalho de Castro and G. Edward Schuh) the use of an economic model to establish research priorities, based on Brazilian data, is illustrated. The role of resource endowments and relative prices in inducing the choice of alternative paths of technical change – that is, labor saving versus land saving – is documented (by Hans P. Binswanger). The final paper in this section (by Alain de Janvry) presents a theoretical model of dialectical interaction between technical and institutional change in agriculture which suggests important directions for future research.

The final section is devoted to discussion of the research strategy and management issues that will affect *the future of the international research system* and the productivity of national research systems. The first paper (by A. T. Mosher) is devoted to a discussion of unresolved issues in the evaluation of the international system. A second essay (by Theodore W. Schultz) focuses on the role of economic policy in influencing the prospects for gains from agricultural research. In the final paper (by J. G. Crawford) the policies and problems facing the Consultative Group on International Agricultural Research and its Technical Advisory Committee in their efforts to strengthen national and international research are reviewed.

The papers included in this volume cannot, of course, reflect fully the excitement and challenge of the Airlie House conference. Particularly significant contributions to the discussions were made also by Richard Baldwin of Cargill Inc.; Joel Bernstein of the U.S. Agency for International Development; John K. Coulter of the Consultative Group on International Agricultural Research; George Darnell, James M. Fransen, Raj Krishna, and Montague Yudelman of the World Bank; M. McDonald Dow of the National Academy of Sciences; Walter L. Fishel of the U.S. Department of Agriculture; Lowell S. Hardin of the Ford Foundation; W. David Hopper of the International Development Research Centre; Richard Nelson of Yale University; Peter Oram of the Food and Agriculture Organization of the United Nations; S. J. Webster of the Ministry of Overseas Development of the United Kingdom; A. M. Weisblat of the Agricultural Development Council, Inc.; and F. R. Wittnebert of the Parker Pen Co.

The Airlie House conference was given considerable impetus and background by the materials presented and discussed at an earlier symposium on Resource Allocation in Agricultural Research held at the University of Minnesota in February 1969. The results of the Minnesota conference, which focused on many of the same issues considered at Airlie House but from a domestic (United States) viewpoint, were published in a book edited by Walter L. Fishel and entitled *Resource Allocation in Agricultural Research* (Minneapolis: University of Minnesota Press, 1971).

Both the Minnesota and Airlie House conferences were oriented to formal research carried out at nonprofit institutions and supported primarily with public or philanthropic funds and, to a lesser extent, to research conducted by private firms. It is recognized, however, that significant technical improvements may also result from informal research by farmers themselves or by small, local firms. Simple selection of improved varieties of seeds has traditionally been an important source of varietal development and yield increase. In the field of mechanical technology, simple but effective mechanical devices invented by farmers or mechanics, such as small irrigation and cultivation equipment, continue to provide many of the models for equipment engineered, manufactured, and marketed by larger firms. The work of both national and international research programs in many cases builds on this indigenous technology. Thus while little is said about informal research in this volume, its potential importance should not go unrecognized.

The structure of public agricultural research differs somewhat between the developed and the less developed nations. In the developed countries, research is conducted both by ministries (or departments) of agriculture and by colleges and universities. In the developing nations, research has traditionally been much more strongly concentrated in specialized arms of the ministries of agriculture or in autonomous, commodity-oriented research institutes. Colleges of agriculture in developing countries have usually concentrated on teaching, although in some cases they have done significant research (the popular rice variety C4-63, for instance, was developed by the College of Agriculture at the University of the Philippines) and may well play a larger role in the future. A paper on "Articulation of the International System with Other Regional Organizations: Ministry and University," by Jose Drilon was originally scheduled for the conference, but unfortunately Dr. Drilon, for reasons beyond his control, was unable to prepare the paper or to participate in the discussions.

We would also like to emphasize that, while this volume focuses on the importance of improved technology in the process of agricultural development, we are quite aware that it is not the only factor. A host of other forces – social, institutional, economic, and technical – both influence the adoption of technology and set the stage for the complex of other changes which must take place in the process of agricultural and economic development. Although improved technology may be the key factor in some societies, in others it may not. Improved technology alone is clearly not tantamount to agricultural development. But if the proper forms of technology can be efficiently generated through research, the development process can often be facilitated. It is our hope that this volume will contribute to a better understanding of the role of research in the process of technical change and of the role of technical change in agricultural development.

The conduct of the Airlie House conference and the preparation of this volume have been possible only through the help of many individuals and several institutions. The conference planning committee consisted of Robert E. Evenson, Walter L. Fishel, and Vernon W. Ruttan. The conference was sponsored by the Agricultural Development Council under its Research and Training Network Program, a program which, in turn, is funded under a contract with the United States Agency for International Development. Additional support for international participation in the conference was provided by the World Bank.

Turning a set of conference papers by an international cast into a book is, as we learned, far from a quick or simple job. Our task was facilitated by the cooperation and forbearance of the individual authors, most of whom found their carefully revised papers further edited and trimmed. Virginia O. Locke played a key role in the copy editing; her professional skills and contributions were of great assistance. We would also like to express our appreciation to Willis L. Peterson for his review of technical portions of the final page proofs and to Raymond D. Vlasin for his very helpful review and evaluation of a preliminary draft of the conference proceedings manuscript.

Thomas M. Arndt
United States Agency for International Development
Colombo, Sri Lanka

Dana G. Dalrymple
United States Department of Agriculture and United States Agency for International Development
Washington, D.C.

Vernon W. Ruttan
The Agricultural Development Council, Inc.
New York and Singapore

January 1976

Contents

Organization and Development of the International Institute System

Organization and Management of Agricultural Research Systems

Economic and Social Factors in Research Resource Allocation

The Future of the International Research System

List of Figures

List of Tables

List of Contributors

MARTIN E. ABEL, Professor, Department of Agricultural Economics; Director, Economic Development Center, University of Minnesota, St. Paul, Minnesota

MASAKATSU AKINO, Visiting Research Associate, East Asian Research Center, Harvard University, Cambridge, Massachusetts; Associate Professor, Department of Agricultural Economics, University of Tokyo, Tokyo, Japan

ELISEU R. A. ALVES, Director, Empresa Brasileira de Pesquisa Agropecuaria (EMBRAPA), Ministerio da Agricultura, Brasilia, Brazil

JORGE ARDILA, Economist, Instituto Colombiano Agropecuario (ICA), Bogotá, Colombia

THOMAS M. ARNDT, The Representative of the United States Agency for International Development (USAID), Colombo, Sri Lanka; formerly Special Assistant, Bureau for Technical Assistance, USAID, Washington, D.C.

H. K. BAL, Ph.D. Candidate in Statistics, Punjab Agricultural University, Ludhiana, India

HANS P. BINSWANGER, Associate, The Agricultural Development Council, New York City; Agricultural Economist, International Crops Research Institute for the Semi-Arid Tropics (ICRISAT), Hyderabad, India; Assistant Professor, Department of Agricultural and Applied Economics, University of Minnesota, St. Paul, Minnesota

NYLE C. BRADY, Director, The International Rice Research Institute (IRRI), Los Baños, Philippines

J. P. RAMALHO de CASTRO, Adviser to the Minister of Agriculture, Federal Ministry of Agriculture, Brasilia, Brazil

JOHN G. CRAWFORD, Chairman, Technical Advisory Committee, Consultative Group on International Agricultural Research, Washington, D.C.

DANA G. DALRYMPLE, Agricultural Economist, United States Department of Agriculture (USDA) and United States Agency for International Development (USAID), Washington, D.C.

ALAIN de JANVRY, Associate Professor, Department of Agricultural Economics, University of California, Berkeley, California

ROBERT E. EVENSON, Associate, The Agricultural Development Council, New York City; Visiting Professor, Department of Agricultural Economics, University of the Philippines, Los Baños, Philippines

JOSEPH C. FITZHARRIS, Assistant Professor, College of St. Thomas, St. Paul, Minnesota

DAVID FRANKLIN, Systems Engineer, Centro Internacional de Agricultura Tropical (CIAT), Cali, Colombia

HALDORE HANSON, Director General, International Maize and Wheat Improvement Center (CIMMYT), Mexico, D.F.

YUJIRO HAYAMI, Agricultural Economist, The International Rice Research Institute (IRRI), Los Baños, Philippines; Professor, Faculty of Economics, Tokyo Metropolitan University, Tokyo, Japan

REED HERTFORD, The Ford Foundation, Bogotá, Colombia; Research Associate, Department of Agricultural Economics, University of California, Berkeley, California

DAYANATH JHA, Senior Economist, Indian Agricultural Research Institute, New Delhi, India

A. S. KAHLON, Dean, College of Basic Sciences and Humanities, Punjab Agricultural University, Ludhiana, India

YOAV KISLEV, Lecturer in Agricultural Economics, Faculty of Agriculture, The Hebrew University of Jerusalem, Rehovot, Israel

JOHN W. MELLOR, Professor, Department of Agricultural Economics, Cornell University, Ithaca, New York

ALBERT H. MOSEMAN, Consultant, Ridgefield, Connecticut

A. T. MOSHER, Associate, The Agricultural Development Council, New York City; Visiting Professor, Faculty of Agriculture, University of Sri Lanka, Peradeniya, Sri Lanka

JOSÉ PASTORE, Professor, Instituto de Pesquisas Economicas, Universidade de São Paulo, São Paulo, Brazil

WILLIS L. PETERSON, Professor, Department of Agricultural and Applied Economics, University of Minnesota, St. Paul, Minnesota

PER PINSTRUP-ANDERSEN, Director, Agro-Economic Division, International Fertilizer Development Center, Florence, Alabama; formerly Leader, Economics Unit, Centro Internacional de Agricultura Tropical (CIAT), Cali, Colombia

ANDRÉS ROCHA, Dean, Division of Social Sciences, Universidad de Tolima, Ibagué, Colombia

VERNON W. RUTTAN, President, The Agricultural Development Council, New York and Singapore

P. N. SAXENA, Assistant Director General, Indian Council of Agricultural Research, New Delhi, India

ANDREW SCHMITZ, Associate Professor, Department of Agricultural Economics, University of California, Berkeley, California

G. EDWARD SCHUH, Staff Economist, Council of Economic Advisors, Executive Office of the President, Washington, D.C.; Professor, Department of Agricultural Economics, Purdue University, Lafayette, Indiana

THEODORE W. SCHULTZ, Emeritus Professor, Department of Economics, University of Chicago, Chicago, Illinois

S. M. SEHGAL, President, Overseas Seed Division, Pioneer Hi-Bred International, Des Moines, Iowa

C. RICHARD SHUMWAY, Associate Professor, Department of Agricultural Economics, Texas A. & M. University, College Station, Texas

BURTON E. SWANSON, Assistant Professor of International Agricultural Education, Department of Vocational and Technical Education, University of Illinois, Urbana, Illinois

CARLOS TRUJILLO, Economist, Centro Internacional de Agricultura Tropical (CIAT), Cali, Colombia

TILO L. V. ULBRICHT, Head of Planning Section, Agricultural Research Council, London, England

DELANE E. WELSCH, Visiting Professor of Agricultural Economics, Kasetsart and Thammasat Universities, Bangkok, Thailand; Agricultural Economist, The Rockefeller Foundation, Bangkok; Professor, Department of Agricultural and Applied Economics, University of Minnesota, St. Paul, Minnesota

STERLING WORTMAN, Vice President, The Rockefeller Foundation, New York City

Resource Allocation and Productivity in
National and International Agricultural Research

1

Valuing the Productivity of Agricultural Research: Problems and Issues

Thomas M. Arndt and Vernon W. Ruttan

The capacity to develop technology consistent with physical and cultural endowments is the single most important variable accounting for differences in agricultural productivity among nations. Yet the process by which this capacity to create and diffuse technical innovations in agriculture is realized has, until recently, received relatively little attention from agricultural or social scientists.

The Airlie House conference was a milestone in effective collaboration between agricultural and social scientists in analyzing the sources and the effects of agricultural research. This interchange had been evolving hesitantly for the past few years, and the success of the conference suggests that it may grow at an accelerated pace as we add to our knowledge of the organization, management, and productivity of agricultural research in developing and developed countries.[1]

This introductory chapter will attempt to illuminate the major issues that activated the dialogue at the conference. These issues are organized under six headings: (a) the productivity of agricultural research; (b) the demand for research and technical change; (c) the generation and diffusion of agricultural technology; (d) the productivity and potential of the international agricultural research institutes; (e) the organization and management of agricultural research; and (f) the improvement of research decision-making. Our primary objective in considering these issues will be to achieve a synthesis of the views of conference participants as expressed both during our discussions and in the subsequent chapters of this book. The concluding section of this chapter will outline some suggestions for further research.

The Productivity of Agricultural Research

The significance of any technical change is that it permits some substitution of a less expensive and more abundant resource—knowledge—for more expensive and often scarce resources—land, water, and the like. In short, technical change releases the constraints imposed upon growth by inelastic resource supplies. Research increases agricultural productivity in several ways—by raising returns to factors of production through lowering costs or increasing output, by improving product quality or introducing new products, and by reducing the cultivator's vulnerability to forces beyond his control. In recent years there has been a proliferation of studies which indicate that returns to a great deal of investment in agricultural research have been two to three times higher than returns to other agricultural investment. Data presented in several chapters of this book (2, 3, 4, 5, 7, 8, and 9) support this suggestion. These data, together with the findings of several other studies available to us, are summarized in Tables 1-1 and 1-2.

Both the theory and the methods on which rate-of-return estimates for agricultural research are based were subject to careful scrutiny at the conference. Webster pointed out that many studies have arrayed gross benefits from research against only direct costs, omitting or reporting only a part of the costs of research implementation. If this deficiency were corrected, Webster maintained, estimated returns would be more comparable to those realized from conventional development projects (for which a 15-20 percent internal rate of return is considered good). On the other hand, it has been argued that the benefits from research are conservatively stated and that indirect effects, such as spillover benefits beyond the country or region originating the research, are not fully captured by the existing data.

Evenson, concurring with the view that exaggerated returns have been reported in some studies, argues for more careful attention to the theory and method of rate-of-return estimates by both producers and consumers of such estimates (see chapter 9). In general, however, the sources-of-growth studies should be less subject to accounting errors than are direct cost-benefit studies.

In support of a somewhat different point, Ulbricht contended that estimated rates of return assume the accuracy of estimates of benefits while in his judgment, assessments of benefits are highly subjective. Furthermore, he pointed out, relatively few studies, other than Schmitz and Seckler's investigations of tomato harvesting in California (see Table 1-1) and Hayami and Akino's research on rice reported in this volume (chapter 2), have taken the

Table 1-1. Summary of Direct Cost-Benefit Type Studies of Agricultural Research Productivity

Study	Country	Commodity	Time Period	Annual Internal Rate of Return (%)
Griliches (1958)	U.S.A.	Hybrid corn	1940-55	35-40
Griliches (1958)	U.S.A.	Hybrid sorghum	1940-57	20
Peterson (1966)	U.S.A.	Poultry	1915-60	21-25
Evenson (1969)	South Africa	Sugarcane	1945-62	40
Ardito Barletta (1970)	Mexico	Wheat	1943-63	90
Ardito Barletta (1970)	Mexico	Maize	1943-63	35
Ayer (1970)	Brazil	Cotton	1924-67	77+
Schmitz & Seckler (1970)	U.S.A.	Tomato harvester	1958-69	
		With no compensation to displaced workers		37-46
		Assuming compensation of displaced workers for 50% of earnings loss		16-28
Hines (1972)	Peru	Maize	1954-67	35-40[a] 50-55[b]
Hayami & Akino (1975)[c]	Japan	Rice	1915-50	25-27
Hayami & Akino (1975)[c]	Japan	Rice	1930-61	73-75
Hertford, Ardila, Rocha, & Trujillo (1975)[c]	Colombia	Rice	1957-72	60-82
	Colombia	Soybeans	1960-71	79-96
	Colombia	Wheat	1953-73	11-12
	Colombia	Cotton	1953-72	None
Peterson & Fitzharris (1975)[c]	U.S.A.	Aggregate	1937-42	50
			1947-52	51
			1957-62	49
			1967-72	34

[a] Returns to maize research only.

[b] Returns to maize research plus cultivation "package."

[c] From papers presented at Conference on Resource Allocation and Productivity in National and International Agricultural Research, Agricultural Development Council, Research and Training Network Program, Airlie House, Virginia, January 26-29, 1975, and which appear as chapter 2 (Hayami and Akino), chapter 3 (Peterson and Fitzharris), and chapter 4 (Hertford et al.) in the present volume.

Table 1-2. Summary of Selected Sources-of-Growth Type Studies of Agricultural Research Productivity

Study	Country	Commodity	Time Period	Annual Internal Rate of Return (%)
Tang (1963)	Japan	Aggregate	1880-1938	35
Griliches (1964)	U.S.A.	Aggregate	1949-59	35-40
Latimer (1964)	U.S.A.	Aggregate	1949-59	Not significant
Peterson (1966)	U.S.A.	Poultry	1915-60	21
Evenson (1968)	U.S.A.	Aggregate	1949-59	47
Evenson (1969)	South Africa	Sugarcane	1945-58	40
Evenson (1969)	Australia	Sugarcane	1945-58	50
Evenson (1969)	India	Sugarcane	1945-58	60
Ardito Barletta (1970)	Mexico	Crops	1943-63	45-93
Evenson & Jha (1973)	India	Aggregate	1953-71	40
Kahlon, Saxena, Bal, & Jha (1975)[a]	India	Aggregate	1960/61-1972/73	63

[a] From paper presented at Conference on Resource Allocation and Productivity in National and International Agricultural Research, Agricultural Development Council, Research and Training Network Program, Airlie House, Virginia, January 26-29, 1975, and which appears as chapter 5 in the present volume.

Sources for Tables 1-1 and 1-2: The estimates presented at the Conference on Resource Allocation and Productivity in National and International Agricultural Research are identified by an asterisk. The other estimates have been summarized by James K. Boyce and Robert E. Evenson, *National and International Agricultural Research and Extension Programs*. New York: Agricultural Development Council, 1975. The sources of the individual estimates are as follows:

Ardito Barletta, N. "Costs and Social Benefits of Agricultural Research in Mexico." Ph.D. dissertation. Chicago: University of Chicago, 1970.

Ayer, H. "The Costs, Returns and Effects of Agricultural Research in a Developing Country: The Case of Cotton Seed Research in São Paulo, Brazil." Ph.D. dissertation. Lafayette: Purdue University, 1970.

Evenson, R. "International Transmission of Technology in Sugarcane Production." Mimeographed. New Haven: Yale University, 1969.

Evenson, R. "The Contribution of Agricultural Research and Extension to Agricultural Production." Ph.D. dissertation. Chicago: University of Chicago, 1968.

Evenson, R., and D. Jha. "The Contribution of Agricultural Research Systems to Agricultural Production in India." *Indian Journal of Agricultural Economics*, 28 (1973), 212-230.

Griliches, Z. "Research Costs and Social Returns: Hybrid Corn and Related Innovations," *Journal of Political Economy*, 66 (1958), 419-431.

distributional effects of technical change into account. Ulbricht stressed that investment in agricultural research in developing countries should be supported not on the basis of the high rates of return that have been estimated but simply because agricultural research has been an important factor leading to increases in agricultural productivity.

These and related arguments have been examined by Hertford and Schmitz (chapter 6), who point out that regardless of the methodology employed accurate estimation of the change in production attributable to research is the most critical step in any effort to measure research productivity. It is clear that few of the available studies are free of methodological or empirical problems. Nevertheless, the overall robustness of the return figures does not appear to be in doubt.

More than once in conference discussions the value for research planning of even the most precise historical estimates of research productivity was questioned. Nevertheless, as Kahlon pointed out with reference to India,

Griliches, Z. "Research Expenditures, Education and the Aggregate Agricultural Production Function," *American Economic Review*, 54:6 (December 1964), 961-974.

*Hayami, Y., and M. Akino. "Organization and Productivity of Agricultural Research Systems in Japan." ADC/RTN Conference on Resource Allocation and Productivity in National and International Agricultural Research, Airlie House, Virginia, January 26-29, 1975 (chapter 2 in the present volume).

*Hertford, R., J. Ardila, A. Rocha, and C. Trujillo. "Productivity of Agricultural Research in Colombia." ADC/RTN Conference on Resource Allocation and Productivity in National and International Agricultural Research, Airlie House, Virginia, January 26-29, 1975 (chapter 4 in the present volume).

Hines, J. "The Utilization of Research for Development: Two Case Studies in Rural Modernization and Agriculture in Peru." Ph.D. dissertation. Princeton: Princeton University, 1972.

*Kahlon, A. S., P. N. Saxena, H. K. Bal, and D. Jha. "Returns to Investment in Agricultural Research in India." ADC/RTN Conference on Allocation and Productivity in National and International Agricultural Research, Airlie House, Virginia, January 26-29, 1975 (chapter 5 in the present volume).

Latimer, R. "Some Economic Aspects of Agricultural Research and Extension in the U.S." Ph.D. dissertation. Lafayette: Purdue University, 1964.

Peterson, W. L. "Returns to Poultry Research in the United States," *Journal of Farm Economics*, 49 (August 1967), 656-669.

*Peterson, W. L., and J. C. Fitzharris. "The Organization and Productivity of the Federal-State Research System in the United States." ADC/RTN Conference on Resource Allocation in National and International Agricultural Research, Airlie House, Virginia, January 26-29, 1975 (chapter 3 in the present volume).

Schmitz, A., and D. Seckler. "Mechanized Agriculture and Social Welfare: The Case of the Tomato Harvester," *American Journal of Agricultural Economics*, 52:4 (November 1970), 569-577.

Tang, A. "Research and Education in Japanese Agricultural Development," *Economic Studies Quarterly*, 13 (February-May 1963): 27-41; 91-99.

there is continual pressure from political leaders for evidence of the productivity of public investment, including investment in agricultural research. And productivity estimates are a useful device for monitoring research program performance.

The studies under discussion, however, provide only partial information on the policy choices involved in research resource allocation. They explain neither the variability in returns to research (e.g., the disparate results of Colombian research, as reported in chapter 4) nor the many unsuccessful research investments around the world. They indicate little about the distribution of the benefits of research among various groups in society. And they do not fully explain the relationship of research investments at home to research done elsewhere. Collectively, though, the studies do serve as a point of departure for exploring these issues in greater depth.

The Demand for Research and Technical Change

The theory of induced innovation served as a launching point for attempts to develop a more complete understanding of the nature of the demand for technical change in agriculture. The 1971 study by Hayami and Ruttan dealing with technical change in agriculture in Japan and the United States indicated that both countries have experienced similar agricultural growth rates despite radically different factor endowments.[2] In Japan land was expensive and labor was cheap. In the United States labor was expensive and land was cheap. The ability of each country to introduce a series of technical innovations which utilized cheap factors while conserving expensive factors was a key source of productivity growth in their respective agricultures.

In both Japan and the United States factor endowments have provided the compass and much of the motive power for technical change in agriculture. Relative factor scarcities have been reflected in relative factor prices which in turn have induced a search for technical innovations to conserve scarce factors. In Japan this led largely to land-saving biological innovations; in the United States it led to labor-saving mechanical technology.

The effectiveness of the process through which technical progress is generated, along a path induced by relative factor scarcities and by final demand, is conditioned by many circumstances. These include the state of scientific knowledge, the capacity of industry to supply inputs and materials, the levels of technical and scientific skill embodied in people, the distortions in the market, and the tugs and pulls of social and political circumstances.

In its simplest form, the theory of induced innovation assumes that all technical innovations are equally possible. Binswanger presents evidence (see chapter 25) that technical change may be more easily produced in some directions (e.g., labor-saving technology) than in others. If his evidence is confirmed, it implies that there are "fundamental biases" in technical change which may offset or neutralize the inducement mechanisms which bring about the conservation of relatively scarce factors. Binswanger concludes that it may take massive changes in relative factor prices to alter the direction of technical change. Such changes have in fact taken place. The inducement mechanisms described by Hayami and Ruttan, though buffeted by counter-currents, have operated in those countries which successfully escaped from the low-productivity trap.

In an attempt to articulate the variety of forces which condition technical change, de Janvry presented the conference with a conceptual model of the process involved in the inducement and diffusion of technical innovations (chapter 26). De Janvry views technical change as occurring in a dialectical interaction with institutional change rather than as the essentially linear process proposed by Hayami and Ruttan.

The key to the de Janvry model is the payoff matrix, which describes the partitioning of the gains from research among particular interest groups in society—commercial farmers, landed elites, subsistence farmers, consumers—who derive income gains or losses from alternative public goods such as research (see Figure 26-1). The supply and demand for research is centered in the payoff matrix and is conditioned by the socioeconomic structure on the one hand and the political-administrative structure on the other. Each social group pressures the political-administrative structure for research depending on the particular payoff such group expects. The relative social power of different groups determines whether and how their demands are translated into the allocation of people and money for particular lines of research. The extent of basic scientific knowledge determines the area within which technical innovation is possible.

The resulting supply of research is filtered through the socioeconomic structure and produces specific payoffs for different social groups. In agricultural research, these payoffs are determined by (1) the physical characteristics of the innovation in terms of its ability to raise yield or reduce costs; (2) the extent of the diffusion of the innovation, which is conditioned by its suitability to varied physical and cultural environments; and (3) the prices of factors and products, which determine the relative profitability of particular technical innovations. High payoffs, of course, induce further demands for new research.

The study on Japan by Hayami and Akino (chapter 2) indicates how agricultural research can prosper where social and economic forces flow together to induce a clear demand for technical change. During the Meiji era when Japan was modernizing, several groups converged in their demand for agricultural research. Farmers sought land-augmenting technology, consumers sought lower food prices, industrial employers wanted low-priced wage goods to keep costs down and save foreign exchange, and governments sought higher land tax revenues. Several features of the social and political structure of Meiji Japan, particularly the breakdown of feudalism and the high degree of social organization at the rural level, were uniquely conducive to the generation and diffusion of agricultural technology.

Similarly, Peterson and Fitzharris's study of United States agricultural research (chapter 3) illustrates how the "more successful, wealthier farmers," abetted by a social structure which encouraged the organization of farm groups and a political structure which enabled them to press their demands on the body politic, succeeded in establishing a favorable environment for the emergence of a highly productive federal-state research system.

In most countries, particularly the developing countries, effective clientele groups capable of serving as an "agricultural research constituency" have not emerged, and the demand for technical change in agriculture is only latent.

The analysis of research in Colombia by Hertford and his associates (chapter 4) indicates that the concentration of rice, soybean, and cotton production in limited areas or in the hands of tightly organized groups was a major factor in inducing research and in effecting the adoption of research results. By contrast, the land tenure arrangements among Colombian wheat growers had a negative effect on the spread of new wheat technology and on the subsequent demand for technical change.

The Colombian study also reveals the importance of prices in inducing or hindering research. Rice research was stimulated by the 82 percent increase in the price of rice which took place in the three years following the imposition of import controls. On the other hand, incentives for wheat research were dampened for a number of years by the availability of PL 480 wheat. It is important to note, as Schultz pointed out in the discussion of this point, that the persistent underpricing of food grains by political authorities in developing countries is a force majeure constraining the demand for research and for the diffusion of its products.

The conference's exploration of the demand side of technical change ended with a trail of question marks leading to unexplored territory. There have been only a limited number of empirical studies in LDC's that describe how the demand for new technology derives from particular groups and how the payoffs are distributed. The studies on returns to research mentioned earlier are based on aggregate estimates of benefits. Research on how various input and output owners appropriate the surplus generated by agricultural research would be an important means to understanding how research is induced, why it takes the direction it does, and whether it will serve broad development goals.

There remains a persistent question: why, despite the evidence of high returns, has there been so little investment in agricultural research in developing countries? Does the answer lie primarily on the demand side, as some at the conference argued? If so, what is the nature, origin, and direction of this demand?[3]

The Generation and Diffusion of Agricultural Technology

Several recent econometric investigations, primarily those by Evenson and Kislev (chapters 8, 9, and 10), have made some major inroads into understanding how advances in agricultural technology are made and diffused.[4]

Applied agricultural research may be understood as a search for new technology within the limits of existing scientific knowledge. Basic knowledge establishes the boundaries within which innovation is possible. If basic knowledge is static, applied research is subject to the principle of diminishing returns and will eventually come to a halt as the cost of successive technical innovations within the existing knowledge boundary rises. Without an increase in basic knowledge, technical change will eventually stagnate as the marginal cost of innovations rises to meet marginal returns.

Advances in basic knowledge extend the frontiers of applied research and make it more productive by providing new opportunities for technical innovations. Kislev argues that the faster the advance of basic knowledge, the greater the productivity of applied research will be. The rate of technical progress thus reflects both the rate of growth in the supply of new knowledge, resulting from investment in basic or supporting research, and the rate of growth in the effective demand for technical change as reflected by investment in agricultural experiment station capacity.

Since basic knowledge does not expand continuously or smoothly, technical progress can be expected to move in cycles or spurts. A breakthrough typically leads to an initial rapid harvest of innovations, followed by a slowing down of innovative activity.

Evenson documents the existence of such spurts in the successive breakthroughs in the development of improved sugarcane varieties throughout the world, beginning in the late 1800s (chapter 8). Similarly, Hayami and Akino show that agricultural research in Meiji Japan, which was based on the development of technology from existing knowledge, was slowing down until it was revitalized by a turn to more basic research in the 1920s (chapter 2).

Given this characteristic of technical progress, the introduction of "miracle" rice and wheat by the International Rice Research Institute (IRRI) and the International Center for the Improvement of Maize and Wheat (CIMMYT) was not miraculous at all. These pioneering institutes filled a gap created by the delay of the developing countries in taking advantage of technical opportunities that were available to them through previous advances in scientific knowledge. The primary reason for this failure, Evenson asserts, was the low level of research investment in LDC's in the 1940s and 1950s and the consequent incapacity of these countries to capitalize on the stock of knowledge. The delay was particularly apparent in many former colonial countries where agricultural research capacity had been developed to facilitate the production of export commodities rather than of domestically consumed food commodities.

If technical innovation is defined as filling the gap between the technology in practice and the technology which is possible given existing knowledge, then innovations are achieved by well-trained scientists who know what is possible and who can design new technology to take advantage of it. This skill—which Swanson, in chapter 15, terms the skill of the "biological architect"—is what made CIMMYT and IRRI so successful.

Productive applied research in LDC's, Evenson argues, is strongly dependent on the availability in such countries of this type of high-order technical skill. The highly trained scientist has an understanding of science and the basic knowledge embodied in existing technology which he manipulates to create superior technology for the production conditions in his country. Evenson demonstrates that the availability of high-order research skills represents an important source of agricultural productivity growth in LDC's. Without these high-order skills, Evenson asserts, LDC's tend to engage in relatively unproductive low-level research, which often unnecessarily repeats work done elsewhere.

Evenson's studies of the international diffusion of sugarcane varieties and other commodities show direct as well as adaptive diffusion processes at work. He demonstrates that the rate of both processes depends on the availability and quality of indigenous research capabilities.

The availability of the capacity to do research on sugarcane made it possible to speed up the importation, testing, and release of sugarcane varieties generated elsewhere. Evenson reveals that in Australia, South Africa, and the Caribbean area this diffusion effect alone justified the countries' investment in research even without considering the benefits from the adaptive research which resulted from the new research capacity.

Hertford's studies of cotton research in Colombia showed a similar result. Initiation of cotton research there facilitated the importing and the testing of United States cotton varieties which yielded high returns, even though the Colombian research itself did not produce varieties superior to United States varieties. Research leading to the introduction or adaptation of new crops or varieties is often accomplished by the private sector too (see Sehgal, chapter 19).

Evenson concludes that countries without the capacity to do significant agricultural research also lack the capacity to benefit fully from the research of others in similar geoclimatic zones. His estimates of the magnitude of these spillover effects are shown in Table 9-5. His conclusions buttress the argument for investment in agricultural research in developing countries. His data imply, though they do not yet prove, that developing countries will need to emphasize the development not only of the capacity for adaptive or applied research but also of high-order conceptual-scientific skills if they are to take full advantage of the potential contribution of agricultural science to national development. In the future, according to Evenson, it will be primarily people with high-level conceptual skills who will break new ground and lead effective national research programs.

The Productivity and Potential of the International Agricultural Research Institutes

In the early 1970s the international agricultural research institutes accounted for about a tenth of 1 percent of world expenditure for agricultural research. Even in the developing countries they accounted for less than 5 percent of agricultural research expenditure. Yet, their impact has been great. Dalrymple estimates that the technology packages derived from the work of the institutes added 1 billion dollars in wheat and rice production in Asia alone during 1972-73 (chapter 7). These technology packages were the joint products of research at the institutes and of original and adaptive research within the LDC's.

In addition, the institutes have had substantial indirect consequences, which are not measurable. As Wortman points out, they have demonstrated the potential of science-based agriculture and thus have stimulated investment in agricultural research in LDC's. Through such example, the institutes have initiated trends in a number of countries toward problem-oriented, commodity-focused, multidisciplinary research.

It is important, however, to place the success of the institutes in perspective. As we have already noted, the fact that the level of investment in applied agricultural research in the LDC's was very low from about 1940 to 1960 provided an excellent opportunity for the new international institutes to convert existing scientific knowledge into technologies which were superior to those in use in the tropics. Furthermore, both CIMMYT and IRRI adopted research strategies designed to develop grain varieties which were usable under a relatively wide spectrum of environmental conditions.

Progress has slowed as the new varieties have spread into less favored lands. The productivity of the international institutes remains high, but it will be difficult to maintain the rate of return that was achieved from the initial investments. Evenson estimates that the "second generation" returns to wheat and rice research, though exceedingly high by conventional standards, have fallen below the "first generation" levels (Table 9-7).

Progress has not been uniform. The return on investment in corn research at the international centers (and predecessor institutions) has been realized more slowly than that for wheat and rice, although private research on corn has been relatively profitable. Moreover, some of the international institutes are in their early stages and have yet to make major contributions.

This does not mean that in the near future the international center will become just another type of research unit. The centers are at present uniquely structured for effective action. They have independent boards of trustees, organizational discipline, established pipelines to the financial resources of donor countries, and the ability to recruit skilled staff from all over the world. More importantly, the institutes have carved out central roles for themselves in the constellation of research institutions working on improved agricultural technology worldwide. They have access not only to scholarly capital but to genetic materials from around the world. They have established communication links with national research systems and with related production programs that focus on commodities on which the international centers are conducting research.

The centers' relationships with developing countries are described in the papers by Brady (chapter 12) and Hanson (chapter 13). The institutes have utilized international collaboration on research and training systems to achieve significant multiplier effects from limited resources. There are, for example, only thirteen wheat scientists at CIMMYT itself. It is this organizational innovation which may in the long run be judged the most outstanding accomplishment of the institutes and which may ensure their continuing relevance.

The established institutes are now vigorous adolescents. What sort of adulthood lies ahead? A rare path to maturity is taken by planning an adult role based on a careful assessment of one's strengths and limitations. Another allows one just to grow, responding appropriately to present circumstances and trusting in one's innate ability to bring one out all right in the end. Mosher sees the institutes as tending to follow the latter path (chapter 27). He notes that at present the institutes seem to be evolving out of research and training centers into research-based institutes of agricultural development. To their core research and training programs they have increasingly added commitments to strengthen the national research capabilities of LDC's. Moreover, some institutes are beginning to involve themselves in production programs in these countries.

These developments have come about because the institutes are conditioned to judge the success of their programs by the actual increases in commodity production in LDC's. This has given them a practical orientation and a sharp sense of purpose. As the spread of high yielding varieties has slowed, and as evidence mounts that farmers are not adopting the whole package of practices and that the yield increments achieved by LDC farmers are low relative to the increments obtained under experiment station conditions, the tendency for IRRI and CIMMYT to be concerned about LDC production programs has increased.

Mosher asked if these developments will dilute the centers' ability to apply high-level conceptual-scientific skills to research problems. Evenson's analysis shows that the productivity of scarce research skills such as those found at the institutes is higher than almost any other agricultural or nonagricultural investment available in developing countries. Evenson also notes that the institutes may need either to establish closer working relationships with other centers engaged in basic or supporting research or to turn themselves to more basic research in order to sustain their productivity. If this is true, Mosher asked, should not the institutes sharply limit their activities outside research and training?

On the other hand, the evidence on the international diffusion of technology supports the view that the institutes should focus on production as well as research and training programs. If a lack of capacity for indigenous research implies a lack of capacity to benefit from international research, then the institutes, the developing countries, and AID donors can expect high returns in both the short and the long term from the centers' outreach effort.

The important question, as Mosher points out, is how the centers will define their role in relation to other actors who are either on stage or in the wings. The former group is composed of LDC governments and traditional aid agencies. The latter includes representatives of the private sector, whose role in production programs and adaptive research may well expand, and various regional institutes—such as the West African Rice Development Association (WARDA) and the South East Asia Regional Center for Graduate Study and Research in Agriculture (SEARCA)—funded by the developing countries themselves and by aid consortia.

Although these questions were considered by the conference, they were not resolved. Representatives from the institutes did not think their involvement with outreach programs would dilute their effectiveness. Moreover, the institutes have yet to confront the budget restraints that would require them to make hard choices. The consensus was that the institutes probably should and will continue to evolve as research-based agricultural development institutes. However, some participants did question whether the institutes' acceptance of a broader charge might weaken their capacity to contribute in the area where their advantage is greatest relative to national institutions: the design of efficient technologies capable of bypassing many of the institutional constraints in production. Crawford, in particular, pointed out that the institutes could not, and should not, try to assume responsibility for changing economic policies that may limit the diffusion of institute technologies or methodologies in certain countries.

The Organization and Management of Agricultural Research

The United States and Japanese agricultural research systems have been more thoroughly studied than any others. The Hayami/Akino (chapter 2) and Peterson/Fitzharris (chapter 3) papers indicate that each country responded successfully to the needs of its farmers. The two nations shared certain attributes. Both, for example, evolved decentralized federal-state systems. The state (prefectural, in Japan) units were able to respond flexibly to changing local circumstances and to develop locally appropriate technologies even for micro-environments.

In each country the state units were backed up by national research systems. In 1926 Japan and the United States independently of each other reinforced their federal-state systems by introducing centrally orchestrated nationwide research programs on specific crops and problems. The initiation of these coordinated research programs coincided with a trend toward more basic research in both countries. The programs were effective in mobilizing scientific talent around specific problems without sacrificing the responsiveness to local problems which characterized the state, or prefectural, units.

The United States formally integrated research, extension, and education in its land grant colleges. In Japan a liaison between these three levels was maintained by a less formalized arrangement. Both countries had the advantage of well-organized groups of farmers, relatively equitable land distribution, high levels of education, growing industries, progressive government, and a sociopolitical structure which favored communication with farm groups.[5] Whether the Japanese and United States experiences will be modeled by the LDC's, however, is an open question.

The study by Pastore and Alves on the reform of Brazilian agricultural research argues that the decentralized model is not applicable to the extent that it implies relatively autonomous, multipurpose, locally responsible institutions such as the United States land grant colleges (see chapter 18). The social and economic circumstances, such as the presence of farmer organizations, that ensured that these institutions responded to farmers' needs are not present in most developing countries.

Pastore and Alves assert, in effect, that autonomous institutions work best in cohesive social structures. They imply that the unorganized, particularistic structures which characterize rural areas in developing countries favor the development of a centrally directed, aggressive research system. A "directed" system has a central planning unit which coordinates the activities of various subunits. An "aggressive" system proceeds in a logical, organized manner to seek information about the farm sector through social research or other means, to orient itself around explicit development and production goals, and to gear its research program to these ends.

Brazil, then, is moving toward a more centralized and coördinated national and state research system. Wortman argues that research should be organized on a multidisciplinary, commodity basis in order to achieve production targets established by governments (chapter 14). Moseman urges governments to take hold of their typically scattered research units and to institute centrally coordinated national programs around specific crops or problems (chapter 16).

Moseman's appeal is, in a sense, the reverse side of Evenson's in respect to the productivity of national and international research. Evenson, from an analytic standpoint, notes that decentralized, small-scale research stations are characteristic of countries with low skill levels. As skill levels rise, concentration occurs in order to take advantage of the economies of scale and higher productivity which the consolidation of high-order skills can achieve. Moseman, from a practitioner's standpoint, urges governments to begin the process of consolidation which will be needed to fructify the high-level skills that the country presumably will be developing.

The argument for coordinated or directed national research programs shifts our focus from the traditional debate over what type of research institute works best in an LDC (single crop versus multi-crop; land grant college versus government research institute) to the development of a national system. It implies that many types of institutions can be productive if their programs are coordinated effectively to work on specified national-regional research goals. But can a research system continue to be productive if research decision-making is highly centralized? For example, Kahlon mentioned that one effect of a productive agricultural research program (as in the Indian Punjab) is to create a local or regional research constituency.

This discussion of the central coordination and planning of research systems touched off some lively debate at the conference. Nelson raised the issue of whether formal criteria should be used in research planning and project selection, questioning what such criteria should be and how the weights could be derived. He reported that studies by the Rand Corporation of the economics of research and development had shown that estimates of the cost of research and development projects were typically inaccurate.

Moreover, Mansfield's studies of industrial research have indicated substantial error in predictions of the amount of time needed for project completion and for effect on output. Ulbricht reported that at a recent OECD conference on the relationship between agricultural research and socioeconomic policy there was a general consensus that attempts to develop weighting criteria generally result in spurious precision. He stressed that reliance on subjective judgment was inevitable but that such judgment could be refined by systematic analysis utilizing technical and economic information. Schultz argued strongly that research is an entrepreneurial activity whose success depends on such relatively rare personal qualities as creativity and insight. Thus the organizational task of research is to build structures where talented individuals, working independently and in teams, can exercise their creativity. He pointed out that the markets are effective transmitters of information about technological needs and that researchers usually read these signals pretty well. In sum the question is, as Nelson put it, do we want to bet on proposals or on people?

The counterargument was that coordination and planning need not be stultifying. On the contrary, they can enhance the effectiveness of individuals and organizations by clarifying goals, increasing the flow of information, and promoting teamwork. Since many LDC's have fewer qualified research scientists than are on staff at a single major United States experiment station, the question of competition among multiple centers arises primarily in larger countries. It was also argued that the entrepreneurial concept of research reflects a Western philosophy which does not provide appropriate guidelines for transitional societies where economic signals may be distorted and where particularism and communalism are still strong social values. There may be a tendency for scientists in developing countries, left to respond on their own, to direct their entrepreneurial talents toward the international scientific market from which rewards and emoluments flow. Or they may respond to highly limited demands, such as those from large landowners.

The discussion of this issue was characterized by Fishel as excessively ideological and inconclusive. He insisted that it is possible to quantify and communicate events that lie in the undiscovered future based on the experience of the past. Ideally the resulting measures should not be neat point estimates but probability functions which incorporate all the information about the future including the uncertainty involved.

Bernstein stressed that the discussion was hampered by a failure to differentiate between the control and rationalizing dimensions of research management. He pointed out, in discussion, that systems for determining the use of scarce and potentially high-yielding research resources can operate with minimal built-in controls. The discussion did underline the need for better understanding of the origin, nature, and direction of demand for agricultural research in developing countries. Such understanding would make it possible to consider whether to stimulate research productivity by altering the market rather than by striving for more comprehensive systems of planning.

The Improvement of Research Decision-making

There are several reasons why optimal allocation of research resources, though important, is extremely difficult. First, research resources in developing countries are scarce. A particularly severe constraint on research capacity in many countries is the scarcity of well-trained scientists. The lack of the organization necessary to permit the existing research capacity to function effectively is often an even more serious constraint. Because research is potentially very productive, the opportunity costs of bad decisions are high.

Second, different kinds of technical change brought about through research have unequal effects on a nation's economic and social goals. Monotheism in development planning, with its primary focus on growth, has been superseded by pantheism in recent years as other social goals have moved into the sanctuary. There are economic growth goals, such as increasing the net income of the agricultural sector or maximizing the contribution of agriculture to the economy as a whole; there are welfare goals, such as increasing employment and the income of labor employed in agriculture, reducing the real price of food for consumers, and improving health and nutrition; and there are equity goals, such as mitigating income inequalities and opening the benefits of growth to particular groups, such as small farmers. The choice of crops, regions, or disciplines for research affects these goals unequally.

Third, choice is further complicated by the element of uncertainty in the process and outcome of research. This uncertainty has two primary sources: (a) the time and resources required to attain stated technical objectives can only be estimated, and the risks of miscalculation are high; (b) benefits, once technical objectives have been attained, may or may not be appropriated. In agriculture, the decision whether or not to adopt new technology flowing from research is in the hands of an independent individual—the farmer—who maximizes his private welfare and acts in accordance with his own assessment of the risks involved.

Research may shift production functions, but many farmers will tend to operate at less than what appears to be the optimum level of production. How much below and for how long is imponderable. This depends on extension, input supplies, access to credit, economic policies, and other circumstances. It also depends on the accuracy with which the research planners have assessed the private welfare criteria of the farmer. Agricultural commodities are produced by farmers—not by planning commissions, research scientists, or extension workers.

Fourth, the marketplace is imperfect as a decision-making guide for research. Society places a value on research, but the marketplace may not reflect this value accurately. New knowledge produced by research enters the public domain, and the benefits of such research cannot ordinarily be captured by the individual or organization that bears the cost of producing it. This is also true of much agricultural technology produced by research (e.g., new seeds or cultural practices). Hence, private profit is often an inadequate incentive for research, particularly in industries such as agriculture that are characterized by many small producing units. The market undercompensates private innovation in agricultural research. Research decision-making receives indirect guides from the marketplace through factor and product prices, but little direct guidance.

Discussion of these uncertainties evoked two types of responses from the conference, neither of them mutually exclusive.

The first was a consensus that there are severe limits on our ability to make quantitative objective assessments of the value of particular kinds of research. This implies that choice must continue to be determined to a great extent by subjective judgment. The objective criteria available to guide research decision-makers through the uncertainty surrounding research decisions are limited. Use of conventional tools such as cost-benefit analysis is limited by the degree of precision with which the research scientist can estimate both the resource requirements and the output of a research project or program.

At the research project and program levels, the judgment involved is essentially scientific. At higher levels, the judgments are partly scientific and partly political. There was agreement among conference participants that the high rates of return from past research imply that the subjective judgment of knowledgeable scientists and science administrators should receive good marks. Given the right institutional and social setting, including efficient markets for inputs and commodities, scientists' judgments of technical constraints and opportunities for increasing production have led to effective research resource allocation decisions. Nevertheless, the second response to this consideration of uncertainties was that the tools of social science can and should be developed as guides to decision-making. The use of such tools becomes even more significant in countries characterized by the absence of efficient input and commodity markets.

On the micro-level, Pinstrup-Andersen described a methodology being developed at CIAT for determining research priorities in respect to a single commodity (see chapter 20). The method proceeds through logical stages:

(1) attempt to identify reasons for low productivity; (2) identify researchable problems which, when solved, will improve productivity; (3) estimate the impact on production of solving each of the problems; (4) estimate the probability of research success, the likelihood of the results being adopted, and the time required for solving the problem; and (5) estimate the impact of alternative research results on product supply, input demand, farm income, and farm size.

The methodology developed by CIAT and the other methods reviewed by Shumway provide research managers with vastly more information than do traditional informal methods (chapter 21). However, the methods also entail costs, in both time and trained people. Whether their margin of advantage over informal methods justifies their cost is not yet certain. One indication of their potential value is CIAT's report that several research agencies in Latin America have shown interest in their systems.

Attempts to introduce social goals, in addition to explicitly economic considerations, into agricultural research planning are relatively new. Data from Brazil were used by de Castro and Schuh to demonstrate a preliminary model for assessing national research priorities in the light of a country's factor endowments and socioeconomic goals (chapter 24). This model adopts the Hayami/Ruttan thesis that the task of research in promoting economically efficient growth is to introduce technical change which conserves relatively scarce factors of production. Using trends in relative factor prices in Brazil, de Castro and Schuh show that even in land-rich Brazil research efficiency has begun to require greater emphasis on land-augmenting technology. This includes, for example, soil research directed at opening problem lands to production and biological research focused on improving yields. The authors note that there is great regional variation within Brazil, however, and that in some parts of the country a labor constraint is emerging which calls for a different technical choice.

Knowing what research is consistent with relative factor scarcities does not answer the question of how to allocate research resources among commodities. In considering this issue, de Castro and Schuh pose the question of whether a nation wishes to favor the welfare of consumers or of producers. Hertford and Schmitz have demonstrated that whether the benefit of technical change in particular commodities redounds to consumers or producers hinges primarily on the relationship between the demand and supply elasticities for the commodity (chapter 6). Crops with low relative demand elasticities (e.g., food grains, beans, manioc) distribute their benefits primarily to consumers in the form of lower food prices. Lowering the price of food grain also releases wage-good constraints and permits the expansion of employment programs. However, as prices fall, acreage devoted to food grain may shift to other crops. Even labor-intensive technical change in crops with low demand elasticities can displace labor, moving it from that crop to other crops or to the nonfarm sector. Crops with high relative demand elasticities (e.g., cotton, sugarcane, export crops, or other crops where the producing region must accept the prices set in national or international markets) return most of the benefits of technical change to producers and thus can stimulate the demand for labor and increase rural incomes. If the goal of planners is to increase income and the employment of agricultural labor, emphasis in research should be given to crops with high relative price elasticities of demand. If the goal is to increase consumer welfare, research emphasis must be placed on food grains and other crops with relatively low demand elasticities.

Mellor has elaborated on these trade-offs in an analysis of research allocation and social goals based on Indian data (chapter 23). He argues for a sequence of agricultural policies to which research should be tied. First, stress must be placed on increasing yields of food grains in productive areas of the country. This increases both the supply of calories, which improves the health of the poor, and the supply of grain, which relaxes the wage-good constraint on employment growth. Relaxation of the wage-good constraint should be followed by employment programs to maintain demand. Otherwise, the incentive for increased production may diminish with falling prices.

As the wage-good constraint is released, the next strategy should be to promote the production of food grains in less productive regions and to expand the production of labor-intensive crops. These usually have relatively high demand elasticities. Promotion of labor-intensive crops needs to be supplemented with policies encouraging the demand for them through either the expansion of exports, which raises domestic incomes, or subsidies.

The implications of the Mellor and de Castro/Schuh papers are clarified by the de Janvry model of induced technological and institutional innovation (chapter 26). And several important points about relating research resource allocation to social goals emerge from the papers in the section on economic and social factors in research resource allocation.

In the first place, the models clearly indicate that the contribution of research and technical change to society's goals is dependent on other policies. Economic policies, such as the systematic reduction of product prices, can weaken the ability of research to contribute to growth or welfare goals. Distortion of input and product prices affects not only farmers' decisions regarding the use of inputs and the selection of commodities to produce, but also decisions regarding research priorities and hence the new technologies that will become available in the future.

By the same token, many social goals may be achieved more effectively through policies other than research, such as, for example, land tenure reform.

Conference participants generally agreed with Mellor's point that biological research is an inefficient instrument for solving problems involving the distribution of rural income. Too much reliance on such research may interfere with both the generation of improved technology and the achievement of desired goals. Yet, as Ulbricht insisted, the design of agricultural research strategies should not ignore the potential effects of technical change on the distribution of income. Crawford noted that it should be the policy of international institutes to make a range of technical options available and not to bias the technical innovation in a capital-intensive direction. It may also be important to provide technologies that are less "management intensive."

A second conclusion is that economic models of the type presented at the conference have a great deal of difficulty in accommodating multiple social goals. The definitions of welfare used in the models discussed above are relatively simple. A country's actual welfare is more complex. Development planning is still in an early stage of specifying what the various social goals of developing countries may be, let alone of understanding the relationships among such goals. However, clear articulation of long-term objectives is critical for making research choices. Consideration of research priorities must proceed from an understanding of the goals of a given country or region.

Another problem with such models is that, because of uncertainties about the production process for research, they are forced to make highly uncertain estimates of returns from alternative research investments. Marginal returns to certain lines of research may be increasing while others may be declining. This would significantly affect the flow of benefits and would condition research choices, but it is difficult to specify in advance. In addition, there are the problems, mentioned earlier, of predicting research cost functions and the distribution of benefits flowing from research.

These uncertainties are stubborn barriers to developing better tools for predicting the consequences of research choices on social welfare. Economic analysis at present yields only gross indications of the consequences from various choices. More data on the appropriation of research benefits and on the research cost function, in addition to further theoretical development and empirical testing of models, are needed to improve decision-making tools.

Conclusion: Areas for Research

Viewing the conference and the chapters in this volume in retrospect raises the proverbial question of whether to measure how far one has come or how far one has to go. From the latter perspective, it is evident that we remain some distance from a full understanding of agricultural research and from a fully convincing theory of technical change. The capacity of scientists to generate new technology outstrips our understanding of the social and economic implications of technical change and our ability to provide guidance for policy makers. Yet the essays on the future of the international system by Mosher (chapter 27), Schultz (chapter 28), and Crawford (chapter 29) provide some clear-cut guidelines for agricultural research resource allocation and for the organization of agricultural research systems. We do not attempt to repeat these authors' summary comments here. We do suggest, in the following paragraphs, some areas in which further research is clearly warranted.

1. A more precise understanding of the sources of demand for technical change in agriculture is needed. This includes further specification of the natural and institutional biases which condition the processes by which technical change is induced. The need for more careful analysis of the incidence of benefits from technical change is particularly important for understanding more about the origin and nature of demand for agricultural technology and for the further development of allocation tools. Study of research "failures," such as the relative lack of payoff to date from wheat research in Colombia, might also be instructive. Furthermore, there is a need to study how the political-bureaucratic process impinges on research allocation and conditions the demand for technology.

2. More analysis of research cost functions and the production process for research is needed. One area for further inquiry is signaled by Evenson's hypothesis on the relative productivity of various levels of research skill in conjunction with different levels of research organization. Thus far, efforts in this direction have had to rely on a relatively small amount of data. Although the results are consistent with economic principles regarding the productivity of scarce resources, they do counter some popular views about the relevance to LDC's of simple adaptive research and low-level skills.

3. There is need for further understanding of the national and international diffusion of agricultural technology and scientific knowledge. We need to know more about the relationships between different types of technology and advances in scientific knowledge. We need more information about the interaction between the productivity of investments in technology-based and/or science-based research within a country or region and similar investments in other agroclimatic regions.

4. There is also need for research on some elements of the process of technical change which were not well covered at the conference. One example is the relationship between technology policy and economic policy (particularly price policy). Another is the relationship between formal schooling (or literacy) and rates of technical change.

Even though all these areas still need further explanation, when we look back at the distance we have traveled, it is clear that advances have been made.

There have been great strides forward in our general understanding of how technical change is induced, in the modeling of the discovery process, and in mapping worldwide diffusion of technology and scientific knowledge.

There is solid evidence that investment in national and international research has been highly productive. The social returns to agricultural research have been high relative to the alternative investments available to most poor countries. It is clear that investment in agricultural research in developing countries by both national and international agencies should expand.

NOTES

1. For additional information on the organization, management, and productivity of agricultural research systems, see I. Arnon, *Organization and Administration of Agricultural Research* (Amsterdam, London, New York: Elsevier, 1968); I. Arnon, *The Planning and Programming of Agricultural Research* (Rome: FAO, 1975); James K. Boyce and Robert E. Evenson, *National and International Agricultural Research and Extension Programs* (New York: Agricultural Development Council, Inc., 1975); Albert H. Moseman, *Building Agricultural Research Systems in the Developing Nations* (New York: Agricultural Development Council, Inc., 1970).

2. Yujiro Hayami and Vernon W. Ruttan, *Agricultural Development: An International Perspective* (Baltimore: Johns Hopkins University Press, 1971).

3. These questions are not confined to agricultural research. There has been a low rate of investment in research by LDC's in all fields. For one interpretation of this phenomenon see the *Journal of Development Studies* (Special Issue on Science and Technology in Development), 9 (October 1972).

4. In addition to chapters 8, 9, and 10 in this book, see Robert E. Evenson and Yoav Kislev, *Agricultural Research and Productivity* (New Haven: Yale University Press, 1975).

5. In regions where these conditions did not prevail, as in the United States "Old South," agricultural research, agricultural productivity, and rural development lagged. For a description of the social and economic environment which conditioned rural development in the United States South see V. O. Key, Jr., *Southern Politics in State and Nation* (New York: Alfred Knopf/Vintage Books, 1949).

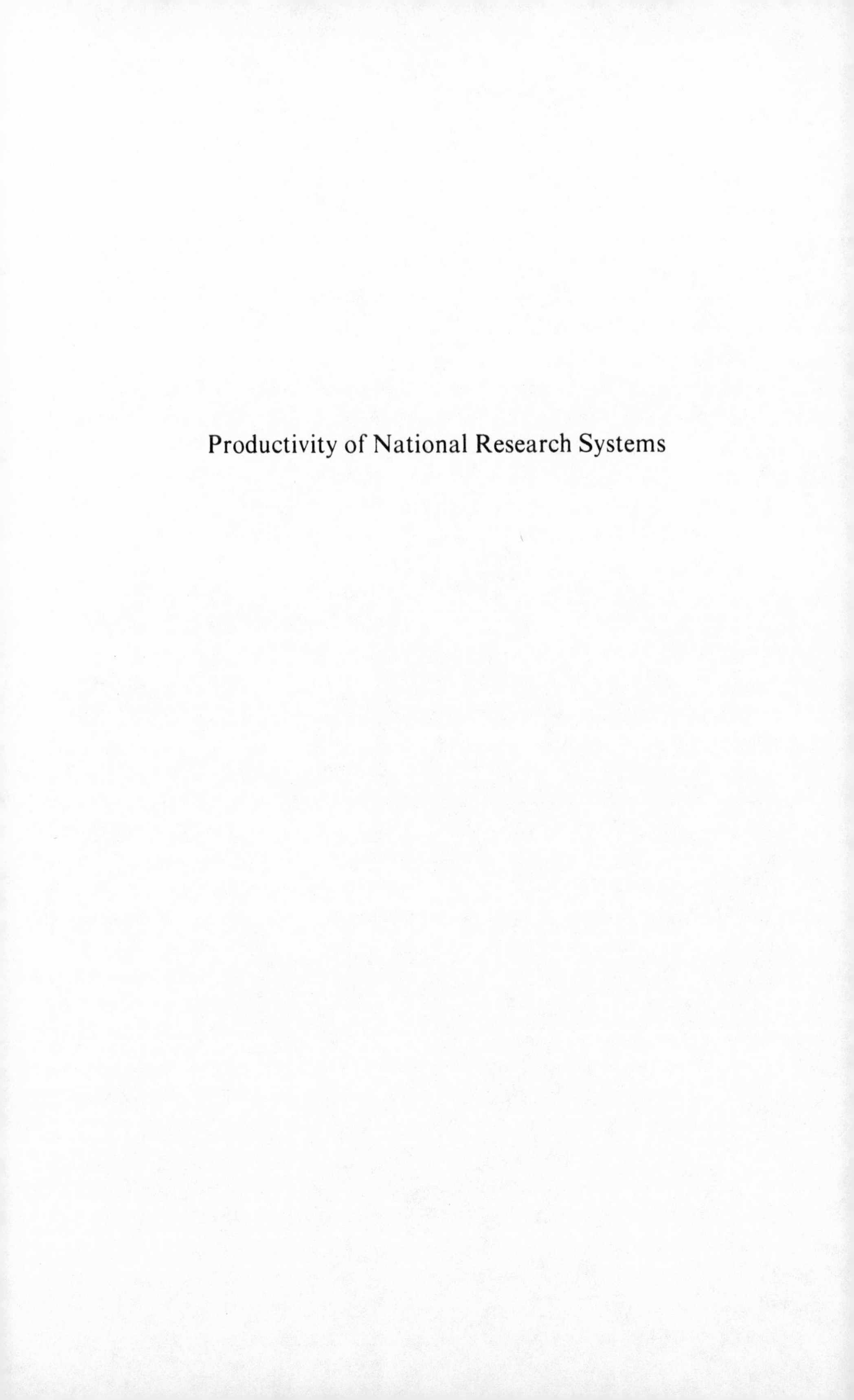

Productivity of National Research Systems

2

Organization and Productivity of Agricultural Research Systems in Japan[1]

Yujiro Hayami and Masakatsu Akino

Agricultural growth in Japan since the Meiji Restoration (1868) is unique in that it was achieved without disrupting Japan's traditional rural structure even though there were strong constraints on the land. Data indicate that Meiji Japan inherited a man/land ratio from the feudal Tokugawa period which was even more unfavorable than that in South and Southeast Asia today (Table 2-1). However, it appears that the unfavorable endowment of land relative to labor was compensated for by higher land productivity. Because of this, labor productivity in agriculture in Meiji Japan was at the level prevailing in Asia today.

From this initially low level, Japanese agriculture has grown to a level of agricultural output per male worker which exceeds Asian standards by a wide margin. A major component in this growth of output per worker was the increase in yield per unit of agricultural land. Meanwhile, there has been little change in rural structure; Japan remains a nation of small farms (Table 2-2).

This clearly indicates that technical innovations in Japanese agriculture were consistent with both its resource endowments and its rural organization. Central to these innovations was the development of biological technology in the form of fertilizer-responsive high-yielding varieties of major cereal crops, especially rice, complemented by improvements in land infrastructure.

The agricultural research system that developed the land-saving and yield-increasing biological technology was essential to the growth in output and

Table 2-1. Agricultural Productivity and Man/Land Ratios in Japan and Selected Asian Countries

Country	Agricultural Output per Male Farm Worker[a] (wheat units per worker)	Agricultural Output per Hectare of Agricultural Land[b] (wheat units per hectare)	Agricultural Land Area per Male Worker[c] (hectares per worker)
Japan			
1878-82	2.5	2.9	0.9
1898-1902	3.4	3.6	0.9
1933-37	7.1	5.5	1.3
1957-62	10.7	7.5	1.4
Asia, 1957-62			
Ceylon	3.9	2.9	1.3
India	2.1	1.1	1.9
Pakistan	2.4	n.a.	n.a.
Philippines	3.8	1.9	2.0

Source: Yujiro Hayami and Vernon W. Ruttan, *Agricultural Development: An International Perspective* (Baltimore and London: Johns Hopkins Press, 1971), pp. 70 and 328. Japan's time series revised to accord with presentation in Yujiro Hayami, in association with Masakatsu Akino, Masahiko Shintani, and Saburo Yamada, *A Century of Agricultural Growth in Japan: Its Relevance to Asian Development* (Minneapolis and Tokyo: University of Minnesota Press and University of Tokyo Press, 1975), pp. 220-230.

[a] Farm workers = economically active male population engaged in agriculture.

[b] Agricultural output in wheat units is the gross agricultural output net of intermediate products such as seed and feed. Individual products are aggregated by the price ratios to the price of wheat per one metric ton.

[c] Agricultural land area includes permanent pasture land.

productivity of Japanese agriculture. This chapter will review the current status and the historical evolution of agricultural research in Japan and will measure the productivity of investments in agricultural research.

Current System of Agricultural Research in Japan

In Japan, research is conducted mainly in public experiment stations under the auspices of national and prefectural governments. The mix of public and private agricultural research is shown in Table 2-3. Measuring both research expenditures and personnel, we find that about 59 percent of agricultural research is conducted at government institutions, nearly 38 percent at universities, and just over 3 percent in the private sector.

This structure contrasts markedly with industrial research in Japan (Table 2-3). It also differs from the United States, where roughly half of all agricultural research expenditure is made by private firms.[2]

Another distinctive aspect of the agricultural research system in Japan is the separation of agricultural experiment stations from education and exten-

Table 2-2. Distribution of Farms in Japan by Size of Cultivated Land Area

Year	Number of Farms, in thousands (% of total in parenthesis)						
	Less Than 0.5 Ha.	0.5-1 Ha.	1-2 Ha.	2-3 Ha.	3-5 Ha.	Larger Than 5 Ha.	Total
1908	2,016	1,764	1,055	348	163	62	5,408
	(37.3)	(32.6)	(19.5)	(6.4)	(3.0)	(1.1)	(100.0)
1910	2,032	1,789	1,048	322	156	71	5,417
	(37.5)	(33.0)	(19.3)	(5.9)	(2.9)	(1.3)	(100.0)
1920	1,935	1,829	1,133	341	154	92	5,485
	(35.3)	(33.3)	(20.7)	(6.2)	(2.8)	(1.7)	(100.0)
1930	1,891	1,892	1,217	314	128	70	5,511
	(34.3)	(34.3)	(22.1)	(5.7)	(2.3)	(1.3)	(100.0)
1940	1,796	1,768	1,322	309	119	76	5,390
	(33.3)	(32.8)	(24.5)	(5.7)	(2.2)	(1.4)	(100.0)
1950	2,531	1,973	1,339	208	77	48	6,176
	(41.0)	(32.0)	(21.7)	(3.4)	(1.2)	(0.8)	(100.0)
1960	2,320	1,923	1,430	233	91	60	6,057
	(38.3)	(31.7)	(23.6)	(3.8)	(1.5)	(1.0)	(100.0)
1970	2,025	1,614	1,286	256	90	71	5,342
	(38.0)	(30.2)	(24.1)	(4.8)	(1.7)	(1.3)	(100.0)

Source: Institute of Developing Economics, *One Hundred Years of Agricultural Statistics in Japan* (Tokyo, 1969), p. 116; Ministry of Agriculture and Forestry, Statistical Research Division, *1970 World Census of Agriculture Report on Farmhouseholds and Population* (Tokyo, 1971).

Table 2-3. Allocation of Expenditures and Staff to Agriculture and Nonagriculture in Public and Private Research in Japan, 1972

Type of Research Institution	Expenditures[a]		Staff	
	Agriculture[b]	Nonagriculture[c]	Agriculture[b]	Nonagriculture[c]
University[d]	35%	14%	40%	24%
Public				
National government	17	4	17	3
Local government	44	2	40	2
Private	4	80	3	71
Total	100	100	100	100

Source: Bureau of Statistics, Office of the Prime Minister, *Kagaku Gijutsu Kenkyu Chosa Hokoku* (Report on the Survey of Research and Development in Japan) (Tokyo, 1972), pp. 62, 150, and 166.

[a] Includes both current and capital expenditures.
[b] Includes forestry and fisheries research.
[c] Excludes medical research.
[d] Includes both public and private universities.

sion. This contrasts with the United States land grant college system, which is characterized by the trinity of education, research, and extension. In Japan, agricultural experiment stations are under the Ministry of Agriculture and Forestry. Agricultural colleges or university faculties of agriculture are under the Ministry of Education. No formal links between experiment stations and universities have been established.

Extension programs are also separate from the experiment stations. They are carried out by the prefectural extension services. However, because both the prefectural experiment stations and the prefectural extension services are under the agricultural departments of the prefectural governments, they operate in close cooperation. In many cases, senior extension specialists are stationed in the experiment stations.

There is also a division of labor between the national and the prefectural experiment stations. National experiment stations under the administration of the Agricultural, Forestry and Fishery Technology Commission within the Ministry of Agriculture and Forestry include the following: the Central Agricultural Experiment Station at Konosu and the seven regional stations for Hokkaido, Tohoku, Hokuriku, Tokai-Kinki, Chugoku, Shikoku, and Kyushu; six specialized experiment stations for horticulture, tea, sericulture, livestock, veterinary medicine, and agricultural engineering; one station for forestry; and nine stations for fisheries.

Under the same administration are the National Research Institute of Agricultural Science, engaged in more basic research in the natural science aspects of agriculture; the National Research Institute of Agriculture, which is concerned primarily with social science aspects; the Food Research Institute for research on food nutrition, chemistry, and processing; the Plant Virus Research Institute; and the newly established Tropical Agriculture Research Center. In addition, there are the Farm Mechanization Research Institute and the Beet Research Institute, which have semipublic status, and more than 300 agricultural experiment stations and research institutes (including those for fishery and forestry research) operated by the forty-six prefectural governments.

The division of labor between the national and local experiment stations is rather broad, and there is considerable overlap. The national stations emphasize research projects for wide areas, while the prefectural stations tend to concentrate on research of local significance. Consequently, there is a tendency for the former to engage in more basic research and the latter in more applied research.

This division of labor is reflected in the differences in staff size between the national stations and local government research stations (Table 2-4). The universities and the national experiment stations, which conduct more basic research, have larger staffs than the prefectural stations, which stress applied research.

Evolution of the Agricultural Research System

This section outlines the historical evolution of agricultural research in Japan.[3] First, we will quickly review the quantitative growth of Japanese agriculture. The trends in agricultural output, inputs, and productivity are shown in Figure 2-1.

For the period 1878-1972, total output, input, and productivity in Japanese agriculture show secular growth trends except during World War II. Over the whole period total output more than tripled. Inputs of the two primary factors, labor and land, changed relatively slowly: labor measured by the number of farm workers declined about 30 percent, and land measured by cultivated land area increased by about 25 percent. To a large extent, the changes in labor and land canceled each other out in the growth in total inputs (aggregate of all conventional inputs by the shares of respective inputs in the total cost of agricultural production). Capital grew relatively slowly in the prewar years, only starting to rise at a rapid pace in the postwar period. The rates of growth in current nonfarm inputs, particularly fertilizers, have been much faster than in other inputs.

Overall, total inputs grew by about 80 percent over the whole period,

Table 2-4. Comparison of Staff Size and Amount of Expenditure for Agricultural Research among Universities and Government Institutions in Japan, 1972

Item	Institutions: Universities (N=63)	National Government (N=30)	Local Government (N=336)
Staff			
Total number	7,174	3,127	7,290
Average number per institution	114	104	22
Expenditures[a]			
Total			
Yen (million)	34,200	16,300	42,400
U.S. dollars (thousand)	129,000	62,000	160,000
Average per institution			
Yen (million)	543	543	126
U.S. dollars (thousand)	2,049	2,049	475
Average per staff member			
Yen (million)	4.8	5.2	5.8
U.S. dollars (thousand)	18.1	19.6	21.9

Source: Bureau of Statistics, Office of the Prime Minister, *Kagaku Gijutsu Kenkyu Chosa Hokoku*. (Report on the Survey of Research and Development in Japan) (Tokyo, 1972).
Note: Agricultural research here includes research in forestry and fisheries.
[a] Conversions from yen to dollars at rate of 265:1.

while the total output grew by as much as 280 percent. Consequently, total productivity, or output per unit of aggregate input, more than doubled. Three phases can be distinguished in the total output and productivity trends: relatively fast growth up to the late 1910s; relative stagnation in the interwar period; and a spurt forward in the post-World War II period.

In the following historical review we will describe the ways in which these growth patterns were related to the evolution of agricultural research.

National Government Initiatives

The national government first sought to develop agriculture by importing farm machinery, plants, and livestock. The Farm Machinery Exhibition Yard was established in 1871 in Tsukiji, Tokyo, to exhibit large-scale farm machinery imported from England and the United States. The machines were demonstrated at the Naito Shinjuku Agricultural Experimental Station, set up in 1872. The government also tried to transplant foreign plants and livestock. The Mita Botanical Experiment Yard (1874), the Shimofusa Sheep Farm (1875), the Kobe Olive Farm (1879), and the Harima Grape Farm (1880) were established for these trials.

The government also established institutions of advanced agricultural education: the Komaba Agricultural School in 1877 (now the University of

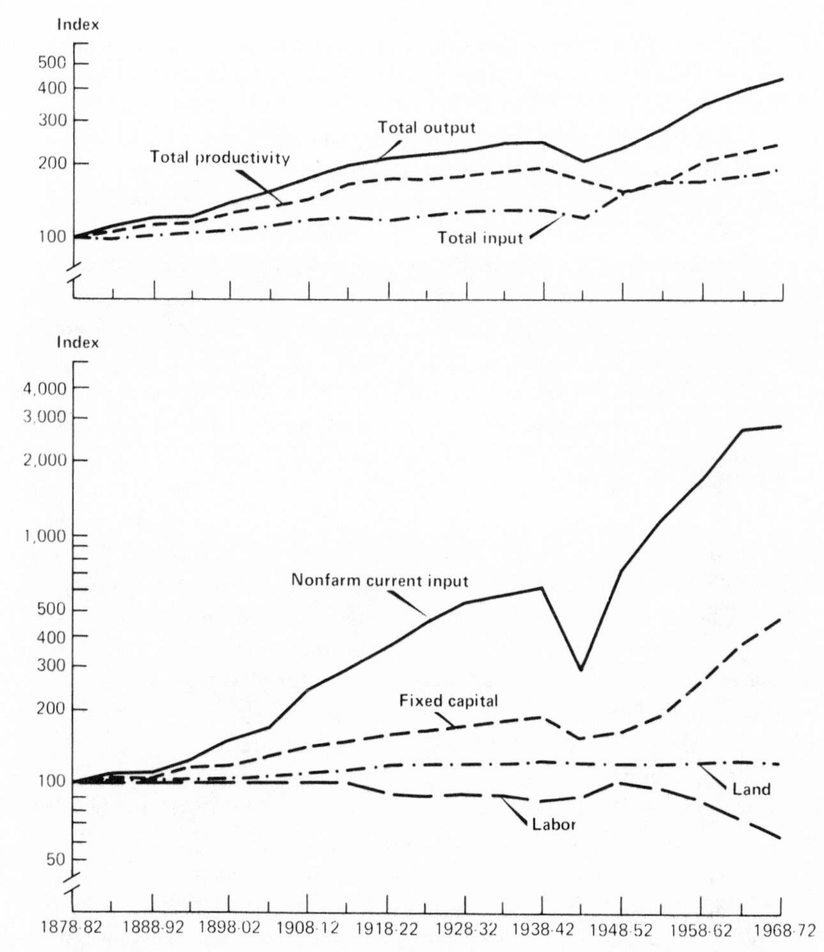

Figure 2-1. Trends in the indices of output, inputs, and productivity in Japanese agriculture (1878-82 = 100), five-year averages, semilog scale.
Source: Yujiro Hayami, et al., *A Century of Agricultural Growth in Japan: Its Relevance to Asian Development* (Minneapolis and Tokyo: University of Minnesota Press and University of Tokyo Press, 1975), pp. 32, 39.

Tokyo, College of Agriculture), and in 1875, the Sapporo Agricultural School which was designed to develop Hokkaido, the last frontier of Japan. British instructors were invited to staff the Komaba School and instructors from the United States were invited for the Sapporo School. The schools taught Anglo-American, large-scale mechanized farming.

This early "technology borrowing" is one example of the broad effort of

Meiji Japan to catch up with Western technology. But, in contrast to similar efforts in industry, this attempt was largely unsuccessful. Factor endowments and farm size in Japan were simply incompatible with large-scale machinery. In most cases the efforts to transplant foreign plants and livestock were not successful because of different ecological conditions.

The Meiji government quickly perceived this mistake and redirected its development strategy toward searching for modern technology that would be consistent with the factor endowments and ecological conditions of Japanese agriculture. In 1881, when their contracts were completed, the British agricultural instructors at the Komaba School were replaced by German scientists. Thereafter agricultural education in Japan placed primary emphasis on agricultural chemistry and soil science of the Liebig tradition. The facilities for demonstrating Western machinery, plants, and livestock were largely discontinued during the 1880s.

The newly founded Ministry of Agriculture and Commerce (1881) established the itinerant instructor system in 1885. Instructors traveled throughout the country holding agricultural extension meetings. In contrast to the earlier emphasis on importing Western technology, the itinerant instruction system was designed to diffuse the best seed varieties and cultural practices already used by many Japanese farmers. Not only the graduates of the Komaba School but also veteran farmers (*rōnō*) were employed as itinerant instructors in order to combine the best practical farming experience with the new scientific knowledge of the inexperienced college graduates.

To provide better information for the itinerant lecturers, the Experiment Farm for Staple Cereals and Vegetables was set up in 1885. By 1893, the farm, considerably expanded, had become the National Agricultural Experiment Station with six regional branch stations across the nation. The itinerant instuction program was subsequently absorbed by the National Agricultural Experiment Station. Meanwhile, the national government encouraged the prefectural governments to set up local experiment stations. However, only a few prefectures had established their experiment stations before the Law of State Subsidy for Prefectural Agricultural Experiment Stations was enacted in 1899.

The development of agricultural experiment stations in Japan was characterized from the beginning by the strong initiative of the national government. This experience in Japan contrasts with the experience in Western Europe. In England the famous Rothamsted Experimental Station was established in 1843, financed by Sir John Lawes. The Edinburgh Laboratory (founded in 1842) was supported by the Agricultural Chemistry Association of Scotland, a voluntary agricultural society. In France the first agricultural experiment station was established by Jean Boussingaullt in his estate at Bechelbrom in

1834. Even in Germany, where the first experiment station was publicly supported, Saxon farmers initiated the movement for the station and drafted its charter.

Why was the government of Meiji Japan (its leaders were primarily from the ex-*Samurai* class) clearly determined from the very beginning to take responsibility for conducting agricultural research? A part of the answer appears to lie in the organization of Japanese agriculture – a host of dwarf-sized, family farms. Individual farms were too small to take advantage of scale economies inherent in research. They were too small to exercise monopolistic power and gain enough benefit from research to cover the costs.

Because land was the factor limiting agricultural production, farmers demanded technology which would save land. Research emphasized the development of biological technology such as improved seed varieties. There was no incentive for private firms to conduct biological research since there were no institutions, such as patent laws, to allow them to profit from the research. Furthermore, because of the low price elasticity of demand for staple foods – the major products of Japanese agriculture – the gains from agricultural research were transferred primarily to consumers through declines in food prices. In this situation, the need for public support of agricultural research was obvious. Why did the national government rather than the local governments take the initiative? Why was it not left to cooperatives of farm producers, especially in the case of export crops?

The answer must be sought in the basic approach of Meiji Japan to economic development, which was to exploit agriculture for industrialization. When it opened its doors to foreign countries shortly before the Meiji Restoration, Japan was in danger of colonialization. The national slogan was *fukoku kyohei*, to "build a wealthy nation and strong army," and to attain this goal it was considered necessary to "develop industries and promote enterprises," or *shokusan kogyo*.

In predominantly rural Meiji Japan, industrialization was necessarily financed from agricultural surplus. Revenue for industrial development was raised by taxing agricultural land. Foreign exchange needed for imported goods was earned by exporting primary products. In a sense, the establishment of the National Agricultural Experiment Station in 1893 was a response to agitation for the reduction of the newly established land tax. The *Konoronsaku* (*A Treatise on the Strategy of Agricultural Development*), which was drafted in 1891 by the Agricultural Science Association and which had an immediate impact on the establishment of the experiment station, denied the argument for a land tax reduction on the basis that it would only contribute to the welfare of landlords and give no benefit to a large number of tenant farmers; it advocated "more positive measures to develop agriculture such as agricultural schools, experiment stations, itinerant lecturing, and agricultural societies" to reduce the burden on farmers.[4]

More fundamentally, increasing the supply of food for the growing urban industrial population was critical for industrialization. If the food supply had not kept up with urban demand, the price of food would have risen. In the early stages of development characterized by a high Engel coefficient, a rise in food prices would have significantly increased the cost of living and wages. This would have reduced profits and depressed industrial accumulation of capital.

Given its aspirations for industrial development, it was natural for the government to undertake measures, including agricultural research, to develop agriculture and to increase the supply of food, a critical wage-good for industrial development. Furthermore, because foreign exchange was a constraint on the importation of capital goods and on economic development, it was rational for the government to use tax revenues to support research on export products such as tea and silk, even though this research might benefit the producers primarily.

Linkage with Extension and Education

The linkage between agricultural research and extension has traditionally been strong in Japan. As mentioned previously, the Experiment Farm for Staple Cereals and Vegetables was designed to provide relevant data for itinerant instructors. The itinerant extension program was later absorbed into the National Experiment Station. As prefectural experiment stations were established, extension programs were transferred to them. When the agricultural associations organized under the Agricultural Association Law (1899) began to develop extension programs with government subsidies, their extension workers were trained in agricultural training centers attached to the prefectural experiment stations.

In contrast, the connection between agricultural experiment stations and educational institutions has not been formally established. The pioneers of agricultural science in the Meiji period had expected otherwise. In the *Treatise on the Strategy of Agricultural Development* they proposed:

> It is advantageous that the agricultural experiment stations and the itinerant instruction system are combined . . . it is advantageous if the agricultural experiment stations belong to agricultural colleges . . . it is highly effective for students to see the projects in the experiment stations. The results of experiment are useful for experiment stations . . . Agricultural associations should encourage the study of students, and encourage farmers to use the established results of experiments.[5]

It appears they had in mind a trinity of research, extension, and education, supported by agricultural associations – a system similar to that of the United States. However, the National Agricultural Experiment Station was established independently from the Komaba School. The system of national and prefectural experiment stations developed separately from institutions of advanced agricultural education.

This development may be explained by the strong (and hasty) demand of the government that experiment stations produce practical results for farmers immediately and by the fact that basic research at universities in the Meiji period was not producing practical techniques.

In the beginning stages, the staff at the agricultural colleges primarily studied principles and theories developed abroad. Although a few produced distinguished research, this was not immediately applicable on the farms. In the short run the more productive approach was to exploit the best indigenous farming practices by simple tests and demonstrations.

The initial research conducted at the Experiment Farm for Staple Crops and Vegetables and at the National Agricultural Experiment Station was primarily applied research. The major projects were field experiments comparing various seed varieties or husbandry techniques (for example, checkrow planting of rice seedlings versus irregular planting). Facilities, personnel, and, above all, the state of knowledge were not adequate for more than simple comparative experiments. Nevertheless, such experiments provided a basis for the rapid growth of agricultural productivity during the latter years of the Meiji period. This was a result of the substantial indigenous technological potential which could be further tested, developed, and refined at the new experiment stations as well as the strong propensity toward innovation among farmers, with whom the research workers effectively interacted.

For 300 years before the Meiji Restoration, farmers were constrained by feudalism. Personal behavior and economic activity were highly structured within a hierarchical system of social organization. Farmers were bound to their land and were, in general, not allowed to leave their villages except for such pilgrimages as the *Ise-Mairi* (Pilgrimage to the Ise Grand Shrine). They were not free to choose which crops to plant or which varieties of seeds to sow. The division of the nation into feudal estates discouraged communication. In many cases, feudal lords prohibited the export of improved seeds or cultural methods from their territories. Under such conditions the diffusion of superior seeds and husbandry techniques from one region to another was severely limited.

The Meiji Restoration removed feudal restraints. Farmers became free to choose which crops to plant, which seeds to sow, and which techniques to practice. The introduction of a modern postal service and of railroads reduced the cost of diffusing information about technology. The land tax reform, which granted a fee simple title to the farmers and transformed a feudal share tax on crops into a fixed land tax, increased the farmers' incentive to innovate. The farmers, especially of the *gōnō* class (landlords who personally farm part of their holdings), vigorously responded to such new opportunities. They voluntarily formed agricultural societies called *nodankai* (agricultural discussion society) and *hinshukokankai* (seed exchange society) and searched for better techniques. Rice production practices, such as the use of salt water in seed selection, improved preparation and management of nursery beds, and checkrow planting were discovered by farmers and propagated by the itinerant instructors. They were sometimes enforced by the sabers of the police. The major improved varieties of seeds, up to the end of the 1920s, also resulted from selections by veteran farmers.

Experiment station research was successful in testing and refining the results of farmer innovations. The *rōnō* techniques (veteran farmers' techniques) were based on experiences in the specific localities where they originated. They tended to require modification when transferred to other locations. Simple comparative tests effectively screened the *rōnō* techniques and varieties, thereby reducing greatly the cost of technical information for farmers. Slight adaptations of indigenous techniques based on experiments often gave them wide applicability.

Given the backlog of indigenous technological potential, the innovative attitude of farmers, and the infant state of university education and research, it might have been more effective at that time to organize the experiment stations and extension services separately from the universities. Also, the loose linkages between experiment stations and universities did not pose a serious problem during the Meiji period when the agricultural research-education-extension complex was small. Since the key personnel in experiment stations, extension services, and agricultural colleges were all graduates of either the Komaba School or the Sapporo School, the interaction among experiment stations and agricultural colleges worked sufficiently well on an ad hoc basis.

Centralization versus Decentralization

It appears that the evolution of the organization of agricultural research in Japan has been marked by a search for an optimum balance between centralization and decentralization. The pioneers of agricultural science in Japan seem to have recognized that agricultural technology is highly location-specific. To produce practical results for farmers, agricultural research must be conducted in various ecological regions. They also appear to have recognized the need for the coordination of central and local experiment stations. In the *Treatise on the Strategy of Agricultural Development* they proposed to establish one central experiment station in Tokyo, five regional stations, and one or more stations in each prefecture.[6] However when the National Agricultural Experiment Station (headquartered in Nishigahara, Tokyo) and the six regional branch stations were established, the director, Atsushi Sawano, considered the system inadequate. He remarked that "regional breakdowns are not sufficient for the variations in soil and climate. If additional budget is available, another forty-five stations must be established."[7]

Local political groups lobbied the National Diet for more branch stations, and the number of these was increased from six to nine in 1895. But the experiment stations were strapped for funds. The initial staff, including that in the branch stations, comprised only twenty research scientists and seven technicians. This increased to thirty scientists and fifteen technicians in 1899. Experiments were always handicapped by insufficient facilities and logistical support.[8] Under such conditions it was hardly possible to conduct more basic research in addition to the simple tests and demonstrations noted previously.

Meanwhile, the backlog of indigenous technological potential was gradually exhausted as it was exploited. Research institutions felt the need to recharge this declining potential by turning to more basic research. The prefectural experiment stations gradually accepted responsibility for more applied research. The rapid expansion of local research during this time is shown by the prefectural government's increasing expenditure for agricultural research (Table 2-5).

In response to the establishment of the prefectural experiment stations, the National Experiment Station reduced its branch stations from nine to three in 1903, with the intention of exploiting scale economies in more basic research by concentrating research resources in fewer stations. The following year, for the first time, the National Experiment Station launched an original crop breeding program at its Kinai Branch. The object of this project was to develop new rice varieties by artificial crossbreeding based on the Mendelian principles rediscovered in 1900. It took almost two decades before new varieties of major practical significance were developed, though the project contributed greatly to the accumulation of experience and knowledge. Another project was started in 1905 at the Rikuu Branch to improve rice varieties by pure line selection. This approach brought about quicker practical results. Thereafter the main efforts of crop breeding in the Taisho era (1912-25) were in pure line selection.

Rice breeding by pure line selection represented the final exploitation of the indigenous technological potential embodied in the *rōnō* varieties. As the purity of those varieties was raised, the potential was exhausted.[9] The exploitation and consequent exhaustion of indigenous potential became evident in the 1910s before more basic research represented by the crossbreeding

Table 2-5. Expenditures for Agricultural Research by National and Prefectural Governments in Japan, 1897-1970 (in thousand yen)[a]

Year	Total	National[b]	Prefectural
1897	616	367 (60)	249 (40)[c]
1902	2,044	930 (45)	1,114 (55)[c]
1907	2,032	718 (35)	1,314 (65)[c]
1912	3,044	822 (27)	2,222 (73)[c]
1918	2,521	849 (34)	1,672 (66)[c]
1923	5,385	1,286 (24)	4,099 (76)[c]
1927	6,561	1,251 (19)	5,310 (81)[c]
1932	8,196	1,686 (21)	6,510 (79)[c]
1955	9,478	4,190 (44)	5,288 (56)[d]
1960	12,300	4,661 (38)	7,639 (62)[d]
1965	38,814	12,257 (32)	26,557 (68)[b]
1970	60,093	17,257 (29)	42,836 (71)[b]

Source: Yujiro Hayami et al., *A Century of Agricultural Growth in Japan* (Minneapolis and Tokyo: University of Minnesota Press and University of Tokyo Press, 1975).

[a] Based on 1934-1936 prices. Parentheses enclose percentages of total expenditures.

[b] Five-year averages ending the years shown, except for the 1918 and 1923 figures which are the five-year averages ending in 1917 and 1922, respectively.

[c] Single-year figures of the years shown.

[d] Estimations: (1) Change in percentage between 1951-55 and 1956-60 was assumed to be the same between 1956-60 and 1961-65. (2) The percentage for 1958 was used for 1956-60.

project produced major breakthroughs. The rate of increase in rice yield began to decline. Japanese agriculture began to stagnate during the interwar period.

Meanwhile, not only the National Experiment Station but also the prefectural stations began crossbreeding projects. The projects were handicapped by a lack of coordination which tended to dissipate the limited research funds. In these circumstances, a nationwide coordinated crop breeding program called the Assigned Experiment System (the system of experiment assigned by the Ministry of Agriculture and Forestry) was established, first for wheat (1926), next for rice (1927), and subsequently for other crops and livestock.

Under the Assigned Experiment System, the national experiment stations were responsible for crossbreeding up to the selection of the first several filial generations. The regional stations, in each of eight regions, conducted further selections to adapt the seeds to regional ecologies. The varieties selected at the regional stations were then sent to the prefectural stations to be tested for their acceptability in specific locations. The varieties developed by this system were called *Norin* (abbreviation of the words "Ministry of Agriculture and Forestry" in Japanese) varieties.

This system was outstandingly successful. The Mexican dwarf wheat, which is revolutionizing Mexican and Indo-Pakistan agriculture, was based on the *Norin* No. 10 wheat variety. The *Norin* numbered varieties successively replaced older varieties in the late 1930s. If the supply of fertilizer and other agricultural inputs had not been restricted during World War II, the second epoch of agricultural productivity growth in Japan – which occurred after the war – would probably have started in the late 1930s.

The Assigned Experiment System was an institutional innovation which economized on research resources – above all, knowledge and experience – while satisfying the requirement for location-specific agricultural research. The system has evolved so that now both the national and the prefectural experiment stations conduct crossbreeding from the first step of artificial crossing. This change reflects the increase in knowledge and experience of the prefectural stations. It has enabled the prefectural stations to conduct research more specifically designed to satisfy local demand.

Social Returns to Agricultural Research Investment

From the historical review in the previous section it seems reasonable to hypothesize that the evolution of agricultural research in Japan was spurred by the benefits which research contributed to society and to national development. To demonstrate the gains from public investment in agricultural research, we will attempt in this section to estimate the social rates of return to rice-breeding research.

Theoretical Framework for Estimating Social Returns

Using Marshallian concepts of social welfare and cost, social returns to rice-breeding research are measured by the changes in consumers' and producers' surpluses that result from the shift in the rice supply curve corresponding to a shift in the rice production function. (Producers' surplus is defined as the total value of output in agriculture minus the payment to the inputs applied to agricultural production that are supplied from nonagriculture; it includes not only the entrepreneurial profit of farmers but also land rent, wages to family labor, and returns to farm capital. For a more detailed discussion see chapter 6.) This relationship is shown in Figure 2-2, in which d and s_o represent the actual demand and supply curves, whereas s_n represents the supply curve that would have existed if improved rice varieties had not been developed.

Assuming market equilibrium and no rice imports, the shift in the supply curve from s_n to s_o would increase the consumers' surplus by (area ABC +

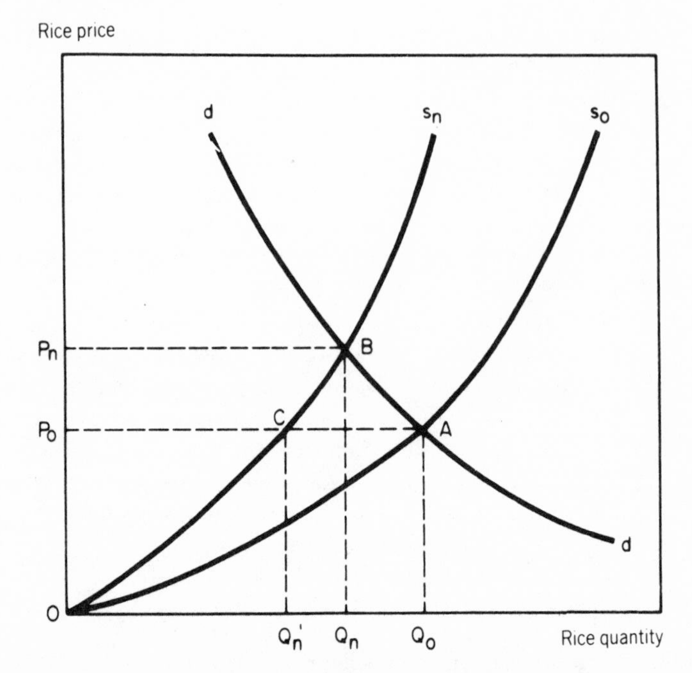

Figure 2-2. Model of estimating social returns to rice-breeding research.
Source: Masakatsu Akino and Yujiro Hayami, "Efficiency and Equity in Public Research: Rice Breeding in Japan's Economic Development," *American Journal of Agricultural Economics*, 57:1 (February 1975), 4.

area BP_nP_oC), the producers' surplus by (area ACO − area BP_nP_oC), and the social benefit by (area ABC + area ACO).

In reality, however, Japan remained a net importer of rice during this period. Rice imports were regulated by government tariffs and quotas.[10] As discussed previously, the government sought to prevent a rise in the cost of living of urban workers. In fact, the government maintained stable rice prices relative to the general price index until around 1960. Rice prices then began to rise sharply because government policy regarding rice shifted to protecting producers (Figure 2-3, lower section).

If basic policy was to secure sufficient rice to prevent a rise in the urban workers' cost of living, and, if increasing domestic rice production owing to varietal improvement and other means did not meet increasing demands, the gap would have to have been filled by imports. Let P_o in Figure 2-2 be the price of rice that the government determined to maintain. If the domestic supply schedule did not shift from s_n to s_o, the government would have in-

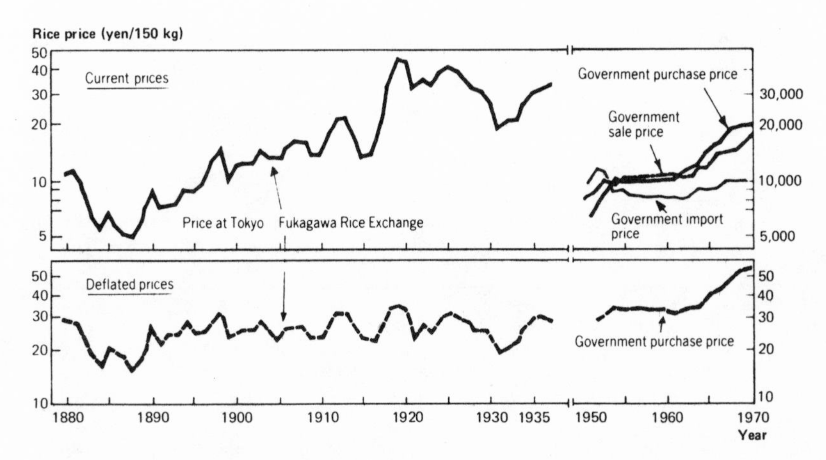

Figure 2-3. Changes in rice prices, both current and deflated by the General Price Index (1934-36 = 100), in Japan (log-scale in brown rice term), 1880-1937 and 1951-70.
Source: Yujiro Hayami, "Rice Policy in Japan's Economic Development," *American Journal of Agricultural Economics*, 54:1 (February 1972), 22.

creased rice imports by $Q'_n Q_o$. Then, the producers' surplus would have been reduced by area BP_nP_oC without being compensated for by area ACO. In this case, foreign exchange would have been reduced by area ACQ'_nQ_n.

If there had been no breeding program to shift the domestic supply from s_n to s_o, the producers' surplus would have been smaller by area ACO. This area may be defined as the producers' gain in economic welfare from the rice breeding research (assuming price stability through rice imports). Since the consumers' surplus would remain unchanged under this assumption, the producers' gain would equal the total social benefit produced from the rice breeding programs. Another contribution of the breeding research to the national economy in the open economy case would be the gain in foreign exchange by area ACQ'_nQ_o.

In reality, in spite of the efforts to shift the rice production function upward, domestic supplies did not keep up with demand. This resulted in rice imports of about 5 to 20 percent of domestic production. Therefore, s_o in Figure 2-2 would have been located somewhere to the left of A if we define A as the point of equilibrium of total market supply and demand. However, this does not require modification of our model. We will now estimate the social benefits of rice research in Japan using this model.[11]

Distribution of Social Benefits

Estimates of social benefits were conducted separately for the breeding programs before and after the introduction of the Assigned Experiment System

Table 2-6. Estimates of Social Benefits from Rice-Breeding Research Programs Begun before the Establishment of the Assigned Experiment System, Japan (million yen in 1934-36 prices)

	Autarky Case			Open Economy Case	
Year	Producers' Gain (1)	Consumers' Gain (2)	Total Social Benefits (3)=(1) + (2)	Producers' Gain (Total Benefits) (4)	Saving in Foreign Exchange (5)
1915 . . .	− 0.30	0.45	0.15	0.15	0.18
1916 . . .	− 0.32	0.48	0.16	0.16	0.19
1917 . . .	− 1.80	2.70	0.90	0.90	1.08
1918 . . .	− 4.20	6.30	2.11	2.10	2.52
1919 . . .	− 7.68	11.52	3.85	3.84	4.60
1920 . . .	− 12.47	18.75	6.28	6.25	7.50
1921 . . .	− 15.69	23.64	7.95	7.88	9.45
1922 . . .	− 26.21	39.51	13.30	13.17	15.80
1923 . . .	− 33.49	50.73	17.24	16.91	20.29
1924 . . .	− 45.90	69.75	23.85	23.25	27.90
1925 . . .	− 60.97	93.03	32.06	31.01	37.21
1926 . . .	− 57.37	87.54	30.17	29.18	35.01
1927 . . .	− 66.07	100.86	34.79	33.62	40.34
1928 . . .	− 67.65	103.38	35.73	34.46	41.35
1929 . . .	− 70.68	108.12	37.44	36.04	43.24
1930 . . .	− 86.68	132.87	46.19	44.29	53.14
1931 . . .	− 80.80	124.23	43.43	41.41	49.69
1932 . . .	− 102.25	157.86	55.61	52.62	63.14
1933 . . .	− 102.56	157.62	55.06	52.54	63.04
1934 . . .	− 77.01	118.41	41.40	39.47	47.36
1935 . . .	− 86.21	132.63	46.42	44.21	53.05
1936 . . .	− 103.53	159.35	55.82	53.12	63.74
1937 . . .	− 104.33	160.67	56.34	53.56	64.27
1938 . . .	− 103.65	159.62	55.97	53.21	63.85
1939 . . .	− 110.97	171.07	60.10	57.03	68.43
1940 . . .	− 95.79	147.52	51.73	49.18	59.01
1941 . . .	− 84.69	130.35	45.66	43.45	52.14
1942 . . .	− 100.18	154.12	53.94	51.38	61.65
1943 . . .	− 91.07	139.95	48.88	46.65	55.98
1944 . . .	− 81.11	124.52	43.41	41.51	49.81
1945 . . .	− 78.69	120.62	41.93	40.21	48.25
1946 . . .	− 73.44	112.35	38.91	37.45	44.94
1947 . . .	− 65.79	100.55	34.76	33.52	40.22
1948 . . .	− 69.27	105.70	36.43	35.24	42.28
1949 . . .	− 52.91	80.62	27.71	26.88	32.25
1950 . . .	− 57.49	87.50	30.01	29.17	35.00
1951 . . .	− 48.99	74.50	25.51	24.84	29.80
1952 . . .	− 48.84	74.15	25.31	24.72	29.66
1953 . . .	− 36.42	55.20	18.78	18.40	22.08

Table 2-7. Estimates of Social Benefits from Rice-Breeding Research Programs Established under the Assigned Experiment System, Japan (million yen in 1934-36 prices)

	Autarky Case			Open Economy Case	
Year	Producers' Gain (1)	Consumers' Gain (2)	Total Social Benefits (3)=(1) + (2)	Producers' Gain (Total Benefits) (4)	Saving in Foreign Exchange (5)
1932 . . .	−0.66	0.99	0.33	0.33	0.39
1933 . . .	−2.32	3.48	1.16	1.16	1.39
1934 . . .	−3.12	4.68	1.56	1.56	1.87
1935 . . .	−4.72	7.08	2.36	2.36	2.83
1936 . . .	−7.40	11.10	3.71	3.70	4.44
1937 . . .	−10.90	16.38	5.48	5.46	6.55
1938 . . .	−15.15	22.80	7.65	7.60	9.12
1939 . . .	−19.61	29.55	9.94	9.85	11.82
1940 . . .	−19.96	30.09	10.13	10.03	12.03
1941 . . .	−19.86	29.94	10.08	9.98	11.97
1942 . . .	−26.97	40.71	13.74	13.57	16.28
1943 . . .	−28.13	42.48	14.35	14.16	16.99
1944 . . .	−29.69	44.88	15.19	14.96	17.95
1945 . . .	−33.30	50.40	17.10	16.80	20.16
1946 . . .	−37.74	57.18	19.44	19.06	22.87
1947 . . .	−40.18	60.93	20.75	20.31	24.37
1948 . . .	−50.49	76.68	26.19	25.56	30.67
1949 . . .	−47.81	72.72	24.91	24.24	29.08
1950 . . .	−59.86	91.20	31.34	30.40	36.48
1951 . . .	−61.52	93.87	32.35	31.29	37.54
1952 . . .	−62.64	95.43	32.79	31.81	38.17
1953 . . .	−48.18	73.32	25.14	24.44	29.32
1954 . . .	−51.04	77.64	26.60	25.88	31.05
1955 . . .	−66.73	101.40	34.67	33.80	40.56
1956 . . .	−47.42	71.88	24.46	23.96	28.75
1957 . . .	−39.98	60.48	20.50	20.16	24.19
1958 . . .	−37.64	56.88	19.24	18.96	22.75
1959 . . .	−34.58	52.20	17.62	17.40	20.88
1960 . . .	−32.79	49.47	16.68	16.49	19.78
1961 . . .	−28.51	42.99	14.48	14.33	17.19

based on the data in Appendix 2-3. The results are summarized in Tables 2-6 and 2-7. The most remarkable aspect of the autarky case results is that the social benefits produced from the research were totally captured by the consumers; the producers were worse off. Such results were derived from the application of low price elasticities of demand and supply. In particular, the demand elasticity plays a decisive role in the distribution of benefits among consumers and producers.

In reality, however, Japan did not operate in the condition of rice autarky during the period of this analysis. If it is assumed that rice was imported, the producers' situation would have been even worse (as shown in Figure 2-2 by area ACO and as measured in Tables 2-6 and 2-7) if there had been no program for rice-breeding research. Thus, rice research preserved a larger share of the Japanese rice market for domestic producers. Without the research the Japanese economy would have lost foreign exchange, by area ACQ'_nQ_o. In fact, as the estimates in Tables 2-6 and 2-7 indicate, the possible loss of foreign exchange owing to the shortage in the domestic rice supply during the 1930s would have amounted to about 5 percent of the total commodity import of Japan. Considering the chronic shortage of foreign exchange in the course of industrialization in Japan, the contribution of rice-breeding research to economic growth was quite significant. (However, it is difficult to estimate, in a term comparable to that of consumers' or producers' surplus, the gain in national economic welfare as a result of the saving of foreign exchange.)

In the open economy case, the producers were made better off by rice-breeding research, while the consumers continued to enjoy the same level of economic welfare without causing a drain on foreign exchange. In reality, however, it would appear that consumers' welfare could not have remained the same in the absence of the rice-breeding research. The constraints on foreign exchange would not have allowed additional rice imports on such a large scale. The autarky and the open economy cases in our analysis represent the polar cases between which reality lies.

The Social Rate of Returns

In order to assess the efficiency of rice-breeding research, both the external and the internal rates of returns are calculated by relating the research costs (data reported in Appendix 2-3) to the estimates of social benefits shown in Tables 2-6 and 2-7. (See accompanying box.)

The social rates of returns for the breeding programs before the Assigned Experiment System are calculated for two alternative cases: Case A assumes that net returns $(R_t - C_t)$ in 1935 would have been maintained forever from that year; Case B assumes that net returns would become zero after 1953. Case A represents a polar case in which the knowledge and experience accumulated in a breeding program would continue to be utilized even after the varieties developed by the program were replaced by varieties developed by subsequent breeding programs, whereas Case B assumes that the life of the varieties ends when they are replaced by new ones.

In calculating the rates of returns in the programs under the Assigned Experiment System, two alternative assumptions about the stream of returns were made: Case A assumes that the net returns would have continued to be maintained forever at the 1951 level since that is the year when the area plant-

The external rate of returns (r_e) is defined as the rate calculated from the following formula:

$$r_e = \frac{100\,(iP + F)}{C}$$

where i is the external rate of interest, P is the accumulation of past returns, F is the annual future returns, and C is the accumulation of past research expenditures. The external rate of interest (i) is applied to the accumulation of both returns and expenditures. In this study 10 percent is assumed for the interest rate.

The internal rate of returns (r_i) is the rate that results in

$$\sum_{t=0}^{T} \frac{R_t - C_t}{(1 + r_i)t} = 0$$

where R_t is the social benefit in year t, C_t is the research cost in year t, and T is the year that the research ceases to produce returns.

ed in the *Norin* varieties reached a peak; Case B assumes that the net returns would have become zero after 1961.

The results of the estimates of the social rates of return for the autarky and the open economy cases are reported in Table 2-8. There are only small differences between the two. Both indicate that the crop-breeding research represents a lucrative public investment opportunity.

The estimates for the programs before the introduction of the Assigned Experiment System are comparable with Griliches's estimates for hybrid corn research in the United States (about 35 percent for the internal rate and 700 percent for the external rate of return); they also compare favorably with estimates for poultry research in the United States reported by Peterson (about 20 percent for the internal rate and 140 percent for the external rate).[12] The estimates of the rate of returns for rice research under the Assigned Experiment System are comparable with those for the cotton research in São Paulo, Brazil, by Ayer and Schuh (about 90 percent for the internal rate) and for wheat research in Mexico by Ardito Barletta (about 75 percent for the internal rate).[13] Judging from these estimates, there has been gross underinvestment in research to improve grain varieties in the world.

The results in Table 2-8 show that the social rate of return increased after the introduction of the Assigned Experiment System. This seems to suggest that efficiency was improved by this institutional innovation. We do not deny the possibility that the increase in the rate of return over time reflects scale

Table 2-8. Estimates of Social Rates of Returns to Rice-Breeding Research in Japan (million yen in 1934-36 constant prices)

Rates of Return	Autarky Case		Open Economy Case	
	Case A	Case B	Case A	Case B
Before Assigned Experiment System				
External Rate				
(1) Net cumulated past returns	985.88	7,660.95	952.52	7,392.64
(2) Past returns expressed as annual flow	98.58	766.09	95.25	739.26
(3) Net annual future returns	44.63	0	42.41	0
(4) Total net annual returns, (2) + (3)	143.21	766.09	137.66	739.26
(5) Cumulated past research expenditures	123.39	783.47	123.39	783.47
(6) Rate of return, 100 x (4)/(5)	116%	98%	112%	94%
Internal Rate	27%	25%	26%	25%
Under Assigned Experiment System				
External Rate				
(1) Net cumulated past returns	487.98	1,639.77	480.11	1,610.65
(2) Past returns expressed as annual flow	48.79	163.97	48.01	161.06
(3) Net annual future returns	31.73	0	30.67	0
(4) Total net annual returns, (2) + (3)	80.52	163.97	78.68	161.06
(5) Cumulated past research expenditures	14.51	46.78	14.51	46.78
(6) Rate of return, 100 x (4)/(5)	554%	350%	542%	344%
Internal Rate	75%	73%	75%	73%

economies inherent in the process of research in producing knowledge and information. However, if there were no organizational improvements to permit better coordination of the enlarged research complex, the increase in the efficiency of rice-breeding research would not have been as dramatic as measured in this study.

Conclusion and Implications

The Japanese agricultural research system evolved under the strong leadership of the government. The current system is dominated by public-supported institutions, including both national and prefectural experiment stations.

We hypothesized that the government's involvement in agricultural research activities was prompted by the high rate of social returns. The case study of rice-breeding research indicates that the investment in agricultural research was indeed lucrative for society. Moreover, we found that financing agricultural research out of tax revenue can be justified because the major gains from the research increased consumers' surplus or contributed to economic development by lowering the cost of living for urban workers and saving foreign exchange.

The government's recognition of the need for agricultural research is one of the keys to understanding how agriculture grew in Japan despite the very unfavorable endowment of land relative to labor. However, it appears that there was gross underinvestment in agricultural research even in Japan. For example, if many more resources had been allocated to agricultural research in the early days of modern economic growth, the interwar stagnation of Japanese agriculture might have been avoided or reduced considerably. Public planners and policy makers in the world should be constantly reminded that there is a tendency to underestimate the returns to research because of uncertainty and long gestation periods. Rarely are sufficient resources allocated to agricultural research. Hence, efficient allocation of research resources among different research enterprises is important.

The conflict between the needs for location-specificity and scale economies poses a critical problem in the allocation of scarce research resources. The Assigned Experiment System in Japan represents a successful attempt in solving this problem. Such organizational innovations should be promoted in developing countries whose research resources, particularly in research and technical staff, are very limited.

Interaction among research administrators, scientists, and farmers is of critical importance to produce information useful to farmers. At the same time, the interaction of agricultural scientists with those in neighboring disciplines is a source of research productivity. The close association of agricultural experiment stations with extension programs and agricultural associations in Japan increased the responsiveness of agricultural research to the needs of farmers.

It must be noted, however, that this system was established without formally linking experiment stations with universities. Perhaps this represents a bad example which should not be repeated in developing countries. Today agricultural science in Japan is far more advanced than it was in the Meiji period. Specialized research with an interdisciplinary approach seems to be required for the transfer of technology developed in advanced countries to developing countries. For this requirement the close linkage between experiment stations and universities is necessary.

At the same time, the close linkage between experiment stations and universities should by no means be established at the sacrifice of the responsiveness of agricultural research to the needs of farmers. How to establish a close association between experiment stations and universities while promoting active interactions among farmers and research workers, either directly or through extension agents, remains a major unsolved issue in organizing agricultural research for agricultural economic development.

APPENDIXES

Appendix 2-1. Specification of Demand and Supply Schedules

The first step to estimate the changes in consumers' and producers' surpluses is the specification of the demand and supply schedules. In this study a constant elasticity demand function is assumed as

$$q = Hp^{-\eta}$$

where q and p are, respectively, the quantity and the price of rice, and η is the price elasticity of demand. Similarly, a constant elasticity supply function is assumed as

$$q = Gp^{\gamma}$$

where γ is the price elasticity of the rice supply. We assume a hypothetical supply curve that would have existed in the absence of improved varieties as

$$q = (1 - h)Gp^{\gamma}$$

where h represents the rate of shift in the supply function owing to varietal improvement. In competitive equilibrium the supply function is equivalent to the marginal cost function derived from the production function. Since the relation between the rate of shift in the marginal cost function (h) and the rate of shift in the production function (k) can be approximated by

$$h \cong (1 + \gamma)k$$

the following approximation formulas hold in equilibrium:

$$\text{area ABC} \cong \frac{1}{2} p_o q_o \frac{[k(1 + \gamma)]^2}{\gamma + \eta}$$

$$\text{area ACO} \cong k p_o q_o$$

$$\text{area } BP_nP_oC \cong \frac{p_o q_o k(1 + \gamma)}{\gamma + \eta} \left[1 - \frac{1}{2}\frac{k(1 + \gamma)\eta}{\gamma + \eta} - \frac{1}{2}k(1 + \gamma)\right]$$

and

$$\text{area } ACQ'_nQ_o = (1 + \gamma)k p_o q_o.$$

For the derivation of the above formula, see Appendix 2-2.

Appendix 2-2. Derivation of the Model of Estimating Social Benefits from Rice-breeding Research

The relationship between h and k

The actual and the hypothetical supply functions that would have existed in the absence of improved varieties are assumed, respectively, as

$$q = Gp^{\gamma} \tag{1}$$
$$q = (1 - h)Gp^{\gamma}. \tag{2}$$

Assuming that the supply curves are equivalent to the marginal cost curves, the marginal costs $(\frac{dc}{dq})$ are

$$\frac{dc}{dq} = p = G^{-1/\gamma} q^{1/\gamma} \tag{3}$$

$$\frac{dc}{dq} = p = (1 - h)^{-1/\gamma} G^{-1/\gamma} q^{1/\gamma}. \tag{4}$$

Total cost curves derived by taking the integrals of the marginal cost curves which are assumed to pass through the origin are

$$C = \frac{\gamma}{(1 + \gamma)} G^{-1/\gamma} q^{(1 + \gamma)/\gamma} \tag{5}$$

$$C = \frac{\gamma}{(1 + \gamma)} G^{-1/\gamma}(1 - h)^{-1/\gamma} q^{(1 + \gamma)/\gamma}. \tag{6}$$

Let q_o and q'_n represent respectively the output levels for a given cost in equations (5) and (6). Then, the relation between q_o and q'_n is represented approximately for a sufficiently small value of h as

$$(q'_n/q_o) \cong 1 - h/(1 + \gamma). \tag{7}$$

Since k is denoted $(q_o - q'_n)/q_o$, the relationship between h and k can be shown approximately as $h \cong (1 + \gamma)k$. This formula implies that h becomes infinite when $\gamma = \infty$. This is due to the approximate nature of the formula. Actually, h is equal to $k/(1 - k)$ when $\gamma = \infty$.

The Formulas of Social Returns, Changes in Consumers' Surplus, and Producers' Surplus

Area ABC is derived as follows:
p_o and p_n in Figure 2-2 are represented respectively as

$$p_o = (H/G)^{1/(\gamma + \eta)} \tag{8}$$

$$p_n = (H/G)^{1/(\gamma + \eta)} (1 - h)^{-1/(\gamma + \eta)}. \tag{9}$$

Hence, $(p_n - p_o) \cong p_o h/(\gamma + \eta)$ for a sufficiently small value of h. Thus, area $ABC \cong \frac{1}{2} p_o q_o h^2/(\gamma + \eta) = \frac{1}{2} p_o q_o [k(1 + \gamma)]^2/(\gamma + \eta)$.

Area ACO is derived as follows:

$$\text{area ACO} = \int_o^{p_o} hGp^{\gamma} dp = p_o q_o h/(1 + \gamma) \cong kp_o q_o.$$

Area BP_nP_oC is derived as follows:

$$\text{area } BP_nP_oC \cong (p_n - p_o)q_o - \frac{1}{2}(p_n - p_o)(q_o - q_n) - \text{area ABC}.$$

Since $(p_n - p_o)$ is approximately equal to $p_o h/(\gamma + \eta)$, and $(q_o - q_n)$ to $q_o h\eta/(\gamma + \eta)$, then

$$\text{area } BP_nP_oC \cong \frac{p_o q_o k(1 + \gamma)}{\gamma + \eta} \left[1 - \frac{1}{2} \frac{k(1 + \gamma)\eta}{\gamma + \eta} - \frac{1}{2} k(1 + \gamma)\right].$$

Appendix 2-3. Parameters and Data of Estimating the Social Rates of Returns to Rice-Breeding Research

Demand and Supply Parameters

The estimate of the price elasticity of demand for rice (η) is available from Ohkawa, whose estimates are based on 1931-38 household survey data for the urban population, and on 1920-38 market data for the rural population.[14] The estimates differ for different occupational, regional, and income groups, but they cluster around 0.2. We will adopt 0.2 for η.

The price elasticity of the rice supply (γ) was estimated by Hayami and Ruttan on the basis of 1890-1937 time-series data.[15] The supply elasticity was also estimated by Yuize on a 1952-62 time-series.[16] The results of the former study indicate that γ was in the vicinity of 0.2; those of the latter indicate that γ ranged from 0.2 to 0.3. Here we will adopt 0.2 for γ.

Shift in Rice Production Function

We estimated the rate of shift in the aggregate rice production function (k) by averaging the yield differences between the improved and the unimproved varieties for the same level of inputs, using the areas planted in the improved varieties as weights. The data for the differences in yield between the improved varieties and the varieties that were replaced by the improved varieties at the same level of inputs are based on the results of the comparative yield tests at various agricultural experiment stations.

A good collection of the results of comparative rice yield tests for the varieties developed before the Assigned Experiment System is available in the reports of a survey conducted by the Ministry of Agriculture and Forestry. These reports have gathered the results of the three years' tests for the 130 improved varieties in comparison with those of the varieties that they replaced. We used these data to calculate the rate of shift in the aggregate rice production function in the t-th year owing to varietal improvement (k_t) by the following formula:

$$k_t = \sum_i \sum_j k_{ij} \frac{A_{ijt}}{A_t}$$

where k_{ij} is the ratio of the increase in rice yield of the i-th variety in the j-th region over the variety that it replaced; and A_t and A_{ijt} are, respectively, the total rice area in the nation and the rice area planted in the i-th variety in the j-th region.

Because of data limitations a cruder method is applied for the estimation of the rate of the production function shift owing to the varieties developed by the Assigned Experiment System. Judging from a limited number of the results of comparative yield tests, we adopted 6 percent as the average rate of yield increases of the *Norin* varieties over the varieties that they replaced. This rate was multiplied by the ratio of the area planted in the *Norin* varieties in order to calculate the rates of shift in aggregate rice production owing to the breeding research under the Assigned Experiment System. The results of the estimation of the k_t's are shown in Table 2-9.

Table 2-9. Estimates in the Rate of Shift in the Rice Production Function owing to Varietal Improvement in Japan

Programs Initiated Before the Assigned Experiment System		Programs Established Under the Assigned Experiment System	
Year	k_t	Year	k_t
1915 . . .	0.01%	1932 . . .	0.02%
1916 . . .	0.01	1933 . . .	0.06
1917 . . .	0.06	1934 . . .	0.11
1918 . . .	0.14	1935 . . .	0.15
1919 . . .	0.23	1936 . . .	0.20
1920 . . .	0.36	1937 . . .	0.30
1921 . . .	0.52	1938 . . .	0.42
1922 . . .	0.79	1939 . . .	0.52
1923 . . .	1.11	1940 . . .	0.60
1924 . . .	1.48	1941 . . .	0.66
1925 . . .	1.89	1942 . . .	0.74
1926 . . .	1.91	1943 . . .	0.82
1927 . . .	1.97	1944 . . .	0.93
1928 . . .	2.08	1945 . . .	1.02
1929 . . .	2.20	1946 . . .	1.13
1930 . . .	2.41	1947 . . .	1.26
1931 . . .	2.73	1948 . . .	1.40
1932 . . .	3.17	1949 . . .	1.56
1933 . . .	2.70	1950 . . .	1.72
1934 . . .	2.77	1951 . . .	1.89
1935 . . .	2.80	1952 . . .	1.75
1936 . . .	2.87	1953 . . .	1.62
1937 . . .	2.94	1954 . . .	1.55
1938 . . .	2.94	1955 . . .	1.49
1939 . . .	3.01	1956 . . .	1.20
1940 . . .	2.94	1957 . . .	0.96
1941 . . .	2.87	1958 . . .	0.86
1942 . . .	2.80	1959 . . .	0.76
1943 . . .	2.72	1960 . . .	0.70
1944 . . .	2.58	1961 . . .	0.63
1945 . . .	2.44		
1946 . . .	2.22		
1947 . . .	2.08		
1948 . . .	1.93		
1949 . . .	1.79		
1950 . . .	1.65		
1951 . . .	1.50		
1952 . . .	1.36		
1953 . . .	1.22		

Value of Rice Output

Data for the value of rice output ($p_o q_o$) are obtained by valuing the physical outputs of rice by the 1934-36 average price in order to estimate the stream of social benefits in real terms. The years generally used as the basis of

Table 2-10. Expenditures on Rice-Breeding Programs by National and Prefectural Governments in Japan (thousand yen in 1934-36 constant prices)

Programs Initiated Before the Assigned Experiment System				Programs Established Under the Assigned Experiment System			
Year	National	Prefectural	Total	Year	National	Prefectural	Total
1904 . . .	135	330	465	1927 . . .	97		97
1905 . . .	136	327	463	1928 . . .	83		83
1906 . . .	137	362	499	1929 . . .	87		87
1907 . . .	130	365	495	1930 . . .	94		94
1908 . . .	162	445	607	1931 . . .	98		98
1909 . . .	158	439	597	1932 . . .	86	11	97
1910 . . .	160	489	649	1933 . . .	79	29	108
1911 . . .	185	502	687	1934 . . .	70	58	128
1912 . . .	142	465	607	1935 . . .	65	86	151
1913 . . .	113	402	515	1936 . . .	58	116	174
1914 . . .	121	468	589	1937 . . .	49	166	215
1915 . . .	134	520	654	1938 . . .	44	198	242
1916 . . .	142	541	683	1939 . . .	36	205	241
1917 . . .	106	483	589	1940 . . .	32	192	224
1918 . . .	94	499	593	1941 . . .	32	193	225
1919 . . .	100	538	638	1942 . . .	30	187	217
1920 . . .	98	657	755	1943 . . .	41	178	219
1921 . . .	130	923	1,053	1944 . . .	37	167	204
1922 . . .	119	834	953	1945 . . .	25	131	156
1923 . . .	150	877	1,027	1946 . . .		108	108
1924 . . .	182	785	967	1947 . . .		194	194
1925 . . .	112	818	930	1948 . . .		298	298
1926 . . .	135	1,035	1,170	1949 . . .		417	417
1927 . . .	126	1,180	1,306	1950 . . .		479	479
1928 . . .	139	1,265	1,404	1951 . . .		624	624
1929 . . .	147	1,140	1,287	1952 . . .		652	652
1930 . . .	163	1,297	1,460	1953 . . .		685	685
1931 . . .	175	1,350	1,525	1954 . . .		729	729
1932 . . .	320	1,450	1,770	1955 . . .		642	642
1933 . . .	243	1,456	1,699	1956 . . .		588	588
1934 . . .	252	1,454	1,706	1957 . . .		527	527
1935 . . .	261	1,536	1,797	1958 . . .		505	505
1936 . . .		1,323	1,323	1959 . . .		480	480
1937 . . .		1,257	1,257	1960 . . .		419	419
1938 . . .		1,150	1,150	1961 . . .		403	403
1939 . . .		1,075	1,075				
1940 . . .		791	791				
1941 . . .		690	690				
1942 . . .		593	593				
1943 . . .		500	500				
1944 . . .		401	401				
1945 . . .		265	265				
1946 . . .		186	186				
1947 . . .		268	268				
1948 . . .		337	337				
1949 . . .		394	394				
1950 . . .		382	382				
1951 . . .		409	409				
1952 . . .		424	424				
1953 . . .		427	427				

index construction are 1934-36 because it is considered that "normal" price relations prevailed during this period. The price of rice relative to the prices of other commodities was somewhat lower during this time because this period was characterized by a large inflow of rice from overseas territories – Korea and Taiwan – although the government tried to support the price of rice by increasing the government inventory. The valuation of output by the 1934-36 average price might result in an underestimation of the stream of social benefit.

Cost of Rice-Breeding Research

Data for the expenditure on rice-breeding research before the Assigned Experiment System are not readily available. There is an estimate that the ratio of expenditures on crop-breeding programs to the total expenditures of agricultural experiment stations in 1927 was 43 percent for the national experiment stations and 45 percent for the prefectural experiment stations.[17] We estimated the annual expenditures for rice-breeding research by multiplying these ratios by the total expenditures of the national and prefectural stations.

All expenditures for research under the Assigned Experiment System were paid for from the budget of the central government, and these data are readily available. In addition to the expenditures covered by the central government, prefectural governments paid for the costs of the tests of local adaptability of the *Norin* varieties and of the multiplication of improved seeds. Those expenditures by the local governments were estimated by multiplying the expenditures for crop-breeding programs in the prefectural experiment stations by the ratios of area planted in the *Norin* varieties to area planted in the total improved varieties.

The time-series of the expenditures on crop-breeding programs thus estimated were deflated by the Consumer Price Index, with 1934-36 = 100. The results are shown in Table 2-10. The estimates of the expenditures on crop-breeding programs include not only the cost of research and development but also the cost of extension, such as the multiplication of seeds. In addition, our cost data include not only the projects on rice but also those on *mugi* (wheat, barley, and naked barley), although rice research projects should outweigh all others in the programs.

NOTES

1. This chapter summarizes a part of the results of the research project entitled "Science and Agricultural Progress: The Japanese Experience," which was supported by a grant of the Rockefeller Foundation to the University of Minnesota Economic Development Center. It draws heavily on Masakatsu Akino and Yujiro Hayami, "Efficiency and Equity in Public Research: Rice Breeding in Japan's Economic Development," *American*

Journal of Agricultural Economics, 57 (February 1975), 1-10; Yujiro Hayami and Saburo Yamada, "Agricultural Research Organization in Economic Development: A Review of the Japanese Experience," *Agriculture in Development Theory*, ed. Lloyd G. Reynolds (New Haven: Yale University Press, 1975); and Yujiro Hayami, in association with Masakatsu Akino, Masahiko Shintani, and Saburo Yamada, *A Century of Agricultural Growth in Japan: Its Relevance to Asian Development* (Minneapolis and Tokyo: University of Minnesota Press and University of Tokyo Press, 1975).

2. Yujiro Hayami and Vernon W. Ruttan, *Agricultural Development: An International Perspective* (Baltimore and London: Johns Hopkins University Press, 1971), p. 144.

3. The historical sketch in this section draws heavily on Hayami and Ruttan, *Agricultural Development*, pp. 153-163. Useful references are Toshio Furushima, ed., "Nogaku" ("Agricultural Science"), *Nihon Kagaku Gijutsushi Taikei* (*The Comprehensive History of Science and Technology in Japan*), vols. 22 and 23 (Tokyo: Daiichi Hoki Shuppan, 1967 and 1970); Nogyo Hattatsushi Chosakai (Agricultural Development History Research Committee), ed., *Nihon Nogyo Hattatsushi* (*History of Agricultural Development in Japan*), 10 vols. (Tokyo: Chuokoronsha, 1953-58); Takekazu Ogura, ed., *Agricultural Development in Modern Japan* (Tokyo: Fuji Publishing Co., 1963); Yukihiko Saito, *Nihon Nogakushi* (*The History of Agricultural Science in Japan*), 2 vols. (Tokyo: National Research Institute of Agriculture, 1968 and 1971).

4. Ministry of Agriculture and Forestry, *Meiji Zenki Kannojiseki Shuroku* (*The Compilation of Measures to Encourage Agriculture*), vol. 2, reprint (Tokyo: Dainihon Nokai, 1939), pp. 1765-1779, 1766-1767.

5. *Ibid*., pp. 1778-1779.

6. *Ibid*., p. 1774.

7. Saito, *Nihon Nogakushi*, p. 121.

8. *Ibid*., pp. 126-134, 161-164.

9. Takamine Matsuo, "Suito Hinshu Kairyo Shijo no Shomondai" ("Problems in the History of Rice Variety Improvement"), Hattatsushi Chosakai Data no. 42, mimeographed (Tokyo, 1951).

10. For a historical review of rice policy in Japan, see Yujiro Hayami, "Rice Policy in Japan's Economic Development," *American Journal of Agricultural Economics*, 54 (February 1972), 19-31.

11. For the quantitative steps used to put the model into practice see Appendix 2-2.

12. Zvi Griliches, "Research Costs and Social Returns: Hybrid Corn and Related Innovations," *Journal of Political Economy*, 66 (October 1958), 419-431; W. L. Peterson, "Return to Poultry Research in the United States," *Journal of Farm Economics*, 49 (August 1967), 656-669.

13. H. W. Ayer and G. E. Schuh, "Social Rates of Return and Other Aspects of Agricultural Research: The Case of Cotton Research in São Paulo, Brazil," *American Journal of Agricultural Economics*, 54 (November 1972), 557-569; Nicolas Ardito Barletta, "Costs and Social Benefits of Agricultural Research in Mexico," Ph.D. dissertation (Chicago: University of Chicago, 1971).

14. Kazushi Ohkawa, *Shokuryo Keizaino Riron to Keisoku* (*Theory and Measurement of Food Economy*) (Tokyo: Nihon Hyoronsha, 1945), pp. 9-34, 77-96.

15. Hayami and Ruttan, *Agricultural Development*, pp. 236-237.

16. Yasuhiko Yuize, "Nogyo Seisan ni Okeru Kakaku Hanno" ("On the Price Responses in Agricultural Production"), *Nogyo Sogo Kenkyu*, 19 (January 1975), 107-142.

17. Takeichi Oda, "Honpo ni okeru Beibaku Hinshu Kairyo Jigyo no Taiko" ("Summary of the Rice and Mugi Varietal Improvement Programs in Japan"), *Dainihon Nokaiho* (November 1929), 14-28.

3

Organization and Productivity of the Federal-State Research System in the United States[1]

Willis L. Peterson and Joseph C. Fitzharris

The agricultural research system of the United States is discussed in this chapter under two broad headings. In part I, we examine the organization of the federal-state system. We investigate how political and physical geography as well as production and input trends in the agricultural sector influenced the search by more efficient farmers for new inputs, techniques, and organizational forms. The origins of the federal system and its structure are detailed. A view of the workings and structural complexities of the federal-state system is gained through an examination of the agricultural research system of Minnesota. The origins of the state system, the resulting organizational structure, and the types of work done at the state level are reviewed.

Part II deals primarily with the productivity of the federal-state research system. After briefly reviewing the relationship between agricultural research and farm productivity, we attempt to explain the absence of productivity growth in United States agriculture until about forty years after the establishment of the federal-state system. Viewing agricultural research and extension as an investment, we then provide rough estimates of the marginal internal rate of return to this investment for specified periods from the 1930s to the 1980s. Finally, some evidence is presented which bears upon the question of whether or not there is an efficient allocation of public agricultural research in the United States.

Organization of the Federal-State Research System

The federal-state agricultural research system of the United States is a decentralized, cooperative system composed of both federal and state agencies. Without central control, the system attempts to allocate resources, solve pressing problems, produce new varieties of crops and livestock, and conduct basic agricultural research.

The United States research system reflects the political dualism, geographic differentials, and historical accidents surrounding its origins. The vast bulk of work is done on the state level by the various autonomous state agricultural experiment stations. For this reason the following case study, in which we examine the development of the Minnesota Agricultural Experiment Station and the Institute of Agriculture in the University of Minnesota, will provide a good basis for understanding the overall system.

The Minnesota Case

The College of Agriculture of the University of Minnesota was founded in 1869, eleven years after the first efforts to develop a college of agriculture proved unsuccessful. The college lacked stability in the early years, having very few students and a rapid turnover in teaching staff. In addition, its experimental farm was inadequate and poorly funded. A new campus and a farm were acquired in 1882, and the Farmers' Lecture Courses, the forerunner of the Agricultural Extension Service in Minnesota, were initiated.[2]

As a part of the movement to gain federal support for agricultural research, the Minnesota legislature authorized a state agricultural experiment station at the university. Established by the university regents on the university farm in 1885, the Minnesota station remained a paper creation until the passage of the Federal Hatch Act in 1887. After Hatch Act funds became available, the station hired a staff and began operation.

Agricultural extension work was initiated in 1910, and the Agricultural Extension Service was established in 1914. In the years after 1893, branch stations were founded to serve diverse sections of the state. Expansion of the college faculty-station staff was followed by the beginning of graduate training. As this system developed, many of the geographic and economic forces that affected the national system also affected the state system.

The state setting. The Minnesota agricultural research, extension, and education system developed out of local, state, and national movements for government aid to agriculture. Farmers' organizations were instrumental in the origins and development of the Minnesota system. Soil differences, production trends, and other problems too great for farmers and farmers' organiza-

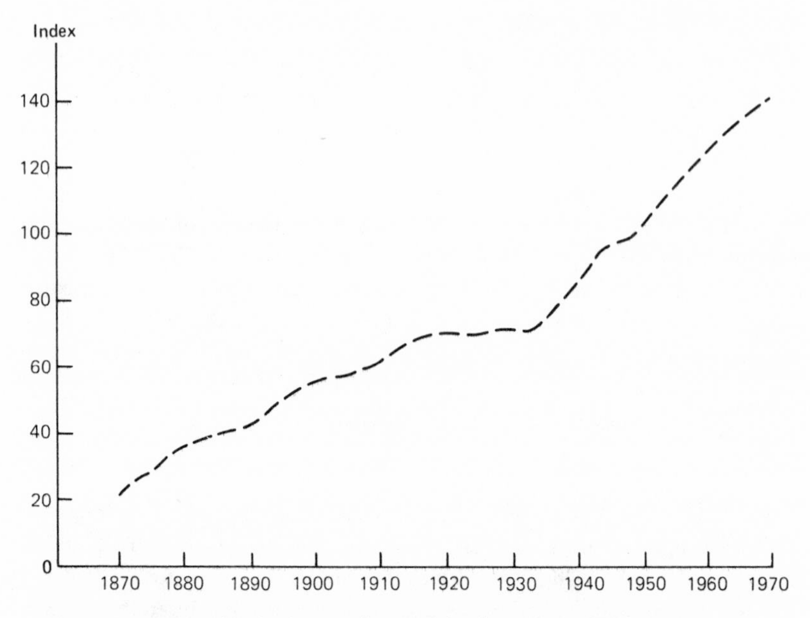

Figure 3-1. Index of United States farm production, 1870-1970.
Source: Bureau of the Census, *Historical Statistics of the United States*, 1960, p. 288 (Series K-190). USDA, *Agricultural Statistics, 1972*, p. 537 (adjusted from a 1967 base). Decennial observations, 1870-1910.

tions alone to solve helped shape the developing system of agricultural research in the state.

Geography. Because no single research unit could begin to deal with the problems posed by the great regional differences in soil and climate in the United States, the country's political structure demanded the creation of a federal-state agricultural research system.

The salient feature of Minnesota political geography is its urban-rural contrast, but this has not seriously affected the development of the Minnesota Agricultural Experiment Station and Institute of Agriculture. More important factors are soil and climatic differences throughout the state. There is considerable variation in growing season, temperature, and average moisture among the diverse regions of Minnesota. Type and quality of soil, which affect type of farming, vary across the state. Because of these physical and climatic factors, branch stations located in the major regions have been important components of the Minnesota agricultural research system.

Production and input trends. United States farm output tripled between 1870 and 1915. Then it remained roughly constant until 1935 when another period of rapid growth began which lasted until 1945. Since 1950, growth in farm output has been rapid (Figure 3-1).

The growth of farm output before 1915 paralleled the increase in improved acreage during the same period. After 1915, this very clear relationship between the expansion of farm land and the growth in farm output ceased to exist, as urbanization increased and farm land dwindled. Between 1915 and 1970, farm labor also declined drastically. Of the traditional factors of production only capital expanded.

The rise in farm output to 1900 can be explained by geographic expansion of agriculture, that is, by expansion in amount of land used for farming, although capital and the number of farmers were also increasing. After 1910, geographic expansion does not contribute appreciably to the expansion of output. Figure 3-2 illustrates the relationship among the expansion of land, labor, and capital. After 1900, capital becomes increasingly important, except during the agricultural depression of the 1920s and 1930s. Clearly, capital (the value of land and buildings) does not account for the amount spent on machinery, fertilizer, disease-resistant strains of crops, and better livestock. And perhaps even more important, capital does not include better farming practices.

Better farming practices alone tend, ceteris paribus, to increase output. Combining better farming methods with hardier and/or disease-resistant crop strains leads to a further increase in agricultural production. Improved health of livestock also increases farm income, or production. After 1880, all these developments resulted from the work of the agricultural experiment stations. These stations worked with existing crops, using a trial selection process. After 1900, some basic research led to cures for various livestock diseases (e.g., hog cholera serum). The objective of this work was to maintain yields and production levels.

Minnesota production trends. In Minnesota the value of agricultural production grew sevenfold between 1880 and 1920. During the 1920s and the early 1930s, output declined slightly. After 1935, production again rose, tripling by 1950. A brief decline in the late 1950s and early 1960s interrupted an otherwise continual increase in the value of agricultural production (Figure 3-3).

Between 1880 and 1930, farms doubled in number and improved acreage tripled. Land in farms continued to expand to 1950, after which it declined gradually (Figure 3-4). The decline in improved acreage after 1950, however, was twice as great as the decline in total acreage. Since 1940, the number of

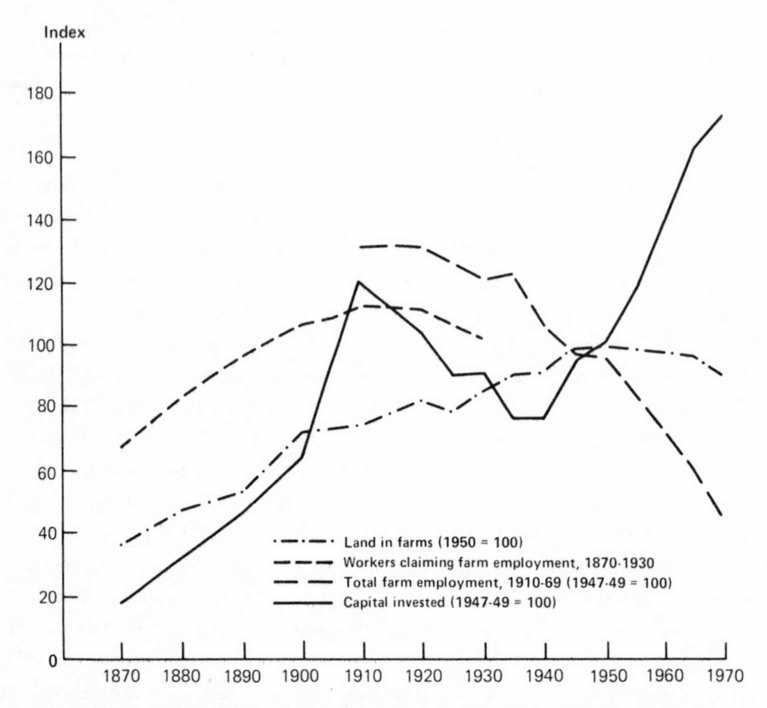

Figure 3-2. Indices of land, labor, and capital in United States agriculture, 1870-1970.
Source: Bureau of the Census, *Historical Statistics of the United States*, 1960, p. 72 (Series D-37), p. 280 (Series K 73-75). USDA, *Agricultural Statistics, 1962*, p. 512; *1972*, pp. 504, 523, 537; *1973*, p. 424.

farms has declined, and total labor employed (both paid and unpaid) in farming fell rapidly between 1940 and 1970. Aggregate capital input (in horsepower equivalents) is the only input that has risen over the entire period (Figure 3-5).[3]

Efforts by the Minnesota station to produce hardier crop varieties with shorter growing seasons resulted in increased land productivity between 1900 and 1920. In the 1930s, and again after 1950, land productivity rose as more fertilizers and pesticides and better disease-resistant crops were utilized. With the exception of the 1890s and the 1930s, labor productivity has risen. The expansion of land per worker has been uneven, varying with the adoption of new methods and machinery. The substitution for human power of animal power, steam power, and the internal combustion engine/diesel engine tractor explains much of the change in the ratio of land to worker (Figure 3-6).

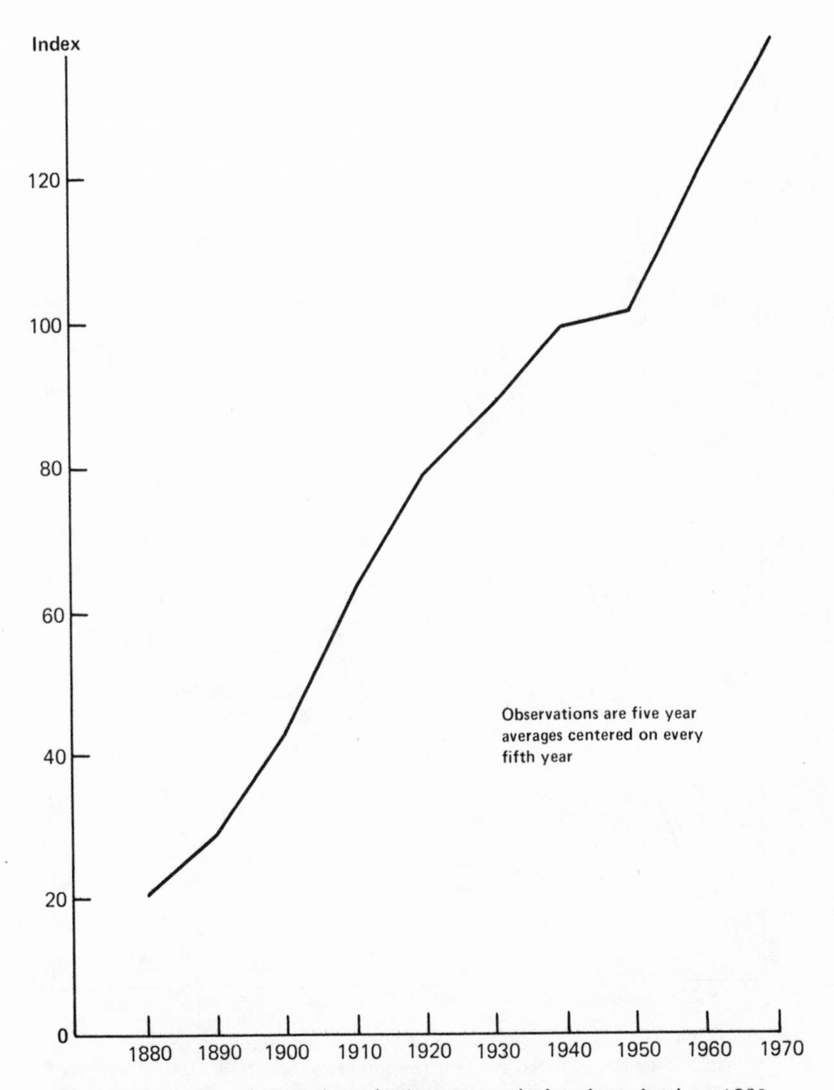

Figure 3-3. Index of the value of Minnesota agricultural production, 1880-1970 (constant dollar values, 1950 = 100).
Source: Joseph C. Fitzharris, *The Development of Minnesota Agriculture, 1880-1970: A Study of Productivity Change*, Department of Agricultural and Applied Economics Staff Paper P 74-20, St. Paul, University of Minnesota, September 1974, p. 5.

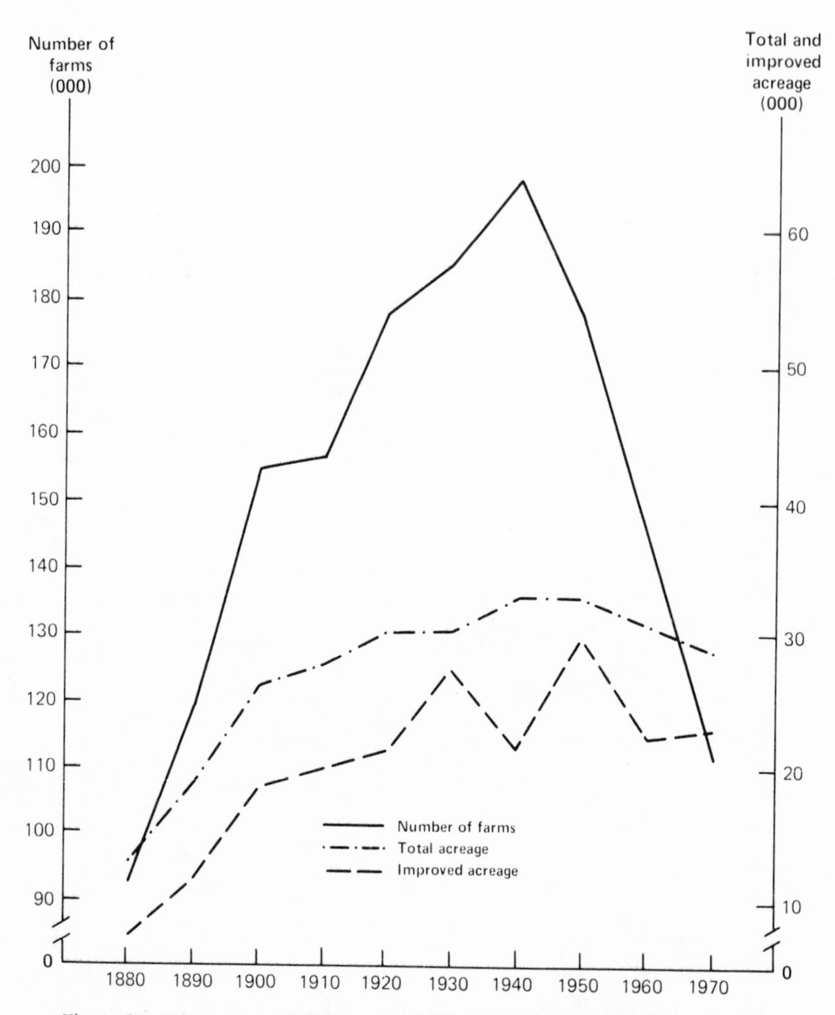

Figure 3-4. Minnesota agriculture, 1880-1970: Farms and farm acreage. Source: Joseph C. Fitzharris, *The Development of Minnesota Agriculture, 1880-1970: A Study of Productivity Change*, Department of Agricultural and Applied Economics Staff Paper P 74-20, St. Paul, University of Minnesota, September 1974, p. 7.

Private efforts. Farmers' organizations in Minnesota from the 1850s to the 1890s attempted to solve many of the problems facing Minnesota farmers. Individual efforts had proved too costly, but group effort, because of the "free rider" problem, also failed. As a consequence, these organizations turned to the state government for assistance.

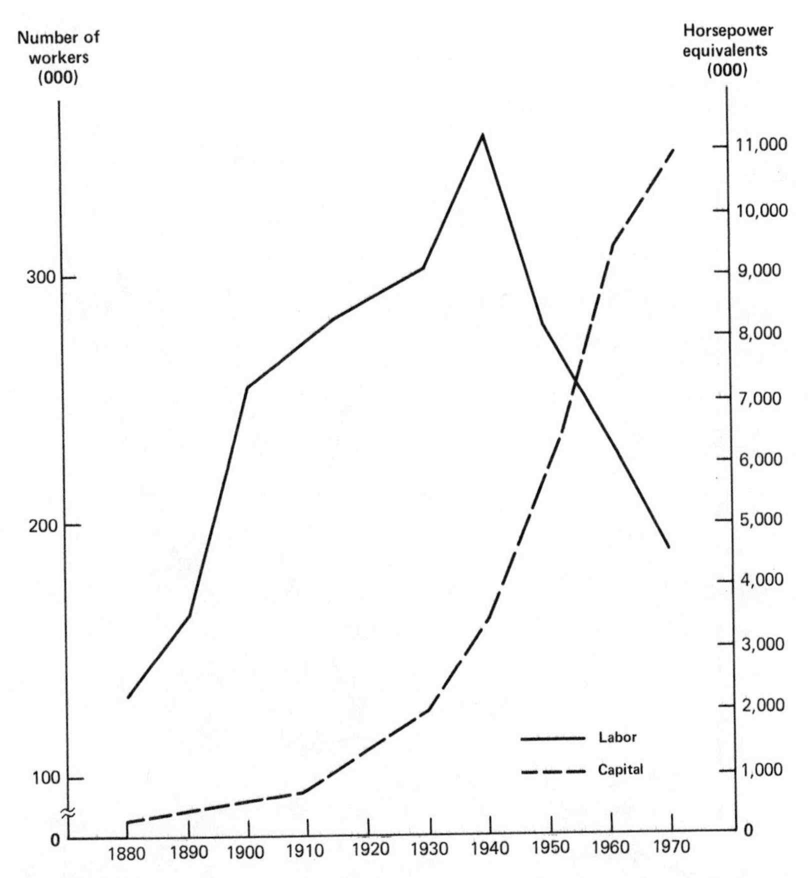

Figure 3-5. Minnesota agriculture, 1880-1970: Labor and capital (horsepower equivalent) inputs.
Source: Bureau of the Census, *Census of Agriculture*, 1880-1970. Joseph C. Fitzharris, *The Development of Minnesota Agriculture, 1880-1970: A Study of Productivity Change*, Department of Agricultural and Applied Economics Staff Paper P 74-20, St. Paul, University of Minnesota, September 1974, p. 9.

In 1885, two years before the Federal Hatch Act was passed, the Minnesota state government authorized an experiment station although it did not provide funding. Federal support was necessary for the development of the Minnesota station and agricultural experiment stations in other states. Yet in Minnesota, farmers' groups were instrumental in developing the agricultural experiment station. In the state elections of the 1880s and 1890s the objectives and accomplishments of the experiment station-college of agriculture

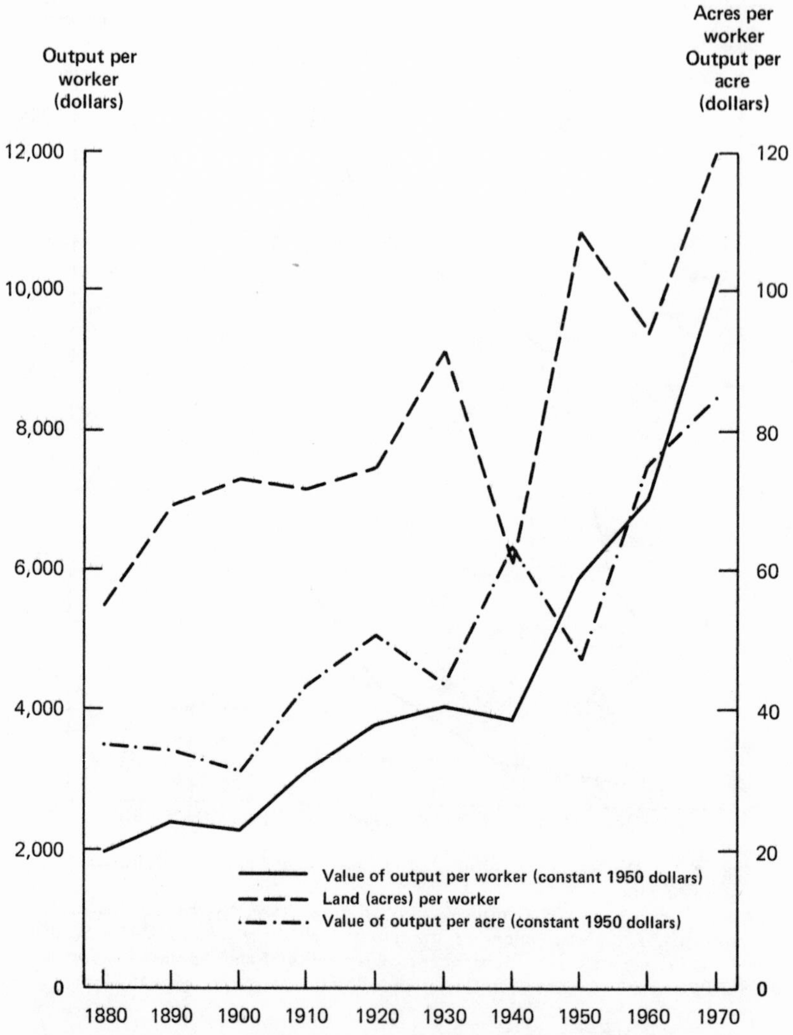

Figure 3-6. Minnesota agriculture, 1880-1970: Labor and land productivity. Source: Joseph C. Fitzharris, *The Development of Minnesota Agriculture, 1880-1970: A Study of Productivity Change*, Department of Agricultural and Applied Economics Staff Paper P 74-20, St. Paul, University of Minnesota, September 1974, p. 11.

were frequent topics of debate. These debates left the station and college administrators and the university regents firmly convinced that the first duty of the agricultural research, teaching, and extension system was to serve the farm sector's immediate needs. Basic research was thus given low priority in the early years.[4]

Origins of the Minnesota system. In 1881, Edwin D. Porter, the fourth professor of agriculture in the College of Agriculture, arrived in Minnesota and proceeded to meet with farmers' organizations, leading citizens, and legislators to determine their views on the role of the college and farm in the service of the state. Obtaining a new campus and farm for the college was Porter's first major accomplishment, and on the new farm he built the foundations for the Minnesota agricultural research and teaching system.[5]

From the early years of the college and station, the staff worked closely with the various farmers' organizations and commodity groups as they were founded. Frequently serving as officers of such groups, staff members gained close contact with the farmers and their problems.

In 1882, a lecture series called the Farmers' Lecture Courses was established, following the example of colleges in other states. Initially well received, the Lecture Courses, which were later expanded and renamed the Farmers' Institutes, went through a period of some uncertainty. In 1914 Congress passed the Smith-Level Act, which provided federal support for agricultural extension work. The Agricultural Extension Division of the experiment station was then separated from the station and became the Agricultural Extension Service. From 1910 to 1917 the Farmers' Institutes were absorbed by this service, which, by law, was supervised by a county farmers' organization called the Farm Bureau. This tie to a single farmers' organization, particularly in years of conflict among the various farmers' groups, had a deadening effect on the Minnesota Agricultural Extension Service (and on those of other states as well). Confidence in the service decreased, and many farmers believed that favoritism was shown to members of the Farm Bureau. In the 1950s, however, the service was formally separated from the farm bureaus and was funded by the state and federal governments, with the assistance of the counties in which state and federal agents were stationed.[6]

Structure. In 1888 the Minnesota Agricultural Experiment Station and Institute of Agriculture were organized in accordance with the requirements of the Hatch Act. The Institute (then the Department) of Agriculture was founded to provide supervision of teaching and research activities, and its dean was also the director of the Agricultural Experiment Station. Academic subject-matter divisions were established in the station and college, and the same staff served both station and college.[7]

After the Agricultural Extension Service was initiated, the director was coequal with the station's vice-director and the associate dean of the college. When the Institute of Agriculture was created out of the Department of Agriculture in 1952, the directorship of the station was separated from the institute deanship. The college deans and the directors of the Experiment Station, the Agricultural Extension Service, and the Office of International Programs in Agriculture were coequal. The College of Veterinary Medicine became an autonomous unit, cooperating with the experiment station in animal research.

Over the years, the station and institute have made cooperative arrangements with agencies of the United States Department of Agriculture and with experiment stations in neighboring states, beginning with North and South Dakota in the 1890s. Several USDA personnel have been assigned to the station and given academic rank in the college. After Professor E. C. Stakman began working as a cooperating federal agent in barberry eradication and cereal rust investigation, numerous federal plant pathologists were assigned to the university, and in the 1950s the USDA Cereal Rust Laboratory was established at the university, cooperating with the Department of Plant Pathology.

Work done at the Minnesota station. In the first years of the Minnesota Agricultural Experiment Station's existence, the staff centered its efforts on disseminating information produced by other stations, adapting that information to Minnesota's soil and climatic conditions. It also began working to develop varieties of crops and shrubs suited specifically to Minnesota agricultural conditions. Later, the station began crop and livestock breeding experiments, conducted research in farm management and agricultural engineering, and worked on plant morbidities and mortalities, emphasizing the cereal rusts.

Much of the work done in the early years was maintenance work, or "applied-developmental" research, as illustrated by the work on cereal rusts. At first, barberry eradication programs were the major emphasis in the station's efforts to combat cereal rusts. Since the barberry plant harbored the wintering parasite, the fastest way to prevent cereal rust was to eradicate the wintering host. Later, as plant-breeding work became more sophisticated and as time permitted, disease-resistant plants were developed. Eradication of the barberry had "bought time" for the station to breed disease-resistant strains. The national effort was relaxed in the late 1940s, however, and in the early 1950s a serious outbreak of cereal rust destroyed much of Minnesota's wheat crop. From that time on, the Minnesota station, in close cooperation with

USDA laboratories and field units in Puerto Rico, on the mainland (particularly the Minnesota Cereal Rust Laboratory), and in Mexico (where new strains of cereal rust have been identified), has made an unremitting effort to breed disease-resistant plants. As a consequence, the problem of cereal rust, like that of blast, has been solved through the joint efforts of state and federal researchers.[8]

A leading and continuous line of work at the station has been crop adaptation. Efforts to move crops northward, adapting them to shorter growing seasons and colder climates, began in the 1890s. The initial work involved trial experiments and the selection of the best varieties. Considerable success was achieved in moving corn northward and in selecting wheat varieties better adapted to the shorter growing season of the northern two-thirds of the state. After the turn of the century, breeding and crossbreeding experiments were initiated. Breeding efforts were even more successful than trial experiments in producing varieties adapted to the rigorous climate and soil conditions in Minnesota. Much of this work has been cooperative, involving the neighboring state experiment stations, the Minnesota branch stations, and various bureaus in the USDA.[9]

Analysis of station publications reveals that applied-developmental work in the first forty years of the station was closely associated with basic research on crops and livestock (feeding trials, breeding, and varietal adaptation) and engineering work. There was a particular emphasis on human and animal nutrition studies in the years before World War I. In the 1920s and 1930s a pronounced trend toward basic-applied work began to develop.[10]

Maintenance research conducted into the 1920s proved to be very useful for the station. Although the station's work did not result in an increase in agricultural productivity, it probably prevented any appreciable decline in productivity owing to crop and animal diseases. Moreover, by conducting adaptation work for both plants and animals, the station produced strains and varieties which could be grown in Minnesota's colder climate and shorter growing season. Shelterbelt and drainage work improved both the soil and the soil retention of the farm. On balance, although the station did not make many new discoveries, it did preserve the status quo.

In the 1920s, the stations began moving more heavily into basic-applied research. The long time-lag between the initial investment in basic work and the beginnings of positive returns helps to account for the relatively constant productivity of the agricultural sector.

The Federal-State System

The federal-state agricultural research system developed in response to forces and factors operative in the American economy in the nineteenth century. Originating in legislative response to the demands of farmers and their organizations, the system is still closely linked to farm groups. Its structure, powerfully influenced by its origins, is not well understood. Americans think of their state stations as autonomous bodies cooperating with other stations and with the USDA. Foreign observers often see the American system as quite centralized, despite some provincial tendencies.[11] Such observers also tend to view the stations' combination of teaching, research, and extension as inefficient.[12]

Origins and early development. In the 1790s, agricultural societies were established in several states. These societies, formed by the more successful, wealthier farmers, encouraged their members to experiment, to collect new varieties of seeds and animals, and to spread this knowledge widely. To this end, the societies published the proceedings of their meetings, sponsored farm journals, and sponsored and were instrumental in the establishment of state agricultural fairs.

As private effort and initiative in agricultural research became increasingly costly, the problem of adequately supporting this research grew. By the 1840s the agricultural societies had turned for assistance to their state governments, several of which responded by founding state departments of agriculture. These departments did not conduct research but served instead as collectors and disseminators of information.

In 1862 Congress authorized the establishment of the United States Department of Agriculture. Although this federal department was not explicitly charged with conducting research, the implication that it should do so was clear. Also in 1862, Congress passed the Morrill Land Grant College Act. This act allocated public lands to the various states to be used to support one or more state colleges of agricultural and mechanical arts. Such colleges were encouraged to maintain experimental farms and to conduct adaptive trials of crops, shrubs, and livestock. These farms, intended to support the teaching function of the colleges, became useful as well in the colleges' efforts to serve farmers' needs for information.

Because in the early 1800s the existing body of agricultural knowledge was inadequate to provide a solid academic curriculum in agricultural education, the colleges worked to extend the scientific underpinnings of agriculture. By the 1870s, the inability of the colleges of agriculture to broaden the frontiers of knowledge and to solve agricultural problems had become apparent. The first agricultural experiment station was established in Connecticut in 1876. Subsequently other states established stations, many of which were separate from the state colleges of agriculture.[13]

The Morrill Land Grant College Act of 1862 and the Hatch Agricultural

Experiment Station Act of 1887 reflect the emergence of a dual federal-state approach to agricultural research. Under these acts, each state received funds for a college of agricultural and mechanical arts and for an agricultural experiment station. This division of effort along state lines had a practical benefit not fully realized by the legislators when the acts were first passed.

In response to the Hatch Act's provision of federal funds, the states authorized the establishment of experiment stations attached to their colleges of agriculture. It may be noted that the American agricultural experiment stations, unlike their German model at Moeckern, Saxony, were and are attached to colleges of agriculture. And in similar fashion, the agricultural extension services of the various states are connected to the agricultural colleges.[14]

Farmers and farmers' organizations have played a central role throughout the developments detailed above: the establishment of state departments of agriculture; the pressure for authorization of a federal department of agriculture; the allocation of federal lands to the states for the support of agricultural colleges; and the movement – at both state and federal levels – for the establishment of state agricultural experiment stations. As we have seen, the more successful, wealthier farmers were instrumental in obtaining government assistance for the agricultural sector. These farmers and their organizations, after helping to initiate the institutional arrangements necessary to utilize government aid,[15] continued to press for such assistance and served as "watch dogs" over the system they had helped to create by criticizing, demanding, and protecting.

Within the federal-state structure, the states have set up research systems in which the college of agriculture forms the base, while research activities are carried out by the staff of the state agricultural experiment station attached to the college. Extension work is the responsibility of the state agricultural extension service, which operates in the counties but is also attached to the state agricultural college.

On the federal level, the USDA maintains a large staff of research workers in the national capital and in laboratories, stations, and other federal installations across the country. Additional federal workers and facilities are located on the campuses of colleges of agriculture in the various states. The states and the federal government cooperate closely on problems that cross state borders or that are national in scope or origin. An example is the problem of cereal rust disease, mentioned earlier, which involved not only federal and state cooperation within the United States but cooperation between the United States and Mexican governments, with some work conducted at research units in northern Mexico.[16]

Productivity of the Federal-State Research System

Agricultural Productivity, 1870-1972

It is helpful to think of research as a production activity in which the inputs consist of scientific man-years, laboratory facilities, and the like, and output is composed of new knowledge. To determine the productivity of research, we need to measure both input and output. Although research inputs can be measured fairly easily, at least in monetary terms, the same is not true for output. Fortunately research output can be evaluated indirectly by measuring the productivity of the industry toward which the research is directed.

In the case of agricultural research, part of the output is transmitted directly to farmers and part is utilized by experiment stations and farm suppliers as an intermediate input. In both situations the new knowledge makes possible the production of new or improved inputs for agriculture. To the extent that improvements in the quality of agricultural inputs are not fully and accurately measured, we *may* obtain an increase in total factor productivity in agriculture. Hence we may use the observed growth in agricultural productivity as a proxy or indirect measure of the output of agricultural research.

As shown in Figure 3-7, the major share of the growth of agricultural productivity in the United States over the past century has taken place since the mid 1930s. Given the establishment of agricultural experiment stations in the late 1880s, it is puzzling why it took over forty years for productivity growth to begin. One possible explanation for the "long dry spell" is that agricultural research simply did not produce any significant results during its early years. But this is too easy an explanation. It does not appear that there was any abrupt change either in the organization of the agricultural research establishment or in the quality of its personnel just before agricultural productivity growth started. If research was not productive in 1900, why should it suddenly have become so in 1930?

Maintenance research. We may shed some light on this puzzle by considering the nature and the absolute amount of research conducted during the early 1900s. As we observed in a previous section, experiment station personnel appear to have devoted the majority of their time to solving immediate and pressing problems faced by farmers. If crop or livestock production was declining or was threatened by disease or some other problem, it was the job of the researcher to come up with a solution to ensure, at least, that agricultural productivity did not decline. In other words, the research effort during the early years of the experiment station system appears to have been aimed largely at maintaining agricultural productivity in the face of a constant surfacing of new problems. It is reasonable to suppose that without this research

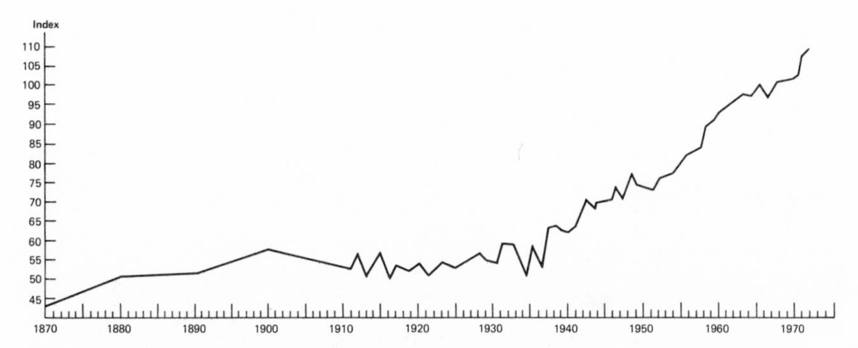

Figure 3-7. Index of output per unit of input, United States agriculture, 1870-1972, 1967 = 100.
Source: USDA, *Agricultural Statistics, 1962*, p. 54; *1972*, p. 31. Decennial observations, 1870-1910.

agricultural productivity would have declined between 1900 and 1935 instead of remaining fairly constant.

Although there can be little doubt that a certain amount of research is required just to maintain productivity in agriculture, two unanswered questions persist: how much research was required for maintenance purposes in the early 1900s and how much is required today? As technology has improved over the years, has the amount of research necessary to maintain productivity increased, remained about the same, or declined? One might argue that as varieties of crops and breeds of livestock are bred up to produce greater yields they lose some of their inherent resistance to disease and pests and thus require an increasing amount of maintenance research. On the other hand, it is probably true that because of both the increase in the stock of knowledge and the creation of new chemical inputs many diseases and pests which represented major problems for farmers fifty years ago are now either nonexistent or routinely controlled. This would imply a decrease in the research required to maintain productivity. In sum, there does not appear to be a strong argument for either a greater or a smaller amount of maintenance research today than was needed in the early 1900s.

The annual expenditures on total agricultural research have, of course, increased greatly over the years. Unless the required maintenance research has increased proportionately with the total, which does not appear likely, the absolute amount of research devoted to technology-producing activities, as opposed to maintenance work, also has increased substantially.

Research deflators. To gauge accurately the growth in real research inputs

Table 3-1. Alternative Research Deflators

Year	Consumer Price Index	Index of Associate Professor Salaries
1915	24	12
1920	48	16
1930	40	23
1940	34	22
1950	58	40
1960	71	57
1972	100	100

Source: For 1915-42, George Stigler, "Employment and Compensation in Education," National Bureau of Economic Research, Occasional Paper no. 33, 1950; for 1948-72, *American Association of University Professors Bulletin*, respective years.

over time it is necessary, because of the increase in the general price level, to deflate the expenditure figures. However, the use for this purpose of a common price deflator such as the Consumer Price Index (CPI) probably will result in a gross underestimate of past research when compared with current figures because professional salaries, which weigh heavily in total research costs, have risen faster than the general price level over the past fifty to sixty years.

To estimate the increase in research costs more accurately, a price index reflecting the average salaries of associate professors in public universities was constructed. As shown in Table 3-1, these salaries multiplied about eight times between 1915 and 1972, whereas the general price level has increased about four times.

Even when we adjust past research expenditures to reflect the change in research costs by using the index of associate professor salaries, the average annual research input (state experiment stations plus USDA) during the 1915-25 period comes to only about 8 percent of the total public research in 1970. As shown in Figure 3-8, annual real research expenditures begin to climb sharply after 1925, increasing by 57 percent between 1925 and 1930. It seems reasonable to assume that at least 5 to 10 percent of total current research is required for maintenance purposes. Unless the amount of research required for maintaining productivity has increased greatly since 1930 (in real terms), it appears safe to say that the bulk of the research input before 1925 was necessary simply to maintain productivity. If this is so, we should expect productivity to show an increase only after 1930, when research inputs began to surpass the maintenance level by a noticeable margin.

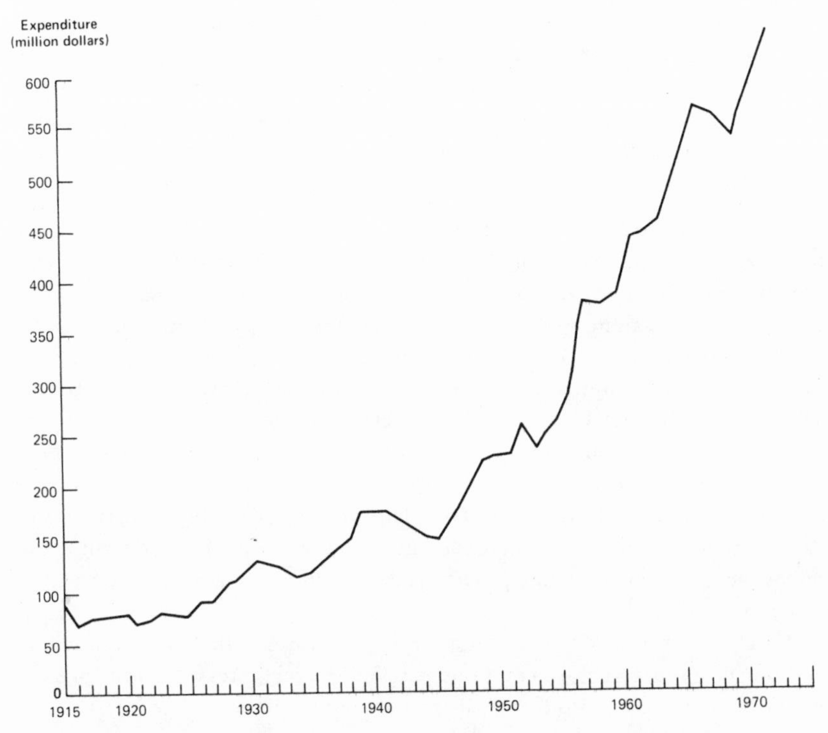

Figure 3-8. United States SAES plus USDA research expenditures, 1915-72, constant 1972 dollars (millions).
Source: See Appendix 3-1. Deflated by Index of Associate Professor Salaries, 1972 = 100.

Marginal Rates of Return to Research and Extension

Methodology. The methodology that has been used to measure the rate of return to investment in agricultural research is reviewed elsewhere in this volume (see chapter 6). In general, two approaches have been utilized. The first, which might be called the index number approach, uses productivity gains to measure the value of inputs saved or consumer surplus stemming from research.[17] The second technique, which might be called the production function approach, involves the use of research as a separate variable in a production function to measure its marginal product and marginal rate of return.[18]

We will use the index number technique in an attempt to measure the value of inputs saved as a consequence of an *increase* in agricultural productivity. To obtain a rough first approximation of the marginal rate of return (as opposed to an overall average), we measure value of *additional* inputs saved over specific periods as a result of the *growth* in productivity and in value of output. We will assume that the growth in productivity (output per unit of input) during a given six-year period is the result of research conducted over the preceding six-year period. This implies a six-year lag between research and its output.

Estimates by decades. Because United States agricultural productivity began its recent long-term growth in 1937, we take 1937-42 as the first of four six-year periods. The average annual value of additional inputs saved during each of these periods, along with the corresponding research and extension inputs, are presented in Table 3-2. For the purpose of computing rates of return, both research and inputs saved are deflated by the Consumer Price Index, 1972 = 100. The fact that research was relatively cheap (compared to inputs saved) in the early years should be reflected in its rate of return. Also, to take account of private research and extension, figures relating to public research and extension are multiplied by two. This implies that private research and extension were equal to public expenditures over the period. We shall argue in the next section that this procedure results in an underestimation of the true rate of return.

Matching the research and extension expenditures with the corresponding additional inputs saved, we compute marginal internal rates of return for each of the four six-year periods (see accompanying tabulation). The internal rate

Period	*Rate of Return*
1937-42	50%
1947-52	51
1957-62	49
1967-72	34

of return is that rate of interest which makes the accumulated research and extension expenditures at the end of the investment period just equal to the discounted present value of the additional inputs saved at the beginning of the payoff period. In calculating the internal rate of return, we assume that the average value of marginal inputs saved over the six-year period will continue into perpetuity. However, because of the high discount rate, these future returns have a small influence on the computed rate of return.

Biases. Although the computed rates of return to agricultural research and extension in the United States turn out to be very attractive, we have reason to believe that these estimates of the true rate of return are biased downward for a number of reasons. First, no return is credited to maintenance research. To capture a return to this activity we would have to know what the productivity

Table 3-2. Average Annual Expenditures for Public Research and Extension with Corresponding Additional Inputs Saved (1972 dollars, in millions)

Research and Extension Expenditures[a]		Additional Inputs Saved[b]	
Period	Annual Average	Period	Annual Average
1931-36	148	1937-42	5,868
1941-46	192	1947-52	6,587
1951-56	322	1957-62	11,747
1961-66	671	1967-72	10,010

[a] Derived from data in Appendix 3-1.

[b] Total inputs saved in year t are obtained by multiplying value of farm marketings plus home consumption by the proportionate change in total factor productivity, 1910-36 = 100. Marginal inputs saved in year t are obtained by subtracting average annual total inputs saved during the preceding six-year period from total inputs saved in year t.

would have been if no such research were performed. Because this information does not exist, we do the next best thing by measuring the gain in productivity from a base period. If productivity declines in the absence of maintenance research, we understate the true productivity gains attributable to research. Our procedure implies a zero return for the years 1921-26 when in fact it is difficult to believe that research during that period was any less productive than it was from 1931 to 1936.

The practice of doubling public research and extension in order to include an estimate of private expenditures should also bias the rate of return downward. We can expect that input prices already include a return to private research and extension. This in turn should increase the input measure and result, therefore, in less productivity gain than would occur if all research and extension were public expenditures. The fact that not all research and extension is aimed at increasing productivity provides a third source of downward bias. For example, most of the extension work in home economics is concerned with improving the quality of life, not only in rural America but also in towns and cities.

On the other side of the coin, one might argue that the rate of return is biased upward because no charge has been made for the increased education of farm people. However, most of the evidence to date suggests that the primary role of education in agriculture is to speed up the adoption of new inputs in order to move more quickly toward an optimum allocation of resources as opposed to a pure "worker effect."[19] In fact the same argument applies to extension, which also yields a return by speeding up the adjustment to new inputs or information.

Future Returns

As we have seen, although the marginal internal rates of return to agricultural research and extension appear to have been relatively high, the figures suggest that this return is beginning to decline. Of course, if research and extension are subject to the law of diminishing returns, it is reasonable to expect a decline in the rate of return to this investment in the absence of additional complementary inputs Moreover, the fact that researchers' and extension agents' salaries probably have increased more rapidly than their marginal productivities in recent years would also imply a decrease in the rate of return to investment in these activities. We might ask, therefore, whether there is a danger that the marginal rate of return to agricultural research and extension will fall below a minimum acceptable level, say 15 percent, in the foreseeable future.

Over the past two decades (1952-72) public agricultural research and extension expenditures (deflated by the CPI) have nearly doubled each decade, for a compound real rate of growth of almost 7.5 percent annually. In 1952 these expenditures totaled $305 million, rising to $509 million by 1962 and $997 million in 1972. If the past twenty-year expenditure growth rate continues over the next decade – which is quite possible in view of the increased concern over world food supplies – public research and extension expenditures would reach $1.336 billion in 1976 and $2 billion in 1982 (1972 dollars).

Predicting future productivity growth is subject to even more uncertainty. If the 1967-72 growth rate continues, the United States total factor productivity index would increase from 109 in 1972 to 124 in 1982 (1967 = 100). Such an increase does not appear unrealistic, particularly if research and extension expenditures continue to grow as much as we have assumed. If we further assume a value of agricultural output of $60 billion per year over the next decade (it was $61 billion in 1972), we can make a rough guess at the expected marginal internal rate of return to 1971-76 research and extension expenditures as they are reflected in 1977-82 productivity growth and resources saved. Utilizing the same procedure by which we computed the marginal internal rates of return presented earlier (see section on "Estimates by Decades") – doubling public research and extension, etc. – we obtain an expected marginal internal rate of return of about 29 percent for 1971-76 research and extension expenditures. Hence there does not appear to be any immediate danger of driving the marginal rate of return to investment in agricultural research and extension in the United States below an acceptable level, at least over the next several years. In fact the rate of return could increase if productivity growth continues at about the same pace and if the value of agricultural output increases because of increased exports and higher farm prices.

Research Allocation

The overall rate of return to all research and extension is a composite of the rates of return to investment in thousands of projects and activities. We know that the overall return will be maximized, for a given expenditure, only if the rates of return on all individual projects are equalized. However, the output of research is very stochastic in nature. Thus it probably is not very fruitful to try to predict expected rates of return on individual projects. At this level, the return is determined largely by the skill (and luck) of the researcher.

On the other hand, as we look at more aggregative groupings of the total research effort, it would seem to be possible to measure ex post marginal rates of return to make predictions about the short-run future. Viewing research as a separate variable in a Cobb-Douglas type production function, we see that its marginal product is equal to e(O/R) where e is the production elasticity of research and O/R is dollars of related output per dollar of research (average product). We know there is great variation among commodities and among states in the average product of research. For example, the average product of corn research in the United States is over four times that of cotton research (Table 3-3). Of course, differences in production elasticities of research between corn and cotton may to a certain extent offset differences in average products, although it is improbable that the ratio of the research elasticities would reach the magnitude of four. Similarly, there is a rather wide divergence between the average products of research in the largest and smallest agricultural states.

Whether marginal products of research exhibit the same variation as the average products is an open question. Preliminary evidence reported by Bredahl suggests that for the most part production elasticities of research are not significantly different among commodities or among large and small states.[20] Therefore it appears fairly certain that the larger the average products of research, the higher the marginal products and hence the marginal rates of return to research. This in turn suggests that, if the objective is to maximize output, growth in agricultural research budgets should take place where the greatest number of dollars of related output can be expected per dollar of research.

This is not to say that marginal rates of return will be equalized if average products or even marginal products are equalized. For one thing, differences in the research lags associated with different commodities will be reflected in different rates of return for the same marginal products. For example, we might expect the lag between research and its output to be longer for livestock than for crops and poultry. If so, an optimum allocation of research would be characterized by higher marginal products for livestock than for crops and poultry.

Table 3-3. Average Products of Research in the United States in 1969

Item	Output per Dollar of Research
Corn	712
Soybeans	672
Wheat	430
Cotton	173
Swine	485
Beef	442
Dairy	323
Poultry	262
Sheep and Wool	76
Ten largest agricultural states	351
Ten smallest agricultural states	97

Source: Howard Engstrom, "Productivity Differences in Agricultural Research Between States," Department of Agricultural and Applied Economics, St. Paul, University of Minnesota, May 1972, pp. 6 and 12.

Differences in lags may also be important among experiment stations. If the large stations engage in more basic research than the small stations, where research may be largely adaptive in nature, we may expect the lag to be longer in the large stations. If so, the large stations would have to exhibit higher marginal products than the small stations in order to have the same marginal rate of return. On the other hand, it is questionable whether differences in lags could justify differences in marginal products or even average products of the order of magnitude of four to five times. Needless to say, we need more information on differences in marginal products and lags of research among commodities and among experiment stations.

APPENDIX

Appendix 3-1. Appropriations for Public Research and Extension (current dollars, in millions)

Year	State Agricultural Experiment Stations[a]	USDA	Extension
1915	4.6	6.0	3.5
1916	3.8	5.2	4.9
1917	3.8	5.8	6.2
1918	4.2	6.3	11.3
1919	4.2	6.9	14.7
1920	5.0	7.7	14.7
1921	5.2	7.8	16.8
1922	6.3	8.2	17.2
1923	7.0	8.5	18.5
1924	7.6	8.4	19.1
1925	7.3	9.3	19.3
1926	8.9	10.2	19.5
1927	9.3	10.5	20.1
1928	11.4	11.7	20.7
1929	12.0	13.8	22.9
1930	13.1	15.5	24.3
1931	12.5	16.7	25.4
1932	12.1	16.1	24.3
1933	11.4	13.1	22.0
1934	11.1	11.1	19.8
1935	11.1	11.4	20.4
1936	12.1	14.4	28.3
1937	12.9	16.4	30.0
1938	14.8	18.0	31.6
1939	15.6	23.3	32.4
1940	16.8	22.1	33.1
1941	16.7	21.4	33.5
1942	17.7	22.0	34.5
1943	17.5	21.8	35.0
1944	18.8	22.0	36.3
1945	19.8	22.9	38.2
1946	23.6	27.6	44.6
1947	28.1	33.2	53.7
1948	35.3	38.2	60.2
1949	39.9	46.0	67.2
1950	48.2	46.8	74.6
1951	50.5	45.1	77.6
1952	56.4	45.0	81.8
1953	60.0	45.3	86.8
1954	68.0	46.0	91.6
1955	73.8	53.4	100.7
1956	85.4	59.6	110.1
1957	92.2	86.6	118.2
1958	105.9	83.7	128.7
1959	110.3	99.0	136.0
1960	120.3	105.2	141.7

Appendix 3-1 – continued

Year	State Agricultural Experiment Stations[a]	USDA	Extension
1961	127.3	128.9	149.4
1962	142.1	126.4	159.2
1963	151.3	136.1	168.6
1964	169.3	149.8	177.9
1965	181.8	192.5	188.9
1966	223.4	212.7	201.2
1967	239.7	218.5	213.7
1968	261.5	219.5	225.5
1969	274.0	213.2	242.0
1970	296.1	238.7	290.7
1971	319.3	263.1	331.9
1972	348.8	294.0	354.4
1973	382.9	303.9	385.1

Source: State Agricultural Experiment Stations: For 1915-60," Report on the Agricultural Experiment Stations," published by Office of Experiment Stations through 1953 and by Agricultural Research Service from 1954 through 1960, Washington, D.C. For 1961-73, "Funds for Research at State Agricultural Stations," Cooperative State Experiment Station Service, Washington, D.C.

USDA: For 1915-53, "Report of the Director of Finance," USDA, Washington, D.C. For 1954-73, "Appropriations for Research and Education," prepared by Office of Budget and Finance, USDA, Washington, D.C.

Extension: For 1915-55, "Annual Report of Cooperative Extension Work in Agriculture and Home Economics," Federal Extension Service, USDA, Washington, D.C. For 1956-73, unpublished data from the extension service.

[a] Federal plus nonfederal funds available. Excludes fees and sales.

NOTES

1. The research reported in this chapter was undertaken by the authors as part of their larger project on "Technology, Institutions and Development: Minnesota Agriculture, 1880-1970," funded by a grant from the Rockefeller Foundation to the University of Minnesota Economic Development Center.

2. Andrew Boss, *Minnesota Agricultural Experiment Station, 1885-1935*, Minnesota Agricultural Experiment Station Bulletin no. 319 (St. Paul: University of Minnesota, 1935); Joseph C. Fitzharris, "Science for the Farmer: The Development of the Minnesota Agricultural Experiment Station, 1868-1910," *Agricultural History*, 48 (January 1974), 202-214.

3. Joseph C. Fitzharris, *The Development of Minnesota Agriculture, 1880-1970: A Study of Productivity Change*, Department of Agricultural and Applied Economics Staff Paper P 74-20 (St. Paul: University of Minnesota, September 1974).

4. Boss, *Minnesota Agricultural Experiment Station*; Fitzharris, "Science for the Farmer"; interview with E. C. Stakman, October 2, 1974.

5. Boss, *Minnesota Agricultural Experiment Station*; Fitzharris, "Science for the Farmer."

6. "Annual Report of the Director of the Minnesota Agricultural Extension Service, 1936, 1953," typescripts (St. Paul: University of Minnesota Libraries Collection, 1889-1965).

7. Interviews with Dean William F. Hueg, Jr., July 23, 1973, and Dean Hubert Sloan, 1973; Boss, *Minnesota Agricultural Experiment Station*; "Annual (Biennial) Report of the Director of the Minnesota Agricultural Experiment Station, 1888/89-1964" (St. Paul: University of Minnesota Libraries Collection).

8. E. C. Stakman, Richard Bradfield, and Paul C. Mangelsdorf, *Campaigns against Hunger* (Cambridge, Mass.: Harvard University Press, 1967); interviews with E. C. Stakman, June 6, 1973, and October 2, 1974; interview with John R. Rowell, August 7, 1973; Sterling Wortman, "International Agricultural Research Institutes: Their Unique Capacities," paper presented at the Tenth Anniversary Celebration of the International Rice Research Institute, Los Baños, Philippines, April 20-21, 1972.

9. Boss, *Minnesota Agricultural Experiment Station*; Herbert Kendall Hayes, *A Professor's Story of Hybrid Corn* (Minneapolis: Burgess, 1963).

10. Fitzharris, "Science for the Farmer."

11. H. C. Knoblauch, et al., *State Agricultural Experiment Stations: History of Research Policy and Procedure*, USDA Miscellaneous Publication no. 904 (Washington, D.C.: United States Government Printing Office, 1962); also, this view is frequently approached by I. Arnon, *Organization and Administration of Agricultural Research* (London: Elsevier, 1968), pp. 5-18.

12. Arnon, *Agricultural Research*, pp. 58-65.

13. Knoblauch, *State Agricultural Experiment Stations*; Albert C. True, *A History of Agricultural Experimentation and Research in the United States, 1607-1925, Including a History of the United States Department of Agriculture,* USDA Miscellaneous Publication no. 251 (Washington, D.C.: United States Government Printing Office, 1937).

14. Knoblauch, *State Agricultural Experiment Stations*; Albert C. True, *A History of Agricultural Extension Work in the United States 1783-1923*, USDA Miscellaneous Publication no. 15 (Washington, D.C.: United States Government Printing Office, 1928).

15. Fitzharris, "Science for the Farmer,"; Douglass C. North and Robert Paul Thomas, *Institutional Change and American Economic Growth* (Cambridge: At the University Press, 1971).

16. Stakman, Bradfield, and Mangelsdorf, *Campaigns against Hunger*; interviews with Stakman, 1973 and 1974; and with Rowell, 1973.

17. See T. W. Schultz, *The Economic Organization of Agriculture* (New York: McGraw Hill, 1953); Zvi Griliches, "Research Costs and Social Returns: Hybrid Corn and Related Innovations," *Journal of Political Economy*, 66 (October 1958), 419-431; Willis L. Peterson, "Return to Poultry Research in the United States," *Journal of Farm Economy*, 49 (August 1967), 656-669.

18. Zvi Griliches, "Research Expenditures, Education, and the Aggregate Agricultural Production Function," *American Economic Review*, 54 (December 1964), 961-974; Peterson, "Return to Poultry Research"; Robert E. Evenson, "The Contribution of Agricultural Research and Extension to Production," Ph.D. dissertation (Chicago: University of Chicago, 1968).

19. See F. Welch, "Education in Production," *Journal of Political Economy*, 78 (January-February 1970), 35-39; Y. Kislev and N. Shchori Bachrack, "The Process of an Innovation Cycle," *American Journal of Agricultural Economy*, 55 (February 1973), 28-37; and Wallace E. Huffman, "Decision Making: The Role of Education," *American Journal of Agricultural Economics*, 56 (February 1974), 85-97.

20. Maury Bredahl, 1975, personal communication.

4

Productivity of Agricultural Research in Colombia[1]

Reed Hertford, Jorge Ardila,
Andrés Rocha, and Carlos Trujillo

This chapter will compare four recent studies of the economic returns to varietal research on rice, cotton, wheat, and soybeans in Colombia. The four studies are part of a larger program of agricultural research, extension, and education administered since about 1950 by the Colombian Agricultural Institute (ICA) and its predecessor agencies, the Department of Agricultural Research (DIA) and the Office of Special Studies (OSS). Our main hypothesis was based on calculations of returns made previously for Colombia, the United States, and several other countries. For Colombia, Harberger had estimated that the average rate of return on all capital ranged from 8 to 10.5 percent between 1960 and 1968 and that the opportunity cost of public funds was about 10 percent during the late 1960s.[2] Studies by Griliches[3] and Peterson[4] for the United States, by Ardito Barletta for Mexico,[5] by Ayer and Schuh for Brazil,[6] and by Duncan for Australia[7] found rates of return to varietal improvement in excess of 10 percent. These investigations also suggested, however, that returns obtained in the United States were exceeded by those obtained in the other countries studied. Accordingly, we hypothesized that the estimated rates of returns for the four Colombian varietal improvement programs would exceed both the opportunity cost of capital within the country and the rates of return previously reported for similar United States programs.[8]

This hypothesis derived additional support from the commonsense notion that, because Colombia began agricultural research after the United States and other developed countries, it should have been able to draw on a large stock of knowledge about plant-breeding techniques and on extensive international collections of plant materials to reduce the gestation periods and development costs of its programs. If we assume that research enterprises in Colombia and the United States are comparable in organization and competence, the cost effectiveness of and returns to research in Colombia should be greater. At the same time, important socioeconomic and structural constraints in Colombia could prevent higher returns to investments in varietal improvement, even if technical breakthroughs had been more easily made.

The foregoing considerations, together with the nature of the data available for analysis, led us to adopt a methodology which could assist us in distinguishing the contributions of biological, socioeconomic, and structural factors to the calculated returns to research. The "social benefits" of varietal research were estimated in the usual way – as changes in consumers' and producers' surpluses resulting from shifts in product supplies generated by the use of improved seeds. (See chapter 6 of this volume.) But the shift in supply itself was taken as the product of two separately estimated variables: the difference in yields between two (average) farm plots of one hectare each, one plot being planted entirely to the improved seeds and the other to the unimproved varieties, multiplied by the percentage of crop land actually planted in the improved variety. We then associated the first of these variables, the "yield advantage" of the improved variety, with the biological determinants of returns and the second variable, "the rate of adoption," with the socioeconomic, structural, and biological determinants, recognizing that a large yield advantage can be a primary cause of rapid and high levels of adoption.

Because it was our impression at the outset that the technical and biological work of the four Colombian varietal improvement programs had been well done, we felt that our main hypothesis would be rejected only if the rates of adoption were low, which would mean that there were major socioeconomic and structural constraints. The only crop of the four studied which evidenced such constraints was wheat. It had been grown under near subsistence conditions by small, traditional farmers in some of Colombia's poorest agricultural areas. Also, for a number of years massive wheat imports, made under PL 480, had depressed the relative price of wheat.

The final step of our analysis concerned the way the yield advantage was to be calculated. We felt that estimates based only on comparisons of yields obtained on plots seeded to new varieties and others seeded to unimproved varieties would be biased upward because of the strong, positive interactions of the new varieties with such inputs as fertilizers and water. Therefore, in comparing yields and calculating the yield advantages of new varieties, we attempted to determine the effects of other inputs by estimating the production relation between yields, seed varieties, and other variables which may have interacted with the seed varieties.

The next four sections of this chapter discuss the returns to research in rice, cotton, wheat, and soybeans, respectively. The final section compares our main results and summarizes our principal conclusions.

Rice[9]

Colombia's rice research program was initiated in 1957 by a predecessor agency of ICA. Its establishment coincided with a sharp rise in rice imports occasioned by an outbreak of the *hoja blanca* disease. (According to *FAO Rice Reports* for relevant years, Colombia usually imported about 2,000 metric tons of rice annually but imported 10,200 tons in 1957; in 1958 and 1959 imports returned to earlier levels.) This is a virus prevalent in Latin America – with symptoms like the stripe disease of Japan – which first caused substantial losses in Venezuela in 1956 and in Colombia in 1956 and 1957.[10] Accordingly, the initial objectives of the research program included varietal selection and breeding for higher yields and resistance to the *hoja blanca* virus.[11]

Rice varieties resistant to the disease were collected throughout Colombia as a first step; in addition, 2,200 varieties were imported from the United States Department of Agriculture's World Collection of Rice in Beltsville, Maryland. By 1959, about 400 of these varieties had shown promising resistance to the virus. Because they were mainly *japonica* varieties, which are not consumed in Colombia, the research program sought to breed the virus resistance of *japonica* into the local long-grain varieties.[12] It was estimated that this might take four to five years. In the interim, the one superior United States variety which had shown some virus resistance, Gulfrose, was multiplied and released in 1961.

Napal, the first improved variety produced by the Colombian research program, was released in 1963. Napal had the long-grain characteristics of Bluebonnet 50, the most popular nontraditional variety, and was resistant to *hoja blanca*.[13] Unfortunately, it was attacked by *bruzone* (rice blast disease) in 1965 and disappeared from commercial use thereafter. In the same year, Tapuripa, earlier imported from Surinam, was distributed to farmers as an alternative to Bluebonnet 50 and Gulfrose. It was long grained and flinty, with some resistance to blast and *hoja blanca*.

In 1966, the Colombian rice research program added an objective derived from the International Rice Research Institute (IRRI): to develop dwarf varieties with a high grain-to-straw ratio and resistance to lodging. About 3,000 additional varieties were imported from IRRI, and an order went out to retain only those varieties already in the Colombian collection which outyielded the most prevalent local variety by 100 percent.

In 1969, ICA joined forces with the rice program of the International Center for Tropical Agriculture (CIAT). Personnel, facilities, budgetary resources, and objectives were shared under informal agreements between the two institutions. This reinforced Colombian ties with IRRI, since the head of CIAT's rice research team had served on IRRI's staff.

In 1968, ICA and CIAT introduced IR-8, which spread quickly even though the medium-type chalky grain sold generally at a 30 percent discount and was susceptible to blast disease. However, it was resistant to *hoja blanca*. Following strong commercial interest, CIAT and ICA also introduced IR-22 in 1970 and recommended it to farmers in irrigated tropical areas.

Between 1966 and 1970, ICA released independently one additional rice variety, ICA-10, which never assumed any commerical importance. Its yields were inferior to the IRRI varieties, although it was superior to and/or less variable than either Gulfrose or Napal. Its grain quality was also less desirable than Tapuripa's.

In 1971, ICA released the CICA-4 variety. Compared with earlier varieties it had improved disease-resistance, was more adaptable to changes in water and air temperature, and it had good grain appearance and cooking qualities. It also produced slightly superior yields. Simultaneously, CICA-4 appeared in Ecuador as INIAP-6, in the Dominican Republic as Advance 72, and in Peru as Nylamp.

Yields recorded in commercial field trials of the seven major rice varieties released by the Colombian and joint CIAT-ICA programs after 1957 are shown in Table 4-1, together with data obtained from the same source on yields of the check variety Bluebonnet 50. The 665 individual trials which are the basis for these yield statistics include all that are available for the fifteen-year period 1957-71. It should be mentioned that ICA's commercial trials, or *pruebas regionales*, are conducted on commercial farms that agree to collaborate with the institute's programs. Farmers run the trials, but materials and instructions are provided by ICA.

The three rice varieties released before 1966 show average yields of 4.1 metric tons per hectare, representing a yield advantage over Bluebonnet 50 of about 33 percent. Varieties introduced after 1966, including ICA-10, double that yield advantage, bringing it to 65 percent above Bluebonnet 50.

In view of these yields, it is interesting to note that the area planted to improved rice varieties did not become a significant proportion of all rice land until the second, or post-1966, stage of the research program (see Table 4-2). Data in the table on the percentage of acreage sown to a given variety were estimated in the following way. First, available information on sales of certi-

Table 4-1. Average Rice Yields from Commercial Trials by Variety, Colombia, 1959-71[a]

Year	Variety (in kilos per hectare)							
	Bluebon-net 50	Gulf-rose	Napal	Tapur-ipa	ICA-10	IR-8	IR-22	CICA-4
1959	1,927							
1960								
1961	2,893	3,071						
1962	2,967	4,065						
1963	3,875	5,391	4,420					
1964	4,336	4,138	5,166					
1965	3,462	2,739	4,343					
1966	1,590		2,436	3,645				
1967	2,893			2,690	4,707	6,098		
1968	3,208		5,356	4,600	4,789	5,890		
1969	3,544		5,110	4,625	5,450			
1970	3,339			4,500	3,852	5,180	5,420	6,125
1971	3,164			3,610	4,234	4,748	5,080	4,600
Average. . . .	3,099	3,880	4,344	4,025	4,441	5,473	5,250	5,362

Source: Jorge Ardila, "Rentabilidad social de las inversiones en investigación de arroz en Colombia," M.S. thesis, Bogota, ICA/National University Graduate School, 1973, Table 5.

[a] Blanks indicate no regional trials were undertaken.

fied seeds by variety were converted to hectare equivalents by dividing by the estimates of seeding rates provided by the ICA National Rice Program director. Second, lacking data on farm-produced seeds of the improved varieties, we assumed that the proportion of all acreage planted to certified seeds of any variety was equal to the proportion planted to later generation seeds produced outside the seed multiplication and certification program. This estimating procedure was followed here, as well as for wheat and soybeans, to estimate total area planted to improved seeds because it produced the simplest and "best fit" between available data on certified seed sales and "expert opinion."

To estimate the shift parameter of each new variety – its yield advantage over Bluebonnet 50 – production functions were fit to the *pruebas regionales* data, using standard least-squares procedures. In the final round of estimation, reported yields (kilos per hectare) were taken as a function of twenty variables: size of the trial plot, seeding rate, seven seed variety variables, two variables to distinguish different time periods, four variables relating to irrigation and its interactions with seed variety, and five variables to differentiate locations and their interactions with variety. Only the first two of these variables entered as continuous arguments. Other continuous variables (relating

Table 4-2. Land Area Planted to Six Improved Varieties of Rice as a Percentage of the Total Area Planted in Rice, Colombia, 1964-73[a]

Year	Variety (in %)						All Improved Varieties (%)
	Napal	Tapuripa	ICA-10	IR-8	IR-22	CICA-4	
1964	2.5						2.5
1965	2.1						2.1
1966		0.1					0.1
1967		3.2	0.1				3.3
1968		21.2	0.6	0.3			22.1
1969 . . .		18.0	0.5	3.7			22.2
1970		12.4	0.2	18.2			30.8
1971		6.9		26.1	3.1	5.0	41.1
1972				18.9	10.1	18.3	47.3
1973				20.1	24.3	12.6	57.0

Source: For 1964-71, Jorge Ardila, "Rentabilidad social de las inversiones en investigación de arroz en Colombia," M.S. thesis, Bogota, ICA/National University Graduate School, 1973, Table 11; for data after 1971, ICA, director of the National Rice Program.

[a] Blanks indicate less than 0.1 percent.

particularly to "cultural practices") were either discarded or respecified in noncontinuous form in the final results presented in Table 4-3.

Because CICA-4 was taken as the check variety, the estimated coefficients on the variety variables are to be interpreted as their "yield disadvantage" in kilos per hectare compared with CICA-4. When results are interpreted in this way, it is evident that the Colombian rice research program has produced through time continuous and substantial improvements in yields. Again, the superiority of the varieties released after 1966 is evidenced.

Results suggest, however, that yields of CICA-4 as well as those of IR-8 and IR-22 are positively influenced by irrigation. The coefficient on the irrigation variable indicates that yields of all varieties are increased by about 1.2 tons with average irrigation practices. Roughly another ton is added when irrigation is applied to IR-8, IR-22, or CICA-4, as indicated by the coefficients on the variables of interaction of those varieties with irrigation. This evidence from the production functions, coupled with data which show that dry land rice yields increased 7 percent during the 1961-72 period while those of irrigated rice increased 133 percent, leads to an inference that the newer varieties have benefited mainly the irrigated rice areas.[14] The other side of the coin, of course, is that the adoption of improved varieties was assisted by the existence of irrigated crop land.[15]

Most of ICA's research has been focused on the irrigated rice areas. Its largest programs have been located at the Palmira and Espinal experiment stations. Although about 75 percent of all Colombian rice land is now irrigated,

Table 4-3. Production Function Estimates for Rice Based on Commercial Trial Data, Colombia, 1957-72

Independent Variable	Estimated Coefficient	Estimated t Statistics
1. Size of trial plot	−0.15	−2.30
2. Seeding rate	2.46	1.58
3. Bluebonnet 50	−1,609.66	−3.45
4. Gulfrose	−1,486.56	−1.31
5. Napal	−1,742.79	−1.83
6. Tapuripa	−884.31	−1.80
7. ICA-10	−536.93	−1.13
8. IR-8	−798.97	−1.54
9. IR-22	−589.97	−0.72
10. Irrigation	1,220.20	5.84
11. Irrigation * IR-8 interaction	1,278.09	3.21
12. Irrigation * IR-22 interaction	700.44	0.89
13. Irrigation * CICA-4 interaction	1,061.87	2.11
14. Location	1,185.26	7.12
15. Location * Gulfrose interaction	991.98	0.94
16. Location * Napal interaction	940.33	1.06
17. Location * IR-8 interaction	428.22	1.24
18. Location * CICA-4 interaction	−1,340.16	−3.14
19. Time I	1,228.03	6.74
20. Time II	−509.78	−2.20
Intercept	2,028.30	3.64
	$R^2 = 0.67$	n = 665

Source: Jorge Ardila, "Rentabilidad social de las inversiones en investigación de arroz en Colombia," M.S. thesis, Bogota, ICA/National University Graduate School, 1973, Table 13.

almost 100 percent has been irrigated traditionally within the areas served by Palmira and Espinal.[16]

This emphasis on the irrigated areas may have been induced by expectations of the sort of variety-irrigation interactions found in the regression results. More plausible is the commonsense explanation that the ICA's creditability would have been seriously threatened had it not produced varieties which yielded well in the irrigated areas of Colombia, since the controlling interests of the rice growers and commercial trade are found there. It has been reported that half of the value of all dues collected by the National Federation of Rice Growers comes from Tolima.

Regression results were used to estimate the overall percentage change in rice supply attributable to the yield advantage of all improved varieties over Bluebonnet 50. It was estimated as a weighted sum – divided by average commerical yields – of the regression coefficient of each improved variety minus the coefficient corresponding to Bluebonnet 50, with weights equaling the percentage of all rice land planted which was sown in each variety.[17] This

Table 4-4. Alternative Values of the Supply Shift Parameter for Rice Attributable to Improved Varieties, Colombia, 1964-71

	Estimate (%)	
Year	Simple (1)	Varietal Effects (2)
1964	1.05	−0.16
1965	1.01	−0.15
1966	0.13	0.03
1967	−0.17	1.07
1968	10.99	5.73
1969	12.81	5.98
1970	14.89	7.42
1971	15.96	10.38

Source: Jorge Ardila, "Rentabilidad social de las inversiones en investigación de arroz en Colombia," M.S. thesis, Bogota, ICA/National University Graduate School, 1973, Tables 5, 11, 12, 13, 17, and 20.

estimate is shown in column 2 of Table 4-4. The "simple" estimate of the yield advantage of the improved varieties is shown in column 1 of the table and is based only on the data for the average annual yield of each variety obtained in the *pruebas regionales* and already presented in Table 4-1.[18]

The fact that the simple estimates of the shift parameter exceed the estimates that are made up only of varietal effects from the regression is consistent with the finding that only the improved varieties of rice interacted with other variables of the production function. Since those interactions were on balance positive and are included in the simple estimate but not in the varietal-effects estimate, the former overstates the shift parameter by as much as seven percentage points.

The simple and varietal-effects estimates were combined with assumed values of the price elasticities of supply and demand to provide upper- and lower-bound estimates of gross social benefits of the new seed varieties for the period 1964-71.[19] Values considered for the price elasticities of supply were zero, 0.2347, and infinity, the intermediate value being derived from the only supply study available for Colombian rice;[20] values considered for the price elasticity of demand were − 0.5, − 1.372, and − 2.0, the intermediate value again having been estimated in another study.[21] Maximum gross benefits resulted from using the simple estimates of the shift parameter and price elasticities of demand and supply, respectively, of − 0.5 and zero; minimum benefits corresponded to price elasticities of demand and supply of − 2.0 and infinity and the varietal-effects estimate of the supply shift parameter. Both estimates of benefits are shown in Table 4-5 for the 1964-71 period.

Table 4-5. Estimated Benefits and Costs of the Rice Research Program in Colombia, 1957-80 (in thousands of 1958 pesos)[a]

Year	Estimated Benefits		Estimated Costs
	Maximum	Minimum	
1957 . . .			15
1958 . . .			193
1959 . . .			235
1960 . . .			286
1961 . . .			429
1962 . . .			441
1963 . . .			252
1964 . . .	3,733	− 563	445
1965 . . .	4,750	− 699	538
1966 . . .	553	127	519
1967 . . .	− 827	5,157	867
1968 . . .	61,659	27,291	937
1969 . . .	60,872	23,675	2,074
1970 . . .	69,444	27,883	2,779
1971 . . .	107,470	52,255	4,165
1972-80	107,543	52,255	4,202[b]

Source: For estimated benefits, preceding tables; for estimated costs, Jorge Ardila, "Rentabilidad social de las inversiones en investigación de arroz en Colombia," M.S. thesis, Bogota, ICA/National University Graduate School, 1973, Tables 44 and 46.

[a] Blanks indicate no benefits.

[b] Figures for subsequent years were estimated by assuming 4,202 grew by 10 percent annually.

Costs of the research program for the same period, also shown in Table 4-5, include direct costs, indirect costs, and complementary costs; these terms were defined earlier by Ardito Barletta.[22] Direct costs of the rice program were available only after 1964. For this reason available cost data were regressed on the number of employees assigned to the rice program and ICA's total expenses for all research programs; the resulting regression coefficients and available data on the two independent variables of the regression were then used to estimate the direct cost data for the missing years, 1957-65. Complementary costs associated with the new program – those it incurred with other collaborating programs – were estimated for this study by the director of the National Rice Program. Included were costs associated with the entomology, plant physiology, plant pathology, soils, and extension programs. Indirect costs included staff training costs, opportunity costs of the services of fixed capital and land, management costs, and the costs of "international cooperation."[23] The latter category comprised the major program cost, the total costs of the CIAT rice program from 1969 to 1972, as estimated by the head of that program, and a prorated share of the cost of Rockefeller Foundation personnel stationed in Colombia from 1958 through 1968. The simple sum of total costs for the 1957-71 period equaled 14.2 million 1958 pesos. The costs of international cooperation were calculated at 5.0 million or 35 percent of the total but are probably understated since benefits derived from the "capital stock" of IRRI and other institutions have not been charged to the Colombian program. (This is discussed further elsewhere in this chapter.) All costs for the rice program in 1971 represented about 12 percent of ICA's total expenses for research.

Recognizing that the current stock of new varieties will continue to produce into the future, we projected the costs and benefits of the Colombian rice program to 1980 using assumptions, when necessary, which would bias downward the estimate of internal rates of return. It was assumed on the cost side, for example, that the real value of the cooperative CIAT-ICA rice program would increase at a rate of about 10 percent a year, primarily on the grounds that programs which are relatively new and reputedly successful tend to grow. Since this assumption was made, ICA's budget has been severely cut, and CIAT's rice program has been phased down. Nonetheless, we hold to the initial assumption to avoid overstating the final estimate of the internal rate of return. On the side of the projection of gross benefits, it was assumed that the value of production during the 1972-80 period will average 581 million pesos, a sum which equals the value of the rather good 1971 crop (in 1958 pesos); that the rate of adoption of the new rice varieties after 1973 will stabilize at the estimated 1973 level of 57.0 percent;[24] that the percentage of rice land planted to IR-8 will trend downward linearly to zero by 1980; that the percentages of all rice land sown to IR-22 and CICA-4 will be equal after 1973; and that the increase in the shift parameter implied by these assumptions will approximately equal increases in commercial yields over the period 1972-80. The increases in yields would be 14 and 11 percent, respectively, for the maximum and minimum values of the shift parameters.

The resulting internal rate of return corresponding to the stream of maximum gross benefits was found to be 82.3 percent; the rate estimated on the basis of the stream of minimum gross benefits was 60.1 percent.[25]

Cotton[26]

Cotton has turned in a striking performance in Colombia. Since the mid 1930s, yields have about quadrupled – in fact, their pattern of change has been broadly similar to that of cotton yields in the United States. Currently,

Colombian cotton yields are comparable to United States yields and roughly twice as high as average yields for all of South America (Table 4-6). In an earlier comparative analysis of changes in cotton yields, it was concluded that Colombia ranked fourth in yield increases in the 1950-60 period among the twenty-four countries that produced 97 percent of all cotton in 1960.[27] Production after the mid 1930s increased at least fifteen times, or from about 30,000 bales in 1937-38 to over 500,000 bales in the early 1970s.[28]

Yields and production advanced most rapidly in two different but not widely separated periods of time: 1951-54 and 1957-59. Yields about doubled in the first period. Although they also increased in the second period, production evidenced a much larger increase of 167 percent. Developments in both periods appear to have been the result of changes in government policies.

The first period of rapid development followed the reopening of the Colombian Ministry of Agriculture in 1948 and the introduction of new policies emphasizing the need to replace imports of food and fibers with local products. For cotton, this meant that the textile industry had to consume stated allotments of national cotton. This produced an 82 percent increase in the farm price between 1948 and 1951.[29] When the local textile industry was faced with the prospect of consuming larger quantities of national cotton, it promoted the establishment in 1948 of the Cotton Development Institute (IFA) for purposes of improving the quality and uniformity of local cotton through both research and the control of ginning.[30] Eventually a government institute with its own budget, IFA also assumed responsibilities for cotton extension, seed distribution, and credit.

The second surge in production, occurring at the end of the 1950s, paralleled changes in exchange policies. The official exchange rate in Colombia was 2.5 pesos per United States dollar from 1951 through May 1957. The free rate was 3.0 to 3.5 pesos through 1954 and then edged up to 6.9 pesos by mid 1957. On June 18, 1957, the official rate was increased to 7.6 pesos. Through the 1951-57 period, the Colombian textile industry was permitted to import raw cotton and capital items at the official exchange rate. As a result, imports steadily built up to a level of 77,000 bales in 1957; production stood at 95,000 bales in that same year. In 1958, following reforms, production jumped to 220,000 bales and in 1959 reached 256,000 bales. Imports fell to 36,000 bales in 1958 and to slightly less than 2,000 bales in 1959, at which time Colombia also showed its first exportable surplus of cotton in several decades. During the early 1960s exports averaged about 100,000 bales and by 1968 had reached a level of almost 300,000 bales.

As producers of cotton attained national prominence and power by satisfying domestic consumption and exporting a growing surplus, the National

Table 4-6. Comparative Statistics for Cotton Yields, 1934-35/1938-39 to 1973-74[a]

Year	All Crops, Index for Colombia	Cotton (pounds per acre)		
		Colombia	United States	South America
1934-35 to 1938-39		133	212	181
1947-48	100	152	267	163
1950-51	115	167	269	175
1951-52	102	150	269	203
1952-53	109	227	280	197
1953-54	119	317	324	212
1956-57	143	319	409	178
1959-60	152	377	462	207
1962-63	176	398	457	231
1965-66	158	352	526	244
1966-67	164	470	480	232
1967-68	176	516	447	267
1970-71		463	437	222
1973-74[b]		470	519	249

Source: For all crops, index for Colombia, *Changes in Agricultural Production and Technology in Colombia*, Foreign Agricultural Economic Report no. 52 (Washington, D.C.: Economic Research Service, USDA, 1969), Table 30; for cotton, Secretariat, International Cotton Advisory Committee, *Cotton-World Statistics*, Washington, D.C., various issues.

[a] Blanks indicate no data available.

[b] Preliminary.

Federation of Cotton Growers (FNA) began to absorb IFA's functions. In 1968, IFA was dissolved completely, and its research and extension activities were passed on to the ICA. Some of its other activities were absorbed by the Ministry of Agriculture.

The inheritance of the IFA was meager when it assumed responsibilities for organized cotton research in 1948. On the advice of an English mission, some cotton research was begun in the Cauca Valley in 1928 but was later suspended when attention there was turned, by a visiting mission from Puerto Rico, to the prospects for sugarcane research. In 1934, some research was established in Armero, State of Tolima, to introduce and test United States Uplands and some Peruvian varieties. Until 1948, however, the most progress had been made in improving the perennial tree cotton. A station on the outskirts of Barranquilla on the north coast is reported to have obtained yields of 350-400 kilos per hectare or at least twice the then prevailing average yield.[31] Nonetheless, one of the first things IFA did was to close that station, since the quality of the tree cotton was considered inferior to imported cotton and tree cotton had become infested with diseases which threatened the introduction of annual varieties.

From the beginning, the institute's sole research objective was the introduction, testing, and multiplication of improved United States cotton varieties. No attempt was made to produce a national variety until 1961, and that effort appears to have languished until IFA's research was absorbed by ICA in 1968.

ICA adopted the same primary objective for its cotton research program as IFA. But, it added a second objective – the development of a national cotton variety through selection and hybridization. It also improved the design of research, expanded experimentation beyond the three locations used by IFA (at Buga, Espinal, and Codazzi), and undertook more trials on a commercial scale as *pruebas regionales*.

Actually, the first United States cotton variety was introduced into Colombia well before the establishment of IFA. Deltapine 12 was imported by cotton producers in 1941 and came into general use in Tolima State during the 1940s. The year before, the Brazilian variety *Expresso do Brazil* had been introduced and it likewise gained acceptance in Tolima during the 1940s. In the late 1940s and the 1950s, Deltapine 15, Earlystaple, Coker 124 B, and Deltapine Smoothleaf were introduced. These so-called "T" type cottons accounted for about 93 percent of all cotton production by 1959; Deltapine was by all odds the most important among them. By 1971, Deltapine 15 was no longer in use but Deltapine 16 accounted for 42 percent of all cotton acreage, Deltapine Smoothleaf for 38 percent, Acala 1517 BR-2 for 8 percent, and Stoneville 213 for 8 percent, with the remaining 4 percent being accounted for by Deltapine 45 and Coker 201.[32]

As noted, the cotton research program emphasized the selection and multiplication of promising United States varieties rather than the development of improved national varieties. The United States varieties sharply increased Colombian cotton yields. To estimate the contribution of the Colombian selection and testing program, two questions need examination.

First, would single farmers or groups of farmers acting without government help have been as efficient or more efficient in selecting and importing United States varieties than IFA and ICA?

Colombian cotton production is concentrated among a small group of farmers. As of 1958, 422 farmers accounted for 61 percent of total production;[33] in 1967, 343 producers were reported to have accounted for 40 percent of all output. At prevailing average yields, this would imply that each large cotton producer was harvesting about 550 acres, given total production for Colombia of 465,000 bales in 1967. The data for 1958 imply that each of the 422 farmers was harvesting in that year only about 200 acres of cotton, suggesting that large cotton producers were major contributors to the increase in production that occurred between 1958 and 1967.[34]

The demand curve facing producers has been highly elastic because of the

Table 4-7. Average Yields of Seed Cotton by Variety Obtained from Commercial Trials Conducted in Colombia, 1953-72

Variety	Number of Observations	Yield (kilos per hectare)
Deltapine 15	193	2,312
Deltapine Smoothleaf	71	2,369
Stardel	18	2,296
Stoneville 213	39	2,375
Coker 124 B	12	2,634
Acala BR-2	48	2,287
Deltapine 45	9	2,693
Deltapine 45 A	40	2,575
Deltapine 16	42	2,457
Coker 201	27	2,568
Total[a]/average	499	2,366

Source: Andrés Rocha, "Evaluación económica de la investigación sobre variedades mejoradas de algodón en Colombia," M.S. thesis, Bogota, ICA/National University Graduate School, 1972, Table 10.

[a] Excludes twenty-four trials on other varieties.

existence of an export market. The elastic demand curve would have served as insurance to individual innovators that prices and profits would not be eroded by the increased production brought about by the diffusion of their innovations. The fact that the industry was composed of a few large farmers makes it more probable that a single individual or small group could have anticipated large enough rewards from search and research efforts to justify undertaking them.

However, it seems unlikely that a farmer-based research effort would have outperformed IFA and ICA. There seems to be no basis for believing that a private research effort would have uncovered other, more effective varieties than IFA or ICA did. Because of their official status, the two institutes probably were able to import new varieties into Colombia more easily and rapidly. Similarly, they were able to control the distribution of improved seeds. For example, IFA controlled all cotton gins. However, it was probably the special privileges and franchises that IFA and ICA possessed as offical government agencies – not "pure differences" in organization of research – which facilitated their success.

The second question needing examination is whether there were, in fact, significant differences in yields among the imported United States varieties when grown under Colombian conditions.

The programs of both IFA and ICA were founded on the premise that such yield differences did exist. The claim was made that it would be worthwhile to identify the size of these differences and to key programs of seed

Table 4-8. Production Function Estimates for Cotton Based on Commercial Trial Data in Colombia, 1953-72

Independent Variable	Estimated Coefficient	Estimated Standard Error of Coefficient
1. Nitrogen	2.58	0.53
2. Irrigation	606.56	48.71
3. Parcel type	− 471.41	61.41
4. Rain deficiency	− 1,150.22	47.60
5. Stardel	− 241.43	89.09
6. Coker 124 B.	− 214.75	109.08
7. Acala BR-2	− 331.64	59.23
8. Location 1.	381.53	88.86
9. Location 2.	189.13	80.78
10. Location 3.	600.73	94.34
11. Location 4.	1,094.14	149.39
12. Location 5.	349.18	91.05
13. Location 6.	− 649.88	116.68
14. Location 7.	− 408.99	181.04
15. Location 8.	991.48	71.77
16. Location 9.	408.49	62.74
17. Location 10	1,167.86	77.85
18. 1953	− 218.53	106.79
19. 1954	− 268.37	81.12
20. 1967	− 428.55	156.78
21. 1970	363.15	61.72
22. 1971	366.96	70.66
23. 1972	452.88	75.20
Intercept	2,081.20	n.a.[a]
	$R^2 = 0.82$	n = 523

Source: Andrés Rocha, "Evaluación económica de la investigación sobre variedades mejoradas de algodón en Colombia," M.S. thesis, Bogota, ICA/ National University Graduate School, 1972. Table 3.

[a] Not available.

multiplication and distribution to them. If, however, yields of all United States varieties harvested in Colombia were equal, then there would be no payoff to a program of varietal selection and distribution. Any individual farmer could import a variety of United States cotton selected at random and expect to obtain as good results as he would have obtained through an organized program of research like IFA's and ICA's. Similarly, such a program would not be useful if the distribution of yields by variety were the same in both the United States and Colombia.

To explore the question of yield differences more carefully, all available data were obtained on the IFA and ICA commercial trials, which were comparable in design to those reported earlier for rice. The trials covered the period 1953-72 and included 523 individual experiments. They are summarized in Table 4-7 as mean values of yields obtained for each of ten cotton varieties.

Additional trial data were destroyed when IFA's research was absorbed by ICA. Presumably, they included information on the two check varieties, Deltapine 12 and *Expresso do Brazil*.

It is evident from the data in Table 4-7 that gross differences in yields are not appreciable and that it would be difficult to reject the hypothesis that they were all, in fact, equal. For this reason, a more refined test was made by estimating production functions from the trial data. In the final round of estimates, twenty-three variables entered the regression; their estimated coefficients and related statistics are shown in Table 4-8. The first variable measures the quantity of nitrogen applied per hectare, the second simply indicates whether or not irrigation was applied, the third adjusts for the fact that some of the trials were undertaken on plots which were "small" by *pruebas regionales* standards, the fourth is an index used by attending agronomists for the lack of rainfall, and Variables 8 through 17 adjust for the location of the experiments, and 18 through 23 adjust results for abnormal years.

Regression results indicated that the only varieties out of the ten tested with yields significantly different from Deltapine 15 were Stardel, Coker 124 B, and Acala BR-2; in each case their adjusted yields were lower than those for Deltapine 15 and lower by rather similar and "small" amounts. On the basis of these results, it is concluded that no significant, positive benefits were derived from the Colombian cotton research programs.[35]

Wheat[36]

In an earlier study of the production trends of Colombia's major crops, it was claimed that "the wheat situation in Colombia contains a number of paradoxes. Despite good experimental development and government programs to expand production, both acreage and output have declined sharply in recent years."[37] Other more refined data now available continue to show that both acreage and production have declined over the past twenty years. The area cultivated in wheat fell steadily from a level of 175,000 hectares in 1953 to about 70,000 hectares in 1973; over the same period production was halved. Yields increased by about 25 percent in the 1953-58 period but stabilized at just above 1,000 kilos per hectare until 1972. In the most recent two years for which information is available (1972 and 1973), yields have increased again, and by about 20 percent.[38] Still, the average yield increase over the entire twenty-year period has been rather unimpressive.

The best explanation currently available for the decrease in acreage planted is that increasing PL 480 sales have dampened incentives for Colombian farmers to devote land to wheat production.[39] According to FAO *Trade Yearbooks*, the quantity of wheat imported has increased over the 1953-73 period from a third to almost three times the quantity of total wheat produced. The modest rise in yields may be attributable to the same forces and to a shift in the regional distribution of wheat production from the State of Boyaca where yields are higher than average to the State of Nariño in southern Colombia where farms are small and poor and yields have traditionally been below the national average.

The wheat improvement program is one of Colombia's oldest programs. It dates from 1926 when the *La Picota* experiment station was established in the central region of the country (State of Cundinamarca) for the purpose of improving yields and certain characteristics of wheat, barley, oats, and rye. Through 1951, at least half the total costs of the activities of *La Picota* were absorbed by wheat research. In 1947, "cold climate" wheat research expanded out from *La Picota* to two additional locations: one at Bonza in Boyaca and another at Isla in Cundinamarca. A few years later, additional locations for research were acquired at Tibaitata (Cundinamarca) and Obonuco (Nariño). After the addition of the Surbatá Station in Boyaca in 1959 for cold climate wheat research, activities were consolidated there, in Tibaitata, and in Obonuco.

Colombia's wheat research has received important assistance from a number of different foreign and national organizations. In 1948, personnel of the program were sent to Mexico to study methods of wheat breeding with the staff of the OSS in Mexico City, then supported by the Rockefeller Foundation. Later, in 1950, Rockefeller personnel were assigned to collaborate with the Colombian research program; the foundation's assistance continued until the mid 1960s. In 1953, wheat seed distribution and multiplication programs received a lift from the Colombian Agricultural Credit Bank which ultimately assumed responsibility for them. Two years later the National Federation of Rice Growers provided some support for research on the potential for wheat production in the warmer tropical areas that traditionally produced rice; similar support was received in the same year from INA, the National Marketing Institute. The federation of barley producers, PROCEBADA, contributed to the wheat program's budget in the 1959-61 period to support expansion of the *pruebas regionales* effort; aid was received for the same purpose from FENALCE, a federation of Colombian cereal producers. In the 1967-71 period, the wheat program was assisted by the University of Nebraska Mission, financed by a consortium of international assistance agencies, including USAID and major United States foundations.

Data compiled on the costs of the wheat improvement program reflect this support from the outside and provide a profile of the development of the program. Table 4-9 presents data on the direct, complementary, and indirect costs of the program – comparable in all respects with the cost data shown

Table 4-9. Cost of the Colombian Wheat Research Program by Major Category, 1927-73 (in thousands of 1972 pesos)

Year	Direct Costs[a]	Complementary Costs[b]	Indirect Costs[c]	Added Costs of New Seeds[d]	Total Costs
1927	184		9		193
1928	236		9		245
1929	287		11		298
1930	338		12		350
1931	389		15		404
1932	441		17		458
1933	492		48		540
1934	543		52		595
1935	598		58		656
1936	653		63		716
1937	708		69		777
1938	709		78		787
1939	709		108		817
1940	710		120		830
1941	711		135		846
1942	670		143		813
1943	630		156		786
1944	589		170		759
1945	548		189		737
1946	508		207		715
1947	467		239		706
1948	416		255		671
1949	365		279		644
1950	315		351		666
1951	403		394		797
1952	492	88	440		1,020
1953	580	117	315	655	1,667
1954	669	74	349	2,638	3,730
1955	758	93	385	242	1,478
1956	1,169	238	511	1,680	3,598
1957	1,117	232	562	282	2,193
1958	996	271	584	1,058	2,909
1959	841	265	483	1,170	2,759
1960	1,202	319	500	2,182	4,203
1961	914	388	510	3,002	4,814
1962	828	277	396	2,599	4,100
1963	615	280	280	2,717	3,892
1964	1,229	487	1,146	2,628	5,490
1965	1,889	658	585	5,548	8,680
1966	2,427	592	957	4,129	8,105
1967	2,150	993	2,307	8,731	14,181
1968	2,919	915	1,261	7,050	12,145
1969	3,045	1,314	2,768	5,623	12,750

Table 4-9 – continued

Year	Direct Costs[a]	Complementary Costs[b]	Indirect Costs[c]	Added Costs of New Seeds[d]	Total Costs
1970	2,352	1,446	3,106	3,343	10,247
1971	2,020	1,603	3,260	2,901	9,784
1972	1,501	1,467	2,407	1,507	6,882
1973	1,570	1,385	2,354	1,626	6,935

Source: Carlos Trujillo, "Rendimiento económico de la investigación en trigo," M.S. thesis, Bogota, ICA/National University Graduate School, 1974, Tables 4.1, 4.2, 4.3, 4.4, and 4.5.

[a] Salaries, supplies, and office materials directly related to the wheat varietal improvement program.

[b] Represents costs of the plant pathology, soils, entomology, biometrics, extension, and plant physiology programs incurred on behalf of the wheat improvement program; for the period 1927-51 these costs were included in direct costs.

[c] Includes costs associated with the use of experiment station facilities, agricultural machinery, and land as well as costs of administration and training of program staff.

[d] Equals the difference between the average price of certified wheat seeds and the price received by farmers times the quantity sold of certified seeds; for the period 1927-49, improved varieties relevant to this study were not planted.

earlier for the rice improvement program – plus data on the additional costs to farmers of the improved wheat seeds which were adopted after 1952. It can be seen that, beginning in 1927, total costs built up slowly and by 1935 had reached a level which was subsequently maintained for about fifteen years. Following the establishment of the joint Colombian-Rockefeller Foundation program, direct, complementary, and indirect costs again built up to a level which was maintained until 1964 with the exception of three years – 1959, 1963, and 1964, the latter two being years during which the research agency was reorganized. Investments began to drift upward after 1964 and then increased sharply during the period of the University of Nebraska Mission's presence, falling off after the mission began to leave Colombia in 1971. From 1968 through 1971, total costs of the program represented only 5 percent of ICA's research budget but fully 3 percent of the value of wheat production. Wheat research had become an expensive program.

Activities of the research program revolved around four kinds of wheat: cold climate, warm climate, *Triticales*, and Durums. The first has been the most important in terms of both the time and the resources devoted to it.

When the research program began in 1927, some promising cold climate wheats were introduced from the United States and tested over a six-year period. Fifteen varieties were released to farmers from the *La Picota* experiment station in 1933. By the early 1940s, the number had increased to twenty-four. Of these the best eight were Klein, General San Martin, Klein 40,

Marzuolo, Pentad, Florence, Barcino Barbado, and Bola Picota. Because the latter two were the most widely used varieties by the mid 1940s, they are considered as check varieties in this analysis. The first reference to the yields indicated that between 850 kilos and 1.5 tons per hectare were expected. The other six of the eight varieties could reasonably be expected to yield 1.8 to 2.8 tons per hectare under experimental conditions.[40]

The vast majority of promising wheats were obtained by crossing local *criolla* varieties. By the early 1940s it was thought that available foreign varieties were inferior to local Colombian wheats and thus the program turned inward until the arrival of foreign personnel in the early 1950s. When the OSS was established in Colombia with Rockefeller Foundation support, 11,000 varieties were immediately imported from the Rockefeller Mexican program. Selections were made from these imports primarily on the basis of their resistance to the yellow, black stem, and leaf rusts. These rusts had become the major preoccupation of the Colombian wheat improvement team because data produced in 1949 showed that wheat yields were being cut 6 percent by leaf and stem rusts and 14 to 41 percent as a result of yellow rust.[41]

The first new commercial variety released after the establishment of the joint Colombian-Rockefeller Foundation program was Menkemen 52. Distributed in 1953, it was the product of a cross of varieties from the Mexican collection, including Mentana and Kenya. It reduced time to maturity by thirty-five days, was somewhat resistant to the major rusts, had strong stems, and outyielded Bola Picota by 30 percent. Two years later a second variety, Bonza 55, was released. It was the product of two Rocamex varieties, Yaqui and Kentana. Because it was especially resistant to the yellow rust of Nariño State, it was most widely distributed there. A third variety, Nariño 59, was released in 1959; it also was particularly well adapted to the State of Nariño, being resistant to its variety of yellow rust. Three years later, a large batch of new varieties produced by the joint Colombian-Rockefeller program was released, including Miramar 63, Bonza 63, Crespo 63, Napo 63, Tiba 63, and Tota 63. At the time this release was made, the industry was advised that the research program would in the future attempt to make "batch releases" (i.e., releases of more than one variety) to reduce susceptibility to new wheat rusts. Millers are reported to have reacted adversely to this announcement on the grounds that a single mill could not handle more than two varieties of wheat; an appeal was made to the research team to revise its strategy. As it turned out, the wheat program for cold climates made only one additional release — in 1968 Sugamuxi 68, Zipa 68, and Samacá 68 were distributed simultaneously.

The Colombian wheat plant began to change as the result of the introduction of dwarf varieties from Mexico in 1958. The effort to incorporate char-

Table 4-10. Comparative Wheat Yields for Thirteen Varieties Obtained Under Experimental Conditions, Colombia, 1970

Variety	Area Planted (hectares)	Reported Yields (kilos per hectare)	Reported Yields Compared with Bola Picota Yield (%)
Menkemen 52	0.095	2,820	427
Bonza 55.	0.123	2,360	358
Nariño 59	0.106	1,700	258
Bonza 63.	0.153	2,220	336
Miramar 64	0.101	2,535	384
Crespo 63	0.163	1,570	238
Napo 63	0.274	2,770	420
Tiba 63.	3.740	3,017	457
Tota 63.	0.200	2,680	406
Zipa 68.	0.134	3,000	455
Samacá 68	0.050	3,700	561
Sugamuxi 68.	0.223	2,300	348
Bola Picota.	0.132	660	100

Source: Colombian Agricultural Institute (ICA), *Informe del Programa Nacional de Trigo* (Bogota: ICA, 1970), Appendix 4.

acteristics of the smaller plant did not, however, gain force and importance until about 1964. By 1970, 60 percent of all materials in the Colombian wheat research program included dwarf wheats. Of the thirteen improved varieties released to farmers after the establishment of the joint Colombian-Rockefeller program, small-plant characteristics were incorporated in nine of them: Bonza 63, Miramar 63 and 64, Napo 63, Tiba 63, Tota 63, and the three varieties released in 1968. In this regard it is important to mention that practically none of Colombia's wheat is irrigated and that the use of fertilizer is negligible.

In 1971, ICA published the data in Table 4-10 comparing yields obtained under experimental conditions on small plots of land for twelve improved varieties and Bola Picota. Reported yield advantages over the Bola Picota variety were in excess of 500 percent for the highest yielding wheats and not less than 250 percent for any improved variety. By international standards, these yields of the Colombian varieties also appeared to be quite good. As Table 4-11 indicates, in the International Wheat Trials of 1968 three of Colombia's most recently released wheats outyielded the best of the Mexican wheats, Azteca 67. The average level of these yields, however, is extremely high even by experiment station standards in Colombia (e.g., those reflected in Table 4-10).

With reference to the second category of wheat research – namely, that undertaken on "warm climate" varieties – it is worth mentioning that large

Table 4-11. Comparative Wheat Yields Obtained for Colombian and Mexican Varieties in the International Nursery Trials, 1968

Variety	Country of Origin	Reported Yields (kilos per hectare)
Sugamuxi 68	Colombia	6,232
Crespo 63	Colombia	6,215
Samacá 68	Colombia	6,217
Azteca 67	Mexico	6,110
Tota 63	Colombia	6,054
Napo 63	Colombia	6,044
Tiba 63	Colombia	5,894
Penjamo 62	Mexico	5,833
Centrifén	Chile	5,755
Norteño	Mexico	5,538
Lerma Rojo 64 x Sonora 64	Mexico	5,349
Sonora 64 x TZ.PP	Mexico	5,249
Nai 60(B)	Mexico	5,249
Bonza 63	Colombia	5,233
Jaral	Mexico	5,166
Zipa 68	Colombia	5,116

Source: Colombian Agricultural Institute (ICA), *Informe del Programa Nacional de Trigo* (Bogota: ICA, 1968), p. 13.

areas of wheat had existed in the warmer regions since the colonial period. However, these wheat areas were practically eliminated in the mid 1930s as a result of attacks by stem rust. Thus in 1955 when the first rust-resistant varieties were available, the Federation of Rice Growers persuaded the research agency to experiment with Bonza and Menkemen in the Cauca Valley and Tolima State. Although the rust resistance of the new wheats was confirmed in these early studies, the experiments were not continued because of the unpromising levels of yields obtained. It was thought at the time that, for wheat to compete with rice, 2,500 to 3,000 kilos of wheat per hectare would be required. Commercial yields averaged 1,500 kilos, and maximum experimental yields did not exceed 2,750 kilos per hectare.[42]

The wheat program first experimented with rye as a rust-resistant, high-protein, water-saving alternative to wheat in 1937, and experiments with *triticales* were initiated in *La Picota* in 1946. However, interest in *triticales* appears to have languished until recently. The wheat program has also evidenced interest in Durum wheats. Work began on Durum in 1952 and was stepped up somewhat in the mid 1960s. However, it was not successful primarily because of the high humidity in Colombia's wheat areas, the short days, and the occasional heavy rainfalls which occur when the grains are maturing.

As with rice and cotton, available data on the *pruebas regionales* were collected in an attempt to quantify the shift parameter and gross social benefits attributable to the Colombian wheat research program. However, the collection of these data was much more difficult for wheat than for rice and cotton because the information had been scattered when the research program changed its affiliation with outside agencies. In the final analysis, data for only 1,016 individual trials were obtained for the 1953-73 period; many more trials had been undertaken on the major improved varieties.

Most of the trials for which data were obtained (about 80 percent) related to six varieties: Bonza 63, Crespo 63, Menkemen 52, Napo 63, Nariño 59, and Tota 63. By region, the bulk of the data related to two states, Cundinamarca and Nariño. Only about 5 percent of the data are from the State of Boyaca. The director of the National Wheat Program stated that this does not reflect any slighting of the Boyaca wheat regions because there are many regions in Cundinamarca and Nariño that are fully representative of the areas in Boyaca. Given that wheat production in Boyaca has declined more sharply than in any other state, the facts here are important, although difficult to establish and qualify. Finally, it should be noted that most of the *pruebas regionales* data obtained (70 percent) were for the years 1963, 1964, 1968, 1971, and 1972.

Table 4-12 summarizes the data collected for thirteen improved wheat varieties and two check varieties, Bola Picota and "150." Mean yields in kilos per hectare are reported by variety, together with the estimated standard error of yields, the range of trial yields corresponding to a 5 percent level of probability, and the coefficient of variation of yields. When compared with the data in Table 4-10, it is apparent that these data assign rather different relative yield ranks to specific varieties. For example, in Table 4-12 the yield of Menkemen 52 puts it in twelfth place among the improved varieties, but its yield in Table 4-10 ranks it in third place. Also, the average level of yields reported in Table 4-12 is lower than the averages of Tables 4-10 and 4-11, and the yield advantage of the improved varieties is noticeably less than indicated by Table 4-11.

The range in yields of all improved varieties in Table 4-12 includes the upper-bound yield reported for Bola Picota; the range in yields for six out of the thirteen improved varieties includes the upper-bound yield for "150." The yield advantages of Samacá 68 and Bonza 63, as a percentage of the average yields of "150" and Bola Picota, are 83 and 75 percent, respectively; the corresponding value for all improved varieties shown in Table 4-12 is 50 percent.

To adjust these estimates of the gross yield advantage of the improved wheat varieties for the effects of other determinants of yields, production functions were estimated from the data on commercial trials. The final ver-

Table 4-12. Comparative Wheat Yields for Fifteen Varieties Obtained in Commercial Trials, Colombia, 1953-73

		Output per hectare (kilos)				
				Range at 5% Probability		
Variety	Number of Observations	Mean	Standard Error	Lower Bound	Upper Bound	Coefficient of Variation of Yields (%)
"150"	8	1,624	1,022	771	2,476	62.9
Bola Picota.	8	1,194	895	448	1,941	74.9
Samacá 68	47	2,584	1,592	2,117	3,051	61.6
Bonza 63.	106	2,460	1,197	2,230	2,690	48.7
Miramar 63	29	2,348	1,218	1,885	2,812	51.9
Zipa 68.	31	2,190	1,382	1,684	2,697	63.1
Bonza 55.	77	2,172	1,504	1,831	2,513	69.2
Crespo 63	129	2,115	1,369	1,876	2,353	64.7
Tiba 63.	51	2,110	1,265	1,754	2,466	59.9
Nariño 59	119	2,106	1,351	1,861	2,352	64.1
Napo 63	136	2,097	1,340	1,869	2,324	63.9
Sugamuxi 68.	12	1,973	1,157	1,238	2,708	58.7
Tota 63.	104	1,893	1,283	1,643	2,142	67.8
Menkemen 52	138	1,836	1,237	1,627	2,044	67.4
Miramar 64	21	1,643	1,496	960	2,326	91.1
All varieties . . .	1,016	2,099	1,340	2,027	2,175	66.1

Source: Carlos Trujillo, "Rendimiento económico de la investigación en trigo," M.S. thesis, Bogota, ICA/National University Graduate School, 1974, Table 5.7.

sion of the production function is shown in Table 4-13. Thirty-nine variables entered: twelve of these represented zero-one adjustment variables for the location of the trials; fourteen adjusted for the effects of variety; four, measured as indices above a certain threshold level and zero otherwise, accounted for major diseases reported (*vaneamiento*,[43] foot and root rot, stem rust, and dwarfing virus); one adjusted for seeding rates of 80 kilos per hectare (only two rates were actually reported – 80 and 111 kilos per hectare); two each were used to adjust for soil type and reported weather; and one variable each adjusted for how well the soil had been worked before planting, for weed growth, and for the application of lime. Coefficients on the noncontinuous variables shown in Table 4-13 need to be read with some care. Since the regression package reparameterized all variables by imposing the restriction that the sum of the regression coefficients equal zero, an estimate of the corrected mean yield associated with a given noncontinuous variable should be calculated by adding its estimated coefficient to the overall mean value of yields, which was 2,099 kilos per hectare.

The statistical significance of the variety variables entering the regression was surprisingly low. Only the estimated coefficients on the Bonza 63 and

Table 4-13. Production Function Estimates for Wheat Based on Commercial Trials in Colombia, 1953-73

Independent Variable	Estimated Coefficient	Estimated t Statistic
1. *Vaneamiento*	− 13.8	− 6.98
2. Foot and root rot.	− 11.1	− 5.75
3. Stem rust	− 8.6	− 3.18
4. Dwarfing virus	− 15.8	− 3.73
5. Plot size	− 48.5	− 5.88
6. Seeding rate	− 225.0	− 3.69
7. Good soils	482.7	8.10
8. Poor soils	− 340.2	− 4.31
9. Poor prior soil preparation	− 157.1	− 2.68
10. Heavy weed growth	− 300.6	− 6.79
11. Unfavorable weather	− 1,333.9	− 10.20
12. Favorable weather	895.4	8.43
13. Lime applied	154.1	2.35
14. Location 1.	− 732.0	− 3.21
15. Location 2.	3,672.8	16.40
16. Location 3.	692.4	7.37
17. Location 4.	1,604.9	9.24
18. Location 5.	− 466.9	− 2.29
19. Location 6.	− 1,084.6	− 4.44
20. Location 8.	− 349.1	− 3.12
21. Location 9.	− 644.9	− 5.38
22. Location 10	− 831.1	− 6.60
23. Location 11	−812.2	− 1.94
24. Location 12	− 856.6	− 3.05
25. Location 13	49.1	0.56
26. Menkemen 62	− 113.2	− 1.20
27. Bonza 55	98.8	0.89
28. Nariño 59	163.8	1.70
29. Miramar 63	− 375.5	− 2.17
30. Bonza 63	340.5	3.27
31. Miramar 64	− 300.3	− 1.46
32. Crespo 63	210.9	2.12
33. Napo 63	131.7	1.44
34. Tiba 63	− 169.2	− 1.28
35. Tota 63	− 80.7	− 0.77
36. Zipa 68	− 152.8	− 0.92
37. Samacá 68	196.9	1.22
38. Sugamuxi 68	210.3	0.80
39. Bola Picota	− 303.5	− 0.91
Intercept	2,460.8	15.73
	$R^2 = 0.53$	n = 1.016

Source: Carlos Trujillo, "Rendimiento económico de la investigación en trigo," M.S. thesis, Bogota, ICA/National University Graduate School, 1974, Table 5.14.

Crespo 63 varieties were positive and significant. In an independent estimate of the partial contribution of the variety variables as a group to yield variance, the significance of the variety variables was found to be less than that of any other single variable or group of variables (e.g., the variables adjusting for location).

Also, the values of the estimated yield advantages of most of the improved varieties are lower on the basis of the regression of Table 4-13 than on the basis of the unadjusted estimates of mean yields presented in Table 4-12. The largest and most significant yield advantage of any improved variety in the regression – that of Bonza 63 – is only 36 percent more than the adjusted yield of Bola Picota, or roughly half the value implied by the unadjusted yield estimates of Table 4-12.

Table 4-14 presents statistics summarizing the use of the improved varieties of wheat. Underlying these summaries are data for each improved variety used in weighting shift parameters taken from the estimated production function to arrive at an average annual estimate of the percentage yield increase of improved wheats over average commercial yields. Two estimates of these weights were considered, and their implications for overall rates of adoption are reflected in Table 4-14 in the "upper-bound value" and in the "most probable value" of the percentage of wheat land planted to improved varieties. The first estimate simply assumes that the total use of an improved variety equaled in any year two times its reported sales in certified form and that the average seeding rate was 120 kilos per hectare for all varieties. As can be seen in the table, this assumption results in levels of adoption in the late 1960s which were high by known standards for unirrigated wheat. The second estimate – the one used in this study – maintained that the total seed use of any variety would equal two times its certified sales and that seeding rates averaged 120 kilos per hectare but set the germination rate of certified seeds at 86 percent and the corresponding rate for seeds retained and planted by farmers from prior harvests at 39 percent. These low rates of germination, based on several ICA studies,[44] were not encountered for rice, cotton, or soybeans. Since the sum of the two germination rates is 125 percent, the effect of this procedure was to assume that the real, postgermination rate of employment of an improved variety was 1.25 times its quantity sold in certified form.

Estimates of the yield advantage of each improved variety taken from the regression, divided by average commercial yields in each year and weighted by the appropriate adoption rate, produced two streams of gross benefits for the 1953-73 period of the Colombian wheat improvement program. In each case it was assumed that the c.i.f. import price of wheat was the relevant "price" at which to value the crop. Because of the overvaluation of the Co-

Table 4-14. Selected Data on Employment of Improved Wheat Varieties in Colombia, 1953-73

Year	Total Certified Seed Sales (tons)	Wheat Land in Improved Varieties(%) Upper-Bound Value	Most Probable Value
1953	147	1.4	0.9
1954	1,039	8.9	5.6
1955	113	0.9	0.5
1956	639	5.2	3.3
1957	599	5.5	3.4
1958	1,610	22.0	13.7
1959	3,050	43.4	27.1
1960	2,149	28.7	17.9
1961	2,830	33.7	21.1
1962	2,470	31.7	19.8
1963	2,100	31.8	19.9
1964	1,864	27.0	16.9
1965	2,782	38.6	24.1
1966	3,113	45.0	28.1
1967	3,795	66.6	41.6
1968	4,494	83.2	52.0
1969	2,809	72.0	44.6
1970	1,694	56.5	35.1
1971	1,641	56.9	35.3
1972	1,528	40.4	25.3
1973	1,429	33.0	20.5

Source: Carlos Trujillo, "Rendimiento económico de la investigación en trigo," M.S. thesis, Bogota, ICA/National University Graduate School, 1974, Tables B.3, 5.17, and 2.10.

lombian peso, this assumption underestimates the gross benefits of research. It was found, however, that the estimated internal rate of return to the wheat improvement program would increase by only 4 percent if the (higher) price received for wheat by farmers was used instead.

In one of the yield advantage estimates it was assumed that the price elasticity of the supply of wheat equaled 0.55 and that the price elasticity of demand was – 0.04. These values of the price elasticity parameters were derived from estimates of two independent studies.[45] For the second estimate of gross benefits, it was recognized that wheat was imported throughout the 1953-73 period and that the value of gross benefits should therefore not include a surplus to consumers.

The two estimates of gross benefits, as well as the total costs shown in Table 4-9, were then projected through 1976 on assumptions similar to those used for rice. The internal rate of return estimated for the "closed economy" case corresponding to the stream of net program benefits (gross benefits minus costs) for the 1927-76 period was 11.9 percent. When allowance was made for the fact that wheat was imported, the estimated internal rate of return was reduced to 11.1 percent.

Soybeans [46]

Soybean production in Colombia has grown very rapidly in recent years. The total area cultivated was only 16,000 hectares in 1962; production stood at 25,000 tons, and yields were 1,500 kilos per hectare in the same year. By 1972, just ten years later, the area harvested had increased to 58,000 hectares, production was 116,000 tons, and yields had risen by a third to 2,000 kilos.[47] This rapid development is attributed to the fact that soybeans are excellent in rotation with several major crops (cotton, in particular) and that the demand for soybeans has been strengthened by a fast-growing poultry industry. The crop is cultivated in Colombia only in the Cauca Valley; this is probably because the feed industry is near there. An equally important explanation, however, is that available high-yielding, disease-resistant soybean varieties produced by the ICA experiment station at Palmira have been adapted to conditions found in the Cauca Valley.

ICA did not begin soybean research until 1960, and then work was restricted to the Palmira Station. In about seven years, however, the research effort produced three new varieties with superior yield potential and resistance to major diseases, principally *cercóspora*, a fungus which attacks and destroys almost all parts of the soybean plant. Table 4-15 summarizes experimental data relating to yields of four soybean varieties obtained for this study. Unfortunately, data generated from commercial fields or *pruebas regionales* were unavailable; thus, the information used in Table 4-15 and elsewhere in this section relates to small experimental plots of the Palmira Station. However, the use of experimental data is somewhat less troublesome for soybeans than for other crops because of the high level of technology and improved practices used by farmers in the Cauca Valley.

The ICA Pelican, Lila, and Taroa varieties were released successively by the experiment station. The Mandarin variety was imported earlier from the United States and by 1967 had come to occupy about four-fifths of all soybean acreage. Compared with the data previously shown for cotton, rice, and wheat, the yield superiority of the improved varieties in Table 4-15 is not particularly outstanding.

Nonetheless, adoption of the new varieties has been nothing short of spectacular. Table 4-16 presents the percentages of soybean acreage planted to each of the four main varieties grown in the 1967-71 period. These data were

Table 4-15. Average Soybean Yields from Experimental Trials by Variety, Colombia, 1967-71 (in kilos)

Year	Unimproved Variety, Mandarin	Improved Varieties		
		ICA-Pelican	ICA-Lili[a]	ICA-Taroa
1967	2,068	2,406		2,490
1968	2,329	2,373	2,700	2,650
1969	1,756	2,138	2,525	2,400
1970	1,751	2,373	2,300	2,500
1971	1,828	2,578	2,410	3,034
Average.	1,946	2,455	2,483	2,622

Source: Gabriel Montes, "Evaluación de un programa de investigación agrícola: El caso de la soya," M.S. thesis, Bogota, University of the Andes, 1973, Table 3.

[a] No experiment reported for 1967.

estimated using a procedure analogous to the one followed for the other crops discussed in this chapter, i.e., total acreage planted to a variety in a given year was taken to be equal to two times certified seed sales of that variety divided by an estimate of the seeding rate. The important point to note about the data in Table 4-16 is that roughly three-quarters of the area planted in soybeans used the Mandarin variety in 1968 and 1969, while by 1971 Mandarin had practically disappeared, and ICA Pelican and Lili varieties had come to be used on 84 percent of all acreage.

There were two major reasons for the rapid and high levels of adoption of the improved varieties. First, there was a severe outbreak of the *cercóspora* fungus on the Mandarin variety in 1969. In 1970 ICA found itself in the en-

Table 4-16. Land Area Planted to Improved Varieties of Soybeans as a Percentage of the Total Area Planted in Soybeans, Colombia, 1967-71[a]

Year	Variety					Total
	Mandarin	ICA-Pelican	ICA-Lili	ICA-Taroa	Other	
1967[b]	89	1			10	100
1968	77	13			10	100
1969	71	18	5		6	100
1970	35	29	24		12	100
1971	2	43	41	2	12	100

Source: Gabriel Montes, "Evaluación de un programa de investigación agrícola: El caso de la soya," M.S. thesis, Bogota, University of the Andes, 1973, Table 6.

[a] Estimates derived from data on certified seed sales, assuming that the total use of a variety of seed equaled two times its sales in certified form. Blanks indicate less than 0.5 percent.

[b] Only data on ICA-Pelican use were available. The Mandarin estimate was derived on the assumption that "other" varieties occupied 10 percent of all acreage planted in 1967 as they did in 1968.

Table 4-17. Production Function Estimates for Soybeans Based on Experimental Data, Colombia, 1967-71

Independent Variable	Estimated Coefficient	Estimated t Statistic
1. ICA-Pelican	268.44	3.06
2. ICA-Lili	418.31	3.93
3. ICA-Taroa	436.95	4.37
4. Number of weedings	86.93	2.43
5. Herbicide and insecticide use	78.45	2.35
6. Rainfall	113.11	1.75
7. *Cercóspora*	– 107.30	– 3.68
8. Plant density/seeding rate	200.80	4.58
Intercept[a]	692.08	
	$R^2 = 0.70$	n = 68

Source: Gabriel Montes, "Evaluación de un programa de investigación agrícola: El caso de la soya," M.S. thesis, Bogota, University of the Andes, 1973, Table 8.

[a] No estimated t statistic was available.

viable position of having two high-yielding, fungus-resistant varieties available for distribution and plenty of seed. Second, it was easy for this news to get around; the only Colombian farmers interested in soybean production are in a relatively small geographic area, where some of the best communication and infrastructure facilities in the country are located. The farmers themselves are among the most modern in Colombia.

As with cotton, rice, and wheat, an attempt was made to generate more refined estimates of the yield superiority of the new soybean varieties by identifying a relation between yields and their major determinants, including seed variety. Final results of this effort are shown in Table 4-17, which reports on a regression of experimental yields on three independent variables for the major improved seed varieties (observations on the check variety, Mandarin, were included in the regression, of course), the number of times the experiment was weeded, kilos per hectare of active herbicide and insecticide ingredients applied, millimeters of rainfall, the presence of the *cercóspora* fungus measured as an index with a range of 0 to 5, and an index (likewise with a range of 0 to 5) which reflected essentially the ratio between the observed plant density and the seeding rate. Signs of all estimated coefficients are those which were hypothesized at the outset, and the significance of most coefficients is seen to be high. One exception, the estimated coefficient for rainfall, reflects the fact that the rainfall variability was limited because most experiments were undertaken in a small geographic area. The statistical strength of the plant density/seeding rate variable is attributable to the fact that it is capturing the effects of several unspecified cultural practices used in the ex-

Table 4-18. Estimated Benefits and Costs of the Soybean Research Program in Colombia, 1960-80 (in 1,000 pesos, 1958)[a]

	Gross Benefits		
Year	Based on "Varietal Effect" Shift Parameters	Based on "Simple" Shift Parameters	Total Costs
1960			40
1961			41
1962			37
1963			39
1964			33
1965			37
1966			40
1967	49	62	57
1968	1,288	2,230	98
1969	3,102	6,187	179
1970	7,847	16,300	463
1971	10,217	28,643	267
1972-80 . . .	10,217	28,643	267

Source: Gabriel Montes, "Evaluación de un programa de investigación agrícola: El caso de la soya," M.S. thesis, Bogota, University of the Andes, 1973, Table 14 and 21

[a] Blanks indicate no benefits during this period.

periments. This variable may thereby have adjusted the estimated coefficients of the improved varieties for the experimental nature of the data – the high levels of technology and intensive use of improved cultural practices. The fact that the coefficients on the improved varieties increase in value from Pelican (the first released) through Lili to Taroa (most recently released) indicates substantial progression in ICA's research program. A test of the null hypothesis that the estimated coefficient on the Pelican variety equaled that of the Lili variety was rejected at the 99 percent level of significance. Similarly, the hypothesis that the coefficients estimated for the Lili and Taroa varieties are equal was rejected at the 95 percent level.

The yield advantage of the improved varieties taken from the production function, divided by commercial yields and weighted by the percentages of the land area planted in each variety, led to a "varietal effect" estimate of the shift parameter. The yield advantage of the improved varieties estimated directly from the data in Table 4-15, also divided by commercial yields and weighted by the percentages of the land area planted in each variety, led to a "simple" estimate of the shift parameter associated with the soybean research program. These two estimates of the shift parameter were combined with plausible values for the price elasticities of demand and supply – respectively, – 0.77[48] and infinity – to give a range of gross benefits in each year for the 1967-71 period. These two streams of gross benefits are shown in Table 4-18 along with estimates of the costs of the soybean research program, which include the same categories of expenses as do the other three commodities considered in this chapter. Costs and benefits were projected nine years beyond 1971 on the assumption that in real terms both would remain about constant. The resulting internal rate of return for the smaller benefit stream was 79 percent, while the rate for the larger one was 96 percent. These rates did not change appreciably when program costs were assumed to increase 10 percent annually after 1971.

Comparisons and Conclusions

At the outset we hypothesized that net internal rates of return to varietal improvement of rice, cotton, wheat, and soybeans in Colombia had been higher than the opportunity cost of public funds (10 percent) and, in fact, even higher than rates of return on the order of 50 percent calculated for similar programs in the United States. Among the four programs, somewhat lower estimated returns were expected for wheat because its domestic price had been under pressure from PL 480 imports, and production had moved to less productive areas.

To examine this latter possibility more carefully – as well as the roles of socioeconomic and structural constraints generally in the estimated returns to research – the total shift in product supplies caused by the use of improved varieties generated through research was divided into two parts: an estimate of the "yield advantage" of the new over the old varieties and an estimate of the rate of adoption of the new varieties. Low returns attributable to socioeconomic and structural constraints were then associated mainly with low rates of adoption; the role of the biological determinants of the return to research was associated principally with the calculated yield advantage of the improved varieties. The yield advantage was estimated with regression techniques which were designed to factor out assumed positive interactions between the improved varieties and such inputs as fertilizers and water.

Our main results are summarized in Table 4-19. Estimated net internal rates of return were found to exceed 50 percent in the cases of soybeans and rice. Returns calculated for the wheat improvement program turned out to be much lower – in fact, well below the 50 percent level; and gross returns to cotton research were found to have been negligible. In all cases, the estimated yield advantage was smallest when interactions of the improved varieties with other variables were factored out.

The very high rates of return estimated for soybean research were explained by a large shift in product supply caused principally by the rapid uptake of the new varieties and their virtual displacement of the unimproved Mandarin

Table 4-19. Selected Comparative Data on the Rice, Cotton, Wheat, and Soybean Varietal Improvement Programs in Colombia[a]

Concept	Unit	Rice	Cotton	Wheat	Soybean
1. Estimated net internal rates of return	Percentage	60-82	0[b]	11-12	79-96
2. Estimated value of the supply shift parameter, 1971	Percentage	10-16		16	17-35
3. Estimated yield advantage, 1971	Percentage	25-39		46	17-36
4. Land area planted to improved varieties, 1971	Percentage	41	100	35	98
5. Average yields, 1971, Colombia/United States	Ratio	0.68	1.03[c]	0.53[c]	1.01[c]
6. Total research costs/ value production, 1968-71	Percentage	0.5	0.1	3.0	0.1

Source: Concepts 1-4 are based on a summary of previous tables in this study. Concept 5, Colombia, is from: Jorge Ardila, "Rentabilidad social de las inversiones en investigación de arroz en Colombia," M.S. thesis. Bogota, ICA/National University Graduate School, 1973; Gabriel Montes, "Evaluación de un programa de investigación agrícola: El caso de la soya," M.S. thesis. Bogota, University of the Andes, 1973; Andrés Rocha, "Evaluación económica de la investigación sobre variedades mejoradas de algodón en Colombia," M.S. thesis. Bogota, ICA/National University Graduate School, 1972; and Carlos Trujillo, "Rendimiento económico de la investigación en trigo," M.S. thesis. Bogota, ICA/National University Graduate School, 1974. Concept 5, United States, is from USDA, *Agricultural Statistics, 1973*, p. 441. For concept 6, see Ardila (1973); Montes (1973); Rocha (1972); and Trujillo (1974).

[a] Blanks indicate no data available.

[b] Since gross benefits were negligible, this net rate should be negative.

[c] 1970-72 average.

seed. The calculated yield advantage of the new varieties was not spectacular. The striking adoption pattern of the improved soybeans was attributed to the strength of product demand, derived in the main from a fast-growing poultry industry, the geographic concentration of production in a small area (the Cauca Valley), which facilitated the rapid diffusion of information concerning the improved varieties, the expected severity of attacks by the *cercóspora* virus, to which the improved varieties were resistant, and the fact that soybean producers figure among Colombia's most modern farmers. That soybean yields have been practically equal in Colombia and the United States in recent years (Table 4-19) reinforces our characterization of the industry as a modern one.

The Colombian cotton industry has evidenced similar characteristics. Yields of cotton in Colombia have not only equaled United States yields but even surpassed them in some recent years. Adoption of the improved United States varieties of cotton was practically instantaneous as a result of the government's ownership of gins and control over seed distribution. Yield increases since the early 1950s, when improved varieties came into widespread use, have been spectacular. Still, in spite of these similarities with the case of soybeans, it was concluded that returns to the cotton research program have been negligible.

This apparent contradiction was explained in terms of the organization of the research effort. The Colombian textile industry, long accustomed to importing United States cotton, partly as a result of a preferential rate of exchange, was compelled "to buy Colombian" by a change in government policy. Textile firms then sponsored the establishment of research which would lead ultimately to the local production of United States varieties of cotton. The final organization of the research program involved merely the importation, local testing, and distribution to farmers of the highest yielding United States varieties. This organization was justified on the premise that yields obtained locally from the United States cotton would vary according to variety; thus, there would be a payoff for identifying those kinds of cotton which yielded best under local conditions.

Our data did not sustain this premise, however. Information compiled on about 500 commercial field trials undertaken in Colombia for over ten varieties of improved United States cotton indicated that differences in yields by variety were minimal. Thus, the main research activity – local testing of imported varieties – appears to have been unnecessary. United States varieties could just as well have been selected at random for distribution in Colombia. Therefore, even though the widespread use of United States cotton increased yields, resulting surpluses were not attributed to the cotton research program.

As mentioned earlier in this section, net internal rates of return found for the rice research program were high by any standard of comparison. Yet, in view of the comparative data of Table 4-19, they are a puzzle. Although the ranges of estimated rates of return for the rice and soybean programs overlap, we see that the range of the calculated supply shift parameter for rice is significantly lower than the corresponding range for soybeans, principally because of differences in the levels of adoption of the improved rice and soybean varieties. Also, it can be observed that estimated rates of return to rice were much higher than those for wheat, even though the calculated values of their supply shift parameters were roughly comparable. Why then were estimated net rates of return to the rice research program so high?

An important answer lies with the cost side of the net rates of return calculations and with the organization of the rice improvement program. We believe that the direct costs of rice research to Colombia were effectively reduced by the program's having tapped into the accumulated stock of plant-breeding capital – general knowledge, improved breeding techniques, and plant materials – available in the two international centers, CIAT and IRRI, and in the World Collection of Rice. Without that accumulated capital, the costs of achieving comparable shifts in the supply of rice would have been higher and the corresponding net rates of return would have been lower.

This characteristic of the rice program was also found in the wheat research program. In fact, wheat had a longer history of using the accumulated foreign stock of plant-breeding capital than did rice. Linkages with the Rockefeller-Mexican program dated from about 1948, and additional collaborative support was provided the program during the late 1960s and early 1970s by the University of Nebraska Mission to Colombia. Judged from a purely technical and biological point of view, these foreign inputs were associated with success as they were in the case of rice. The estimated yield advantages of the improved wheat varieties were found to be large, even after the effects of variables which interacted with the new wheat varieties had been factored out. If they were included, the improved wheats could be shown to outyield the unimproved varieties by considerably more than 250 percent. Also, in international nursery trials the Colombian wheats easily outyielded the Mexican wheats from which they were largely derived.

Thus, the low estimated returns to the wheat research program were not the result of technical failures in plant breeding. Part of the explanation for the low returns lies in patterns of on-farm adoption of the improved seeds. The uptake of the new wheat varieties was notoriously slow. Fully twelve years elapsed from the time the first improved varieties of wheat were sold commercially in 1953 until they were in use on roughly one-quarter of all wheat land. Rates of adoption peaked at 50 percent in 1968 and then began a downward trend. Levels of use of the improved varieties in 1974 were estimated to include barely a fifth of all crop land planted to wheat. The slow uptake of the new seeds and the low levels and distressing trends in their use were attributed primarily to socioeconomic and structural constraints on production, especially the depressed domestic market resulting from continued PL 480 imports at levels which represented a large multiple of national production.

Two additional explanations for the low estimated rates of return to wheat research should also be stressed. One is that it became a very expensive program in the middle and late 1960s. Annual investments averaged fully 3 percent of the total value of wheat production, a figure which was not even remotely approximated by investments made in the other three varietal improvement programs (Table 4-19). A second explanation relates to the program's long gestation period. The Colombian wheat program dates from 1927. Yet our review of that history indicated that a well-organized research effort probably did not get underway until 1948, and the first improved varieties were not released on a major scale until 1953. As a consequence, investments (albeit at reduced levels) were being made for almost a quarter of a century before offsetting benefits were realized, and this had an adverse effect on the calculated net rates of return for wheat research.

NOTES

1. An earlier version of this chapter was presented in a workshop on Methods Used to Allocate Resources in Applied Agricultural Research in Latin America which was held at the International Center for Tropical Agriculture (CIAT), November 26-29, 1974. The current version of the chapter benefited from comments of participants in that workshop and from suggestions subsequently made by Norman R. Collins and Alain de Janvry.

2. A. C. Harberger, "La tasa de rendimiento del capital en Colombia," *Revista de Planeación y Desarrollo*, I:3 (October 1969).

3. Zvi Griliches, "Research Costs and Social Returns: Hybrid Corn and Related Innovations," *Journal of Political Economy*, 66:5 (October 1958), 419-432.

4. Willis Peterson, "Returns to Poultry Research in the United States," *Journal of Farm Economics*, 49:3 (August 1967), 656-669.

5. Nicolas Ardito Barletta, "Costs and Social Benefits of Agricultural Research in Mexico," Ph.D. dissertation (Chicago: University of Chicago, 1971).

6. Harry W. Ayer and G. Edward Schuh, "Social Rates of Return and Other Aspects of Agricultural Research: The Case of Cotton Research in São Paulo, Brazil," *American Journal of Agricultural Economics*, 54:4 (November 1972), 557-569.

7. R. C. Duncan, "Evaluating Returns to Research in Pasture Improvement," *Australian Journal of Agricultural Economics*, 16:3 (December 1972), 153-168.

8. See, for example, Willis Peterson, "The Returns to Investment in Agricultural Research in the United States," *Resource Allocation in Agricultural Research*, ed. Walter Fishel (Minneapolis: University of Minnesota Press, 1971), p. 160..

9. This section draws heavily on the work by Jorge Ardila, "Rentabilidad social de las inversiones en investigación de arroz en Colombia," M.S. thesis (Bogota: ICA/National University Graduate School, 1973).

10. S. H. Ou, *Rice Diseases* (England: Commonwealth Mycological Institute, 1972), pp. 28-33. Apparently, the attack was least severe where the Colombian red rice was grown; see Philippe Leurquin, "Rice in Colombia: A Case Study in Agricultural Development," *Food Research Institute Studies*, 6:2 (1967), 231.

11. The means by which the disease was transmitted were not identified with the insect, *Sogotodes*, until 1958.

12. Chemical control of the virus, though expensive, proved somewhat effective among partially resistant varieties; see G. E. Galvez, "*Hoja Blanca* Disease of Rice," *The Virus Diseases of the Rice Plant* (Baltimore: Johns Hopkins Press, 1968), pp. 35-49.

13. Bluebonnet 50 was first imported to Colombia in 1954. The history of its introduction and rapid adoption is discussed by Leurquin, "Rice in Colombia," pp. 250 and 251.

14. *Informe de Gerencia al XIII Congreso Nacional* (Bogota: FEDEARROZ, 1971), p. 32.

15. In 1948 Raul Varela Martinez, *Industria y comercio de arroz en Colombia* (Bogota: Ministerio de Agricultura y Canaderai, 1949), p. 15, estimated that 18 percent of all

rice land was irrigated and 16 percent was partly irrigated. By 1974, it is estimated that the percentage of rice land irrigated had increased to about 75 percent.

16. See, for example, Leurquin, "Rice in Colombia," Tables 5 and 11.

17. The exception to this procedure was taken in estimated CICA-4's yield advantage since, given the specification of the regression, the yield advantage of CICA-4 simply equals the negative value of the coefficient on Bluebonnet 50.

18. The varietal-effects estimate shown here evidences lower values for the shift parameters than the source (Ardila, "Arroz en Colombia," Tables 17 and 20) by reason principally of the omission of adjustments for "time." Ardila originally added appropriate values of the coefficients on the time variables in each year to the yield differentials estimated from the production functions. Since those coefficients are on net strongly positive for the 1964-71 period, the effect was to increase estimated values of the shift parameters.

19. Calculations based on the varietal effects plus interaction effects shift parameters are not included here but can be inspected in Ardila, "Arroz en Colombia," Table 27.

20. Nestor Gutierrez and Reed Hertford, *Una evaluación de la intervención del gobierno en el mercado de arroz en Colombia*, Technical Pamphlet no. 4 (Palmira: Centro Internacional de Agricultura Tropical, 1974), Table 3.

21. *Ibid*., Table 4

22. Ardito Barletta, "Agricultural Research in Mexico."

23. One staff member of the rice program held a doctorate and four others held masters degrees. Costs of training personnel of the National Federation of Rice Growers at the ICA/National University Graduate School were also included.

24. This is a more moderate assumption than that used by Ardila, "Arroz en Colombia," Table 35, who assumed that the percentage of area sown in improved rice would fall between 72 percent and 84 percent by 1980. The data of Dana G. Dalrymple's *Development and Spread of High-Yielding Varieties of Wheat and Rice in the Less-Developed Nations*, Foreign Agricultural Economic Report no. 95 (Washington, D.C.: Economic Research Service, USDA, 1974), Table 38, indicating that adoption rates in the Philippines, Malaysia, and India are topping out at rates well under 60 percent, plus the analysis by Robert Evenson, "The 'Green Revolution' in Recent Development Experience," *American Journal of Agricultural Economics*, 56:2 (May 1974), 387-394, suggest that the use of improved varieties may never reach 70 percent in Colombia and that an assumption of 57 percent is more conservative and probable.

25. Ardila, "Arroz en Colombia," placed a range on the internal rate of return of 53.0 to 53.4 percent. The minimum (and the maximum) reported here would have been lower, and lower than Ardila's estimates, had the additional costs of the new varieties to farmers been netted out of "gross benefits." Data on those costs were simply judged to be too weak to include.

26. This section is based largely on the work by Andrés Rocha, "Evaluación económica de la investigación sobre variedades mejoradas de algodón en Colombia," M.S. thesis (Bogota: ICA/National University Graduate School, 1972).

27. Philippe Leurquin, "Cotton Growing in Colombia: Achievements and Uncertainties," *Food Research Institute Studies*, 6:2 (1966), 145.

28. These and other production and trade data cited in this section were obtained from the relevant number of *Cotton-World Statistics* published by the Secretariat of the International Cotton Advisory Committee, Washington, D.C.

29. *Changes in Agricultural Production and Technology in Colombia*, Foreign Agri-

cultural Economic Report no. 52 (Washington, D.C.: Economic Research Service, USDA, 1969), p. 76.

30. Leurquin, "Cotton Growing," pp. 158-159.

31. *Ibid*., p. 154.

32. *Cotton in Colombia*, FAS-M (Washington, D.C.: Foreign Agricultural Service, USDA, 1971), p. 11.

33. *The Cotton Industry of Colombia*, FAS-M-113 (Washington, D.C.: Foreign Agricultural Service, USDA, 1961), p. 9.

34. *Agriculture Production and Trade of Colombia*, ERS-Foreign 343 (Washington, D.C.: Economic Research Service, USDA, 1973), p. 31.

35. This is not to deny the role of improved United States varieties in increasing Colombian cotton yields. There are several references to their importance in the available literature. One of the strongest is International Bank for Reconstruction and Development, *The Agricultural Development of Colombia* (Washington, D.C.: IBRD, 1956), p. 86: "It is the policy and program of the Cotton Institute to provide the full seed requirements for the entire cotton crop annually. Largely as a result thereof, the average yield per hectare of the cotton crop has doubled in four years."

36. This section is based on the research of Carlos Trujillo, "Rendimiento económico de la investigación en trigo," M.S. thesis (Bogota: ICA/National University Graduate School, 1974).

37. USDA, *Changes in Agricultural Production*, p. 12

38. Data on production, land area planted, and yields of wheat before 1971 are from Roger Sandilands, *Algunos problemas en la selección de datos estadísticos para trigo, cebada y papa*, Technical Report no. 14 (Bogota: ICA, 1974). Data for 1971-73 are from the National Department of Statistics, DANE, Bogota.

39. Many of the important issues involved are discussed in Roger Sandilands and Leonard Dudley, "The Side-Effects of Foreign Aid: The Case of P.L. 480 Wheat in Colombia," *Economic Development and Cultural Change*, 23:2 (January 1975), 325-336.

40. Estación Agrícola Experimental La Picota, *Informe de labores de la sección experimental e industrial realizados en 1936* (Bogota, 1937); also, Estación Agrícola Experimental La Picota, *Informe de labores de la sección experimental e industrial realizados en 1941* (Bogota, 1942).

41. Estación Francisco José de Caldas, *Influencia de las royas sobre el rendimiento del trigo*, Project I-SF-1949 (Bogota, 1951).

42. Mario Zapata, "Informe general de programa de trigo en clima cálido," mimeographed (Bogota, 1969).

43. Vaneamiento is a disease in which sterility results from sharp changes in temperature.

44. Trujillo, "Investigación en trigo," Table 2.10.

45. The price elasticity of demand estimate is from *Un método de proyectar la producción y demanda para productos agrícolas en Colombia*, Research Bulletin no. 15 (Bogata: ICA, 1971). The price elasticity of supply estimate is from Sandilands and Dudley, "Side-Effects of Foreign Aid."

46. Based on the research by Gabriel Montes, "Evaluación de un programa de investigación agrícola: El caso de la soya," M.S. thesis (Bogota: University of the Andes, 1973).

47. *Ibid*., Table 4.

48. The price elasticity of demand estimate was suggested by the results of James P. Houck, "A Statistical Model of Demand for Soybeans," *Journal of Farm Economics*, 46:2 (May 1964), 371-372.

5

Returns to Investment in Agricultural Research in India

A. S. Kahlon, P. N. Saxena,
H. K. Bal, and Dayanath Jha

The problems of measuring returns to investment in agricultural research are beset with serious difficulties, both conceptual and practical. Even with advanced methodology, researchers in the less developed countries face a serious problem: the lack of a suitable information base on key variables. Still, the growing realization of the importance of research in agricultural growth has stimulated the estimation of likely payoffs to investment in agricultural research. These estimates have become vital because of the extremely critical resource position of the LDCs and the concomitant necessity of making the most judicious investment decisions. Recognizing the significance of this issue, the Indian Council of Agricultural Research has recently formed an expert panel to investigate this problem.[1]

Two general approaches, namely, the index number approach and the production function approach, have been used for *ex-post* evaluation of agricultural research. The pioneering works of Schultz and Griliches have been followed by a fair amount of empirical and theoretical work in this area.[2] But these studies have covered mainly the developed countries.

There is very little empirical evidence to facilitate measurement of the contribution of Indian agricultural research systems to real productivity growth in Indian agriculture. Two studies, however, have recently been conducted. Evenson and Jha estimated the return to investment in Indian agriculture, using the total factor productivity approach, and obtained an internal rate of return that exceeded 50 percent a year.[3] Karam Singh estimated returns to investment in agricultural research in the Punjab and concluded that investment of one rupee in agricultural research gives a return of twenty-seven rupees.[4] In both studies research expenditures were directly incorporated into the function.

These estimates appear to be somewhat crude because they ignore qualitative improvements that are embodied in technological changes. Moreover, a technical change, which is a proxy for research expenditure, improves the quality of inputs and outputs, changes the combination of the inputs, and has factor-augmenting effects. The importance of qualitative changes and factor-augmenting technical change was emphasized by Evenson and Jha. However, these factors were not fully incorporated in their final analysis.

The Evenson and Jha study went on to investigate the share of annual output growth accounted for by the growth in different inputs. Sawada used a similar approach in his study of the sources of the growth of aggregate production in Japan.[5] In both cases, the assumption of constant returns to scale – "homogeneity of degree one" in the production function – was made. This does not seem entirely realistic for Indian agriculture.

When the residual approach is employed, it is possible only to include selected major variables in the specified relationship. The residual will include the effects both of the omitted variables and of any measurement errors in the variables that are included in the output measure.[6] Therefore, using the residual to measure the effect of technology has some serious limitations. Technology is not the only variable included in the residual. It is difficult to isolate the effects of other excluded variables, such as management. Moreover, since few of the variables contained in the model are independent of the effect of technology, it becomes even more difficult to attribute all of the residual effect to such technology.

A further difficulty is that the magnitude of the residual is dependent upon the proper specification of the model, i.e., the specification of the variables and the specification of the type of relationship. The principle of the least-squares technique minimizes the sum of the squares of the residuals. The best specification of the model would lead to residuals of small magnitude; incorrect specification would lead to residuals of larger magnitude.

We try to avoid some of these problems in the approach presented in this chapter. Our study attempts to develop a model wherein the assumption of constant returns to scale is not considered necessary. An attempt is also made to weigh the factor share in the growth of output in Indian agriculture. The use of dummy variables makes it possible to analyze the shifts in production functions and the change in the quality of different inputs, both of which may affect the factor combinations.

The chapter is divided into two parts. The results for the country as a whole, comprising fifteen major states, are presented in the section entitled "The All-India Analysis." The second section is devoted to "State-Level Analysis." In this section, estimates of the contribution of agricultural research to gross agricultural output are presented for the four states representing the four major regions of the country: Andhra Pradesh, Bihar, Maharashtra, and Punjab. Since levels of technological progress vary considerably among states and regions, this type of analysis is expected to provide a closer understanding of the underlying relationships.

The All-India Analysis

In studying the returns to research expenditure, we used two different approaches. Each covered two periods: the pre-green revolution period (1960-61 to 1964-65) and the post-green revolution period (1967-68 to 1972-73). The first approach incorporates research expenditure (with suitable lags) as an independent variable in the types of functions already discussed. The coefficient of this variable in the production function for the two periods (with D = 0 and D = 1) gives the return to one rupee invested in agricultural research for a linear function and the percentage return to a 1 percent increase in the research expenditure for a double-log function. The use of a dummy for this variable also allows for the possibility of a shift in returns owing to the green revolution.

The second approach excludes research expenditure from the production function analysis.[7] The estimates of output for the two periods were obtained at fixed levels of the inputs. The difference between these two estimates can be attributed to the additional investment in agricultural research. This approach loses information on both the level of returns to investment in the first period and the way in which these returns increased in the second period.

The model used in the all-India analysis is outlined in Appendix 5-1. The data used and their sources are described in Appendix 5-2.

Growth Rates in Indian Agriculture

During the pre-green revolution period, agricultural output in value terms, using constant prices, increased at a compound rate of 2.66 percent annually, and the average annual increase was 61.01 million rupees (Table 5-1). These growth rates were lower than those of the post-green revolution period, when the compound growth rate was 5.80 percent a year and the linear growth rate was 182.98 million rupees a year.

The growth rate of net sown area increased from 2.20 percent in the first

Table 5-1. Output and Input Growth Rates in Indian Agriculture, 1960/61-1972/73

Variable	Linear Growth Rate[a]		Compound Growth Rate[b]			
	First Period 1960/61-1964/65	Second Period 1967/68-1972/73	First Period 1960/61-1964/65	Second Period 1967/68-1972-73	Mean	Standard Deviation
Y : Output value in million rupees	61.01	182.98	2.66	5.80	5573.28	2760.42
X_9 : Net sown area in thousand hectares	47.92	52.03	2.20	2.70	9064.99	4827.01
X_2 : Human labor in thousand persons	1073.04	148.52	5.63	2.32	7258.31	710.68
X_3 : Bullock labor in thousand numbers	1196.88	– 0.50	6.39	– 0.17	5692.15	7497.83
X_4 : Fertilizer consumption in million rupees	34.06	52.70	45.00	16.97	244.85	2704.70
X_5/X_9 : Proportion of irrigated area	0.20	0.0047	6.54	1.81	0.3360	.42
X_6 : Tractor population in thousand numbers	0.16	1.31	31.23	15.62	5.22	6.79
X_7 : Expenditure in agricultural research in thousand rupees	329.12	395.33	7.58	6.08	5030.36	3311.37

a $\left(\frac{dX_i}{dt}\right)$.

b $\left(\frac{1}{X_i} \times \frac{dX_i}{dt} \times 100\right)$.

period to 2.70 percent in the second period. However, the growth rates for human labor declined from 5.63 percent in the first period to 2.32 percent in the second period. Bullock labor declined from 6.39 percent to − .17 percent per annum.

Fertilizer consumption increased at a rate of 45.0 percent in the first period and 16.97 percent in the second period. However, this does not imply that the increase in fertilizer consumption was lower in the second period. In fact the average annual increase in the consumption of fertilizer was 34.06 million rupees in the first period and 52.70 million rupees in the second. The high rate of growth in the use of fertilizer during the first period was attributable to a very low base.

Contrary to expectation, the growth rate of the irrigated area declined during the second period. Tractor population increased from the first period to the second period, but the compound rate of growth declined from 31.23 percent to 15.62 percent.

The rate of growth in agricultural research was steady over the two periods: 7.58 percent in the first period and 6.08 percent in the second period. Annual research expenditure increased by 329.12 thousand rupees and 395.33 thousand rupees in the first and second periods, respectively.

Shift in Production Function and Qualitative Effects

The preceding section has described the growth of agricultural output and of the specified inputs. The importance and the impact of these inputs have varied from time to time and therefore it was necessary to examine the change in the contribution of various factors over time.

Owing to the problem of multicollinearity, all the factors shown in Table 5-1 could not be included in the production function. Bullock labor (X_3) was highly intercorrelated with human labor (X_2). The inclusion of both these variables in the production function resulted in large sampling variances of the estimates of the coefficients, and it was decided to omit bullock labor from the production function. Because they added nothing to the results, two other variables – number of tractors (X_6) and research expenditure (X_7) – were also excluded from the production function. The structural equation used is provided in the accompanying box.

The significance of the coefficient of the period dummy variable (0.2298) indicates an important shift in the production function associated with the green revolution. The coefficients for D log X_9 and D log (X_5/X_9), though positive in signs, were not significant in the trial functions and were, therefore, deleted from the final function. This would suggest that the impact of the net sown area (X_9) and the proportion of irrigated area to net sown area (X_5/X_9) did not change over time; hence, the elasticity coefficients of 0.3879

$$\text{Log } Y = a + b_9 \log X + b_2 \log X_2 + b_4 \log X_4 + b_5 \log (X_5/X_9) + CD + C_2 D \log X_2 + C_4 D \log X_4$$

where Y, X_9, X_2, X_4, and X_5/X_9 are the variables indicated in Table 5-1.

$$\begin{aligned} \log Y = {} & 1.5717^* + 0.3879^* \log X_9 + 0.1875^* \log X_2 \\ & (0.1477) \quad (0.0601) \qquad\qquad (0.0581) \\ & + 0.0492^* \log X_4 + 0.2715^* \log (X_5/X_9) \\ & (0.0170) \qquad\qquad (0.0423) \\ & + 0.2298D^* - 0.1695^* D \log X_2 + 0.1957^* D \log X_4 \\ & (0.1045) \qquad (0.0355) \qquad\qquad (0.0371) \end{aligned}$$

where D = 0, for pre-green revolution period

D = 1, for post-green revolution period

$R^2 = 0.7252^*$

$N = N_1 + N_2 = 75 + 90 = 165$

* Significant at .01 level.

Standard errors of the estimates are in parentheses.

for the net sown area and 0.2715 for the proportion of irrigated area remained the same over the two periods. It may be noted that these elasticities were significantly different from zero, which means that these factors were important for expanding output, although their impact did not change from the first to the second period.

The elasticities for human labor (X_2) and fertilizers (X_4) were positive and significant in the first period. The impact of these variables underwent a significant change, however, from the first to the second period. In the second period, though still positive, the elasticity for human labor declined to 0.0180, and it was not significantly different from zero. Fertilizer, on the other hand, experienced a positive shift of 0.1957 (significant) in the elasticity coefficient. One effect of the green revolution was to increase the productivity of fertilizer.

Factor Shares

The relative share of each factor in the growth of output for the two periods and the change in factor shares associated with the green revolution are presented in Table 5-2. These shares were calculated by utilizing the production function estimated above and the expressions (5) and (10) in Appendix 5-1.

During the first period, fertilizer and irrigation contributed 37.44 and 30.15 percent respectively to the growth of output experienced. Next in im-

Table 5-2. Factor Shares in the Output Growth Rate

Factors of Production	Pre-Green Revolution Period	Post-Green Revolution Period
X_9 : Net sown area	14.49%	18.27%
X_2 : Human labor.	17.92	0.73
X_4 : Fertilizers	37.44	72.43
X_5/X_9: Proportion of irrigated area.	30.15	8.57
Total	100.00%	100.00%

portance were human labor and net sown area, contributing 17.92 percent and 14.49 percent, respectively.

In the second period, fertilizers accounted for 72.43 percent of the output growth, thereby making fertilizer the most important factor in the growth of agricultural output in that period. The growth rate of net sown area increased in the second period from 14.49 percent to 18.27 percent. A decline in the growth rate of the irrigated area, with elasticity remaining the same over the two periods, contributed to a decline in its relative share in output growth. Human labor remained an important factor of production, as indicated by the significance of the elasticity coefficient, but the results show that growth of labor inputs contributed very little to growth of agricultural output in the post-green revolution period.

Change in Factor Combinations

It is important to know whether the factor combination for a given level of output changed because of changes in technology. For purposes of this analysis, the output level was fixed at the overall mean of 5573.28 million rupees in both time periods. The variables that finally appeared in the production function were net sown area (X_9), human labor (X_2), fertilizers (X_4), and proportion of irrigated area (X_5/X_9).

Since the elasticity coefficient for X_9 and X_5/X_9 did not change over time, the levels for these variables were fixed and held at the mean levels indicated in Table 5-1. The levels of variables such as human labor (X_2) and fertilizer (X_4) need to be determined for given values of other variables. This again gives one equation with two unknowns, X_2 and X_4, in each time period. It was possible, therefore, to determine X_4 for varying levels of X_2 and vice versa. For example, when X_2 was held at the mean level, X_4 was estimated at 3.16 million rupees in the first period and 68.29 million rupees in the second.

Obviously, a considerable shift in the combination of factors was required to produce a fixed level of output. There was more use of human labor and less use of fertilizers in the first period. Fertilizers became a dominant factor in the second period.

Return to Investment in Agriculture

To study the contribution of research to output growth, we considered two alternative approaches at the model-building stage. However, the first approach (directly incorporating research expenditure in the function) could not be used because research expenditure had a high multicollinearity with the other independent variables. The second approach (measuring returns to autonomous public investment in agricultural research) was used to estimate the research contribution. The results are shown in Table 5-3.

Table 5-3. Return to Investment in Agricultural Research in India[a]

Output and Investment	First Period (1)	Second Period (2)	Difference (2) – (1)
Estimated output (million rupees)	6592.00	6945.00	353.00
Average investment in agricultural research (thousand rupees).	3372.05	6412.28	3040.23

[a] Return to 1 rupee invested = $\frac{\text{353.00 million rupees}}{\text{3040.23 thousand rupees}}$ = 11.61 rupees.

An investment of one rupee in agricultural research gave a return of 11.61 rupees, with a lag of five years between research expenditure and returns. This was converted into an internal rate of return, assuming a five-year lag. The internal rate of return worked out to be 63.3 percent per annum.

Evenson and Jha also estimated the return to investment in agricultural research (within a state).[8] Their estimates gave a return of 10,650 rupees from a 1,000-rupee increase in research spending. The internal rate of return in this case was 50 plus percent per annum over a time lag of about eight years.

Owing to the weakness of the data and the estimation problems that were encountered, we do not regard the two rates of return as essentially different. Their similarity, in spite of differences in estimation procedures, tends to reinforce the conclusion that the rate of return to agricultural research in India has been high.

State-Level Analysis

Both linear and log-linear models were utilized in analyzing the data for individual states. The use of slope (or elasticity) dummies was not feasible in the state-level analysis because of the short time series available. All the period dummies, however, were included in the pooled analysis of data for all selected states. Regional dummies were also included in this analysis to account for interstate differences. Finally, a time variable (t) was introduced into the model explicitly to account for trend components. It was generally found to improve the fit.

Data and Methodology

Four states – Andhra Pradesh, Bihar, Maharashtra, and Punjab – were selected for the state-level analysis as a representative cross section of the state of agricultural development in India. These states cover the major agroclimatic zones as well as the important cropping regions for rice, wheat, sorghum, and pearl millet. It is these crops which have been most affected by the seed-based technological revolution in the country.

Ten variables were included in the production function analysis. (See the accompanying tabulation.) The research expenditure variable was used in the equations with lags varying from zero to six years, and the variable was then denoted by $R_t, R_{t-1}, R_{t-2}, \ldots, R_{t-6}$.

Variable	*Unit of Measurement*	*Symbol*
Agricultural production index	1956-57 base	Y or X_1
Annual rainfall	Millimeters	X_2
Fertilizer	Thousand tonnes	X_3
Male agricultural workers	Thousands	X_4
Literacy in rural males	Percentage	X_5
Draft bovines	Thousands	X_6
Tractors	Numbers	X_7
Total cropped area	Thousand hectares	X_8
Gross irrigated area	Percentage	X_9
Research expenditure	Thousand rupees	X_{12} or R_t

State-level time series data on these variables were used for the period 1956-57 to 1972-73. A period dummy variable (X_{10}) with the value zero up to 1965-66 and unity for subsequent years was also introduced in an attempt to capture the effects of technological change. The time trend variable was denoted by "t."

In the pooled analysis of data for all the four states, the index numbers of agricultural production were converted to value terms (in millions of rupees) at constant 1956-57 prices and used collectively as the dependent variable. Regional dummy variables, S_1, S_2, and S_3, were also introduced to account for interstate variations. In this analysis, slope dummies were also used to examine the period-specific changes in the effects of individual variables. Further particulars about these variables, along with the sources and limitations of data, are given in Appendix 5-3.

Results and Discussion

The means and coefficients of variation (C.V.) for all the ten variables under consideration for each state are given in Table 5-4. The mean values reveal the wide interstate variability which characterizes Indian agriculture. For example, there was wide variation in the rise in the agricultural production index. From a 1956-57 baseline of 100, the index rose in Maharashtra to only 103.10 by 1972-73, but in Punjab it reached a level of 154.90. Similar differences are noticeable with respect to other variables. The coefficients of variation were the lowest, as expected, for such variables as gross cropped area, draft bovines, and male agricultural workers – the conventional inputs. Rainfall, literacy, and irrigated area lie in the medium range. For others, such as fertilizers, tractors, and research expenditure, a high temporal variability is indicated, mainly on account of substantial secular increases. The linear annual growth rate of agricultural output over the seventeen-year period was – 0.98 percent for Maharashtra, 1.30 percent for Andhra Pradesh, 4.36 percent for Bihar, and 9.73 percent for the Punjab.

To estimate the relative contribution of the input variables to agricultural productivity, both linear and log-linear regression equations, using all nine variables, were fitted to the data for each of the four states individually. For each state, seven equations were fitted, corresponding to the lags of zero, one, two, three, four, five, and six years in the research expenditure variable R_t. Out of these equations, the one showing consistent signs and a relatively lower standard error for the R_t variable was chosen for subsequent analysis. Thus, this variable had different lags, ranging between one to four years, in different states. Although there is no logical explanation for this, the crude nature of the data on research expenditures made this compromise necessary. The equations containing all nine variables are given in Appendix 5-4.

Using the usual criteria of consistency in sign, significance of regression coefficients, multicollinearity, and the closeness of fit of the model, we utilized the equations from Appendix 5-4 for a second run to arrive at the final regression functions. The results for each state are discussed separately in the following paragraphs. It was observed that with one exception (Bihar), the log-linear form was not appropriate for depicting the relationships among the

Table 5-4. Variables and Summary Statistics Used in the State-Level Analysis, 1956/57-1972/73

Variable	Unit	Parameter	States			
			Andhra Pradesh	Bihar	Maharashtra	Punjab
Index of agricultural production (Y)	Index	Mean	111.94	135.81	103.10	154.90
		C.V.[a]	10	23	14	34
Annual rainfall (X_2)	Millimeters	Mean	920.70	1,179.41	814.29	712.12
		C.V.	16	17	18	20
Total fertilizer consumption (X_3)	Thousand metric tonnes	Mean	149.12	49.41	102.41	96.06
		C.V.	68	84	78	98
Male agricultural workers (X_4)	Thousands	Mean	7,437.94	10,267.88	7,074.35	1,973.12
		C.V.	8	12	8	15
Literacy in rural males (X_5)	Percentage	Mean	24.28	26.31	36.50	29.55
		C.V.	15	8	13	14
Draft bovines (X_6)	Thousands	Mean	6,561.12	7,837.00	6,559.12	1,855.88
		C.V.	5	3	4	4
Tractors (X_7)	Numbers	Mean	2,964.94	2,794.53	2,646.88	15,088.59
		C.V.	49	67	55	95
Total cropped area (X_8)	Thousand hectares	Mean	12,713.94	10,763.76	19,081.59	5,204.66
		C.V.	5	5	1	6
Gross irrigated area (X_9)	Percentage	Mean	29.85	21.24	7.16	62.58
		C.V.	5	13	14	14
Research expenditure (R_t)	Thousand rupees	Mean	5,800.23	4,607.76	7,704.35	5,326.71
		C.V.	35	24	56	46

[a] Coefficient of variation.

variables. Hence only the linear model is discussed for the remaining three states.

Andhra Pradesh. The accompanying box indicates the way in which the final linear regression equation relating the index of agricultural production to the other input variables for the State of Andhra Pradesh was obtained. The

$$Y = -249.5412 + \underset{(0.0871)}{0.2117 X_3^*} + \underset{(0.0186)}{0.0551 X_6^{**}} + \underset{(0.0071)}{0.0086 R_{t-3}} - \underset{(3.7884)}{9.1019 t^*}$$

$R^2 = 0.6672$

$$\bar{R}^2 = 1 - \frac{(1 - R^2)(N - 1)}{N - K - 1} = 0.5563$$

$F = 6.01$

Durbin-Watson $d = 1.97$

* Significant at .05 level.
** Significant at .01 level.
Standard errors are in parentheses.

Durbin-Watson test revealed an absence of serial dependence in the residuals. The value of R^2 was rather low, for the model explains only 55 percent of the observed variation in Y. In this case, the period dummy variable (X_{10}) was found to be nonsignificant and thus could not be entered in the final equation. These findings indicate the lack of any positive contribution by the green revolution to the agricultural productivity of the state.

Other variables included in the equation with the corresponding standard partial regression coefficient values (β_i) are given in the accompanying tabulation. It can be seen that fertilizer consumption made the greatest contribu-

	Variable	β_i
X_3	: Total fertilizer consumption	1.8886
X_6	: Draft bovines	1.5687
R_{t-3}	: Research expenditures	1.5543

tion to the dependent variable, followed by draft animal power and lagged research expenditure. The last variable, however, was not statistically significant. The marginal product of R_{t-3} came to 32.40 rupees. This can be interpreted as the external rate of return to investment in agricultural research.

Bihar. The accompanying box indicates the results of the final regression

Linear

$$Y = -1053.8380 + \underset{(0.0381)}{0.0814}\ X_4^{**} + \underset{(0.0073)}{0.0417}\ X_8^{***}$$
$$+ \underset{(14.7080)}{9.3249}\ X_{10} + \underset{(0.0120)}{0.0187}\ R_{t-4}^{*} - \underset{(11.4729)}{20.6002}\ t^{**}$$

$R^2 = 0.9161$ $\bar{R}^2 = 0.8780$

$F = 24.02$ $'d' = 2.02$

Log-linear

$$\text{Log } Y = -19.0693 + \underset{(0.6890)}{1.2076} \log X_4^{*} + \underset{(0.4817)}{3.8297} \log X_8^{***}$$
$$+ \underset{(0.0424)}{0.0111}\ X_{10} + \underset{(0.2109)}{0.2855} \log R_{t-4}^{*} - \underset{(0.0818)}{0.1547} \log t^{**}$$

$R^2 = 0.9405$ $\bar{R}^2 = 0.9135$

$F = 34.82$ $d = 2.09$

* Significant at .10 level.
** Significant at .05 level.
*** Significant at .01 level.
Standard errors are in parentheses.

equations for Bihar. The values of 'd' were not significant in either equation. Fairly high R^2 values suggested reasonably good fit. The period dummy variable (X_{10}) here, as in Andhra Pradesh, was found to be nonsignificant. This means that the post-1966 period did not reveal any positive impact on agricultural production.

For the linear equation, the other variables included in the equation and their corresponding standard partial regression coefficient values (β_i) are given in the accompanying tabulation. It is clear that the highest contribution to

	Variable	β_i
X_4	: Male agricultural workers	3.1282
X_8	: Total cropped area	0.7621
R_{t-4}	: Research expenditure	0.6556

the dependent variable was made by male agricultural workers, followed by total cropped area and the lagged research expenditure. Converting the marginal product of R_{t-4} in the linear equation into value terms, we found that every rupee invested in research gave a return of 63.75 rupees. The variable in question is significant at the 10 percent level of probability.

Maharashtra. The accompanying box indicates the way in which the final re-

$$
\begin{aligned}
Y = &- 2065.9382 - \underset{(0.0187)}{0.0464}\, X_2{}^{**} + \underset{(0.0155)}{0.0671}\, X_6{}^{***} \\
&+ \underset{(0.0211)}{0.0760}\, X_8{}^{***} + \underset{(19.5667)}{72.0594}\, X_9{}^{***} + \underset{(7.3289)}{24.0126}\, X_{10}{}^{***} \\
&+ \underset{(0.0021)}{0.0034}\, R_{t-1}{}^{*} - \underset{(4.1988)}{26.3015}\, t^{***}
\end{aligned}
$$

$R^2 = 0.8932$ $\quad\bar{R}^2 = 0.8101$

$F = 10.74$ $\quad d = 2.80$

* Significant at .10 level.
** Significant at .05 level.
*** Significant at .01 level.
Standard errors are in parentheses.

gression equation relating the index of agricultural production to input variables was obtained. The value of 'd' was found to be nonsignificant at the 5 percent level, indicating the absence of any autocorrelation in the residuals. The values of R^2 were quite high, the model explaining nearly 81 percent of the variation in Y. The period dummy variable (X_{10}) was found to be highly significant, indicating marked differences in agricultural productivity in the state owing to changes in technology in the two periods.

Other variables included in this equation and their corresponding standard partial regression coefficient values (β_i) are provided in the accompanying tabulation. Here irrigation percentage made the greatest contribution to the

	Variable	β_i
X_2	: Annual rainfall	− 0.4628
X_6	: Draft bovines	1.1615
X_8	: Total cropped area	1.3771
X_9	: Irrigation percentage	4.9152
R_{t-1}	: Research expenditures	0.9725

dependent variable, followed by total cropped area and draft bovines. The research expenditures variable was also significant, at the 10 percent level of probability. Converting the marginal product of R_{t-1} into value terms, we find that every rupee invested in agricultural research in the state yielded a return of 14.28 rupees.

Punjab. The final regression equation for this state gave the results shown in the accompanying box. The value of 'd' suggested lack of serial dependence. The value of R^2 in this case was the highest among all states, accounting for nearly 97 percent of the total variation in Y. Moreover, the period dummy

$$Y = -891.4295 + \underset{(0.1527)}{0.2341\, X_3^{*}} + \underset{(0.2398)}{0.2509\, X_4}$$
$$+ \underset{(10.5703)}{22.1758\, X_5^{**}} + \underset{(1.4941)}{2.2428\, X_9^{*}} + \underset{(9.9899)}{20.4666\, X_{10}^{**}}$$
$$+ \underset{(0.0048)}{0.0113\, R_{t-2}^{**}} - \underset{(15.8339)}{37.2630\, t^{**}}$$

$R^2 = 0.9864$ $\bar{R}^2 = 0.9758$

$F = 93.53$ $d = 1.96$

* Significant at .10 level.
**Significant at .05 level.
Standard errors are in parentheses.

variable (X_{10}) was significant at the 5 percent level, indicating significant change in the state productivity level in the two periods as a result of the introduction of new agricultural technology.

The standard partial regression coefficient values (β_i) are provided in the accompanying tabulation. It can be seen that the highest contribution to the

	Variable	β_i
X_3	: Total fertilizer consumption	0.4613
X_4	: Male agricultural workers	1.3753
X_5	: Literacy in rural males	1.6809
X_9	: Irrigation percentage	0.3833
R_{t-2}	: Research expenditures	0.5279

dependent variable was made by literacy, followed by labor and research. The last variable (R_{t-2}) was significant at the 5 percent level of probability. The results indicate that every rupee invested in agricultural research in the state gave a return of 15.93 rupees.

Pooled Analysis. The results of the pooled state-level analysis, which aimed at providing additional information on the impact of the pre- and post-green revolution periods through the incorporation of slope dummies, were totally unsatisfactory. The regression coefficients turned out to be both nonsignificant and inconsistent. The results could not, therefore, be interpreted meaningfully. The results presented and discussed above reveal the inherent weakness of aggregative analysis of this sort. It is quite clear that the great heterogeneity of the environment of agricultural production makes it very risky empirically to estimate a unique functional specification for all states. Indeed, even a state is too heterogeneous a unit. In view of this fact and the usual errors of measurement and aggregation of inputs, the relatively poor results are not surprising. We can, however, draw from the individual state-level analyses some conclusions regarding the importance of different input variables in different states with implications for *ex-ante* assessments.

Conclusions from State-Level Analysis

In analyzing the rainfall and irrigation results, it is helpful to recognize that Maharashtra and Punjab are relatively dry regions while Andhra Pradesh and Bihar are areas of generally sufficient rainfall. Rainfall appeared as a significant variable only in Maharashtra, and even there it had a negative sign. At the same time, for Punjab the contribution of irrigation was positive and significant. The explanation for this apparent contradiction may lie in the high sensitivity of the dry region crops to the distribution of rainfall and to the improvement in water management following the development of irrigation.

In Andhra Pradesh and Bihar, irrigation did not appear to be significant. This means that in areas of sufficient rainfall irrigation is more an adjustment mechanism to control periodic insufficiences in the amount of rainfall than a direct stimulant to output. It also underscores the importance of the development of irrigation in the relatively drier regions.

In Bihar and Maharashtra, fertilizer did not appear to be significant. Literacy was significant only in Punjab, whereas a positive contribution by labor was indicated in Bihar and Punjab. The draft animals variable was significant in Andhra Pradesh and Maharashtra. Tractors, used as a representative measure for machines in general, did not appear to be significant in any state. Bihar and Maharashtra had a significant positive coefficient for total cropped area. The research expenditures variable was found to have a significant positive impact on output in all states except Andhra Pradesh.

The period dummy variable was significant only in Maharashtra and Punjab, indicating an upward shift in production function in the post-1965/66 period in both cases. The traditional rice-growing states of Bihar and Andhra Pradesh did not show this effect. The trend variable emerged significant in all cases and carried a negative sign.

The overall impression one gets from these results is that the conventional factors of production play a more dominant role in states like Andhra Pradesh, Bihar, and Maharashtra. Only in the Punjab do the nonconventional input factors seem to be of relatively greater importance.

The research expenditures variable is the primary focus of this chapter. It is, as stated earlier, difficult to explain the differences in lags between different states. Still, we do have enough basis to suggest that agricultural research does make a positive and substantial contribution. The state-level analysis tends to support the results of the national analysis in confirming that agricultural research is a productive investment. The range of the estimated marginal returns is 14.28 rupees in Maharashtra, 15.93 rupees in Punjab, 32.40 rupees in Andhra Pradesh, and 63.75 rupees in Bihar. It seems apparent that whether one normalizes research investment on a per worker or on a per hectare basis, there is not enough investment in agricultural research in Bihar relative to Punjab.

It is important to bear in mind that these figures are, at best, very rough indicators. Besides errors which crop up as a result of data limitations, another important source of bias is our inability to account for the pervasiveness of research findings by the diffusion of research information from one region to another. We have avoided attempting to derive internal rates of return because we lack knowledge of lag structures. The data limitations are obvious. Further improvements in the quality of the data and specifications of the type of relationship among the variables should improve the quality of the results.

The returns to investment worked out in this study are, in fact, returns to both research and development expenditures. Since the data on development expenditure were not available, the returns were estimated on expenditure in agricultural research only. To the extent that the development expenditure is ignored, the estimated returns to investment in agricultural research are inflated. The results, however, are sufficiently promising and the issue of research productivity is so important that an effort to push this line of work forward with better data and improved models should be pursued.

APPENDIXES

Appendix 5-1. The Model

We start with a general production function in which Y is the output and X_1, $X_2, \ldots, X_n$ are n different inputs:

$$Y = f(X_1, X_2, \ldots, X_n) + e \qquad (1)$$

where the e's are stochastic residual terms with standard least-squares properties.

The assumption that the intercept term, as well as the input coefficients, may change over time was incorporated by introducing a dummy variable D with the value zero to represent the pre-green revolution period (1960-61 to 1964-65) and the value one to represent the post-green revolution period (1967-68 to 1972-73). With the incorporation of dummy variables, the functional form (1) takes the form:

$$Y = f(X_1, X_2, \ldots, X_n, D, DX_1, \ldots, DX_n) + e. \qquad (2)$$

Here Y and all the X's are functions of the time variable t. The total derivative of Y with respect to t can be written as

$$\frac{dY}{dt} = \sum_{i=1}^{n} \frac{\partial f}{\partial X_i} \frac{dX_i}{dt} + \sum_{i=1}^{n} \frac{\partial f}{\partial (DX_i)} \times \frac{d(DX_i)}{dt} . \quad (3)$$

Estimates of $\partial f/\partial X_i$ can be used as weights in the summation of input growth rates dX_i/dt on the right-hand side of (3). Expressed in percentages, the relative share of different input growth rates can be obtained from the rate of growth of output. This was done for the two periods separately by substituting the values of the dummy variables. The difference in the growth rates of output over the two periods can be attributed to the change in the level of technology. Relative change in the factor share over the two periods can be considered to be the effect of technological changes on the factor share.

Assuming that the distribution can be either normal or log normal, we used both linear and Cobb-Douglas forms of functions.

Linear Relationship

With two subperiods, the general linear production function may be written as:

$$Y = a + b_1X_1 + b_2X_2 + \ldots + b_nX_n$$
$$+ C_oD + C_1D\ X_1 + \ldots + C_nDX_n \quad (4)$$

so that,

$$\frac{dY}{dt} = \sum_{i=1}^{n} b_i \frac{dX_i}{dt} + \sum_{i=1}^{n} C_iD \frac{dX_i}{dt} . \quad (5)$$

Here b_i's are the partial regression coefficients, and C_i's are the coefficients of dummy variables which represent the shift in the coefficients of X_i's owing to qualitative and technological change. The coefficient C_o measures the change in the intercept of the production function, and the coefficients C_i measure the shifts in the slopes of the linear production function (or shifts in the elasticities of the double-log function). Further, the b_i's may be treated as weights in estimating the factor share in the output growth rate of the first period and $(b_i + C_i)$ as the weights for the second period. Substituting the values for the dummy variable, we get the expressions for the two periods in the following form:

$$\left(\frac{dY}{dt}\right)_I = \sum_{i=1}^{n} b_i \left(\frac{dX_i}{dt}\right)_I \quad (6_a)$$

and

$$\left(\frac{dY}{dt}\right)_{II} = \sum_{i=1}^{n} (b_i + C_i) \left(\frac{dX_i}{dt}\right)_{II} \qquad (6_b)$$

The relative share of each growth factor was worked out as:

$$b_i \frac{dX_i}{dt} \Big/ \sum_{i=1}^{n} b_i \frac{dX_i}{dt}$$

for the first period and

$$(b_i + C_i) \times \frac{dX_i}{dt} \Big/ \sum_{i=1}^{n} (b_i + C_i) \frac{dX_i}{dt}$$

for the second period.

The average annual increase in output owing to the green revolution is given by

$$\left(\frac{dY}{dt}\right)_{II} - \left(\frac{dY}{dt}\right)_{I} . \qquad (7)$$

The significance and positive sign of the coefficient C_o of the dummy variable in (4) will indicate an upward shift in the intercept term, whereas the significance of the coefficients C_i indicates the shift in the slope owing to technological change. For the input X_i, the shift is expected to be of the type shown in the following manner:

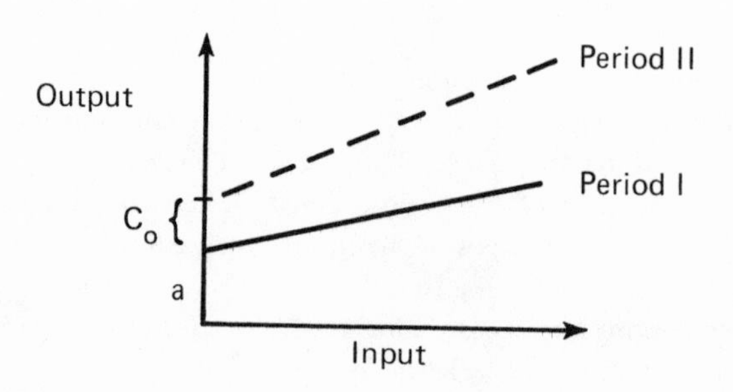

Double-Log Relationship

$$\log Y = \alpha + \beta_1 \log X_1 + \ldots + \beta_n \log X_n + \gamma D + \gamma_1 D \log X_1 + \ldots + \gamma_n D \log X_n . \qquad (8)$$

Differentiating with respect to time on both sides:

$$\frac{1}{Y} \times \frac{dY}{dt} = \sum_{i=1}^{n} \beta_i \frac{1}{X_i} \cdot \frac{dX_i}{dt} + \sum_{i=1}^{n} \gamma_i D \cdot \frac{1}{X_i} \cdot \frac{dX_i}{dt} \quad (9)$$

$$G_Y = \sum_{i=1}^{n} \beta_i G_{X_i} + \sum_{i=1}^{n} \gamma_i D G_{X_i} \quad (10)$$

where G_Y and G_{X_i} are the compound rates of growth for output and inputs. Here the elasticities for the first period β_i and $(\beta_i + \gamma_i)$ for the second period serve as weights in pooling the growth rates of inputs. The relative factor share of X_i can be expressed as

$$\beta_i G_{X_i} \Big/ \sum_{i=1}^{n} \beta_i G_{X_i}$$

in the first period, and

$$(\beta_i + \gamma_i) G_{X_i} \Big/ \sum_{i=1}^{n} (\beta_i + \gamma_i) G_{X_i}$$

in the second period.

The coefficient γ for the dummy variable gives the shift in the production function, whereas γ_i's are the shifts in the elasticities. The shift in the compound growth rate (G_Y) of output can be written as

$$(G_Y)_{II} - (G_Y)_I . \quad (11)$$

Qualitative changes in the inputs can be studied with the help of either of the two forms of production function. New farm technology, which leads to qualitative changes in the various inputs, can be interpreted through the coefficients C_i's and γ_i's, which were earlier used as shifts in weights but here represent the shifts in the slopes of the linear function and shifts in the elasticities of the double-log function. The change in the quality of variables leads to change in factor combinations. A comparison of factor combinations that produce the same output in the pre- and post-green revolution periods can represent the factor-augmenting effects.

Appendix 5-2. All-India Data

The dependent variable Y in the production function analysis is the agricultural output in value terms. The output indices for the fifteen states (excluding the newly formed states of Himachal Pradesh, Meghalaya, and Nagaland) were taken from *Growth Rates in Agriculture* for 1965. The data on output by state for later years were obtained from *Agricultural Situation in India* and *Estimated Area and Productivity of Principal Crops in India.*[9]

These indices were converted to output values (at constant prices for 1956-57) in absolute terms. This step was necessary to account for interstate variation. Whereas year-to-year variation is maintained in the use of index numbers, the interstate variation is lost because of the difference in the base for each state. Evenson and Jha used state indices not only for output but for inputs as well. Insofar as the results are presented at the state level, the use of index numbers or the product in absolute terms does not affect the results. But in the second part of the study, where the major objective is to estimate the returns to investment, the use of index numbers does not seem appropriate.

The data for the input variable of net sown area (X_9) and percentage of irrigated area ($X_5/X_9 \times 100$) were taken from *Statistical Abstracts, India.*[10] Fertilizer data were collected from *Fertilizer Statistics*, and N, P, and K were pooled by using prices as weights.[11]

Human labor (X_2) comprises the population of male cultivators + male agricultural laborers + 0.67 of female agricultural laborers. The census figures for 1961 and 1971 were used to estimate the human labor for the remaining years by working out the compound growth rates. Female cultivators were not included because the definitions and concepts used in the census of 1971 had changed from those used in the census of 1961.

The cattle used only for work formed the bullock labor (X_3). These figures were taken from the *Livestock Census* data for 1961 and 1966 and from the *Agricultural Census* in India for 1972 (see note 9). The data on the number of tractors (X_4) were obtained from the same source. The estimates for X_3 and X_4 were worked out for the remaining years.

The figures for the expenditure on research in agriculture (X_7) were taken from Evenson and Jha. An important point in the use of this variable was the time-lag. Evenson reports that the mean time-lag between expenditures on research and its effect on production in the United States was between six and seven and a half years. For our study, we have considered the time-lag to be five years, which became obvious from the jump in research expenditure during 1961-62 and the jump in agricultural production in 1966-67. In the states that did not experience a sudden jump in the investment in agricultural research, the time-lag did not matter, but for the sake of uniformity we used a five-year lag for all states.

Appendix 5-3. Data Used and Sources for State-Level Analysis

The dependent variable Y (or X_1) in the production function analysis is the usual Laspeyres Index of Agricultural Production with constant base year prices used as weights. In the present series of state indices, the farm harvest prices for 1956-57 were used as weights. The data were taken from the publication *Growth Rates in Agriculture, 1949-1950 to 1964-1965* and other related sources for subsequent years (see note 9). For the purpose of pooled analysis, these output indices were converted into absolute value terms to take into account the interstate variations in the value of production in the base year. For the reorganized state of Punjab, the relevant information was collected from *State Statistical Abstracts* for 1972 and earlier years.[12]

Data on X_2, the average annual (agricultural year) rainfall, were compiled from the rainfall figures for subdivisions published by the India Meteorology Department. They were weighted by the geographical area of the subdivisions to arrive at the rainfall values for the states.

Data on total fertilizer ($N + P_2O_5 + K_2O$) consumption (X_3), total cropped area (X_8), and gross irrigated area were available from published reports of the Ministry of Agriculture. The irrigated area was converted into the percentage (X_9) of the total cropped area. In some cases extrapolation on the basis of trend became necessary.

The variable X_4, male agricultural workers, is composed of the population of male cultivators and male agricultural laborers as reported in the census data for 1951, 1961, and 1971. The female working force had to be excluded because of changes in the definition of a "worker" in the three population censuses. Data on X_5, the literacy percentage in rural males, is based on 1 percent sample tabulations of the census information. Values for intermediate years were estimated on the basis of observed compound growth rates.

Data on the number of draft bovines (X_6), including both cattle and buffalo, and on the number of tractors (X_7) were compiled from *Livestock Census* for the years 1956, 1961, 1966, and 1972 (see note 9). For the reorganized state of Punjab, the data had to be compiled from the detailed tables for individual districts. The remaining values were estimated by means of interpolation in the census figures.

Data regarding expenditure on agriculture research (X_{12}, or R_t) were taken from the paper by Mohan, Jha, and Evenson.[13] This series extends only to 1968. Figures for subsequent years were estimated on the basis of trends in each state in the number of agricultural science publications selected for abstraction in *Indian Science Abstracts*.[14]

Appendix 5-4. First Run Linear Equations with All Variables[a]

Variables/ Parameters	States			
	Andhra Pradesh	Bihar	Maharashtra	Punjab
Constant	– 1,732.0878	– 6,125.1833	– 845.0994	– 1,508.2228
X_2	0.0021	– 0.0167	– 0.0382	0.0094
	(0.11)[a]	(0.67)	(1.62)	(0.40)
X_3	0.1072	– 0.6844	0.0666	0.2036
	(0.83)	(1.41)	(0.62)	(1.02)
X_4	– 0.0860	0.2858	– 0.0052	– 0.1770
	(0.86)	(1.23)	(0.02)	(0.51)
X_5	– 44.3513	– 103.3886	–28.2030	11.1348
	(1.34)	(1.27)	(0.70)	(1.10)
X_6	0.4838	1.0323	0.0423	0.5826
	(1.27)	(1.23)	(0.70)	(0.79)
X_7	0.0105	– 0.1786	0.0146	– 0.0048
	(0.26)	(1.17)	(0.29)	(0.58)
X_8	0.0183	0.0566	0.0697	0.0601
	(2.19)	(3.87)	(2.67)	(1.49)
X_9	0.2503	0.1952	48.5763	4.8419
	(0.06)	(0.04)	(1.81)	(2.48)
D[b]	– 0.4341	53.6893	15.1439	7.0628
	(0.04)	(1.36)	(1.36)	(0.57)
$R_{t-\tau}$[c] or X_{12}	0.0167	0.0522	0.0049	0.0056
	(2.26)	(1.46)	(1.54)	(1.04)
R^2	0.88	0.94	0.88	0.99
F	4.31	10.12	5.12	68.29

[a] Figures in parentheses are t values for the regression coefficients.
[b] D is the period dummy variable.
[c] $R_{t-\tau}$ is the lagged research expenditures variable with τ = 3, 4, 1, 2 respectively for the four states.

NOTES

1. *Progress Report of the Special Panel for the Study of Return on Investment in Agricultural Research* (New Delhi: Indian Council of Agricultural Research, 1974).

2. T. W. Schultz, *The Economic Organization of Agriculture* (New York: McGraw Hill, 1953). Z. Griliches, "Research Costs and Social Returns: Hybrid Corn and Related Innovations," *Journal of Political Economy*, 66 (1958), 419-431. For other excellent accounts of work in this area see A. R. Prest and R. Turvey, "Cost-Benefit Analysis – A Survey," *Surveys of Economic Theory*, vol. III (London: Macmillan, 1968); Walter L. Fishel, ed., *Resource Allocation in Agricultural Research* (Minneapolis: University of Minnesota Press, 1971); and R. E. Evenson, "Technology Generation in Agriculture," Bellagio Conference, Yale University, 1973.

3. R. E. Evenson and D. Jha, "The Contribution of Agricultural Research Systems to Agricultural Production in India," *Indian Journal of Agricultural Economics*, 28 (1973), 212-230.

4. Karam Singh, "Returns to Investment in Agricultural Research in the Punjab,"

Progress Report of the Special Panel for the Study of Return on Investment in Agricultural Research (New Delhi: Indian Council of Agricultural Research, 1974).

5. Shujiro Sawada, "Technological Change in Japanese Agriculture: A Long-Term Analysis," *Agriculture and Economic Growth: Japan's Experience*, eds. Kazushi Ohkawa, Bruce F. Johnston, and Hiromitsu Kaneda (Princeton, N.J.: Princeton University Press, 1970), pp. 136-154.

6. J. Johnston, *Econometric Methods* (New York: McGraw Hill, 1963).

7. Singh, "Returns to Investment."

8. Evenson and Jha, "Contribution of Agricultural Research Systems."

9. Publications of the Directorate of Economics and Statistics, Ministry of Agriculture and Irrigation, Government of India, New Delhi: *Growth Rates in Agriculture, 1949-1950 to 1964-1965*; *Agricultural Situation in India* (a journal); *Estimated Area and Productivity of Principal Crops in India*; *Livestock Census*; *Agricultural Census*.

10. Central Statistical Organisation, Ministry of Planning, *Statistical Abstract, India* (New Delhi: Government of India).

11. *Fertilizer Statistics* (New Delhi: Fertilizer Association of India).

12. *State Statistical Abstracts*. Issued by Bureaus of Economics and Statistics, individual state governments, India.

13. Rakesh Mohan, D. Jha, and R. E. Evenson, "The Indian Agricultural Research System," *Economic and Political Weekly*, 8:13 (March 31, 1973), A21-A26.

14. *Indian Science Abstracts* (New Delhi: Indian National Scientific Documentation Centre).

6

Measuring Economic Returns to Agricultural Research[1]

Reed Hertford and Andrew Schmitz

Economists have devoted a considerable amount of effort to estimating the economic payoff from agricultural research. Many of their early studies focused on agricultural activities in the United States. Results showed that past research had yielded impressive rates of return. Similar conclusions are now available for a number of other countries.

The framework used in estimating the rates of return to agricultural research was introduced almost a century ago by Alfred Marshall.[2] It involved the concept of an economic surplus. Since Marshall's time, numerous debates have arisen among economists over the theoretical validity of this concept. Unfortunately for the practitioner, there is no consensus at present on either its validity or its usefulness in economic analysis.

The purpose of this chapter is to review briefly the methods currently used by economists to estimate the rates of return to agricultural research and to suggest some modifications in those methods which would expand their usefulness. We do not attempt to resolve the debate over the validity of the concept of economic surplus. Also included in the chapter are some comments about why those rates of return to research estimated in the past have been so high.

Methods

The concept of economic surplus underlies the methods used by economists to estimate the benefits of agricultural research.[3] Such methods involve deriv-

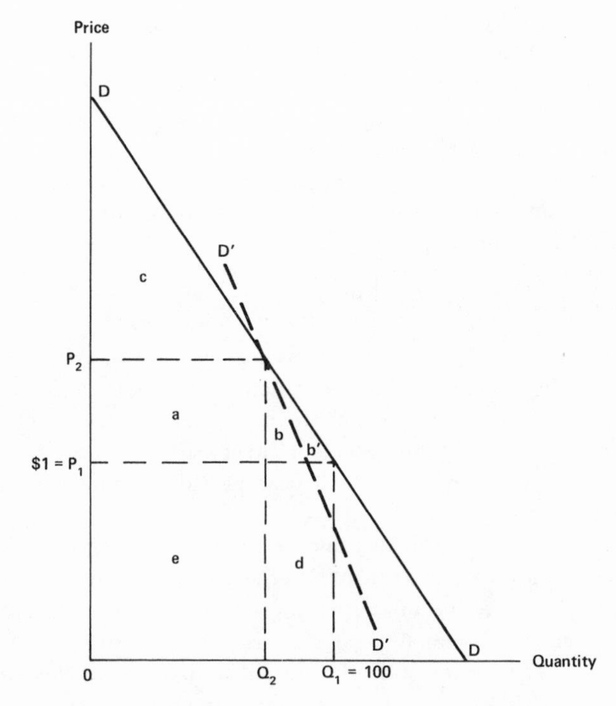

Figure 6-1. Consumers' surplus.

ing a total annual benefit from research by aggregating the separate surpluses of consumers and producers. Annual benefits are then compared with the annual costs of research by finding that rate of interest (the internal rate of return) which equates the discounted value of benefits and costs.

Consumers' Surplus

The concept of consumers' surplus was introduced by Dupuit[4] over a century ago, later popularized by Marshall,[5] and ultimately extended by Hicks[6] in the 1940s in ways which expanded and clarified its range of applicability. Consumers' surplus is measured with reference to demand curves.

Consider Figure 6-1, where D represents a compensated demand curve (CDC) showing the maximum prices a consumer would be prepared to pay for successive, additional units of a commodity. If he were to pay such maximum prices until he had obtained, say, 100 units, he would have made a total expenditure equal in value to the area under the demand curve and left of Q_1. If, on the other hand, he were to purchase the 100 units on the market at a single average price of \$1.00 per unit, he would save a value equal to the area under the demand curve above the \$1.00 price line (area a + b + b′ + c). It is this saving which is termed "consumers' surplus." For a group of consumers, or a "market," the consumers' surplus is generally the sum of the individual surpluses of consumers.

Any factor which increases the market price of the good from P_1 of \$1.00 to, say, P_2 would reduce the consumers' surplus by the area a + b + b′ in Figure 6-1. This area can be shown to equal

$$KP_1Q_1 \quad (1 - \frac{1}{2}Kn) \qquad (1)$$

where K is $(P_2 - P_1)/P_1$ and n is the absolute value of the price elasticity of demand.[7] If K is associated with the reduction in the average costs of a commodity resulting from a new yield-improving technique generated through research, then this formula can provide a direct estimate of the value to consumers of that research in a particular year, provided estimates are also available for P_1Q_1 and n.

This method of attributing a value or product to a research activity has been criticized on the grounds that the value of n generally estimated is not exactly equal to the value of the price elasticity of demand on the CDC.[8] When the price of a commodity falls, the consumer with a given money income is better off in terms of total utility or welfare because his real income rises. Since the CDC abstracts from this increase in real income, although an "ordinary" demand curve does not, the price elasticity of demand as usually estimated can be expected to overstate the CDC price elasticity. If, for example, the solid sloped line in Figure 6-1 were actually an ordinary demand curve, instead of a CDC as it has been represented to this point, the CDC would correspond to the broken line with less slope which passes through the point (P_2Q_2). The implication would be that any estimate of the value of research to consumers based on equation (1) would be biased upward by the value of the triangle b′. Although this is the usual situation, the actual direction of the bias, as well as the slope of the CDC in relation to the ordinary demand curve, depends on the sign of the income elasticity of demand for the commodity. One could argue that the bias is probably not substantial. First, the bias is small in the sense that the value of the triangle b′ is small in relation to the value of the total change in consumers' surplus, a + b. Second, the magnitude of the bias is even smaller when the commodity in question either has a small income elasticity of demand or represents a small proportion of total expenditures on all commodities.[9] It has been argued that the kinds of agricultural commodities for which research returns have been quantified using the consumers' surplus concept have, in fact, low income elasticities and represent a small proportion of total consumer expenditures.[10]

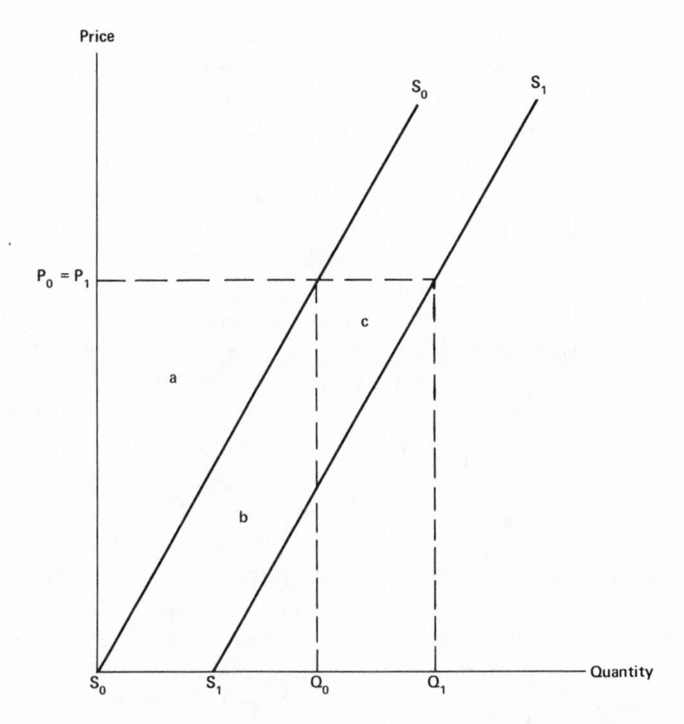

Figure 6-2. Producers' surplus.

Producers' Surplus

The concept of producers' surplus is analogous to that of consumers' surplus and refers, in general, to a difference between what is actually received from the sale of a good and the minimum amount required to induce a seller to part with it. Consequently, producers' surplus has been equated with an area – like area a in Figure 6-2 – between the prevailing price of a good and its supply curve since the latter has been traditionally defined as a locus of minimum prices at which quantities will be sold. If the supply curve shifts to a position like S_1 as a result of the adoption of a new production technique generated through agricultural research, then the benefit of that research to producers has been taken equal to area b + c.

Because there are sellers of input services, for example labor and capital, as well as sellers of final products, all of whom could theoretically earn a "surplus," some confusion has arisen over what producers' surplus and changes in it really measure. Is only the surplus of producers included, as the term implies, or are surpluses earned by factors of production also included? Marshall, the author of the concept, was himself not clear on this point. It is of crucial practical significance, since any estimate of the returns to agricultural research would be incomplete if it included only benefits to consumers and producers and overlooked those accruing to factors of production.

As it turns out, the composition of producers' surplus is dependent upon the specification of the supply curve. When variable inputs of production are available to a competitive industry at prices that are independent of the level of output, it is well known that (1) the supply curve of the industry is the lateral sum of the marginal cost (or supply) curves of the individual firms, (2) the area under the industry's (firm's) supply curve represents the total costs of all variable inputs of production, and (3) the area above the industry's (firm's) supply curve and below the prevailing price of the product is a return to all other inputs – namely, all fixed factors of production including inputs physically associated with producers, like their labor and "entrepreneurial capacity," as well as other fixed inputs. By definition these fixed inputs are specific to the industry or firm under consideration in the sense that they cannot produce anything else in the economy. Therefore, on a broader view, there are no real costs associated with their employment, and any return to them would be an economic rent or surplus. Since producers' surplus has been given the same definition, we may conclude that producers' surplus is, in fact, a return to the producer as well as to other fixed factors of production, and that there is a direct correspondence between the concept at the level of the firm and the industry when prices of variable inputs are invariant with respect to changes in output.

When some variable factors of production are supplied at prices that are not constant but increase with the level of an industry's output because their supply curves slope upward, this simple interpretation of producers' surplus disappears. Assume that the sum of the marginal cost curves of an industry, S_0 in Figure 6-2, for example, shifts down to S_1 at the level of output Q_0 as a result of the adoption of a cost-saving, improved technique generated through research. The value of costs saved, given our definition of the supply curve, would be the area b. These savings, of course, would permit the industry to expand its output beyond Q_0. However, any output expansion achieved through increased employment of variable factors of production would now tend to bid up their prices to the industry and make their costs exceed the area under the new industry sum of marginal costs curve, S_1, at any level of output above Q_0 since the supply curves depicted in Figure 6-1 assume that unit costs of variable inputs are constant. Therefore, as output expands in the industry, the sum of marginal costs curve, S_1, begins to slip back to the position of initial supply curve, S_0. Its final position of equilibrium would approximate that of the initial supply curve if all variable inputs of production were subject to very inelastic supply curves. In that case the initial saving created by adoption of the new production technique would have accrued mostly to the variable inputs of production as a surplus. Area b of Figure 6-2 would then approximately equal the change in surplus obtained by variable factors of production from the cost-saving improvement in production techniques resulting from research. The surplus of producers and other fixed factors of production would be practically unchanged as would the level of industry output (Q_0).

The final position of the supply curve of the industry would normally result in an output somewhere between Q_0 and Q_1 of Figure 6-2; that is, some variable inputs of production would be available at somewhat higher prices to the industry and, as output expanded, the supply curve would shift equivalently from S_1 but not return to its initial level, S_0. The area between this final position of the curve and the sum of marginal costs curve, S_1, up to the final level of industry output would then measure the increase in expenditures by firms in the industry on variable inputs as a result of increases in their unit prices.

Although such increased expenditures would represent very real increases in resources devoted to production by firms in the industry, they would not all be increases in real resources devoted to production from the point of view of the economy at large since a part would accrue to some sectors of the economy – namely, to variable inputs or variable input suppliers – as surpluses above their positively sloped supply curves and below their prevailing prices. A surplus, of course, should be accounted for as such and should be added to other estimated products or returns to research. To estimate such surpluses exactly, however, would require data on the demand and supply curves for variable inputs employed by the industry. Since they are not generally available, the most that can usually be said is that the area between the final position of the sum of marginal costs supply curve of the industry up to the new level of output and the sum of marginal costs curve that would have prevailed had input prices been invariant with respect to output represents the increase in returns to variable inputs and their suppliers associated with employment of the improved technique of production. The area between the initial and the final sum of marginal costs curve up to the product price line would then be a measure of the returns of the new technique captured by producers and other fixed factors of production.

If the point of intersection of the vertical line through Q_0 with the supply curve S_1 were connected with the point of intersection of the new equilibrium level of industry output with the price line P_1, the resulting line would define a new supply curve which was adjusted for the kinds of input price changes just discussed. This "adjusted" industry supply curve would have a smaller slope than the "unadjusted" sum of marginal costs curve of the industry or the relevant curve when prices of all variable factors of production are invariant with respect to output. If there are only two factors of production, the elasticity of supply approximately equals $a_1e_1 + a_2e_2$ where $a_1 = 1 - a_2$ is the first input's share in total production costs and e_1 and e_2 are the two inputs' price elasticities of supply. From this expression, it follows directly that the supply curve will be less elastic when input supply curves are less elastic.

To recapitulate: Area b + c in Figure 6-2 is an unambiguous measure of the increase in the surplus of all fixed factors of production, including the surplus of producers from their self-employment, when variable inputs of production are available at prices that do not change with output. If variable input prices change with changes in output, then area b becomes a reasonable measure of increases in surpluses of variable inputs when their supply curves are very inelastic. If variable input supply curves are neither very elastic nor inelastic, then the area b + c is normally an upper-bound estimate of the changes in surpluses of fixed and variable factors of production since it includes some increases in real resource costs to the economy. Finally, two industry supply curves have been identified: (1) the unadjusted sum of marginal costs curve relevant when prices of all variable inputs are invariant with respect to output and (2) an adjusted supply curve, normally more inelastic, connecting points on unadjusted supply curves after adjustments for input price changes have taken place.

Combining Surpluses

To estimate the returns to research, consumers' and producers' surpluses are combined. Figure 6-3 contains a normally sloped demand curve, D, and an initial supply curve, S_0, which shifts to the position S_1 as the result of the on-farm employment of a new or improved input developed through research. Before this shift, consumers' surplus equaled area a; afterward, it is given by a + b + c. Thus, the net gain to consumers from the research-induced shift in supply is b + c. Similarly, before the supply shift, the surplus b + d corresponded to producers and to other production inputs; afterward, their surplus corresponds to f + d. Their net gain from the supply shift is then f − b. The sum of the gains in surplus to both groups is b + c + f − b = c + f. It can be shown that this latter area approximately equals

$$kP_1Q_1 \left(1 + \frac{1}{2}\frac{k}{n + e}\right) \qquad (2)$$

where k is defined as the percentage increase in production attributable to research (the horizontal distance between the two supply curves divided by the value of final production Q_1); P_1 is the price of the commodity after the sup-

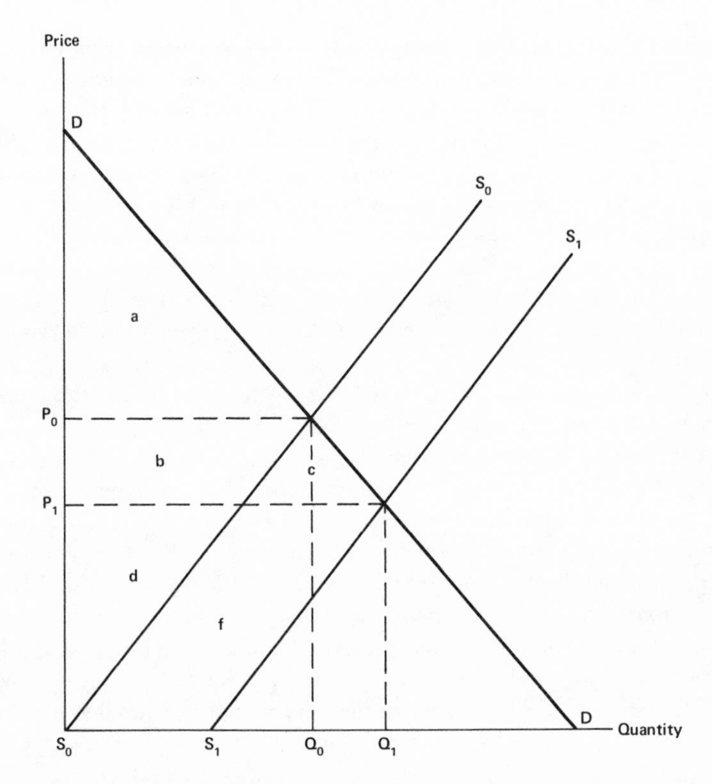

Figure 6-3. Combined consumers' and producers' surplus.

ply shift; and n and e are, respectively, the price elasticities of demand and supply. This formula, of course, can be disaggregated into its primary components, consumers' and producers' surplus, for purposes of quantifying the distribution of gains accruing to each group. Corresponding to (2) are the following expressions:

$$\text{Consumers' surplus} \quad \frac{kP_1Q_1}{n+e}\left(1 - \frac{1}{2}\,\frac{kn}{n+e}\right) \tag{3}$$

$$\text{Producers' surplus} \quad kP_1Q_1\left\{1 - \frac{1}{n+e}\left[1 - \frac{1}{2}k\left(\frac{2n+e}{n+e}\right)\right]\right\} \tag{4}$$

These reveal that the larger the consumers' surplus, ceteris paribus, the smaller the price elasticity of demand, and that the larger the producers' surplus, the larger the price elasticity of demand. Such results support the notion that research on basic subsistence commodities would particularly benefit consumers, while research on more price elastic commodities – for example, rubber, cotton, and perhaps coffee – would be especially remunerative for producers, input owners, and certain factors of production. (See chapter 24 for elaboration of this point.) Note further from (3) and (4) that, although consumers' surplus will always be positive-valued because $2(n + e) > kn$, producers' surplus could be negative-valued if $e + n < 1$. Nevertheless, the sum of the surpluses is always positive-valued according to (2). Expressions for the surplus generated by research in the special cases in which one or both of the price elasticity parameters assume extreme values can easily be derived from the general cases just developed.

More complicated formulas for the values of the combined surplus of producers and consumers have been suggested by Ardito Barletta and Peterson to allow for the possibility that the demand and supply relationships may not always be linear.[11] However, the differences in the estimates of benefits provided by the more complicated formulations and those presented here are small for usual values of the key parameters. The main reason is that *in all formulations the critical determinant of the value of the benefits derived from research is simply* kP_1Q_1 *or the percentage change in the value of production attributable to research*. Other parameters combine to produce a value which is small by comparison.[12] This is an important point to keep in mind when considering the empirical relevance of theoretical controversies that have been sparked by the concept of economic surplus.

The Theoretical Controversy

The concept of economic surplus occupies a controversial place in economic theory. For example, Samuelson has said that consumers' surplus is of "historical and doctrinal interest with a limited amount of appeal as a purely mathematical puzzle."[13] More recently, Bergson has stated: "Despite theoretic criticism, practitioners have continued to apply consumer's surplus analysis through the years. As some have urged, that must already say something about the usefulness (as well as the use) of such analysis, but just what it says has remained more or less in doubt."[14] However, Hicks has disagreed:

> But enough has been said to show that consumer's surplus is not a mere economic plaything, a *curiosum*. It is the foundation of an important branch of Economics, a branch cultivated with superb success by Marshall, Edgeworth, and Pigou, shockingly neglected in recent years, but urgently needing reconstruction on a broader basis. Beyond all doubt it is still capable of much further development; if economists are to play their part in shaping the canons of economic policy fit for a new age, they will have to build on the foundations of consumer's surplus.[15]

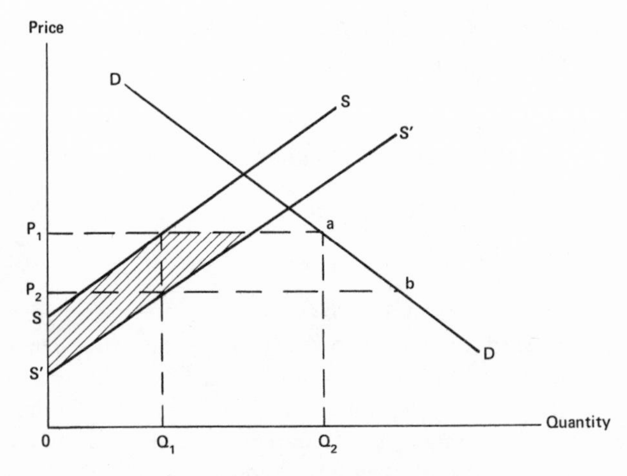

Figure 6-4. Research: The case of a traded good.

Although we acknowledge the place of this controversy in economic theory, we believe that most shortcomings of studies of returns to research arise not from the concept of economic surplus but from overlooking or mistreating practical characteristics of the real world. Such characteristics imply a need to introduce some modifications in the methods used to calculate the returns to research, and to those we now turn.

Modifications

Methods used to estimate the returns to agricultural research should (1) treat products, and research itself, as traded goods if appropriate, (2) consider problems of unemployment, (3) distinguish between intermediate and final goods, and (4) account for the income distributional effects of research. This section briefly discusses these problems and how they can be handled.

Traded Commodities

Consider Figure 6-4 where, at a price of P_1, imports are Q_1Q_2. If the supply curve shifts from S to S′ and the "small country" assumption is used – that is, the price remains unchanged – there is no gain to consumers, but producers gain the crosshatched area. The result is a smaller surplus, ceteris paribus, attributable to research than had the commodity not been traded at a constant world price. However, if the country under question is a large trader and the price drops to P_2 owing to the increase in supply, producers actually lose while consumers gain area P_1abP_2. Thus, the effects of research on a traded good are critically dependent upon the extent to which its world price is affected by changes in supply. One can easily work out the case for a good that is exported.

A related point to consider is that technology can be developed by one country but transferred to another at nominal cost. Such international spill-over effects may enhance the estimated returns to research of those countries with newer programs which can benefit from the prior investments of older programs in other countries.

Resource Unemployment

Research can result in the unemployment of agricultural resources. If, for example, a new technology generated through agricultural research favors certain producers of a commodity, others may be forced out of business as supply expands and prices decline. The resources released may not find employment elsewhere; if this is not taken into account, it can lead to errors in estimating the benefits of research.

Suppose an industry is made up of producers on nonirrigated land (Type 1) and irrigated land (Type 2). In Figure 6-5, S_1 represents the supply of the good produced on Type 1 land and S_2 the supply of the good produced on Type 2 land. In equilibrium, Q_1 of the commodity is produced at price P_1. Suppose now that a new technology is applied to Type 2 land which causes the supply curve to shift to S_2'. The corresponding equilibrium price and quantity are P_2 and Q_2. However, note that although the net gain (measured in producers' surplus) is positive, there is a loss to producers on the Type 1 land because the good cannot be grown there at P_2. If the resources (except land) were employed elsewhere and if all the surplus originally accrued to landowners, the crosshatched area would represent the loss to the owners of Type 1 land.

Now let us assume that resources formerly equal in value to the area under the unirrigated supply curve up to Q^* fail to find alternative employment when the unirrigated farms go out of business. These resources do not include such industry-specific fixed factors of production as producers, family members, and farmland since the losses they sustain are fully reflected in the area of producers' surplus. Rather, they refer to other "normally variable" factors of production like hired labor, machinery, implements, other forms of farm power, livestock capital, and perhaps even such things as fertilizers, insecticides, and seeds. Were such resources to become unemployed, their equivalent value – the area under the unirrigated supply curve up to Q^* – would have to be subtracted from the benefits of research estimated in the usual way where the assumption is made that all "normally variable" factors of production find employment elsewhere in the economy.

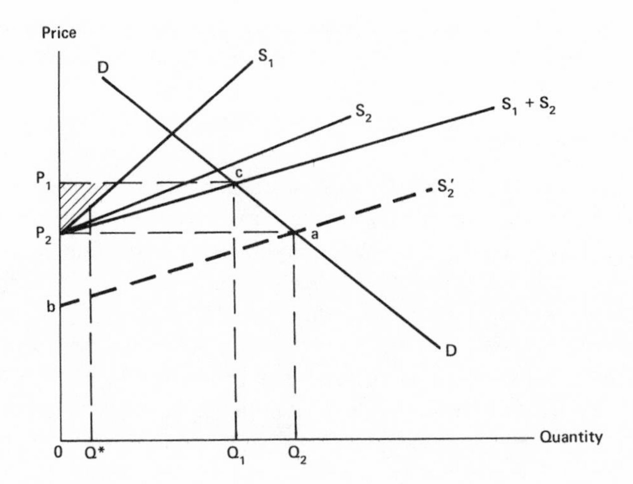

Figure 6-5. The resource unemployment effect from research.

Of course, all that has really happened is that "normally variable" factors of production have become equivalently fixed factors in the face of falling prices. This has altered the division of total revenues between fixed and variable factors. In particular the loss of returns to fixed factors specific to the industry and the lack of employment alternatives elsewhere have turned out to be larger than the crosshatched shaded area above the supply curve, and the value of variable factors of production finding employment elsewhere in the economy is smaller than the area under the supply curve. The implication is that the supply curve corresponding to Type 1 land is more inelastic than that shown in Figure 6-5.

A related case was analyzed by Schmitz and Seckler.[16] They looked at the mechanical tomato harvester as an innovation of agricultural research which had resulted in the unemployment of farm workers. They first estimated the benefits of the research on the harvester in the usual way. A side calculation was then performed in which the returns to laborers who were unemployed because of the introduction of the harvester were deducted from benefits. An alternative approach would have been to reduce the area of surplus generated by the introduction of the harvester by decreasing the elasticity of the new supply curve of tomatoes.

Because past studies have not sufficiently emphasized the "unemployment" effect from research activities, the previous discussion has neglected to indicate that research leading to technological change may, under some circumstances, result in an increase in the employment of resources. For example, many developing countries that export agricultural commodities may add to the employment of labor through technological change made possible by research.

Derived Demand Curves

The demands for farm workers, mechanical harvesters, and hybrid corn involve largely derived, not final, demand curves; yet past research has not made this distinction. The demand for hybrid corn, for example, is largely derived from the demand for grain-fed beef – in this instance, a principal final product for corn.[17] In Figure 6-6A the derived demand for corn is shown, and it is assumed that all the corn is fed to cattle. As the result of a new technique generated by research, the price of corn drops from P_0^c to P_1^c. This induces a downward shift from P_0^B to P_1^B in the price of beef in Figure 6-6B. Are the shaded areas in the two diagrams of equal value as has been implicitly assumed?

If there is one other input in beef production besides corn and its supply is elastic, then it can be easily shown that

$$\frac{P_0^B - P_1^B}{P_1^B} = a_c \left(\frac{P_0^c - P_1^c}{P_1} \right)$$

where a_c is corn's share of the costs of producing beef. An implication is that the rectangles between the two price lines in both diagrams up to the new equilibrium quantities are of equal value. The size of the triangles between the two price lines below the demand curves could still be different, however, as a result of different price elasticities for corn and beef.[18] This would imply different values of consumers' surplus measured with reference to the final and derived demand curves. Also, at the macro-level the supplies of inputs are usually not perfectly elastic. When they are not, the relationship between surpluses estimated under final and derived demand curves is further complicated.

Distributional Effects

How realistic is it to aggregate all consumers and producers of a given product and to look at only these two groups when estimating the rates of return to research? If, for example, the welfare consequences of the well-known Russian wheat transaction were considered, a meaningful analysis would include its effect on grain producers, landowners, livestock producers, consumers of various farm products (both high- and low-income groups), machine manufacturers, fertilizer producers, and so on. A partial analysis

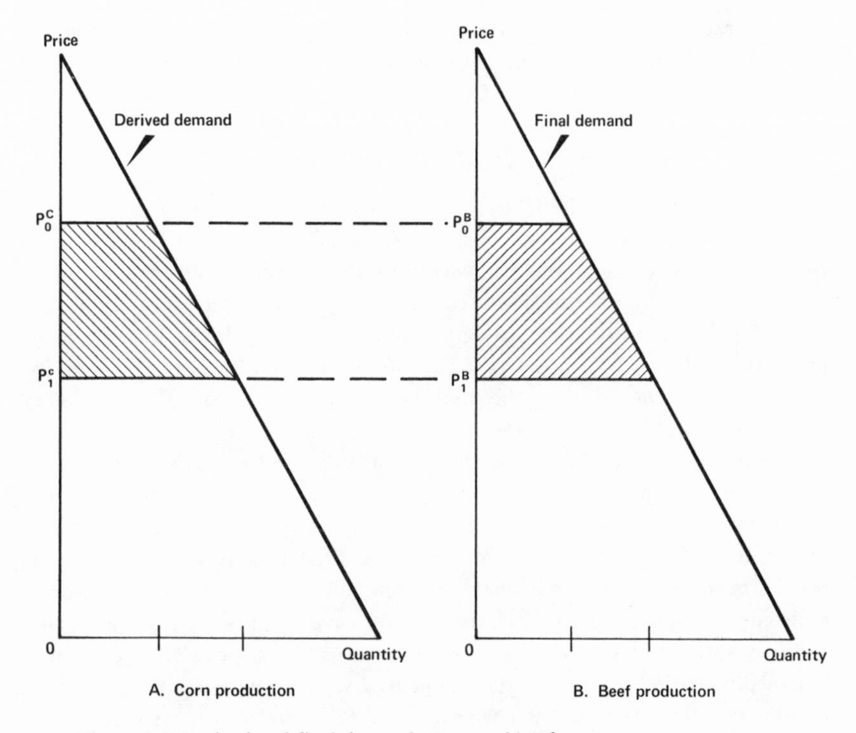

Figure 6-6. Derived and final demand: Corn and beef.

would include only grain producers and consumers. But this would leave out a great deal. Similarly, for research, a great deal is overlooked if the effects on farm workers, profits, farm size, and land values are not considered.

Because of a tendency to deal with aggregative models, the distributional effects from research have not commonly been considered. For example, there are many types of producers of a given commodity – small-scale farmers, large-scale farmers, landowners, sharecroppers, and farmers with unmechanized and mechanized units. An aggregate producers' surplus sheds little light on how research affects each group. It is not enough to know that the rate of return is high or low. Generally we also want to know who will benefit and who will lose within the producing and consuming sectors.

One reason is that a dollar of benefits need not equal a dollar of losses in the context of economic welfare. There is no reason why a 1:1 welfare scheme is superior to any other weighting system. For example, 2:1 weights might be assigned losers and gainers, respectively, or perhaps 4:1 weights. Obviously, *negative* rates of return could result from assigning different welfare weights to losers and gainers even though, with the usual 1:1 arrangement, rates of return are highly positive.

Why Are Returns So High?

Economic studies now available for a number of countries point to high rates of return to programs of agricultural research.[19] One cannot help but view this evidence as unusual and ask, why? In fact, this issue did engender a heated discussion at the Airlie House conference; S. J. Webster pointed out, for example, that the literature on agricultural research is almost unanimous in presenting rates of return that are astronomically high by normal standards. This fact alone should bring the methods employed in arriving at the results under the closest possible scrutiny.

The preceding section of this chapter has suggested some possible answers to this question when: the fact that agricultural products are traded is not taken into account; international spillover effects of research are overlooked; the effects of research on intermediate and final products have been confused; the costs of resource unemployment induced by research have been omitted; and inappropriate welfare weights have been assigned to the gains and losses underlying calculations of the returns to research. Not mentioned, but of possible relevance, are the problems associated with properly accounting for research projects that have produced "dry holes" and with determining how much money has actually been spent developing particular technological or biological breakthroughs. The costs of administration and of other complementary, supporting programs are extremely difficult to impute to individual research projects.

The purpose of this section, however, is to suggest that two factors associated with the agricultural research "delivery system" – the mechanism that assists in transferring new technology from research centers to farms – may explain more generally why estimated returns to agricultural research have been so high. Although it is hypothesized that the influence of these two factors has been pervasive, there is no intention of leaving an impression with the reader that past estimates of the returns to research have been grossly inflated. The evidence in this regard appears to be overwhelming: agricultural research does pay handsome rewards, indeed.

Traditionally, in making calculations on the cost side of the equation, past economic studies of the returns to research have included outlays incurred in developing and sustaining the research program under appraisal plus the differential costs to farmers of purchasing the resulting technology. A great deal of care has generally been taken to include among costs of the program the direct expenses of the research (mainly salaries and supplies), the costs of complementary inputs from such collaborating programs as agricultural extension, and the costs of indirect inputs like staff training and management.[20] Not included, however, have been the costs of transferring technology to farmers which are borne either by organizations largely unrelated to the research program or by farmers themselves.

In the case of developing countries, the first of these two omissions may have been of minor consequence. Most of the relevant costs of transferring technology to farmers are probably captured by imputing to the research program some portion of the costs of the agricultural extension service. In the case of developed countries, however, the omission may have been of some greater consequence because of the existence of a complex and sophisticated network of organizations and facilities in rural areas which form part of the delivery system. Those of us from rural areas of developed countries tend to take for granted such things as agricultural banks, newspapers, radios, television, city libraries, and programs of continuing education. All, however, assist in their own way to transfer new technology to farms at costs which may not be fully reflected in the expenses of research and extension programs, in the prices paid for new techniques of production, or in the area below a product supply curve. The implication, of course, would be that the real costs to society of research programs in developed countries may have been much larger than was estimated by past studies of the returns to agricultural research.

Boyce and Evenson have recently provided some data which relate very directly to this proposition.[21] They show in a multicountry study that there is a strong *positive* correlation between the level of development of a nation and the proportion of the total agricultural product devoted to agricultural research, whereas there is an equally strong but *negative* correlation between the level of development and the proportion of the agricultural product spent on extension. Noting that extension workers are paid generally less than researchers, Boyce and Evenson conclude that developing countries may have been lulled into thinking they can increase agricultural production more cheaply by emphasizing extension activities. An equally plausible explanation, however, is that agricultural programs in developing countries have been forced to pay part of the price of deficient delivery systems by investing in agricultural extension services. Programs in developed countries, on the other hand, have been able to avoid such costs by piggybacking on a delivery system constructed at a substantial cost to society at large.

A second component of the costs of the delivery system not usually taken into account in reckoning the returns to agricultural research includes the farmer, his family workers, and hired laborers.[22] In most settings of dynamic change, some labor time must be allocated to searching for, learning about, experimenting with, and ultimately adopting new techniques generated through research. Needless to say, time spent in these ways decreases when the productivity of the research establishment itself is low. It may also be lower, however, precisely where the delivery system is highly developed and functioning efficiently. In developed countries, technology is practically laid at the farmer's doorstep in prepackaged and pretried forms. By contrast, in developing countries, where levels of formal schooling are low and the delivery system is not well developed, a substantial share of total labor time may be required to adjust to new technologies released by the research establishment. Such time cannot be ignored among the costs of research from society's point of view unless agricultural labor at the margin is unemployed.

In most countries the proportion of the value of agricultural production spent on research and extension is quite small – less than 2 percent.[23] Agricultural labor's share of the value of production, on the other hand, is large, commonly falling into the 40 to 60 percent range.[24] This means that a small fraction of total farm-level labor time bulks large in relation to research expenditures. If it is assumed, for example, that the labor time of searching, learning, experimenting, and adopting activities associated with new agricultural technologies absorbs just 2 percent of total labor time each year, the costs of research could be increased by 50 percent were labor's share of final output 50 percent.

It should be pointed out that both of the usually omitted costs of the delivery system discussed in this section – those associated with the activities of organizations largely unrelated to the research program, especially in the developed countries, and those associated with the labor time absorbed by adopting improved technologies, especially in developing countries – would not be included in assessments of the returns to research done from the more restricted point of view of the research administrator or the research agency. The reason is that the delivery system, including the labor time of farmers, is not usually an item in the budget of agricultural research programs. However, precisely because the probabilities of success of a research program are enhanced by a delivery system that hastens the adoption of improved practices, strategies chosen for research may capitalize on better delivery systems through the particular selections of commodities to be studied and/or geographical areas to be emphasized. Such strategies are obviously at variance with the social interest by not taking into account their full cost to society. The remedy would be to internalize the costs of the delivery system in the budget of the research program and thus close the gap between returns to research as viewed by the research agency and by society at large. Internalizing costs in this way could lead to some interesting changes in the allocation of resources for research. For example, a research agency might choose to initiate work in a depressed area of small farmers rather than in an irrigated region with a more highly developed delivery system.

Concluding Remarks

Estimates of the economic returns to past agricultural research can be most helpful in highlighting commodities and programs that have represented low risk, high payoff endeavors. Rates of return calculations are also useful in pointing up and appraising the value of some unanticipated side effects of agricultural research that might merit additional study. An example is the effect of the tomato harvester on the unemployment of farm labor. Further, where major funds have not yet been committed to a national agricultural research program – as is the case in some developing countries – estimates of the returns to past research may be of assistance in securing financing by demonstrating the objective value of research to society. Consumers in such settings do not typically pressure for agricultural research because they lack firsthand knowledge of its effects, and pressures exerted by producers are viewed as self-serving and are frequently ignored. For each of these reasons, calculations of the returns to agricultural research are an important means of helping to produce a better allocation of society's scarce resources.

This chapter has reviewed the methods used to estimate such returns and has alerted the reader to modifications which should be introduced in particular situations. It was suggested that calculated returns could be biased upward if these modifications are not made where appropriate. It was also suggested that the returns to research may be generally overstated if the full costs of getting technology from the research center to farms are not taken into account. Emphasized in this regard were the costs of using a highly sophisticated infrastructure of delivery services in rural areas of developed countries and the costs of farmers' time absorbed in finding and adopting a new technology in developing countries. Further applications of the methods discussed in this chapter to appraisals of research programs should point out other refinements needed to improve estimates of the economic returns to research.

NOTES

1. This paper has benefited especially from comments and suggestions made by Vernon Ruttan of the Agricultural Development Council, Tilo L. V. Ulbricht of the Agricultural Research Council in London, and S. J. Webster of the Ministry of Overseas Development in London.

2. Alfred Marshall, *Principles of Economics*, 8th ed. (London: Macmillan, 1930).

3. The development and use of the concept of economic surplus has been dealt with in detail by J. M. Currie, J. A. Murphy, and A. Schmitz, "The Concept of Economic Surplus and Its Use in Economic Analysis," *Economic Journal*, 81:324 (December 1971), 741-800.

4. J. Dupuit, "On the Measurement of the Utility of Public Works," *Annales des Ponts et Chaussees*, 2nd series, 8 (1844), 14-28.

5. Marshall, *Principles of Economics*.

6. J. R. Hicks, "The Rehabilitation of Consumers' Surplus," *Review of Economic Studies*, 8:2 (February 1940), 108-116.

7. The area $a + 2\ (b + b')$ in Figure 6-1 equals KP_1Q_1 and the area $b + b'$ equals $½(Q_1 - Q_2)\ (P_2 - P_1) = ½K^2P_1Q_1n$. The difference between these two quantities is equal to (1) and the area $a + b + b'$ in Figure 6-1. Note that this area is always positive-valued.

8. For a recent example of this criticism, see Wolfgang Bonig, "Comment," *American Journal of Agricultural Economics*, 56:1 (February 1974), 177.

9. This statement reflects the exact expression for the price elasticity of demand of an ordinary demand curve, $n_{xp}{}^* - k_x n_{xI}$, where the terms are, respectively, the price elasticity of demand along the CDC, the proportion of income spent on the commodity "x" in question, and the income elasticity of demand for x.

10. See, for example, Harry W. Ayer and G. Edward Schuh, "Reply," *American Journal of Agricultural Economics*, 56:1 (February 1974), 175-176.

11. Nicolas Ardito Barletta, "Costs and Social Benefits of Agricultural Research in Mexico," Ph.D. dissertation (Chicago: University of Chicago, 1971), Appendix C has provided the following formula for the surplus:

$$kP_1Q_1 \left\{1 + \frac{k}{2n}\left[1 - e\ \frac{(1-n)^2}{n+e}\right]\right\}.$$

Willis Peterson, "Returns to Poultry Research in the U.S.," *Journal of Farm Economics*, 49:3 (August 1967), 656-669, has suggested:

$$kQ_1P_1 + k^2\ \frac{P_1Q_1}{2n} - \frac{Q_ok^2P_1}{2}\left(\frac{P_1}{P_o}\right)\left(\frac{en}{n+e}\right)\left(\frac{n-1}{n}\right)^2.$$

12. If, for illustrative purposes, we assume $k = 0.1$, $n = 0.3$, and $e = 0.5$, then the value of the term that multiplies kP_1Q_1 in Ardito Barletta's formula, as well as the formula defined here by equation (2), is 1.1.

13. Paul A. Samuelson, *Foundations of Economic Analysis* (New York: Atheneum, 1967), p. 195.

14. Abram Bergson, "A Note on Consumer's Surplus," *Journal of Economic Literature*, 13:1 (March 1975), 43-44.

15. Hicks, "Rehabilitation of Consumers' Surplus," p. 116.

16. A. Schmitz and D. Seckler, "Mechanized Agriculture and Social Welfare: The Case of the Tomato Harvester," *American Journal of Agricultural Economics*, 52:4 (November 1970), 569-578.

17. The research on rates of return from hybrid corn research was done by Zvi Griliches, "Research Costs and Social Returns: Hybrid Corn and Related Innovations," *Journal of Political Economy*, 66:5 (October 1958), 419-432.

18. Under the assumptions made to this point, it can be shown that the price elasticity of demand for corn equals

$$- [a_1 n + a_2 \sigma]$$

where $a_1 = (1 - a_2)$ is the first input's share of production costs, $-n$ is the price elasticity of demand for beef, and σ is the elasticity of substitution between corn and other inputs in beef production. (This elasticity of substitution is equivalent to the percentage change in the inverse ratio of their prices. Thus, the elasticity is positive-valued and ranges in value from zero to infinity.) We see that this elasticity would equal $-n$ if $n = \sigma$. Although the equality is unlikely to obtain exactly, it seems reasonable to expect that $n - \sigma$ would usually be small. If this is true, consumers' surplus measured under final or derived demand would be quite similar.

19. Several of these are listed and discussed by James K. Boyce and Robert E. Evenson, "Agricultural Research and Extension Systems," Department of Agricultural Economics, mimeographed (Los Baños: University of the Philippines, 1975), Tables 6.1 and 6.2.

20. These are outlined, for example, by Reed Hertford, et al., chapter 4 in this volume.

21. Boyce and Evenson, "Agricultural Research," chapter II.

22. We are indebted to our colleague, Carlos Benito, for suggesting this point. See his "Peasants' Response to Rural Development Projects—With Special Reference to the Puebla Project," Department of Agricultural Economics, Rural Development Project Working Paper no. 1 (Berkeley: University of California, 1974).

23. Boyce and Evenson, "Agricultural Research," chapter I.

24. For the United States, a labor "share" of roughly 50% was implied by the regression results of Zvi Griliches, "Research Expenditures, Education, and the Aggregate Agricultural Production Function," *American Economic Review*, 54:6 (December 1964), 961-974. For a developing country, see Reed Hertford, *Sources of Change in Agricultural Production, 1940-65*, Foreign Agricultural Economic Report no. 73 (Washington, D.C.: Economic Research Service, USDA, 1971).

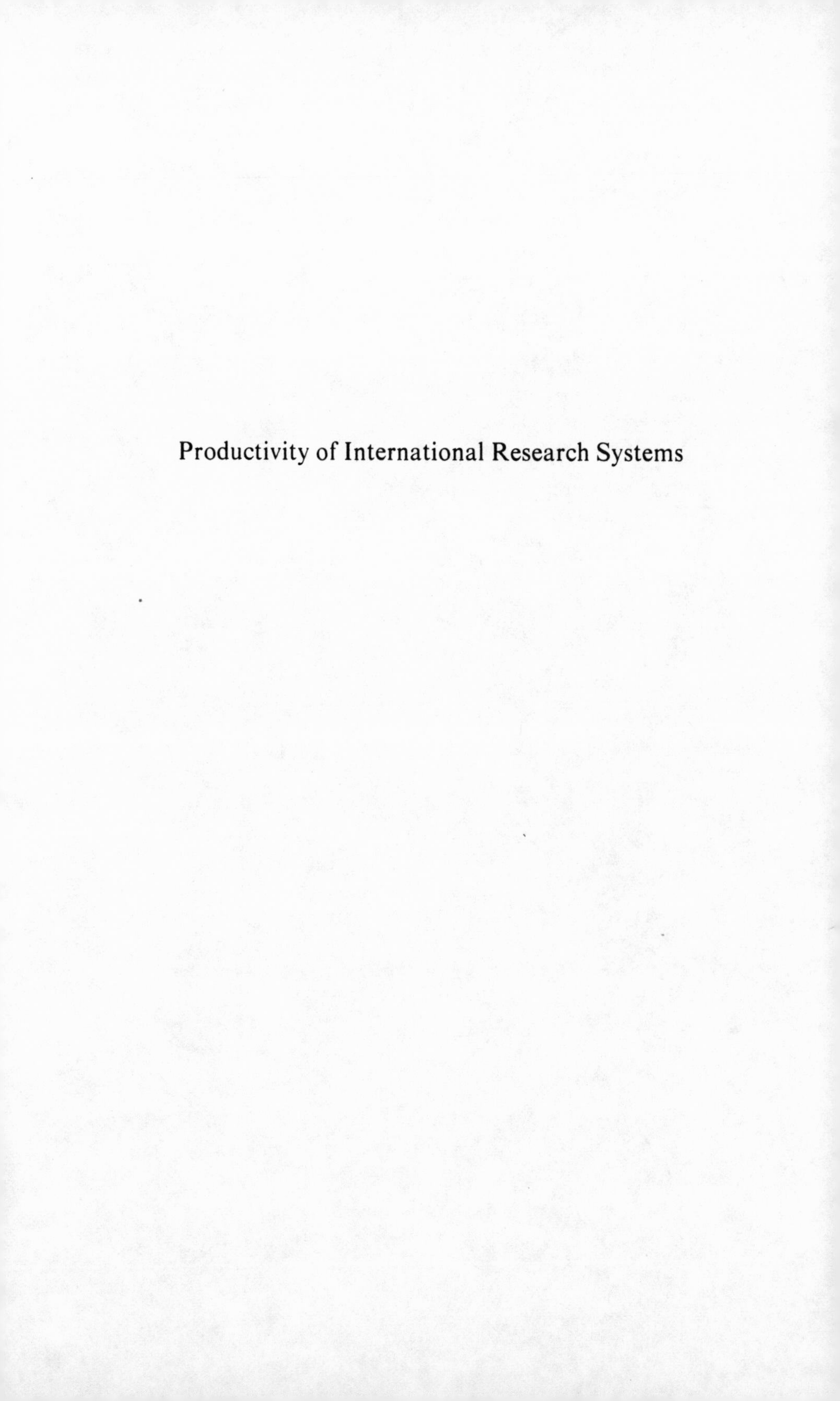

Productivity of International Research Systems

7

Evaluating the Impact of International Research on Wheat and Rice Production in the Developing Nations[1]

Dana G. Dalrymple

Research on food crops in or for the less developed countries (LDC's) is relatively new. For decades, much of the agricultural research in LDC's focused on plantation or export crops. Food crops for domestic consumption, with a few exceptions, were largely ignored. The situation began to change in the years following World War II, but even then national research on food crops was usually given low priority and limited funding.

There were some exceptions. Perhaps the best known exception is the cooperative program on food crops begun by the Rockefeller Foundation and the Mexican government in 1943. This work led to new research programs in other Latin American countries in the 1950s.[2] Some other international cooperative research activities were carried out in the same decade, such as the rice hybridization project sponsored by the Food and Agriculture Organization in India. And a few developed nations supported scattered institutional development and research programs in the LDC's. But most of the research on food crops continued to be done in the developed nations. Although precise figures are not available, data compiled by Evenson suggest that, of the total investment in *agricultural* research in 1958, about 90 percent was in the developed nations and approximately 10 percent was in the less developed nations.[3] The proportions spent on *food* crops in the developing nations may have been even less.

A significant change took place in the early 1960s with the establishment of two international crop research institutes: the International Rice Research Institute (IRRI) in the Philippines and the International Maize and Wheat Improvement Center (CIMMYT) in Mexico. These two institutes were located in LDC's and oriented to their food problems. Their early successes led to the establishment of a number of other international research activities as well as to a rebirth of interest in improving and expanding national research programs. All these activities were enhanced by earlier and concurrent programs of human and institutional development.

As of the mid 1970s, research on food crops in and for the LDC's is finally coming of age. The Consultative Group on International Agricultural Research (CG) – composed of nations, international organizations, and foundations – has been established. The annual investment in international research through this group reached about $47.3 million in 1975. The United States Agency for International Development (AID) contributes up to 25 percent of the costs of CG-sponsored activities and invested $10.655 million in 1975. In addition, AID is actively stepping up financial support for national research programs within LDC's.

Although the funds involved in such projects are substantially greater than they were a few years ago, they are miniscule for the job that has to be done. They are also relatively small in comparison with global expenditures for agricultural research in the developed nations or for other items of public expenditure. Data compiled by Evenson suggest that the total expenditure on agricultural research in 1970 was $1.32 billion in the developed nations and $236 million in the developing nations, or a total of $1.56 billion.[4] The international research funds, however, do represent a significant addition to the total expenditure on agricultural research for developing nations.

Such an investment is likely to spur interest in measuring results. The technical products are abundant and are presented in considerable detail in the annual reports and other publications of the institutes. Economic and social aspects of the resulting technologies are also beginning to be studied in greater detail.

But the quantitative effect of the institutes' efforts on actual production in the LDC's has not yet been closely examined. There are good reasons for this lag: the centers are new, such an analysis is difficult, and few resources have been devoted to the task. Nevertheless, the field is not entirely unexplored. Some studies have been carried out in the past on the effect of national agricultural research programs in both developed and less developed countries. Generally, the results have shown high rates of return to investment in research.[5]

The next step will be a more specific evaluation of the effects of international agricultural research. But to do this efficiently will require more than knowledge of economics and quantitative tools. It will also require theoretical and empirical knowledge of (1) the nature of the international centers and the associated international agricultural research system; (2) the adoption process at the farm level for resulting agricultural technology; and (3) available statistical data which help measure both the input into research and the effect of the product.

Some such knowledge exists at present, but it tends to be in fragmentary form. Dr. Robert Evenson and I have been separately involved in analyzing certain components for several years. His attention has been focused on fairly quantitative and aggregative analysis of agricultural research in general.[6] I, on the other hand, have been more concerned with analyzing specific technologies and most recently have been involved in documenting the development, spread, and influence of the high-yielding varieties of wheat and rice.[7]

Both approaches are necessary but not sufficient for evaluating the impact of international research on crop production. There is a need to find a middle ground where quantitative concepts and tools of measurement are more closely woven with empirical knowledge of the technology. And there is a need to blend highly aggregative analysis with studies that are somewhat more local. This chapter moves toward this middle ground.

First we will examine the general question of the various effects of research that must be considered in evaluating its impact, and then we will offer more specific and narrow quantitative analyses of the direct effects on yield and production. A precise and definitive measure of the effect of international research on wheat and rice production is not attempted; this, as will be demonstrated, is most difficult. Rather, conceptual and methodological problems involved in the process are introduced. Empirical data are used largely for illustrative purposes.

Though production changes can have important effects on economic and social factors, these matters were simply beyond the scope of this study. In any case, they have been discussed elsewhere.[8]

Much more work will be needed before the effects of international agricultural research can be comprehensively assessed. This chapter introduces some of the major considerations involved and, it is hoped, will encourage further study of this most important subject.

The International Research Institutes

International agricultural research as defined here consists of work carried out under the aegis of the Consultative Group on International Agricultural Research (CG). As of early 1975, the CG was sponsoring six active international agricultural research institutes, three other institutes in varying stages of development, and three related programs.

Of the six active institutes, only IRRI and CIMMYT have been in operation for more than ten years. Because of the newness of the other four institutes, it is too early to assess their impact on crop production.[9] Consequently, this study focuses on two of the three crops covered by the first two institutes: rice and wheat. Corn is excluded. Research on this crop, for a variety of reasons, has not been as successful as the work on the other two.[10] Any general study of the payoff to research should, of course, include the full range of efforts.

Work leading to the establishment of CIMMYT began in 1943 with the founding of a grain program in Mexico by the Rockefeller Foundation, in cooperation with the Office of Special Studies of the Mexican Ministry of Agriculture. In 1959, Dr. Norman Borlaug became director of the Rockefeller Foundation's International Wheat Improvement Project. The wheat program was merged with a comparable corn program in October 1963 to form the International Center for Corn and Wheat Improvement.[11] By early 1966, "the growing demands on this program by the ever-widening food gap around the world indicated the need for a restructuring and expansion of activities. As a result, the center was reorganized on April 12, 1966, in accordance with Mexican law, as a nonprofit scientific and educational institution . . . to be governed by an international board of directors."[12]

The new board held its first meeting in September 1966 and approved programs for 1967. Major financial support was at first provided by the Ford and Rockefeller foundations. In 1969, AID became a contributor. A new headquarters and laboratory facility were completed at El Batan (forty-five kilometers northeast of Mexico City) and dedicated on September 21, 1971. The initial construction cost of $3.5 million was provided by the Rockefeller Foundation;[13] through 1974, the total capital costs were $6.4 million. (The CIMMYT capital investment did not include housing for the staff. Also, when CIMMYT was legally constituted in 1966 it had acquired a number of vehicles and a fair amount of field equipment; the replacement of this equipment has been charged to operating costs and not to capital.)[14]

In 1959, the Ford and Rockefeller foundations decided jointly to establish a rice research institute in the Philippines, and on April 13 and 14, 1960, when its trustees met for the first time, IRRI was formally organized. Construction was finished in January 1962, and the institute was dedicated on February 7, 1962. By that time the research program was underway. The capital cost was $7.5 million. (This included housing for staff.)[15] Initially, Ford provided the physical plant and Rockefeller the operating funds; in 1965 they began to split the operating costs. Support from AID was added in 1970.

The growth in the total expenditures of the two institutes is depicted in Table 7-1. One should not, of course, total the columns for the individual institutes without at least making an allowance for inflation. Barker, in cumulating expenditures at IRRI from 1960 to 1972, used a GNP deflator and then went on to include a discount factor equivalent to an interest rate charge for the use of money over the period. The unadjusted total was $24.3 million; allowance for inflation raised it to $28.6 million, and addition of the discount factor increased the total to $51.6 million.[16]

Since the establishment of the centers, their programs have grown somewhat beyond the crops indicated in their titles. On the other hand, some regional rice work has been taken up by the International Center for Tropical Agriculture (CIAT) in Colombia and the International Institute for Tropical Agriculture (IITA) in Nigeria. The total amount proposed for actual expenditure on wheat and rice research in 1975, exclusive of related or overhead costs, is given in the accompanying tabulation.[17] Even if a prorated portion of the other

Institute	*Wheat*	*Rice*	*Total*
CIMMYT	1,166		1,166
IRRI		2,380	2,380
IITA		225	225
CIAT		153	153
Total	1,166	2,758	3,924

Amounts are given in thousands of dollars.

costs were assigned to the two crops and special projects were added, the totals would probably not be over $10 million. The annual total would have been less in previous years. As noted earlier, the work on wheat in Mexico goes back to 1943, but the annual expenditures by Rockefeller were relatively modest. The total annual expenditures on wheat research by the Office of Special Studies for 1954 to 1960, converted from 1958-60 pesos, ranged from $345,000 to $203,000.[18]

Hence, when the impact of the international centers on wheat and rice production in the LDC's is evaluated, the benefits can be compared with a relatively small investment over a short period. In an overall view, the expenditures on research in relation to the annual values of the crops involved are miniscule indeed.

Throughout their history, IRRI and CIMMYT have been closely involved with national LDC programs. As Hardin and Collins have noted, these centers "were not designed to supplant country efforts, but indeed were developed to

Table 7-1. Annual Total Expenditures, both Core and Capital, for IRRI and CIMMYT, 1959-75[a]
(in thousands of dollars)

Year	IRRI	CIMMYT[b]
1959	250[c]	
1960	7,060[c]	
1961	229[c]	
1962	405[c]	
1963	875[c]	
1964	625[c]	
1965	1,055	
1966	1,125	457
1967	1,164	843
1968	1,641	1,427
1969	1,955	2,053
1970	2,135	5,017
1971	2,676	4,836
1972	2,960	4,942
1973	3,084	6,231
1974 (est.)	4,557	5,563
1975 (proposed)	8,520	6,834

Source: For 1959-64 (IRRI), letter from Faustino M. Salacup, executive officer and treasurer, IRRI, August 28, 1974. For 1965-69 (IRRI), Werner Kiene, Ford Foundation, August 1974. For 1966-71 (CIMMYT), "This is CIMMYT," CIMMYT Information Bulletin no. 8, March 1974, Chart 15/2, Tables 1 and 2. (Table 1 lists donors but really means expenditures [letter from Robert D. Osler, deputy director general and treasurer, CIMMYT, September 11, 1974].) For 1970-75 (IRRI), 1972-75 (CIMMYT), budget submissions or presentations for each center for 1974-75, Table III.

[a] Except as noted, data refer to actual total expenditures. In most of the source tables for 1970-75, this category is referred to as "application of funds" (exclusive of funds carried over to the following year). It includes, in addition to funds obtained from the Consultative Group (CG) or individual donors before 1972, three other sources of "income": earned, indirect, and unexpended balances from the previous year. The totals therefore exceed, by these amounts, the annual funding requested from the CG. The totals exclude working capital and funds received and spent on special projects.

[b] The International Center for Corn and Wheat Improvement was first formed in cooperation with the Mexican government in late 1963 but was then reorganized and reestablished on an international basis as CIMMYT in 1966.

[c] Grants received for capital and operating costs; not actual expenditures.

complement and stimulate national research programs."[19] The nature of these institutional ties is amply described in the annual reports of the centers and in other papers.[20]

Relating Research Results and Production Changes

It is a long way from the international agricultural research institute to the farmer's field. Relating the activities of the institute to actual changes in crop production requires an understanding of (1) the potential effects of research and (2) the reasons for the gap between potential and reality. To judge the results of international research in terms of farmers' yields is to judge many other aspects of the rural economy as well. It is a severe test.

Potential Effects of Research

The major product of the international institutes is new technology, which, in turn, brings about changes in the production process for the commodity involved. The direct quantitative effects are that (1) output is expanded at the same overall cost, (2) the same output is produced at lower cost, or (3) there is some combination of these two results. Direct effects may also be accompanied by indirect effects.

Direct effects of the HYV's. High-yielding varieties (HYV's) of wheat and rice are best known for their effect on the quantity of output. In addition, they may also influence the quality of the product.

HYV's usually bring about increased output per unit of land. While yields are increased, so are the total costs per unit of land, because a package of associated inputs is needed. However, if HYV's are properly sited and used, returns per unit of product are usually increased. A recent example is Sidhu's study of wheat in the Punjab of India, which revealed that unit costs of production with the new varieties declined about 16 percent.[21] This increased profitability is, of course, largely responsible for the widespread adoption of the HYV's.

Yield potential is increased largely because of the semidwarf characteristics of the varieties. Additional fertilizer tends to be reflected in increased yields rather than in increased vegetative growth. The short, stiff straw of these varieties also means that they are less likely to lodge (fall over).

Although HYV's, given the proper package of inputs, usually have a clear yield advantage over traditional varieties, it is difficult to measure the difference precisely. The improvement is not the same for wheat and rice. And advantages vary widely within each crop, depending on the degree to which the recommended level of inputs is used, the quality of the land base, and a host of other factors.

In the late 1960s, multiples of two or three times the traditional yield were claimed for the HYV's. These were largely measures of *potential* taken from experiment station trials or supervised demonstration plots. In itself, this increased potential could be considered one possible measure of the fruits of international research. Actual farm yields, however, have been lower. Some of the reasons for this difference will be outlined later in this chapter.

The yield effect has taken two different patterns in the breeding programs for wheat and rice.[22] Semidwarf wheat varieties were developed in the second stage of the Mexican breeding program and were first released in the early 1960s. The semidwarf characteristics were part of the IRRI rice-breeding program from the outset. As a result, the yield potential of the newer Mexican wheat varieties, which incorporate the dwarfing characteristic, is greater than that of the earlier improved varieties, while the maximum yield potential of the IRRI varieties has not increased greatly since the introduction of IR-8.

These different patterns were partly related to problems of disease. The major problem for wheat was rust (a moldlike fungus). Development of resistant varieties was considered to be the only answer, and Borlaug took up this work in 1945. By 1949, four new varieties were developed which were soon widely planted. The battle is a continuing one, however, because rust is extremely persistent, appearing repeatedly in new strains.[23] In 1974, CIMMYT reported that, although the wheat varieties that moved out of Mexico in the 1960s showed good resistance, "resistance to some of the rusts is now breaking down. New varieties with different genetic resistance are urgently needed. It appears that 10 years may be the longest period that a variety can withstand the constantly changing attack of the three rusts."[24]

Disease was not an important factor in the early IRRI activities, but it soon became a serious concern. Other problems receiving major attention include insect resistance and tolerance for such stress factors as drought, cold, deep water, and soil problems.

In addition to looking for increased yield *potential*, the institutes are placing considerable emphasis on achieving yield *stability*. Resistance to insects and disease and tolerance for environmental stress factors play a major role in reducing year-to-year fluctuations in production. In pursuing yield stability, CIMMYT is making a number of crosses between spring and winter wheats and between wheats and other cereals. IRRI has established the Genetic Evaluation and Utilization Program which seeks to develop varieties with improved resistance and tolerance. As a result of this search for yield stability, the potential geographic area of varietal use may be broadened.

Some of these research efforts will produce higher average farm yields, but other research will be needed just to maintain these higher yields in the face of ever changing attacks from insects and disease. Such maintenance research, although absolutely necessary, may not show up well in conventional measures of productivity.[25] Increased yield stability, however, may be viewed by farmers as a reduction in risk and hence could lead to a subsequent increase in agricultural output.[26] Since maintenance research may become increasingly important as agriculture becomes more complex, it is vital that further attention be given to its measurement.

The new varieties differ qualitatively from traditional varieties in two primary ways: consumer acceptance and nutrient composition. Some of the early institute wheat and rice varieties achieved only limited acceptance in certain areas because of color, appearance, or taste differences. The result was a lower price. Most of these problems were taken care of in subsequent breeding programs, though traditional varieties still may be preferred in some places.

The question of relative nutrient quality is more difficult to assess. It depends on an involved interplay of genetic makeup, quantity and timing of nitrogen applications, and environmental factors. Although on balance there may not be much of a difference between the HYV's and the traditional varieties, an attempt is being made, particularly with rice, to breed in higher protein levels or quality.[27] The challenge is to find varieties which have both higher yields and higher nutrient levels.

Indirect effects of the HYV's. The indirect effects of the HYV's, like the direct effects, may have important quantitative and qualitative dimensions. Both of these dimensions are often overlooked.

One of the major biological features of the HYV's, especially rice, is their photoperiod insensitivity, which often shortens the time needed to reach maturity, thereby providing greater flexibility in planting dates. This helps make it possible to grow an extra crop a year in some regions. Several rice-eating nations in Southeast Asia have recently requested CIMMYT's help in introducing a wheat crop during the winter season. And Pakistan is studying the possibility of growing two crops of wheat a year. For these reasons, multiple cropping usually increases in green revolution areas. Castillo notes that in Asia adoption of the modern varieties "is almost synonymous with the adoption of multiple cropping" and that in some cases where their yields were not superior to local varieties "they were adopted nevertheless because of the shorter growing period."[28] Perhaps, in the long run, this indirect effect on output will be as important as or even more important than the direct influence on yield.[29]

A second indirect effect is that higher yields may free resources for other uses. This was recently reported to be the case in Uttar Pradesh in India, where "the coming of the new technology has freed the small farmer from the less profitable cropping patterns on which he could always depend to provide minimum quantities of such staples as wheat and animal fodder for home consumption. If he grows high-yielding varieties, the small farmer can supply his home consumption needs and still have land remaining to grow high-yielding cereals for market or other high-profit crops like sugarcane."[30]

To take these and other effects into account we should increasingly turn our attention from yields *per crop* to yields *per unit of land per year*. This will be particularly true as more work is devoted to developing improved farming systems.

The research on wheat and rice can have many economic and social effects, in addition to its effects on production. But measurement of the effects of research on output – detailed in later sections of this chapter – is a necessary and often missing link in the chain of analysis.

The Gap between Potential and Reality

High-yield technology developed at the research level simply reveals the potential for improved yield; this potential must be transformed into reality in actual farmers' fields in the LDC's. However, many factors outside the control of the experiment station – such as biological and economic constraints or traditional farming methods – may interfere with the optimal use of HYV's.

Nature of the institute product. The new varieties are generally high yielding only if accompanied by a package of inputs. Chief among these are fertilizer and improved management, but both water and control of insects and diseases may also be vital. The international center may provide, along with the seed, a set of recommendations for such inputs but these must actually be applied by the farmer at the local level. Many forces beyond the farmer's control can affect the availability of some of these inputs, as has recently been shown for fertilizer. And other factors, such as the availability of credit, influence the farmer's willingness to actually use the inputs.

In many cases, the HYV provided by the institute is only raw material which needs to be refined for local use by national research programs. It is instructive that CIMMYT does not release varieties as such but rather "distributes germ plasm to national programs" leaving "governments . . . free to release them as varieties under local names or . . . [to] use CIMMYT germ plasm in their own breeding programs. Either way, the national programs take responsibility for what is selected and released."[31] Similarly, IRRI varieties have been reissued under other names and/or extensively crossed with local varieties in national programs.[32]

Another complicating factor in measuring research efforts is that some varieties included in the HYV category were developed in national programs either before the centers were established or independently of them. In fact, the IRRI and CIMMYT varieties are not wholly new varieties; in most cases, they build on generations of breeding efforts which have gone on before at the national and regional levels.[33] For these reasons, the new wheats and rices should be viewed as joint products of national and international research efforts. This makes it difficult to distinguish the particular contributions of the institutes and hazardous to attempt, as Evenson has done, to sort the HYV's into the three groups of "institute-bred," "joint institute-national," and "other independent" (see chapter 9). Such a breakdown is further hindered by the lack of information on discrete varieties in HYV data from many countries.

Constraints on realizing potential. The yield potential of HYV's determined on experiment stations is often several times higher than that obtained in practice. In the Philippines, for instance, the potential rice yield is in the neighborhood of eight metric tons per hectare, whereas actual overall yields (traditional and HYV) are slightly less than two tons.[34]

What accounts for such differences? First, the HYV's are not planted on all of the crop land. In Asia in 1972-73, the HYV's accounted for about 35 percent of the total wheat area and 20 percent of the total rice area. In a few nations the proportions were relatively high: for wheat the HYV proportion was 55.9 percent in Pakistan and 51.5 percent in India; for rice the HYV proportion was 56.3 percent in the Philippines and 43.4 percent in Pakistan.[35] Data on trends are provided in Figure 7-1.

Second, even with local breeding efforts, there are biological limits on the proportion of crop area suitable for the HYV's. For instance, much of the wheat area in Turkey is suited only for winter wheats, whereas the Mexican HYV's are spring wheats. Within an area planted to HYV's, numerous other biological problems restrain output. A breakdown of the constraints reported in one small sample rice survey in the Philippines in 1972-73 suggests the variety of possible limitations that face the farmer.[36] (See accompanying tabulation.) Some other factors restraining HYV adoption may be classified as institutional/economic and risk/uncertainty.[37]

Limiting Factor	*Dry Season*	*Wet Season*
Insects and diseases	35%	70%
Water	26	
Nitrogen	21	6
Weeds	9	18
Seedling	9	6
Total	100%	100%

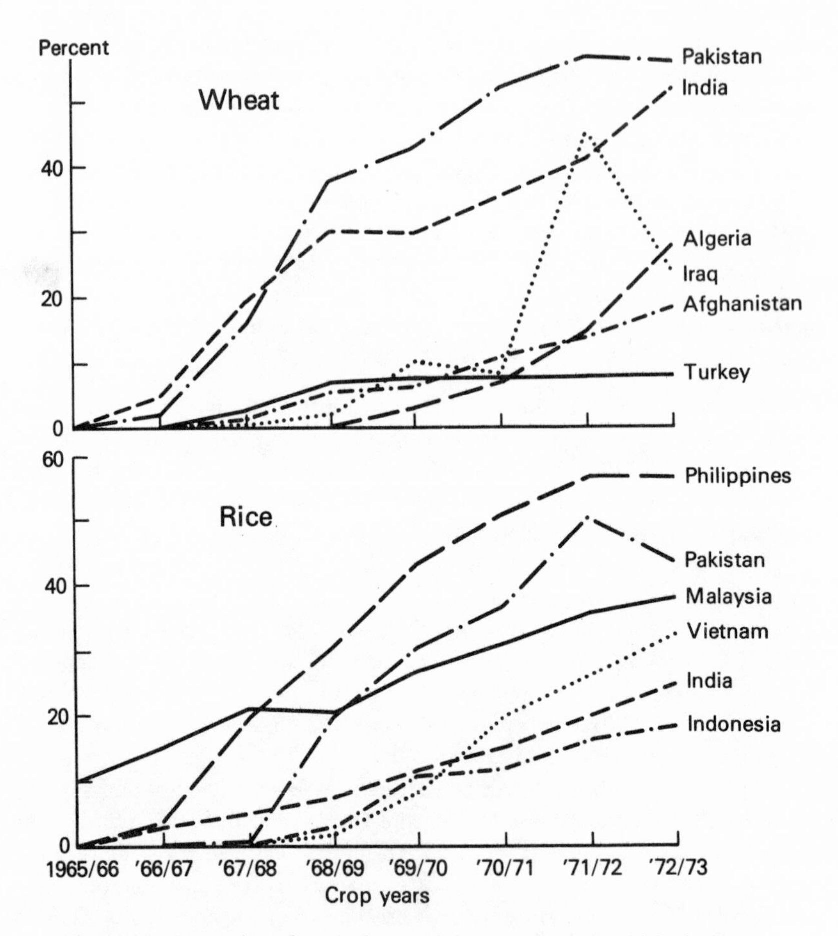

Figure 7-1. Proportion of total wheat and rice area planted to high-yielding varieties, 1965/66-1972/73.

But even if these factors are taken into account, HYV yields are often not as high as might be expected. This is partly because many farmers do not follow the recommended practices for levels of input use. The Philippine survey noted above illustrates the difference in rice yields (in metric tons per hectare) owing to farmers' practices.[38] (See accompanying tabulation.) A num-

Practices	*Dry Season*	*Wet Season*
Recommended	7.3	5.0
Farmers	3.9	3.3
Difference	3.4	1.7

ber of other studies have shown that many farmers either do not use recommended practices or do not use them at recommended levels.[39] There are many reasons for this less than complete usage; in some cases continuation of traditional practices represents a rational allocation of resources under the financial, price, and other conditions at the farm level. Moreover, in measuring increased yield and production at the national level it is impossible to know precisely to what extent the recommended inputs have actually been used.

We see, then, that the gap between potential and reality may be partly reduced by greater use of improved practices. And the effects of some biological factors can eventually be modified through research – for example, by developing greater insect and disease resistance. But there are technical and economic limits to how far this process will go, and there will always be some gap between potential and reality.

Thus, beyond the varieties themselves there are many factors involved in the realization of higher yields at the farm level. To measure the productivity of the international institutes themselves on the basis of productivity at the farm level necessarily involves the measurement of such other factors, which range from the effectiveness of the national research agency, to the price of fertilizer, to the weather.

Changes in Area and Yield

Changes in crop production are usually a function of changes in area and/or yield. Improvements in technology are reflected, for the most part, in increased yield. New technologies are less often needed for expansion of area. Thus, in initially evaluating the effect of the HYV's on production, it is useful to determine the relative importance of changes in area and yield.

Increased yields may be caused by many factors. Technology is only one such factor, and the HYV's are only one form of technology. Still, we can gain an impression of the importance of HYV's by comparing changes in HYV adoption and changes in production and by examining relative yield levels of the HYV's and of the traditional varieties. Examining relative yields will also provide the basis for the more sophisticated analysis of the effect of the HYV's on production which we undertake below (see "Measuring Impact on Production").

Data on area planted to HYV wheat and rice in developing nations go back to 1965-66, the first year the varieties produced by the research institutes began to be used internationally to any degree. The data now available extend through 1972-73. It is often not possible to separate the institute varieties in direct use from their progeny and from other improved varieties, so they are all generally lumped together.

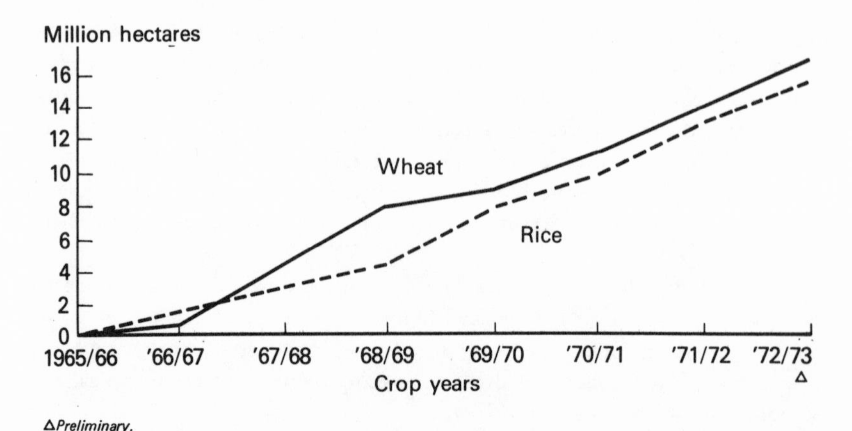

Figure 7-2. Estimated high-yielding wheat and rice area, Asia and North Africa, 1965/66-1972/73 (excluding communist nations).

HYV data for noncommunist LDC's are depicted in summary form for the 1965/66-1972/73 period in Figure 7-2. Area devoted to the HYV's has expanded sharply, but it is still concentrated in Asia, with some HYV wheat in North Africa and some HYV rice in Latin America. Comparable data are not yet available for communist nations.[40]

Total area planted to all types of rice can be obtained for these countries from data compiled by the Foreign Agricultural Service of the USDA or by the Food and Agriculture Organization of the United Nations. Deducting HYV area from the total area gives us, of course, the area planted to regular varieties.

Information can be found on total wheat or rice output for nearly all countries. If the area planted to wheat and to rice is known, it is obviously possible to calculate the average yield for all varieties. However, calculation of relative yields of the HYV's is more difficult. In a few cases, the production and yield of HYV's is reported separately. But more often HYV yields have to be pieced together from a variety of sources.

Effect of Changes in Area and Yield

In assessing the impact of HYV's, some observers look merely at trends in total wheat or rice production in a particular LDC. This procedure alone is inadequate for the measuring of impact because it does not take into account relative changes in area and yield.

Nature of area and yield expansion. There is little information available about the effect of the HYV's on the total cropped area. Considering their biological requirements, it is unlikely that much new land has been cleared for their use. Instead, they have probably substituted for existing crops on the better land. The question then is whether they have substituted for a traditional variety of a like crop or for other crops. It appears that they generally replace like crops, but this is not always the case, especially on irrigated land.

Area trends in India from 1967/68 to 1973/74 reveal different patterns for wheat and rice. For wheat, there was fairly significant expansion of the total area. On the other hand, total rice area expanded only slightly.[41] This suggests that the expansion of HYV wheat involved some replacement of other crops, while the HYV rice area appears to have substituted largely for traditional varieties. Much of the new wheat area would otherwise have been left fallow or planted to chickpeas or other crops. The specific sources of wheat area in 1970-71, compared with those in 1963-65, are presented in the accompanying tabulation.[42] In the Punjab barley, gram, and cotton were the crops replaced by wheat.[43]

	Percentage
Land already in wheat, 1963-65	68.3
Land shifted out of gram (chickpeas)	14.7
Land from fallow or other crops	17.0
Land in wheat, 1970-71	100.0

Relatively little analysis has been made of comparative yield data at the national level. The catch here is the word *comparative*: while we have data on yields where HYV's and traditional varieties are planted, we usually do not have a comparison of the resource base. HYV's are normally planted on the best land. But when they are more widely planted, presumably expanding into less suitable land, yields drop off.

Differentiating area and yield effects. The first step in differentiating the effects of changes in area and yield might be to calculate these changes for countries that have adopted HYV's to a significant extent over a given period of time. For our purposes, averages of two four-year periods, 1960-63 and 1970-73, have been tabulated. The comparisons are conservative in that 1972 was generally a poor year. Countries selected were those where 12 percent or more of the wheat or rice area was planted to HYV's from 1970/71 to 1972/73. Two countries in this classification were omitted: Nepal, because estimates of total wheat areas, yield, and production vary, and South Vietnam because of the influence of the war.

Both area and yield were expanded in each country (see Table 7-2), but in every case except that of Malaysia the relative increase was greater for yield

Table 7-2. Relative Increases in Production, Area, and Yield for Wheat and Rice, 1960-63 to 1970-73

Crop/Country	HYV Proportion 1970/71-1972/73	Increase in 1970-73 Average over 1960-63 Average		
		Area	Yield	Production
Wheat				
Pakistan	52.3 to 55.9%	+ 22.3%	+ 45.2%	+ 77.8%
India	35.5 to 51.5	+ 38.2	+ 56.1	+ 115.7
Rice				
Philippines	50.3 to 56.3	+ 0.4	+ 33.9	+ 34.2
Pakistan	36.6 to 43.4	+ 22.8	+ 73.3	+ 112.9
Malaysia	30.9 to 38.0	+ 43.7	+ 16.5	+ 67.2
India	14.9 to 24.7	+ 4.6	+ 13.8	+ 19.3
Indonesia.	11.2 to 18.0[a]	+ 18.8	+ 29.1	+ 53.4

[a] Government programs only. Additional HYV area planted in private plots.

than for area. The increase in yield ranged from 1.5 times higher than the increase in area for Indian wheat and Indonesian rice, to 2 times for Pakistan wheat, and to 3 times for Pakistan and Indian rice. In the Philippines, virtually all the increase was in yield.

Given this data, it is possible to assess the relative importance of area and yield expansion more formally, as is done in Table 7-3. Increases in yield accounted for a significant portion of the expansion in production in six of the seven cases cited and were of moderate importance in the seventh. Yield increases accounted for virtually all the expansion in rice production in the Philippines and from 50 to 74 percent in the other five cases. Malaysia is the only country where area expansion was more important, and this may have been the result of the addition of some major irrigation projects.

Thus, although both area and yield expansion were involved in production increases in seven cases (five countries) where substantial areas were planted to HYV's, growth in yields generally appeared to be more important.

Annual Changes in Yield

It seems that yield increases were an important factor in production increases in areas where HYV's were planted. What, then, did the annual changes in overall yield patterns look like? How did they differ between HYV's and traditional varieties?

Overall changes in yield. Changes in national wheat and rice yields for the countries noted in the previous section are depicted in Figure 7-3. The following trends are apparent.

Yields in wheat were relatively steady in India and Pakistan through 1967

Table 7-3. Roles of Area and Yield in Production Expansion, 1960-63 to 1970-73

Crop/Country	Production Increase[a] Owing to Expansion	
	Area	Yield
Wheat		
Pakistan	35%	65%
India	42	58
Rice		
Philippines	1	99
Pakistan	27	73
Malaysia (W).	70	30
India	26	74
Indonesia.	40	60

Source: Formula and calculations by Robert Niehaus of the Economic Research Service, USDA.

[a] Calculated according to the following formula:

$$1 = \frac{\log(1 + a)}{\log(1 + p)} + \frac{\log(1 + y)}{\log(1 + p)}$$

where a, y, and p are the percentages in Table 7-2 (but carried out several decimal places in some cases).

and then rose sharply in 1968. Yields in India continued to rise through 1972 but dropped in 1973. Pakistan's yields moved up more slowly but continued to rise in 1973, exceeding those of India.

Yields in rice either remained about the same or rose only gradually through 1966 and 1967. After 1968 Pakistan and Indonesia showed the sharpest and most persistent gains. Though yields in the Philippines appear to have increased only very gradually, changes in accounting and reporting systems may have influenced some of these data. India has shown only a gradual increase over the period. Yields dropped in three of the four countries in 1972 but increased in all of them in 1973. Malaysia was not included on the chart simply because its yield levels averaged above the upper bound. (Malaysia showed no particular trend from 1960 to 1967, but levels moved up substantially in 1968 and 1969; more moderate increases were registered in 1971 and 1973.)

Not surprisingly, these yield trends coincide roughly with the expansion of the HYV area in each country shown in Figure 7-1 (except for the drop in the Philippine rice yields in 1971 and 1972). The impact, however, seemed to be least for rice in India – probably because the HYV area represented only a small proportion of the total area, and because the HYV's used in India have

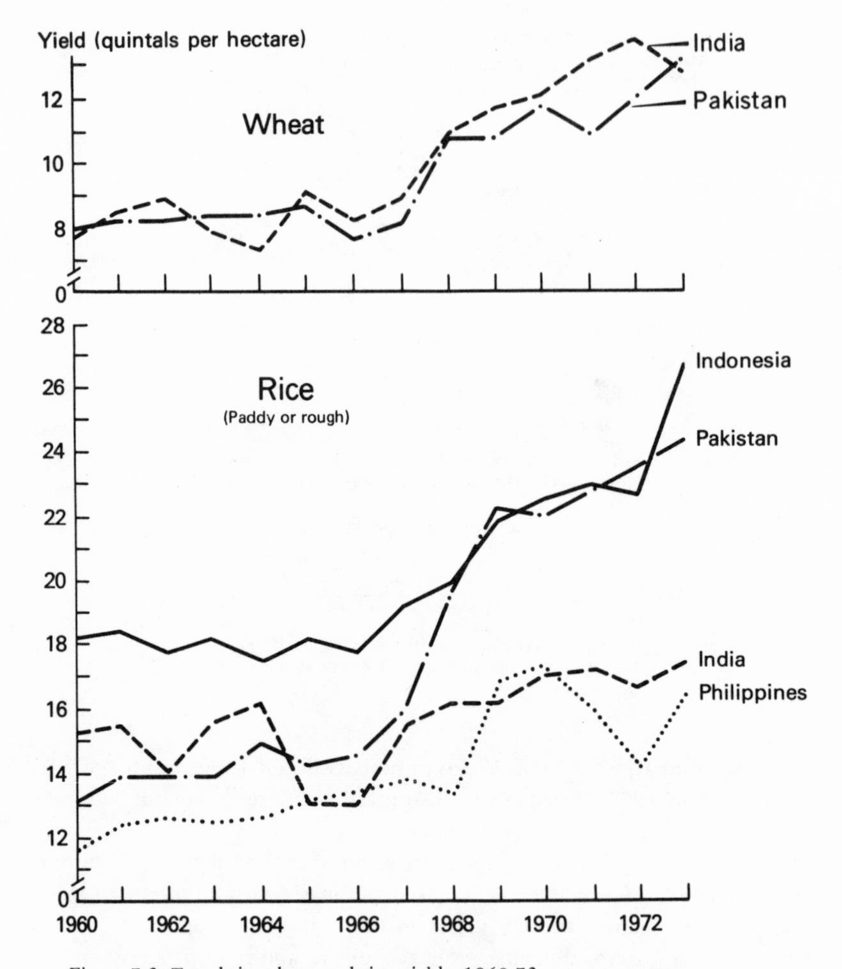

Figure 7-3. Trends in wheat and rice yields, 1960-73.

not yet proved to be well suited to local monsoon conditions. Other factors beside the HYV package, of course, may well have had some influence.

Comparative yield levels. Some national data are available which give an idea of the yield levels of the HYV's compared with those of traditional varieties. These data can be misleading because, as noted earlier, the HYV's are usually planted on the better land. Even so, it may be of interest to review the official statistics and to compare them with other measures.

A few official national statistics have been gathered. One USDA report summarized these figures for wheat from 1966 to 1970 for India, Pakistan, and Turkey.[44] It revealed that HYV yields were substantially above local varieties – from 1.77 to 3.70 times as great; that as area planted to HYV's expanded, their yield levels dropped, though not evenly; and that as HYV area expanded, national yield levels increased. These relationships would be expected. Because they produce higher yields, HYV's account for a larger proportion of total production than of total area. The difference in proportion, however, decreases as the average HYV yield level decreases over time.

Similar data are available for wheat and rice in India for the period from 1966/67 through 1973/74 (Figure 7-4).[45] They show the same general trends noted above, with a few variations. In India, yields for HYV's ranged from less than two to more than three times as high as those for traditional varieties. The wheat multiple was consistently higher than the rice multiple, though the difference narrowed later in the period. These ratios of HYV to traditional yields were fairly consistent through 1970/71 and then dropped. (See accompanying tabulation.) In the Philippines, official estimates for rice

Crop Year	*Wheat*	*Rice*
1966/67	2.87	2.58
1967/68	3.70	2.18
1968/69	3.49	2.05
1969/70	3.68	2.26
1970/71	3.44	2.27
1971/72	2.50	2.03
1972/73	2.35	1.76
1973/74 (prelim.)	2.59	1.71

over the 1968-72 period suggest that HYV yields averaged from 1.30 to 1.35 times higher than those of traditional varieties (including upland).[46]

If the land base were standardized, the comparative yield levels cited above would be somewhat lower. Several years ago I assumed – when pressed for a rough estimate – that the HYV package in irrigated areas might result in a relative yield ratio of 2.0 for wheat and 1.25 for rice. The ratios would be lower in unirrigated areas.

Unfortunately, it has not been possible to review enough studies to provide a good empirical check on these estimates. Two recent investigations, however, provide both larger and smaller multiples for rice, suggesting that the above figure may not be far off the mark as an average. A study of rice production at the village level in six Asian nations in 1971-72 revealed that the overall multiple for both wet and dry seasons was somewhat higher: 1.32 to 1.33.[47] Somewhat lower ratios were obtained in the Philippines for the period from 1968 to 1972 when the national data reported previously were sorted out by type of land base. The HYV advantage was 1.14 on irrigated

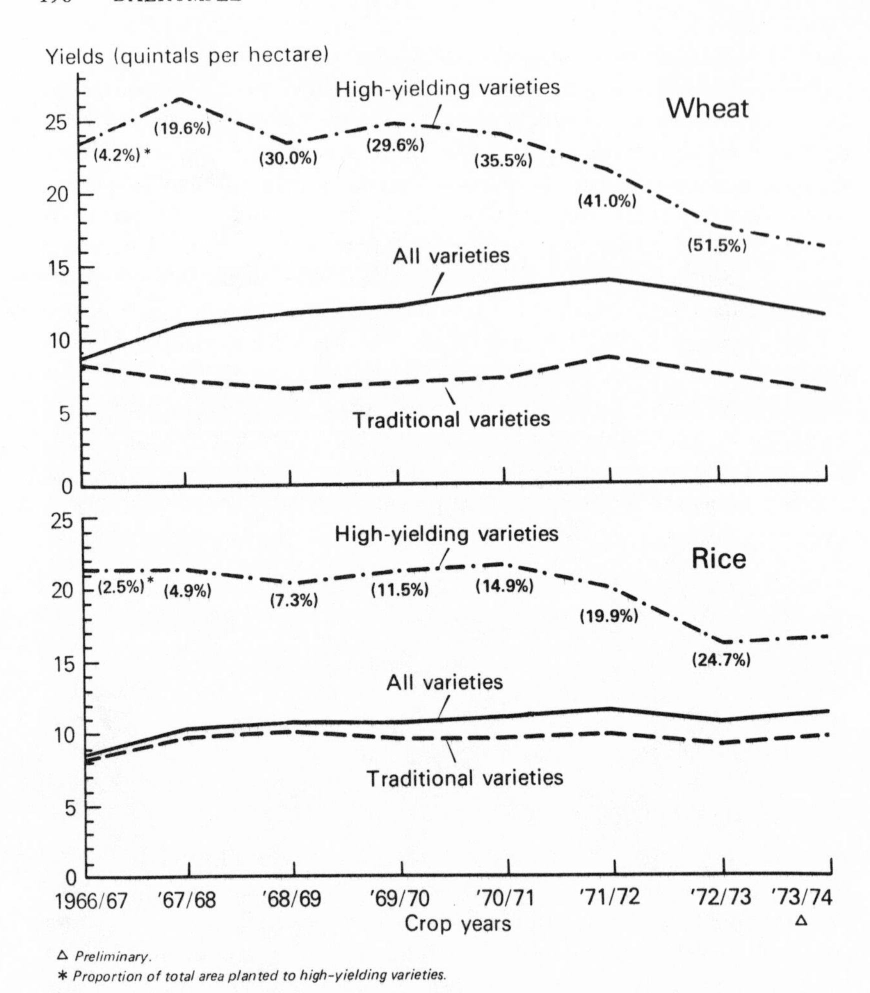

Figure 7-4. Trends in yields for traditional and high-yielding varieties of wheat and rice, India, 1966/67-1973/74.

land and 1.03 on rainfed lowland.[48] Most HYV's are raised in irrigated areas. The multiple did not show any pronounced decline over the period; perhaps the arrival of improved varieties compensated for the possibility that lower quality land may have been planted to HYV's.

For wheat, the countries cited have made extensive use of irrigation. A preliminary review of the data for dryland wheat production in North Africa and the Near East does not yet show a clear pattern of yield increase. This may be because levels of adoption are still relatively low, but may also reflect (1) the impact of lower water levels and of variations in rainfall, and (2) the fact that the traditional varieties in some of the North African nations really are improved varieties which were introduced over the twentieth century and in some cases have characteristics and ancestry similar to the Mexican varieties.[49]

Numerous other data could undoubtedly be found; the difficulty is to distill a meaningful average from them.

Obviously we need to know much more about actual yields at the farm level before we can make precise evaluations of the contribution of the HYV's or of the HYV package to increased yields. And we need to know much more about the influence that various purchased inputs, the weather, and other factors have on production.

Measuring the Impact of Technology on Production

The next step in analyzing the impact of the new technology is to evaluate its effect on production. The main problem we face is that a great many different factors influence changes in production. Furthermore, we do not know precisely what production would have been in the absence of new technology.

To measure production changes, most economists would use (1) a production function, or (2) an index number approach.[50] Each technique has its advantages and limitations. This section presents first a brief review of both techniques in the context of wheat and rice production and then a simplification of the index number technique. Finally, the findings of these two approaches are compared.

Production Function Analysis

A production function is a form of multiple correlation (or regression) analysis in which changes in production are treated as a function of variations in a number of input variables. The variables might include, as Evenson has suggested, (1) utilization of land, (2) fertilizer, (3) irrigation, (4) other agricultural inputs, and (5) some measure of the introduction of new technology, such as the percentage of the crop produced from the new varieties.[51]

Data requirements. Although a logical functional form can be fairly easily laid out, the problem is to obtain statistical data for each of the input variables. This can be accomplished at local or regional levels by farm surveys, but it is a very difficult task at the national level. About the only information readily available is the HYV area. Fertilizer is of critical importance, yet no LDC reports regular national data on the amount of fertilizer applied to individual crops such as wheat or rice, let alone to HYV's. All that is reported on an annual basis is the amount of fertilizer apparently consumed on all crops (these data are presented in FAO's annual *Fertilizer Review*). Some export or nonfood crops are large users of fertilizer. The use of insecticides and pesticides is even less clear. Irrigation is not such an unknown, but it varies a great deal in quality and we have only a vague idea of the amount of irrigated land devoted to HYV's.

Even if these data were available, we would have to take other variables into account. Perhaps the most difficult to measure is weather. While there have been sharp changes in weather since the mid 1960s – and 1972 was particularly bad – there are apparently no indexes which adequately measure the total yearly changes in weather. Perhaps over a long enough time period these changes would balance out, but the period at hand is only eight years long. Some national data are available which make a start possible, such as the all-India rainfall indexes,[52] but they are only a partial means of measuring the weather.

A more easily measured variable is the change in the price of both the product and the various inputs. Increased product prices and lower input prices would be expected to increase the adoption of innovations. Such changes have taken place in the prices of rice and urea. The cost of irrigation water depends on the source but so does quality (in terms of when it is available); canal water is usually much cheaper than tubewell water, but the timing of the application of tubewell water can be regulated much more closely.

All these factors, as well as others, should be considered in specifying a production function – but this is much easier said than done.

Two recent analyses. Despite these problems, many production function analyses have undoubtedly been conducted. Two recent studies on wheat and rice may be representative. One was done at a very broad level, whereas the other was conducted at a regional level within one country. Both used Cobb-Douglas production functions.

Robert Evenson recently reported on a highly aggregated analysis for wheat and rice for Asia and the Middle East.[53] He first considered a country-by-country analysis, but because the data were limited he focused on a regional grouping, using one group of countries for wheat and another for rice. Fertilizer was measured by its total use on all crops, and the HYV areas were based on my earlier area compilations.

The analysis was carried out in two steps. In the first stage, production was expressed as a function of crop area, total use of fertilizer, and the proportion of crop area planted to HYV's. In total, these variables explained nearly all the variation in wheat and rice production. Though each variable was signifi-

Table 7-4. Increase in Production and Value Associated with the Use of High-Yielding Varieties, Asia and Mideast

	Production (%)		Value (million dollars)	
Crop Year	Wheat[a]	Rice[b]	Wheat[c]	Rice[d]
1965/66[e]				
1966/67	0.5	0.5	30	76
1967/68	2.8	1.6	170	233
1968/69	5.2	2.8	325	420
1969/70	5.4	4.6	340	695
1970/71	5.9	6.0	403	905
1971/72	6.7	7.7	445	1,155
1972/73	7.4	9.0	523	1,359

Source: Revised data reported by Robert Evenson, chapter 9 in this volume, Table 9-2; letters from Evenson, September 29, 1975, November 19, 1975.

[a] Thirteen countries.
[b] Twelve countries.
[c] Wheat priced at $75 per metric ton.
[d] Rice priced at $100 per metric ton.
[e] Figures negligible.

cant, crop area was the most important. It was surprising that such a crude measure of the use of fertilizer was significant, but it was not surprising that overall crop area was more important than the HYV area, since the latter was of some magnitude only late in the period. In the second stage of Evenson's analysis, he introduced a number of other measures of research. The results with respect to the variables discussed above were roughly similar.

From this two-stage analysis Evenson concluded that "while the high-yielding varieties did contribute very significantly to increased production, they were by no means the sole source of productivity gains in LDC agriculture."[54] Besides the HYV's and fertilizer, other important reasons for growth in productivity were indigenous research findings and borrowed research discoveries. Whereas two studies revealed (as has been suggested earlier) that the superiority of the HYV's declines as their portion of the total area planted increases, a subsequent and more refined analysis indicated that this decline could be offset to a considerable degree by indigenous research which modifies the technology to suit local conditions.[55]

Evenson went on to calculate the increase in wheat and rice production in the countries studied and then converted this to value terms (Table 7-4). His calculations have gone through three stages of refinement; the third stage is reported here.[56] Even if the figures are only roughly accurate, they suggest that the increased production owing to the use of the HYV's was substantial.

Surjit Sidhu has recently reported the results of a study on wheat in the Punjab of India for the four-year period from 1967/68 to 1970/71.[57] Production, again, was the dependent variable; the independent variables were crop land, capital services, fertilizer/manure, and labor. All independent variables proved to be significant except, in some cases, labor. When production functions were run for HYV and non-HYV farms in 1967/68, it was found that the new varieties used more of all inputs on a per unit of *land* basis; however, "a unit of *output* of new wheat consumes less of all inputs, including land, than old wheat" and this "is of crucial importance as a source of growth."[58]

For the year 1967/68, the "magnitude of the natural upward shift in the wheat production function resulting from the introduction of new wheat" was 22.85 percent.[59] In a subsequent paper, using a somewhat different formulation, Sidhu found an increase in efficiency of 44.79 percent.[60] These two figures form, he feels, the lower and upper limits of the actual change in productivity.[61]

For the other three years of one study, analyses were carried out for HYV's only.[62] The results suggested a downward shift in the production function after 1967/68. Sidhu thought that this drop may have been the result of the weather, the deterioration in seed quality (owing to mixing), and the addition of marginally "inferior lands" but noted that "an assessment of their relative influences seems impossible." The downward shift in the production function, however, was to some extent reversed in 1970/71. Sidhu was not sure whether the downward movement "was a temporary phenomenon or is a long-run technological regression in the production of new wheats."[63]

Sidhu remarks that "during farm visits in 1970 and 1971 Punjab farmers generally complained of defective seed quality after 1967/68. . . . I think mixing of lower quality seed with better seeds occurred at more than one level of the seed distribution channel."[64] If he is right in suggesting that the declining quality of seed may be caused by mixing – and some other recent references from India indicate that he might be – we have another complex and largely unmeasurable variable which should be considered. Forms of "technological regression," however, can be corrected to some extent in national research programs, as Evenson's analysis (cited above) has indicated.

Production functions, though they provide an analytically attractive approach, do have severe data problems unless they are based on farm surveys. And even if they are, there is the problem of extrapolating the results to the national or international level. Is there a way to get around these problems? The index number approach is one possibility.

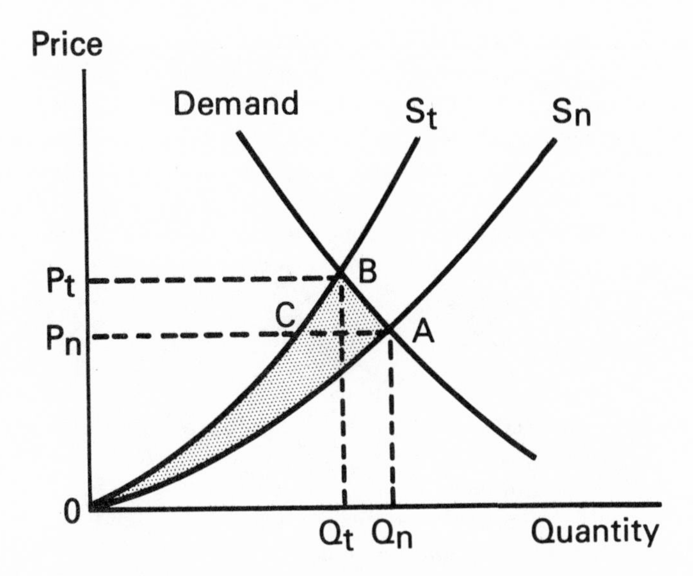

Figure 7-5. Effect of a new technology in shifting supply curves.

Index Number Analysis

New technology usually results in an increase in output for a given set of resources. Through use of the index number approach, it is possible to measure the magnitude not only of this increase but of its value to society. A number of economists have used this approach at the national level.[65] The index number technique can build on some of the results of production function analysis. Although the index number approach does have some limitations, these can be partly avoided by linking this approach with production function analysis.

The general formulation. In economic terms, the introduction of a new technology leads to a shift in the supply curve (graphically shown in Figure 7-5). Curve S_t represents the supply situation with traditional technology. Curve S_n represents the supply situation if the new technology is utilized. With the introduction of the new technology, the quantity of product is increased and the price is reduced. This change results in a gain to society, which is indicated by the shaded area, OAB. (Here I have adopted the simplified depiction of social benefits used by Hayami and Akino in chapter 2.) Since only part of the farming area may utilize the new technology, the actual supply curve would lie somewhere between S_n and S_t.

The usual index number analysis involves a three-stage process, including estimation of (1) gross benefits, (2) research costs, and (3) rate of return over time. Obviously, a full-blown index number study could be rather involved and would demand much data. It also goes beyond the scope of this chapter, which is to evaluate effects on production. Therefore we will focus on step (1), the measurement of gross benefits.

Even the estimation of gross benefits, however, is a rather complex process. The major components and their functional form may be summarized as follows:[66]

$$B = PQK\,(1 + K/2\,E_D)\,(1 - [(1 - E_D)^2\,E_S/(E_D - E_S)])$$

where

B = gross benefits
P = price of the product
Q = quantity of the product
K = shift in supply curve owing to research
E_D = elasticity of product demand
E_S = elasticity of product supply.

The most difficult factor to measure, in turn, is K, since it is hard to separate out the many other factors which may influence productivity. Production function analysis can be very helpful in this process. E_D and E_S may also be difficult to determine over broad areas.

Is it possible, for introductory purposes, to get around some of the data problems by simplifying step (1)? A look at three previous studies provides some help with K, E_D, and E_S.

Several types of estimates of K have been utilized. In his classic study on hybrid corn, Griliches simply assumed, using some industry estimates, that yields were 15 percent higher than for open-pollinated varieties (a shift which he identified as K).[67] A subsequent study, by Ardito Barletta, of the effects of crop research in Mexico made use of three different estimates of K: (1) experiment station results, (2) a weighted average from regression analysis, and (3) a figure obtained by assigning all productivity increases to the new wheat and subtracting the additional costs.[68] Hertford et al. used the results of farm-level experimental trials (see chapter 4). In terms of effects, measures close to the farm level would be most desirable; in terms of measuring potential, experiment station results might be most useful.

How necessary is it that elasticity estimates, E_S and E_D, be included? When Griliches postulated various supply and demand elasticities, he found that "these elasticities have only a second-order effect, and hence different reasonable assumptions about them will affect the results very little."[69] In a concurrent investigation of the returns to research on a disease-resistant cotton in Brazil, Ayer and Schuh found, in calculating internal rates of return, that the results were changed only a little by different assumptions about the respective price and supply elasticities.[70] In reviewing these three papers, as well as Ardito Barletta's, the Statistics Division of the Ministry of Overseas Development in the United Kingdom summarized calculations which suggested that, when the elasticity of demand is within the range of − 0.5 to − 1.85, changes in the elasticity of supply make little difference (less than 5 percent) in the amount of benefit.[71]

All told, then, these findings suggest that it is possible to be flexible and pragmatic in obtaining estimates of K, and that introductory analyses might leave out estimates of E_S and E_D. Clearly, more precise analyses should include the elasticities.

Contribution of the HYV package. Considering data available for wheat and rice, and the possible simplifications suggested in the previous section, we can readily estimate the gross contribution of the HYV package to production by a sequence of a few simple formulas. Several different values for K, the shift owing to research, will be assumed.

The available and required data are described in the following algebraic notation:

Varieties	*Area*	*Yield*	*Production*
Traditional	A_t	Y_t	Q_t
HYV	A_{hyv}	Y_{hyv}	Q_{hyv}
All varieties	A_T	Y_T	Q_T

K is the equivalent of $\frac{Y_{hyv}}{Y_t}$. Five of the nine variables are known: A_t, A_{hyv}, A_T, Y_T, and Q_T. The variables that need to be calculated are Y_t, Y_{hyv}, Q_t, and Q_{hyv}. Q_t and Q_{hyv} as used here, however, are not simply the production from each type of variety: rather Q_t is the quantity that would be produced if all of the area were planted to traditional varieties, and Q_{hyv} is the additional production owing to the HYV package. Four different levels of K have been postulated: 1.25, 1.50, 1.75, and 2.0.

The estimating process is composed of three steps, each of which utilizes a formula.

(1) Estimated yield of traditional varieties (Y_t)

$$Y_t = \frac{Q_T}{A_t + (A_{hyv} \times K)} .$$

(2) Total production if total area planted to traditional varieties (Q_t)

$$Q_t = Y_t \times A_T.$$

(3) Additional production owing to HYV package (Q_{hyv})

$$Q_{hyv} = Q_T - Q_t.$$

The derivation of formula (1) is

$$Q_T = (A_t \times Y_t) + (A_{hyv} \times Y_{hyv})$$

$$Q_T = (A_t \times Y_t) + (A_{hyv} \times (Y_t \times K))$$

$$Q_T = Y_t(A_t + A_{hyv} \times K)$$

$$Y_t = \frac{Q_T}{A_t + (A_{hyv} \times K)}.$$

This is, as suggested, a fairly simple estimating process. It is also flexible; it can be used at any level for which data are available. The main limitation is, as with the index number approach generally, the derivation and specification of K.

Although a range of assumptions on the value of K has been specified, which one appears to be most realistic? In the past, as noted previously, I have used a rough estimate of 1.25 for the HYV rice package and 2.00 for wheat in Asia. Data from several countries suggest that ratios for wheat range from 1.77 to 3.70 and for rice from 1.10 to 2.58. Sidhu's production function analysis indicates farm-level figures ranging from 1.23 to 1.45 for wheat in the Indian Punjab in 1967/68. Research in Colombia placed the yield advantage in 1971 as 1.46 for the improved wheat varieties and between 1.25 and 1.39 for rice (see chapter 4). Clearly there is a wide variation in the ratios.

One explanation for this range of estimates is that they may describe different things. The HYV package is purposely referred to throughout this chapter. The varieties alone may not have a significant effect on overall production because other elements of the package are needed, particularly increased fertilization. On the other hand, without the improved variety the full utility of the other inputs may not be realized. Although some of these factors may be sorted out at the local level through the use of production function or regression analysis, this is much more difficult to do at the national or international level.

While pure varietal effects have reportedly been sorted out for Colombia, some rather exceptional data were available (see chapter 4). The use of more traditional national data on area planted to varieties and estimated quantity of fertilizer may produce a high degree of intercorrelation. In a study underway at the Brookings Institution in 1975, for instance, Roy and Sanderson found such a correlation (r = 0.97) between the area planted to all HYV grains and the estimated use of fertilizer on all grains (including HYV's) in India between 1966/67 and 1973/74. In some cases fertilizer might appear to have a higher correlation with output than the HYV's. If so, this may indicate that the fertilizer figure usually reflects use on both traditional and HYV crops (no one knows how much was actually used on HYV's at the aggregate level) and that the area of the HYV's in the early years in each country is quite small (as shown in Figure 7-1). Hence the HYV figure may be swamped by the fertilizer figure.

Of the various K factors postulated, the most likely for the Asian region as a whole might be 1.25 for rice and 1.50 for wheat. The wheat figure is less than that used a few years ago, partly because of the declines in HYV yields as they are planted more widely within nations (as shown in Figure 7-4 for India) and the fact that some of the newer wheat plantings are in the Near East, where water supplies may be even more limited than they are in South Asia.

When the index number approach is applied to wheat and rice in the non-communist developing nations of Asia for the 1972/73 crop year, the calculations produce the results given in column 2 of Table 7-5. (Column 1, the percentage increase, is simply calculated from some of the original data.) Obviously the results vary considerably, depending on which yield or K factor is utilized. If K factors of 1.25 for rice and 1.50 for wheat are selected as most realistic, the calculations suggest that in 1972/73 the HYV package added 8.7 million metric tons of wheat and 7.7 million metric tons of rice. In terms of the total crop, overall wheat output was increased by 18.3 percent and rice output by 4.9 percent.

These figures may be more meaningful when converted to value terms (column 3 of Table 7-5), though this is a hazardous step since it is difficult to select appropriate prices to use for a broad geographic area. If, to facilitate comparison, one applies the prices used by Evenson ($75/ton for wheat and $100/ton for rice), the gross value of the increased output in 1972/73 is striking: $656 million for wheat and $769 million for rice, or a total of $1.425 billion.

These prices, however, may be on the high side. They are close to international levels and do not reflect the fact that the HYV's, despite improvements in taste and color, still are not exported in quantity and do not bring a premium domestic price.[72] They also do not reflect the effect of increased output on local prices. An increase in output would, of course, result in a decrease in price. The amount of decrease would depend on the price elasticity of demand as well as other factors. Although the price decline reduces the

Table 7-5. Estimated Increase in Wheat and Rice Production in Asia under Different HYV Yield Assumptions, 1972/73 Crop Year[a]

HYV Yield as Multiple of Traditional Yield	Proportion (%)		Quantity (million metric tons)[b]		Value (million dollars)	
	Wheat	Rice	Wheat	Rice	Wheat[c]	Rice[d]
1.25.	9.1	4.9	4.2	7.7	314	769
1.50.	18.3	9.8	8.7	13.8	656	1,379
1.75.	27.4	14.7	11.8	18.4	881	1,841
2.00.	36.6	19.6	14.4	23.5	1,080	2,354

Source: HYV area based on background data for Figures 7-1 and 7-2. Other area, yield, and production data derived from statistics compiled by Foreign Agricultural Service. Prices same as those used by Evenson (see Table 7-4, notes c and d).

[a] Excluding People's Republic of China, North Vietnam, Japan, and Israel.

[b] Calculated according to formulas (1), (2), and (3) in text.

[c] At $75 per metric ton.

[d] At $100 per metric ton.

valuation of the added output, it is at the heart of the social benefits arising from the innovation (as shown in Figure 7-5). The introduction of the improved wheats in Mexico, for instance, had a major effect in lowering prices to consumers.[73] In some countries, on the other hand, farm prices are held artificially low, which unduly lowers the valuation of the impact at the national level.

If for these reasons prices are arbitrarily reduced by a third (to $50 per ton for wheat and $67 per ton for rice) to reflect these factors better, the results are still most impressive: an increase of $435 million for wheat and $513 million for rice, or a total gross value of about $950 million. Overall, it is reasonable to suggest that the gross value of the HYV wheat and rice package in 1972/73 was about $1 billion for Asia alone.

Even though the overall output increases – of 18.3 percent for wheat and 4.9 percent for rice – are not great, the areas involved in noncommunist Asia alone are so vast that the total figures are inevitably significant. The monetary values would be even higher if North Vietnam, North Korea, Latin America, and Africa were included. However, if the additional cost of inputs were subtracted from the gross figures, they would of course be lowered.

Comparison of Results

How do the results obtained using index number analyses compare with those obtained by Evenson for 1972/73 using production function analysis (reported in Table 7-4)? The statistical findings, using the same prices, are summarized in the accompanying tabulation. Although the data cannot be

Analytical Method	*Crop*	*Increase in Total Production*	
		Percentage	*Gross Value (million dollars)*
Production function	Wheat	7.4	523
	Rice	9.0	1,359
	Total		1,882
Index number	Wheat	18.3	656
	Rice	4.9	769
	Total		1,425

precisely compared because the specific countries and regions involved differ slightly as between Evenson's analysis and mine, the production function analysis appears to have led to more conservative estimates for wheat, while the index number approach provided a more conservative estimate for rice (Evenson, however, includes Latin America in his computation). The total values were not greatly different, and both estimates easily exceeded $1 billion.

Just as Evenson has done, I could present estimates on production increase and value for the previous years (Table 7-4). But since the yield ratio between HYV's and traditional varieties has changed over time and has generally declined, it might be appropriate to use different yield assumptions for past years. And perhaps the effect of some lower ratios (such as 1.20 for rice) should also be calculated.

The yield advantage may, of course, vary from season to season if there are widespread changes in the weather. It may be significantly reduced where, as has been the case recently, fertilizer supplies are scarce and prices high. On the other hand, lower yields may be offset by higher grain prices in calculating gross returns.

The index number procedure outlined here seems a promising initial measure of the effects of the HYV package. It is simple and flexible. It is reasonable in its data requirements. It can make use of production function analysis. It does not require any arcane skills (or computation equipment).

But these factors may also be its weakness. It is only an introductory process. To be at least reasonably accurate, it requires a more systematic and thorough evaluation of the yield ratios between the HYV package and the traditional practices than we have at present for many areas. And even then, as is typical of the index number approach, it does not separate the precise effect of the HYV's themselves from other factors influencing productivity. Additional production function analyses could be most helpful in resolving these points.

There are several further steps which should be taken to complete the index number analytical package. These include, as noted earlier in this chapter, estimated research costs as well as the calculation of social rates of return. The procedure for the rate of return computations has been well demonstrated by Griliches, Evenson, Ardito Barletta, Ayer and Schuh, Hertford, Ardila, Rocha, and Trujillo, Hayami and Akino, and others cited in this chapter.

This study will not detail these further steps. However, it should be recalled that the total annual investment in wheat and rice research at the international institutes in 1975 was probably no more than $10 million. The counterpart national investment is not known, but if it was approximately the same, the total research investment was still relatively small. It would appear even smaller if a lag effect were added, and the 1972/73 crop value figures linked to the research investment of several years before.[74] In comparison, the increased value of production was somewhere on the order of $1 billion. Thus the returns to investment are probably very high.

In any case, it is important to remember (as suggested earlier), that only part of the benefits are being evaluated. Even when direct effects have been evaluated, the potential influence of the HYV's in communist nations and in developed nations has not been considered.[75] And the expanded base that the improved varieties provide for future improvements has not been valued. Much remains to be measured.

More sophisticated analysis of the direct and indirect effects of the international institutes on crop production must await further study. It will not be an easy task, but the integrated use of production functions and the index number approach can help to provide a more complete evaluation of these effects.

Conclusion

This chapter has outlined the main conceptual and empirical considerations in evaluating the impact of international agricultural research on crop production in developing nations. The process has been applied to high-yielding varieties of wheat and rice.

The task of evaluation is complex. Although the immediate research product can be readily identified, there are many problems involved in linking this product to actual changes in production in the farmers' fields. Moreover, the HYV package may have a number of indirect and qualitative results in addition to the direct and quantitative effects.

This study, after reviewing all these considerations, focused on only one measure: the direct quantitative effect. Changes in area and yield were first examined. This was followed by an analysis of the effect of the HYV's on yield, using production function and index number techniques. Even this relatively narrow focus encountered a number of analytical difficulties. Some can be solved by using the techniques in combination rather than separately as in the past. Others are more intractable.

Despite these problems, the task is not an impossible one. Crude measures or approximations have been made, and it is certainly possible to make further improvements in evaluation. But to do so will require improved data and analytical techniques. Whether these will be forthcoming will in part depend on the need for improved analysis.

For the moment, the accomplishments of the early centers are well known. They have produced striking technologies whose worth is readily understood. Past studies have shown that investment in research yields high returns. And indeed this preliminary study, while not carried through to the point of calculating an actual cost-benefit ratio, suggests that the returns to international research in wheat and rice must have been very high. Perhaps these findings will be adequate for the near future.

At some point, however, it is likely that more quantitative evidence will be requested. Of all aid recipients, a research organization should be in a good position to provide some measure of its worth. It should be realized that these measures cannot be turned out overnight. Appropriate data must be available. Where data are not available arrangements must be made well in advance for their gathering and assembly. And analytical techniques must be tailored to the job at hand.

Financial resources will be needed to carry out these tasks. Perhaps one or more of the members of the Consultative Group will provide funds for this purpose in the future. Should support become available, the research could be administered in a variety of ways. The newly established International Food Policy Research Institute might play a role in this process (though this institute is not sponsored by the CG at present).

In pursuing a more precise estimate of the effects of technologies, we have recognized several key points. First, the measurement problems, as indicated, are severe. Sponsors need to have some understanding of what can and cannot be readily measured. Second, some research activities might show considerably less quantitative effect than others. Such results might not always be well received, but they ought to be known if resources are to be allocated most effectively.

It should be realized, of course, that quantitative techniques cannot measure everything. Some research programs can be justified on other grounds. And social goals beyond productivity should certainly be considered. Rural equity issues, for example, are becoming increasingly important in the planning process.

The evaluation task, therefore, is broad and challenging. But an enlightened and effective program of international agricultural research requires research on the system itself. It is time to consider a modest but enduring organizational mechanism that can carry out the job.

NOTES

1. The original manuscript of this chapter was titled "Impact of the International Institutes on Crop Production." Following the Airlie House conference, it was revised and published as *Measuring the Green Revolution: The Impact of Research on Wheat and Rice Production*, Foreign Agricultural Economic Report no. 106 (Washington, D.C.: Economic Research Service, USDA, in cooperation with the United States Agency for International Development, 1975). The material presented here is an abridged and slightly revised version of the published bulletin.

2. The origins and dimensions of this work are well reported in E. C. Stakman, Richard Bradfield, and P. C. Mangelsdorf, *Campaigns against Hunger* (Cambridge, Mass.: Belknap Press of Harvard University Press, 1967); and Lennard Bickel, *Facing Starvation: Norman Borlaug and the Fight against Hunger* (New York: Readers Digest Press, 1974).

3. Robert Evenson, "Investment in Agricultural Research: A Survey Paper," prepared for the Consultative Group on International Agricultural Research (October 1973), p. 3.

4. *Ibid*., p. 3.

5. See Tables 1-1 and 1-2 in this volume.

6. Robert Evenson (with Y. Kislev), "Research and Productivity in Wheat and Maize," *Journal of Political Economy*, 81 (November-December 1973), 1309-1329; Robert Evenson, "International Diffusion of Agrarian Technology," *Journal of Economic History*, 34 (March 1974), 51-73; and Robert Evenson, "The Green Revolution in Recent Development Experience," *American Journal of Agricultural Economics*, 56:2 (May 1974), 387-394. Also, Robert E. Evenson and Yoav Kislev, *Agricultural Research and Productivity* (New Haven: Yale University Press, 1975).

7. Dana G. Dalrymple, *Development and Spread of High-Yielding Varieties of Wheat and Rice in the Less Developed Nations*, Foreign Agricultural Economic Report no. 95 (Washington, D.C.: Economic Research Service, USDA, 1974).

8. A listing of some of the more important works is provided in Dalrymple, *High-Yielding Varieties*, p. 2, footnotes 2 and 3. The following more recent studies might also be added: Keith Griffin, *The Political Economy of Agrarian Change: An Essay on the Green Revolution* (Cambridge, Mass.: Harvard University Press, 1974), and *The Social and Economic Implications of Large-Scale Introduction of New Varieties of Foodgrain* (Geneva: United Nations Research Institute for Social Development, 1974).

9. Evenson places the mean time-lag between expenditures on research and effect on production in the United States at about six and a half years (Evenson, "Investment in Agricultural Research," p. 18).

10. Corn, in many ways, is a more difficult plant to work with. For a discussion of the main problems, see Delbert T. Myren, "The Rockefeller Foundation Program in Corn and Wheat in Mexico," *Subsistence Agriculture and Economic Development*, ed. Clifton R. Wharton, Jr. (Chicago: Aldine, 1969), pp. 438-452.

11. Stakman, Bradfield, and Mangelsdorf, *Campaigns against Hunger*, pp. 5, 12, and 273.

12. *1966-67 Report, CIMMYT*, p. 9.

13. See "CIMMYT's New Headquarters at El Batan," *CIMMYT Report*, 1:1-6 (November-December 1972), 1.

14. From a letter from Robert D. Osler, deputy director general and treasurer, CIMMYT, September 11, 1974.

15. Robert F. Chandler, "IRRI – The First Decade," *Rice, Science and Man*, papers presented at the Tenth Anniversary Celebration of the International Rice Research Institute, Los Baños, April 20 and 21, 1972, pp. 5-7.

16. Unpublished table provided by Randolph Barker, November 29, 1973.

17. For the six centers in 1975, about 46 percent of the proposed core budget would actually go to research. Of the total proposed budget for the six centers in 1975, 27.7 percent would be allocated to wheat and rice. "Draft Integrative Paper," Consultative Group on International Agricultural Research (July 24, 1974), p. 4, annex A.

18. Computed from data provided by Nicolas Ardito Barletta, "Costs and Social Benefits of Agricultural Research in Mexico," Ph.D. dissertation (Chicago: University of Chicago, 1971), p. 74.

19. Lowell S. Hardin and Norman R. Collins, "International Agricultural Research: Organizing Themes and Issues," *Agricultural Administration*, 1 (1974), 14.

20. See particularly chapters 12, 13, 14, and 15 in this volume.

21. Surjit Sidhu, "Economics of Technical Change in Wheat Production in the Indian Punjab," *American Journal of Agricultural Economics*, 56 (May 1974), 217-226. On the profitability of HYV rice in six Asian nations, see Randolph Barker, "Changes in Rice Farming in Selected Areas of Asia: Some Preliminary Observations," mimeographed (Los Baños: International Rice Research Institute, February 15, 1973), p. 7.

22. For further details on the matters discussed here, see Dalrymple, *High-Yielding Varieties*, pp. 9-20.

23. Stakman, Bradfield, and Mangelsdorf, *Campaigns against Hunger*, pp. 74-88.

24. *CIMMYT Review, 1974*, p. 7. (The three types are stem rust, leaf rust, and stripe rust.)

25. For more detailed discussion of maintenance research in the United States context, see Peterson and Fitzharris, chapter 3 in this volume.

26. For further and more sophisticated discussion of this matter, see Richard E. Just, "Risk Aversion under Profit Maximazation," *American Journal of Agricultural Economics*, 57 (May 1975), 347-352.

27. Dana G. Dalrymple, "The Green Revolution and Protein Levels in Grain," Unpublished manuscript, USDA, Economic Research Service, Foreign Development Division (Washington, D.C., May 5, 1972). See, for example, *IRRI Research Highlights for 1973*, pp. 22-24.

28. Gelia T. Castillo, "Diversity in Unity: The Social Components of Changes in Rice Farming in Asian Villages," *Changes in Rice Farming in Selected Areas of Asia* (Los Baños: International Rice Research Institute, 1975), p. 349.

29. See Dana G. Dalrymple, *Survey of Multiple Cropping in Less Developed Nations*, Foreign Agricultural Economic Report no. 91 (Washington, D.C.: USDA, 1971).

30. Ian R. Wills, "Projections of Effects of Modern Inputs on Agricultural Income and Employment in a Community Development Block, Uttar Pradesh, India," *American Journal of Agricultural Economics*, 54 (August 1972), 457-458.

31. *CIMMYT Review, 1974*, p. 7.

32. See Dalrymple, *High-Yielding Varieties*, pp. 17-21.

33. See Robert Evenson, "Consequences of the Green Revolution," Department of

Economics, mimeographed (New Haven: Yale University, July 1974), p. 387; Dalrymple, *High-Yielding Varieties*, pp. 9-20.

34. *IRRI Research Highlights for 1973*, p. 46.

35. Dalrymple, *High-Yielding Varieties*, pp. 71, 72.

36. *IRRI Research Highlights for 1973*, p. 45. The data may overstate the importance of insects and diseases in the Philippines as a whole (letter from Robert W. Herdt, agricultural economist, IRRI, September 30, 1974).

37. William Jones (of World Bank) and I have reviewed these categories in greater detail in "Evaluating the 'Green Revolution,'" processed draft, USAID, Bureau for Program Policy and Coordination (Washington, D.C., June 18, 1973), pp. 33-37.

38. *IRRI Research Highlights for 1973*, p. 45.

39. A summary of some of these studies and factors is provided in Dalrymple and Jones, "Evaluating the 'Green Revolution,'" pp. 37-39; and Dana G. Dalrymple, "The Green Revolution: Past and Prospects," processed draft, USAID, Bureau for Program Policy and Coordination (Washington, D.C., July 22, 1974), pp. 13-16, 45-47.

40. For a summary of available information, see Dalrymple, *High-Yielding Varieties*, pp. 73-77.

41. Based on review of statistics compiled by John Parker, Economic Research Service, USDA.

42. Carl C. Malone, *Indian Agriculture: Progress in Production and Equity* (New Delhi: Ford Foundation, 1974), p. 99, Table 20.

43. Sidhu, "Technical Change," p. 221.

44. Sheldon K. Tsu, *High-Yielding Varieties of Wheat in Developing Nations*, Economic Research Service-Foreign 322 (Washington, D.C.: USDA, 1971).

45. Based on statistics compiled by John Parker, Economic Research Service, USDA, May 20, 1974.

46. Mahar Mangahas and Aida R. Librero, "The High-Yield Varieties of Rice in the Philippines: A Perspective," Discussion Paper no. 73-11, School of Economics, Institute of Economic Development and Research (Los Baños: University of the Philippines, June 15, 1973), p. 23.

47. Calculated from Teresa Anden and Randolph Barker, "Changes in Rice Farming in Selected Areas of Asia," mimeographed (Los Baños: International Rice Research Institute, December 1, 1973), Table 8.

48. L. J. Atkinson and David Kunkel, "HYV in the Philippines: Progress of the Seed Fertilizer Revolution," unpublished manuscript, USDA Economic Research Service, Foreign Development Division (Washington, D.C., December 10, 1974), Appendix Table 1. Other computational variations are presented in the appendix and discussed in the text, pp. 5-7.

49. Further detail on the latter point is provided in Dalrymple, *High-Yielding Varieties*, pp. 9-15.

50. These approaches are introduced and described by Willis L. Peterson, "Return to Poultry Research in the United States," *Journal of Farm Economics*, 49 (August 1967), 653-669, and Per Pinstrup-Andersen, "Toward a Workable Management Tool for Research Allocation in Applied Agricultural Research in Developing Countries," International Center for Tropical Agriculture, mimeographed (Palmira, Colombia, June 1974), pp. 3-6.

51. Evenson, "Green Revolution in Recent Development Experience," p. 388.

52. As of early 1975, these indexes were being used by Shyamal Roy and Fred

Sanderson in a study at the Brookings Institution. The indexes for 1951/52 to 1968/69 and the methodology used in their calculation are provided in R. W. Cummings, Jr., and S. K. Ray, "1968-69 Foodgrain Production: Relative Contribution of Weather and New Technology," *Economic and Political Weekly* (New Delhi, March 29, 1969).

53. Evenson, "Green Revolution in Recent Development Experience," pp. 387-394. Also see chapter 9 in this volume.

54. Evenson, "Green Revolution in Recent Development Experience," p. 393.

55. Evenson, "Consequences of the Green Revolution," p. 13.

56. The first set was reported in Evenson, "Green Revolution in Recent Development Experience," p. 393. The second set was reported in Evenson, "Consequences of the Green Revolution," p. 14, and in the preliminary version of his Airlie House paper; these data were also summarized in Dalrymple, *Measuring the Green Revolution*, p. 31, Table 6. The main difference between the second stage and the third stage in Table 7-4 was the use of geometric rather than arithmetic means and the use of somewhat more accurate production costs.

57. Sidhu, "Technical Change," pp. 217-226.

58. Letter from Surjit Sidhu, University of Dar es Salaam, Tanzania, October 2, 1974.

59. Sidhu, "Technical Change," p. 219.

60. Surjit Sidhu, "Relative Efficiency in Wheat Production in the Indian Punjab," *American Economic Review*, 64 (September 1974), 743-744.

61. Letter from Surjit Sidhu, University of Dar es Salaam, Tanzania, November 12, 1974.

62. The comparative analysis of old versus new varieties was carried out only for 1967/68 "because during the subsequent years, the number of farms growing old wheat and the area planted to it were substantially reduced." Sidhu, "Technical Change," p. 217.

63. *Ibid*,. pp. 222-223.

64. *Ibid*., p. 223, n. 11. Some other references to seed quality are summarized in Dalrymple, *High-Yielding Varieties*, pp. 32-34.

65. The seminal application was by Zvi Griliches in his study on hybrid corn. This work was reported in several journals; here I refer to "Research Costs and Social Returns: Hybrid Corn and Related Innovations," *Journal of Political Economy*, 66 (October 1958), 418-431. Major studies of the LDC's have included Nicolas Ardito Barletta, "Agricultural Research in Mexico"; Harry W. Ayer and G. Edward Schuh, "Social Rates of Return and Other Aspects of Agricultural Research: The Case of Cotton Research in São Paulo, Brazil," *American Journal of Agricultural Economics*, 54 (November 1972), 557-569; chapter 4 in this volume; chapter 2 in this volume.

66. This formulation is taken from Pinstrup-Andersen, "Research Allocation," pp. 3-6.

67. Griliches, "Research Costs and Social Returns," pp. 419-431.

68. Ardito Barletta, "Agricultural Research in Mexico," pp. 79-89.

69. Griliches, "Research Costs and Social Returns," pp. 419-431.

70. Ayer and Schuh, "Social Rates of Return," p. 561.

71. "A Note on the Use of Commodity-Based Studies in Estimating the Pay Off to Investment in Research," Ministry of Overseas Development, Statistics Division, London, September 1974.

72. See, for example, Anden and Barker, "Changes in Rice Farming," pp. 6-7.

73. This matter is discussed by Jones in Dalrymple and Jones, "Evaluating the 'Green Revolution,'" pp. 15-31. Evaluation of returns to research must give considerably more attention to the price effect.

74. Recall, from n. 9, Evenson's use of a lag figure of six and a half years in the United States. The interval would probably be even greater in the LDC's.

75. A study of the influence of the HYV's in Israel, for instance, was recently completed. It suggested that the influence of the first imports was minimal but that they did become significant when crossed with local varieties. See Yoav Kislev and Michael Hoffman, "Research and Productivity of Wheat in Israel," Center for Agricultural Economic Research (Rehovot: Hebrew University, February 1975).

8

Cycles in Research Productivity in Sugarcane, Wheat, and Rice

Robert E. Evenson

The recent green revolution in wheat and rice production is not unique in the history of the improvement of agricultural productivity. We can identify a number of similar episodes in which a distinct cycle of productivity gains has occurred, attended by an associated pattern of interregional and international diffusion of the primary technology. The improvement of winter wheat in the 1920s, of European alfalfa varieties in the 1930s and 1940s, and of spring wheats and barley varieties in the 1940s and 1950s in the United States are cases in point. This chapter discusses major productivity sequences in sugarcane. It also briefly considers rice and wheat production and attempts to identify some of the elements common to the development of research in all three crops.

In the first section the stylized cycle of productivity development is discussed. The sections to follow provide a historical treatment of the sugarcane productivity cycle and a brief history of the rice and wheat cycles. An attempt is then made to set forth at least a partial theoretical framework capable of providing an explanation for the cyclical phenomenon and the associated diffusion pattern.

The Stylized Cycle

The salient features of the cycle are as follows.

At first, a state of relative or "quasi"-technology equilibrium exists. It may be based on strictly traditional technology or on a relatively stationary or stagnant phase in the improvement of technology. (Technology can refer to crop varieties, agronomic techniques, mechanical implements, or other aspects of production. Similarly, commodity quality characteristics can be considered part of technology.)

New technology (or a set of closely related technologies) which has a high degree of superiority over the initial technology is then discovered. This discovery may itself be described as having occurred over a long interval, but the final development of the technology occurs in a short period of time.

In the next stage, improvements to the new technology are made, but at a diminishing rate over time. Even with increased research, the incremental rate of improvement declines over time and approaches zero (in some cases it may become negative, as crop varieties, for example, become susceptible to insect and disease problems).

Whereas in the initial quasi equilibrium a wide range of technologies are utilized by producers who confront varying environmental and economic conditions, the discovery of a significant new technology leads, in the next stage of our cycle, to diffusion of that technology directly from the region of its origin to other, similar regions.

A considerable amount of screening and testing, sometimes necessitating sophisticated equipment and skills, is then required to enable efficient diffusion of the new technology. Even with perfect information, however, the direct diffusion of technology is ultimately limited by environmental and economic conditions.

In the next stage the new technology begins to provide incentives for indirect diffusion through "adaptive" research. This can be considered to be a diffusion of technical and engineering *knowledge* as opposed to the diffusion of technology. Adaptive research extends the geographical impact of the technology by tailoring it to specific environmental conditions. This research, like the originating research program, is subject to diminishing returns.

Finally, a new quasi equilibrium with characteristics similar to the original equilibrium is reached. Numerous forms of the basic technology are in use, and the initiating technology improvement occurs in all regions. This stylized cycle does not necessarily hold for all types of technology improvement sequences. In fact, as will be noted in a later section, efficient technology improvement should generally be a relatively smooth process through time. Much of the realized productivity gain from technology improvements in modern agriculture does not exhibit marked cyclical activity. Part of this is more apparent than real, however, because we tend to observe aggregate blocks of technology. Aggregation of cyclical series can easily mask the cycles.

In a later section, we will develop a rationale for expecting greater cyclical activity in commodities where production is based on more primitive or traditional technology and in regions where relatively little research capability exists.

Sugarcane Varietal Improvement[1]

The several stages in the stylized productivity cycle can be clearly discerned from the history of sugarcane development.

The Initial Quasi Equilibrium

From the sixth century to the seventeenth century a single variety of sugarcane, the "Creole" (a hybrid with sterile flowers and thus incapable of sexual reproduction), was produced throughout the world. During the seventeenth century a second variety, the "Bourbon" or "Otaheite" cane, was discovered on the island of Tahiti in the Pacific and later introduced to all cane-growing areas of the world. It proved to be superior to the Creole variety and eventually replaced it as the dominant cane in most producing countries. It is of interest to note that it was not introduced to the British West Indies, a major cane-producing area, until 1785-86, more than a hundred years after it was first known to have been commercially produced in Madagascar and on Bourbon (or Réunion) Island. Produced under a variety of names (Lahania, Vellai, etc.), it dominated world production until it became subject to disease in 1840 in Mauritius, in 1860 in Puerto Rico, in the 1890s in the British West Indies, and in the early twentieth century in Hawaii.

A third major set of wild canes, the "Batavian" canes, were discovered in Java about 1782. These canes were eventually produced in many countries (for example, as the "Crystalina" in Cuba, "Rose Bamboo" in Hawaii, and the "Transparent" canes in the British West Indies) but were not always superior to the Bourbon cane. After the disease epidemics in the Bourbon cane, the Batavian varieties became dominant. However, they were later subject to the Sereh disease in many parts of the world. Other wild varieties were discovered in the late 1800s, including the "Tanas" from New Hebrides, "Badila" from New Guinea, and "Uba," probably from India. Badila and Uba became important varieties because of their resistance to the cane diseases which became increasingly prevalent from 1890 to 1925.[2]

The First Major Cycle

The original sugarcane varieties undoubtedly arose as seedlings from rare cases of natural sexual reproduction. Several reports of seedling growth were made after 1858,[3] but it was not until the 1887-88 cane-growing season – when the fertility of the cane plant was firmly established – that a basis for the deliberate use of seedlings for producing new varieties existed. In the early part of that crop year Soltwedel in the Proefstatien Oost Java (POJ, the experiment station in Java which later became the world's leading producer of important varieties) demonstrated that the sugarcane plant could produce seedlings. Later that same year Harrison and Bovell in the newly established experiment station in Barbados, British West Indies, independently made the same discovery. The researchers at both stations recognized that each individual seedling could be grown and allowed to reproduce asexually, thus creating an entirely new variety having the same genetic characteristics as the seedling.

The inducement of flowering in the cane plant depends on temperature and light control. Thus, the production of seedlings was difficult. Only a few experiment stations, including the two pioneer stations in Barbados and Java, were able to establish breeding programs before 1900. The stations in Barbados, Java, and British Guiana (where Harrison made his home shortly after his discovery of cane fertility in Barbados) had produced new varieties which were of commercial importance by that date. The stations in Hawaii, Mauritius, and Réunion produced commercial varieties shortly thereafter.[4] The Indian station at Coimbatore, which later assumed major importance, did not release its first variety until 1912.

These early "noble" sugarcane varieties were all members of the eighty-chromosome species *Saccharum officinarum*. Breeding methods were relatively simple, although as breeders gained experience advances were made. A certain amount of adaptive research appears to have been undertaken during this first cycle, as a number of new experiment stations initiated breeding programs. Bovell, for one, developed in 1900 a breeding program of "selfing" which, by inbreeding, identified the characteristics of progeny of specific varieties and thereby determined their value as breeding stock.[5] By 1910 or so most of the important varieties of this cycle had been developed.

This first cycle resulted in considerable international diffusion, for many countries introduced the noble varieties which initially outyielded the native varieties. However, in a great many cases (e.g., South Africa), this initial superiority was not maintained. A number of serious disease epidemics among the new varieties eliminated their advantage over the disease-resistant native varieties, and considerable retrenchment of the initial diffusion occurred.

The Second Cycle: Interspecific Hybridization

Cane breeding achieved a major advance with the introduction of additional cane species to the breeding program. The term *nobilization* was used to describe the breeding work in Java which sought to improve the wild species of cane (hardy and disease-resistant, but otherwise inferior) by succes-

Table 8-1. Varietal Production of Various Sugarcane Experiment Stations, 1940-64 (in million metric tons)

Experiment Station	Varieties in Production		Variety Parents		Variety Grandparents	
	Production[a]	Rank	Production[a]	Rank	Production[a]	Rank
Coimbatore, India	64.7	1	75.4	2	53.8	3
Java (POJ)	63.4	2	102.3	1	113.6	1
Hawaii	24.9	3	18.1	4	16.8	4
Cuba	20.4	4				
Barbados, B.W.I.	10.8	5	18.8	3	59.4	2
Canal Point, Florida	10.3	6	4.5	6	4.2	7
Queensland	9.1	7	3.3	7		
South Africa	7.3	8	0.3	10	0.3	9
Taiwan	4.2	9				
Mauritius	4.2	10	1.7	9	6.0	6
Brazil	3.9	11	1.8	8	1.8	8
British Honduras	3.9	12				
Puerto Rico	3.8	13				
Peru	2.2	14				
British Guiana	0.2	15	8.5	5	12.2	5

Source: *Yearbook of Agriculture*, USDA (Washington, D.C.: Government Printing Office, 1936), pp. 561-624; *Proceedings of the Twelfth Congress*, International Society of Sugarcane Technologists, New York, 1967, pp. 844-854; *Agricultural Statistics*, USDA (Washington, D.C.: Government Printing Office), various issues.

[a] Total production of 96 percent sugar, 1940-64 (million of tons).

sive crossing and backcrossing with the noble canes. The breeders in Java introduced the species *Saccharum spontaneum* (chiefly a wild variety, "Kassoer") to their breeding program, obtaining important results by 1920. In 1921 the variety POJ 2878 was produced by this program. It proved to be both disease-resistant and high-yielding. More than 50,000 acres were planted to this variety in Java alone by 1926. By 1929, 400,000 acres were in production, with an estimated 30 percent yield increase owing to this single variety. It later was planted in every producing country in the world.

The Coimbatore Experiment Station in India developed a series of trihybrid canes (the CO varieties) by using the noble *S. officinarum* and the vigorous *S. spontaneum* species and introducing a third species, *S. barberi*. The *S. barberi* canes were local varieties which possessed characteristics that afforded adaptability of the resultant new varieties to environmental and economic conditions. The CO and POJ varieties were eventually diffused to almost every producing country (see Table 8-1).

The expansion of the genetic base for varietal discovery was a very im-

Table 8-2. Percentage of Total Sugarcane Acreage Planted to Varieties Developed by Experiment Stations of Selected Countries, 1930-65

Region	1930	1940	1945	1950	1955	1960	1965
Australia	20	20	33	54	83	85	85
Hawaii	50	65	82	100	100	100	100
South Africa. . .	0	0	0	3	49	78	n.a.
Taiwan	0	32	46	56	10	4	42
Puerto Rico . . .	0	9	12	10	3	35	50
Mauritius.	0	8	53	98	93	78	n.a.
Louisiana.	0	23	52	77	65	65	n.a.

Source: *Annual Report*, Bureau of Sugar Experiment Stations, Queensland, Australia, various issues, 1928-64; *Proceedings of the Twelfth Congress*, International Society of Sugarcane Technologists, New York, 1967, pp. 867, 1041; *Culture of Sugar Cane for Sugar Production in Louisiana*, USDA Agricultural Handbook 262, Washington, D.C., 1964.

portant feature of this second cycle. It not only resulted in slower rates of diminution, or "exhaustion," of the technology potential in each country, but it broadened the scope for direct varietal diffusion. In addition, it formed the basis for effective adaptive research programs utilizing local native species. The second cycle thus developed into a full cycle in the context of the stylized cycle.

During the 1920s and 1930s virtually all sugarcane-producing countries established experiment stations as they recognized the potential gains to be had from (1) the varietal screening activity to facilitate direct transfer and (2) adaptive breeding programs. These "second cycle" research programs generally began to release adapted varieties in the late 1930s, but their major contributions were seen in the varietal releases of the 1940s and the early 1950s. Of course, the first cycle stations continued in the second cycle to be the major technology discovery institutions.

Interestingly, almost all the second cycle experiment stations were successful in coming up with adapted varieties even though a large number of improved varieties in the international market were readily available. This near unanimous success is partly reflected in Table 8-2, which shows the increasing percentage of "home-grown" varieties planted in several countries. Of the countries included in the table, South Africa, Taiwan, and Puerto Rico can be roughly categorized as second cycle stations. The remainder were active in the first cycle.

The Third Cycle: The Modern Experiment Station

In a somewhat crude sense, it is possible to identify a third cycle associated with the development of modern research programs based on sophisticated scientific knowledge. These systems are undertaking work in genetics, physiology, and related fields as well as in plant breeding and agronomy. Improved experimental design and screening methods are utilized. These research systems are adapting their technology discovery effort not only to environmental and changing economic conditions (for example, development of machine harvesting technology — a major adaptive economic response) but to advances in scientific knowledge as well.

Given the relatively long lag between the conduct of research and realized productivity gains, this third cycle has been important only in the 1960s and 1970s. It represents an important institutional development, however, which will be touched upon later.

International Diffusion and Yield Patterns

As noted earlier, the direct diffusion of the first cycle varieties, while extensive, was limited by the susceptibility of the noble varieties to disease. The second cycle was characterized by very extensive international varietal diffusion. A rough picture of the extent of this diffusion can be obtained from Table 8-1, which summarizes international production of sugar (from cane) for the period 1940-64 in terms of the experiment stations which developed the varieties grown. In addition, a computation of the origin of parent and grandparent varieties by experiment station is reported. The production figures show that the experiment station in Coimbatore, India, had produced varieties which accounted for almost 28 percent of the world's sugarcane production even though India accounted for only 8 percent of the world sugar production during the period.

Indonesia (Java) also produced varieties accounting for roughly 28 percent of the world's sugarcane production while producing only 2 percent of the world's sugar from cane. Cuba, on the other hand, produced 22 percent of the world's cane sugar but Cuban varieties accounted for only 8½ percent of the world's production.

The bulk of the varieties in production during this period were second cycle varieties. Some first cycle varieties were still in production in a few countries, however. If data were available for a later period, say the early 1970s, a number of third cycle varieties would be present, but most production in the world in this period would be the result of adapted second cycle varieties.

The table provides some evidence for the pattern of adaptive research. The dominant parental and grandparental role of the POJ varieties from Java identifies these as key originator varieties. Likewise, the Barbados grandparent varieties were important source material for adaptation.

Table 8-3 presents historical production and yield data for sugarcane for major producing countries. These data, it should be noted, are subject to

Table 8-3. National Sugarcane Yield and Production Averages for Selected Five-Year Periods

Area	Production[a]	Yield[b]							
		1910-14	1923-24	1928-32	1938-42	1948-52	1958-62	1963-67	1968-72
Brazil	5,329	n.a.	n.a.	n.a.	17.1	17.4	18.8	19.8	21.0
Cuba	4,950	14.5	19.3	18.6	17.2	17.0	17.0	16.2	21.0
India	4,515	11.3	11.0	12.4	11.5	13.1	15.3	20.8	21.7
Mexico	2,319	n.a.	30.2	20.5	22.4	23.1	26.4	27.0	28.6
Australia	1,643	17.3	16.8	16.9	20.3	23.6	27.5	36.2	36.2
Philippines	1,584	n.a.	n.a.	20.4	22.6	20.3	26.7	19.3	20.3
Argentina	1,422	11.6	13.2	13.6	13.4	14.9	17.0	23.0	21.7
U.S. (Hawaii)	1,275	40.7	43.3	60.1	65.1	76.4	90.0	98.7	97.0
U.S. (Louisiana, Florida)	1,104	15.8	9.4	15.0	19.3	19.5	24.5	25.3	26.0
Taiwan	1,100	11.8	16.1	29.3	n.a.	27.3	33.7	38.6	37.2
South Africa	1,002	n.a.	8.8	20.5	26.3	25.1	35.3	35.8	37.5
Puerto Rico	897	n.a.	16.6	25.3	32.4	29.8	30.4	30.6	31.5
Peru	882	22.4	24.3	40.5	52.6	60.0	70.4	64.2	79.6
Indonesia (Java)	854	41.2	46.5	56.4	61.5	40.0	49.5	39.1	34.5
British West Indies	801	n.a.	9.6	24.0	17.6	38.1	39.4	36.4	33.7
Dominican Republic	800	n.a.	n.a.	n.a.	19.0	n.a.	19.5	22.5	27.5
Mauritius	732	15.6	14.5	15.2	19.8	n.a.	24.6	24.0	30.6
Egypt	465	18.8	n.a.	35.0	n.a.	32.8	42.6	40.0	41.9
Mean yields		20.09	19.97	26.48	25.17	29.90	33.81	34.31	35.97
Coefficient of variation		.503	.763	.532	.679	.475	.562	.548	.550
Change in mean yield[c]		n.a.	1.35	6.34	.03	3.88	5.38	.50	1.66

Source: *Yearbook of Agriculture*, 1925-35, *Agricultural Statistics*, 1936-72, and *International Sugar Situation*, 1904, USDA, Washington; *Production Yearbook*, 1948-72, FAO, Rome; *Annual Report*, 1900-64, Bureau of Sugar Experiment Stations, Queensland, Australia; *South African Sugar Yearbook*, South African Sugar Journal, Durban, 1935, 1948-49, 1961-62.

[a] Average annual production in thousands of short tons of 96-degree sugar in 1963-67.

[b] In short tons of cane per acre per year.

[c] Based on common observations.

some error but, on the whole, they serve to identify major trends. The mean yield data and the coefficient of variation in yields provide a crude measure of the overall cyclical effects. The 1910-14 period is early in the first cycle. The noble varieties were beginning to be diffused from the originating countries. By 1923-24, the first cycle diffusion was well under way and by 1928-32 it had been completed. Note that the coefficient of variation of yields reflects the unequal rate of diffusion. It increases in the mid-cycle (1923-24) period and decreases at the end of the cycle.

The second cycle is actually divided into two phases. The 1938-42 period is roughly the mid-period of the direct diffusion of the second cycle varieties. By 1948-52 this direct diffusion was completed, but the adapted varieties were now beginning to increase production. By 1958-62, the adapted second cycle varieties had increased yields substantially. The third cycle is crudely reflected in the 1968-72 data which show increasing yields and yield variability.

This simple comparison of yields and yield variability over time is not intended to be a thorough analysis of productivity change. Yields are not ideal indexes of productivity, and the variability in yield levels internationally is obviously related to many factors besides the underlying technology diffusion pattern. The main purpose of these comparisons is to note the broad consistency of these data with the cycle interpretation of the historical varietal data.

The Role of the Experiment Station in Variety Development and Diffusion

The role of the sugarcane experiment station was not confined to the production of new varieties. It served in an important way to facilitate the international diffusion of the first and second cycle varieties. Experiment stations for sugarcane research were established in many countries where public support of general agricultural research was limited or nonexistent. The support for many stations came from organizations of private growers. The private growers were aware of the changes in the comparative advantage that new varieties (and related technology) would give them in the international market. They were also aware of the comparative disadvantage resulting from improved yields and lowered costs of production which other countries might realize. Although some first cycle cane breeding was undertaken by large private plantations in Hawaii, Cuba, and Java, it soon became clear that it was not profitable to make large investments in private effort because the plantation was unable to capture more than a small fraction of the benefits. It is also true, of course, that sugar producers will not capture the full benefits from improved varieties. In fact, most benefits are likely to be realized by consumers. A recognition of this has resulted in general public support for modern cane experiment stations.

The initial establishment of experiment stations was clearly based on the colonial interests of the British, Dutch, French, and Portuguese. They well understood the principle that Hertford and Schmitz have elaborated in chapter 6: that the gains from improved technology tend to be passed on to the consumers of the product. It was very much in their interests to support research stations in the sugar-producing colonies. From their point of view, these investments paid off quite handsomely.

The South African case is instructive in this regard. The sugar industry in South Africa was established in 1849. Before 1880 several wild varieties imported from Java, Mauritius, and India were cultivated. A wild variety, Uba, was introduced in 1883 and proved to be more disease-resistant than the other varieties. For a period of fifty years it was the only important variety grown.[6]

During this fifty-year period some experimentation was carried on by planters to find new varieties. A number of potentially important first cycle (and some second cycle) varieties actually existed and were widely planted in many countries. However, it was not until an experiment station, financed by the growers, was established at Mount Edgecumbe in 1925 that varieties from Java and India were introduced to the South African growers. The accomplishments of this station, from 1925 until 1945, were confined to the introduction of new disease-resistant second cycle varieties, mostly from Java and India.

The portion of the South African crop composed of these varieties rose from 3.3 percent in 1933-34 to 19.5 percent in 1942-43. An analysis of yield increases based on a direct comparison of Uba and non-Uba yields indicated that by 1945 the new varieties outyielded the Uba variety by a factor of 27 percent. The South Africa station released the first variety from its own breeding program in 1947 (N:Co:310). This variety was the result of cooperative effort with the Indian Experiment Station at Coimbatore. The actual crossing was completed in the Indian station, and the South African station conducted the growing and selection processes. The experiment stations in Australia (at Queensland), Taiwan, Mauritius, Puerto Rico, and several other countries were also instrumental in the testing and introduction of first and second cycle varieties from other countries into their local economies.[7]

It is possible then to distinguish between several different products of the experiment station systems. The discovery and development of the first cycle varieties were of immense value, as was the development of the basic second cycle varieties. In addition, however, the experiment station provided screening and other vital assistance to the basic diffusion of second cycle varieties.

Table 8-4. Variety Adoption Analysis: Twenty-one Diffused Varieties of Sugarcane[a]

Regression Number	Dependent Variable	Peak percentage (K)	Research (R/P)	Constant	R^2
1	Years from introduction to peak (N)	.127 (5.95)	– 124.0 (5.27)	7.20	.73
2	Average adoption rate (K/N)		112.6 (4.13)	1.80	.48

Source: The *1936 Yearbook of Agriculture* (U.S. Department of Agriculture, 1936) reports the results of a survey of sugarcane research stations throughout the world. The research variable was constructed from these data and from other information on the dates of establishment of stations in Queensland, Puerto Rico, and South Africa.

The countries and states to which varieties were diffused, the names of varieties, and the dates of their introduction are as follows: Puerto Rico: D109 (1910), D625 (1913), BH10 (1920), SC 12/4 (1922), POJ 2878 (1930), M336 (1944); Queensland, Australia: POJ 2878 (1933), Co. 290 (1937), CP 29/116 (1945); South Africa: Co. 231 (1934), Co. 331 (1938), Co. 310 (1936); Louisiana: POJ 2878 (1925), POJ 213 (1927), Co. 281 (1930), Co. 290 (1933); Cuba: POJ 2878 (1932), Co. 213 (1932), Co. 281 (1932); British Guiana: POJ 2878 (1934); Jamaica: POJ 2878 (1932).

Abbreviations:

N = Number of years from introduction to peak percentage (sample mean 9.05).

K = Peak percentage of variety (sample mean 34).

R = Number of senior researchers in recipient country (sample mean 5.92).

P = Production of sugar (at time of introduction) in thousand tons of crude sugar (sample mean 997).

[a] t values are given in parentheses.

At a somewhat later point, many stations contributed through adaptive research, and in the modern setting experiment stations are discovering new, third cycle type technology.

Some evidence is available regarding the contribution of research to a number of these dimensions. It should be interpreted in light of the underlying complexity of the technology discovery process. For example, estimated rates of research productivity should be considered as short-run rates and cannot be expected to remain constant. And, of course, the apparent value of research will depend on the stage of the cycle.

Consider first some evidence regarding the role of research stations in speeding up the adoption of second cycle varieties. Table 8-4 presents a simple regression analysis of the speed of adoption of twenty-one first and second cycle varieties in seven different countries. The analysis shows that the rate of adoption after initial introduction (defined as the date when 1 percent of the country's acreage was planted to the variety) is speeded up by research activity in the recipient countries.

Regression 1 indicates that when the peak adoption percentage is held constant, the number of years required to reach the peak percentage decreases in proportion to the research undertaken. In elasticity terms, it indicates that a 10 percent increase in research shortens time for adoption by 3 percent.

Regression 2 measures the relationship between the average adoption rate per year and research activity per unit of output. It is shown later that the economic value of speeding up adoption, even by a small percentage, is significant; this, however, requires an estimate of the relationship between the rate of introduction of new varieties and sugarcane production.

Yield levels by variety are reported annually for South Africa. This allows a direct comparison between yields of the old varieties and of the new set of varieties. Additional data on fertilizer use and on the age structure of the varieties are also available. The latter information is particularly important for this type of comparison. Cane is typically "ratooned" for several years. The first crop in a life cycle is produced from planted cane. Subsequent crops in the life cycles are simple regrowths of the same plants after cutting and are referred to as ratoon crops. Depending on climate and other factors, yields decline with each ratoon crop. In South Africa, for example, the index of relative yields by ratoon crops is as follows: plant, 109.8; one year, 108.5; two years, 100.5; three years, 91.0; four years, 86.7; and five years, 82.3. It eventually becomes profitable to plow the fields and replant the cane.

Consequently, simple comparisons of yields by variety, uncorrected for the age differences in the new and old varieties, are misleading. The actual yields in South Africa of the varieties introduced during the 1930s were 43 percent higher than the yield levels of the native variety Uba. This calculation was based on data for the 1935-39 period when both types of varieties were grown. The correction for the age distribution of the canes reduces this to a 27 percent advantage for the second cycle varieties. A comparable set of variety-yield-ratoon data for the period of 1954-57 allowed a comparison to be made of the advantage of adapted second cyle varieties over the original second cycle varieties. This was calculated to be 28 percent. (Note that all computations were made on the basis of sucrose yield, not cane yields. In fact, much of the advantage of the second cycle varieties over the first cycle varieties was in sucrose content.)

Varietal data reported for other countries are not as detailed as they are for South Africa, although data on variety composition of production are generally available. An analysis of yield changes in several countries with regard to the adoption periods of second cycle and adapted second cycle varieties is shown in Table 8-5.

Table 8-5. Yield-Variety Relationship Estimates[a]

Regression Number	Region	Period	Variety "turn-over" (DV)	Changes in Fertil-izers (DF)	Changes in Rain-fall (DR)	Constant	R^2
			Original Second Cycle Varieties				
1	Caribbean Islands	1935-46	.188 (2.21)	4.66 (2.07)	.156 (1.90)	− 3.17	.22
2	Australia	1936-45	.777 (2.72)	n.a.	n.a.	− 4.58	.56
3	South Africa	1933-44	.163 (1.99)	n.a.	.183 (1.87)		.15
			Adapted Second Cycle Varieties				
4	Australia	1945-58	.465 (1.94)	n.a.	n.a.	− 4.83	.24
5	South Africa	1945-62	.222 (1.90)	n.a.	n.a.	− 1.69	.18
6	India (Andhra Pradesh)	1954-61	.345 (1.43)	n.a.	n.a.	− .758	.33

Source: Caribbean Research Council, "Sugar Industry of the Caribbean," Washington, D.C., 1947; J. W. Suryandrayana and P. Sethuraman, "A Decade of Sugarcane Development in Andhra Pradesh," *Indian Journal of Sugarcane Research and Development*, 7:4 (1963); *Annual Report, Bureau of Sugar Experiment Stations*, Queensland, Australia, 1920-64; *South Africa Sugar Yearbook*, South African Sugar Association, Durban, 1934-62.

In all regressions but no. 3, variety change was the change in varietal mix for all varieties:

$$DY = A + b_1 D + b_2 DF + b_3 DR,$$

where

Y = yield
V = a measure of varietal composition based on area planted
F = application of fertilizers
R = rainfall
D = first difference operation, e.g.: $Dy = y_t - y_{t-1}$.

In regression no. 3, the variety variable is the percentage of non-Uba varieties in total production; it is not the percentage change from year to year.

[a] t values are given in parentheses.

The variety change from year to year may be regarded as a kind of measure of "turnover." It is essentially the sum of the changes in percentage of acreage planted to varieties which have increased (or decreased) their share of planted acreage; the results of the statistical analysis are presented in Table 8-5. Fertilizer and rainfall data were not available in all cases. The variety turnover variable has the expected positive sign, and the intercept term has a negative sign. A negative intercept term reflects two phenomena, a natural deterioration of yields owing to disease and other factors, and changing age distribution. As the South African data show, the younger the cane, the higher the yields. The varietal change measure may be correlated with sugarcane age in these data. A consequence of such a conclusion would be that the intercept term is biased downward and that the varietal change coefficient has an upward bias. Although major changes in average age do not appear to have occurred over the period, we do not have cane age distribution data that would allow us to check this point.

The data in Table 8-4 and 8-5 allow two calculations. The first is the economic value of the activities of experiment stations in speeding up second cycle varietal diffusion. From Table 8-4 we can compute the acceleration or increase in turnover achieved from an investment of $7 thousand (approximately the cost of adding a senior researcher to the staff in 1936). This increase would be approximately .11 percent per year. The value of an increase in variety turnover of .11 percent computed from Table 8-5 is approximately .03 tons per acre in Australia and .02 tons per acre in the Caribbean and South Africa. The total value of this acceleration, resulting from a $7 thousand investment, would be about $50 thousand in Australia and from $12 thousand to $15 thousand in South Africa and the Caribbean area. A reasonable downward adjustment for bias owing to changing age distribution would reduce these estimates by between one-third and one-half.[8] It appears that the "extension type" side benefits from accelerating second cycle varietal diffusion justified much of the investment in experiment stations during the long gestation period before they began producing adapted varieties.

The second computation is the value of adapted varietal output. Research cost data for 1950 can be used to compute the average cost of research per percentage of varietal turnover. These costs were $21 thousand in South Africa, $25 thousand in Australia, and $7 thousand in India. The estimates in Table 8-5 indicate that the varietal change produced by a one-dollar increase in research investment is worth roughly $15 in South Africa, $25 in Australia, and $35 in India (calculations based on estimated coefficients from Table 8-5 and a sugarcane price of $5.50 per ton). These estimates compare current research with current output value. The actual research that results in varietal change requires considerable time. For sugarcane research the average lag is probably at least eight years. Assuming no yield deterioration and an eight-year average lag between investment and benefit realization, the "internal" rates of return are approximately 40 percent in South Africa, 50 percent in Australia, and 60 percent in India. The South African data allow an adjustment for age bias based on a comparison of the Table 8-5 estimates and the comparative data by variety. A downward adjustment of 30 percent is indicated. This adjustment reduces the internal rates of return only slightly.

Rice Varietal Development

Our concern in this section will be with the cyclical nature of rice varietal improvement and with the characteristics shared by both the rice and sugarcane experiences. We draw heavily on the more detailed presentation by Dalrymple in chapter 7. It is useful to consider essentially two major rice economies, one located in the tropical climate zones of the world, the other in the more temperate Mediterranean and Marine climate zones. The two rice economies had independent histories until the 1960s. This independence is primarily the result of the high degree of sensitivity of the rice plant to soil and climate factors.

The temperate zone rice economy, centered in Japan, has a long history of varietal improvement and a consequent record of productivity gains. In some respects it parallels the sugarcane experience with its own cycles and diffusion patterns. The tropical rice zones in Asia's "rice bowl" have quite a different history. For the most part, this region was dominated by colonial relationships of varying types until after World War II. Investment in research directed toward rice improvement was much less intense than it was in the temperate rice zones. And the established research institutes were often designed to serve the interests of the colonial bureaucracy instead of a constituency of rice producers or consumers in the LDC's.

Thus for many years the two rice economies evolved along different paths. With the development thrust of the 1950s and 1960s, several of the new research programs had reached the point where they were capable of incorporating into their own breeding projects the basic scientific advances made in the temperate rice economy. They were slow to exploit the varieties actually produced in their research programs and thus missed the chance to lead the green revolution.

The International Rice Research Institute had a large initial advantage over the fledgling and in some cases bureaucracy-ridden national research programs. It was able to bring together a combination of high-quality intellectual capital (from the temperate zone experience) and to provide the intellectual and moral incentives to direct the work of this group. The resultant discovery of new varieties initiated a major technology cycle in the tropical rice economy.

Cycles in the Temperate Rice Economy

Hayami and Akino offer a useful discussion of productivity in Japanese agriculture (see chapter 2). Hayami and Yamada show that total factor productivity, which is based primarily on rice production, witnessed a distinct period of rapid growth after the Meiji Restoration until about 1920.[9] Then a period of little or no real growth for ten years or so occurred. This was followed by more rapid growth in the 1930s. Of course, post-World War II Japan has realized a further cycle of productivity gains. (Hayami and Akino compute a specific series on varietal improvement which also shows cycles. This is based on experiment station data, however, which generally has not proven to be an accurate index of real producer gains.)

As several studies of Japanese productivity gains reveal, the early (pre-1920) productivity cycle was based to a substantial degree on the *rōnō* (veteran farmer) rice varieties and production techniques. This is an especially interesting case of technology discovery by highly motivated and inventive farmers which did not depend on sophisticated scientific training. (Interestingly, United States agriculture experienced a slightly earlier productivity cycle based on mechanical invention of a similar nature.)[10] These studies also identify an exhaustion of these potential technology gains from the *rōnō* research methods as the primary factor in the slowdown in growth.

The experiment station system in Japan actually facilitated the diffusion of many of the techniques of production developed by the *rōnō*. In fact, it was primarily through the efforts of research institutions that the technology was diffused to Taiwan. (Again, see chapter 2 for a discussion of the role of the early experiment stations.)

The early experiment station system appears to have lacked the basic knowledge needed to compete with the highly motived and more inventive veteran farmers in the first cycle. By the 1930s, however, the experiment stations had acquired the scientific foundation to stimulate a second productivity cycle in rice production. This cycle was based on varieties and agronomic techniques discovered in the research system. Hayami and Akino (chapter 2) conclude that this second cycle was restricted by the diversion of fertilizer and other input resources for military purposes. They attribute the increased productivity of the experiment station system to an organizational change in which coordinated crop-breeding programs (the Assigned Experiment System) were developed. An alternative hypothesis for the basis for the improvement, which will be explored later, is that researchers acquired an improved understanding of the scientific aspects of crop production.[11]

The diffusion of the Japanese "Japonica" varieties to Taiwan and the subsequent development of the "Ponlai" varieties in Taiwan is an interesting case of the diffusion and adaptation of the Japanese technology. The simple introduction of the Japanese varieties during the 1910-25 period did not result in improved productivity or widespread direct diffusion. Some diffusion was implemented by altering the seedling age at transplanting and by screening for blast resistance, but it was not until the late 1920s that new varieties bred in Taiwan began to have an impact on production. These new ponlai varieties were then subject to a cycle of their own, which preceded the second Japanese cycle by a few years. Hayami and Ruttan[12] place considerable emphasis on the price-depressing effects of imported rice from Taiwan in slowing down and partly aborting the second rice cycle in Japan.

The diffusion of improved Japanese rice varieties to Taiwan was the forerunner of the more recent cycle in the tropical zone popularly termed the green revolution. In fact, efforts in Taiwan after World War II to breed shorter season varieties and more nitrogen-responsive varieties (the earliness was important in the two-crop rice pattern in Taiwan but not in the single-crop culture in Japan) essentially provided the genetic stock on which the green revolution was based.

Productivity in the Pre-Green Revolution Tropical Rice Economy

N. Parthasarathy, in his review of rice breeding in tropical Asia up to 1960, points out that most Asian countries had some form of breeding and selection work before World War II.[13] The program in India was especially large. It appears that these programs were reduced in effectiveness by breeding work being organized in small isolated stations and by breeding objectives which failed to assign highest priority to yielding ability. That is, breeders attempted to improve seed quality and disease resistance without having real yielding ability to work with. It also appears that breeding work during this period did not even consider fertilizer responsiveness (actual fertilizer use in the region was essentially nil).

After World War II, most Asian countries revitalized and developed expanded rice-breeding programs. An important project to hybridize the Japonica and Indica rice varieties was undertaken in a number of countries in the 1950s. These efforts did produce a number of significant varieties and probably led to some productivity gains. Certain varieties coming from this work are classified as "high-yielding varieties" in some regions today (e.g., Mahsuri which is an important variety in India, though it is actually a Malaysian variety).

In the 1950s and the early 1960s breeders in the tropics began to draw on the technology and scientific base of the more temperate zone experience. It seems incredible now that the great bulk of the plant breeders in tropical Asia missed the opportunity to exploit the fertilizer responsiveness of the plant. Breeders in the Philippines were, however, working on the shorter season, stiff-strawed, fertilizer-responsive varieties. (It is certainly worth noting that superior varieties, particularly C4-63, were produced in the Philippines at approximately the same time as the first IRRI variety.)

The post-World War II breeding programs in Taiwan had developed short-season semidwarf varieties to suit the multiple cropping technology of the area. This work had markedly reduced the photoperiod sensitivity of the plant. In 1956 the variety "Taichung Native 1" was developed and it has since been diffused to many tropical Asian countries. It became known as a high-yielding variety some ten years after its first release.

The IRRI-Induced Rice Cycle in Tropical Asia

The setting for a major advance in rice-yielding potential and international diffusion of rice varieties in tropical Asia could hardly have been more ideal. The temperate and subtropical research programs, especially in Taiwan, had developed the genetic materials and breeding methods required. The tropical rice programs were bound by tradition and had been slow to learn from the developments in Taiwan and slow to respond to the relative decline in fertilizer prices of the past two decades.

With the establishment of IRRI and the bringing together of a small group of scientists (from the temperate and subtropical zone experience) with the resources and fresh viewpoint of a new institution, it was inevitable that knowledge diffusion would occur. Of particular interest for the stylized cycle model is the apparent diminution in the yield-incremental discoveries by IRRI researchers after IR-8, the first IRRI variety. Essentially, no gain in real yielding ability over IR-8 has been forthcoming to date. The improvements since IR-8 have instead been in three areas: improved grain quality, improved disease resistance, and, most recently, improved insect resistance. Accompanying these varietal gains have been significant advances in related technology.

Another clear characteristic of the IRRI-induced cycle is the diffusion of varieties and genetic materials which has marked similarities with the second cycle in sugarcane productivity. The early diffusion of IR-8, IR-5, and IR-20 was widespread. It was accelerated because a number of international agencies wanted to get on the green revolution "bandwagon." There was then a major push to diffuse the green revolution varieties.

However, in spite of their day-length insensitivity, the IRRI varieties were quite sensitive to soil and climate factors, especially water availability and control. These factors would probably have been sufficient to limit the diffusion of the new varieties to perhaps 20 percent or so of the Asian tropical rice-producing region. The "adaptive" research programs *and* the continued development of the national research programs, which were well underway before IRRI, have expanded the set of high-yielding varieties and extended the green revolution. By 1972-73, some 20 percent of the Asian region had been directly affected. Adaptive research currently under way indicates that the potential area might be in the neighborhood of 50 percent or so.[14]

Wheat Varietal Development

Bread wheat production is effectively undertaken in three relatively independent economic regions: the steppe climate zone spring wheat regions, the steppe climate zone winter wheat regions, and the subtropical-desert climate zone spring wheat regions. Geographically the winter wheat zones tend to be between the two major spring wheat zones. It is not surprising, then, that the steppe spring wheat varietal technology was developed quite independently of the subtropical technology.

No attempt will be made here to detail productivity cycles in the steppe wheat zone or the winter wheat zones. These cycles, especially in the United States, are quite marked. The principal cycle or productivity episode of concern here is the cycle associated with the dwarf wheat varieties in the subtropical zones. This cycle bears certain obvious parallels with the Asian tropical zone rice productivity cycle.

There are, however, a number of important differences in the rice and wheat cases. First, before World War II, very little work on subtropical wheat production had taken place. And developments elsewhere (such as in the steppe spring wheat zone) were not as directly relevant to subtropical production of wheat as was the case with rice. True, the *Norin* 10 dwarf wheats were available by 1935 in Japan, but they were not given much attention until the late 1940s by breeders outside Japan.

The Rockefeller program in Mexico had to support much of the long development process of incorporating desired genetic properties into subtropical wheats. The development of the breeding methodology took time. Because the program has produced a relatively sustained flow of new varieties, it may not be possible to identify real cycles in the Mexican data.

The cyclical aspects of the new wheats in terms of international diffusion closely parallel those of rice, however. The initial degree of superiority of the Mexican wheats over traditional varieties was higher for wheat than for rice, but the same limitations to diffusion imposed by soil and climate factors hold. And the same interactive process in which local or national research efforts have extended the diffusion pattern has taken place in wheat.[15]

Toward Understanding the Basis for Productivity Cycles

Although the foregoing discussion has been heavily descriptive, it has served to raise questions of theoretical importance. Specifically, one is pressed to ask, why the cycles? Why not relatively smooth rates of technology discovery and productivity? Does this reflect inefficiency in the conduct of research?

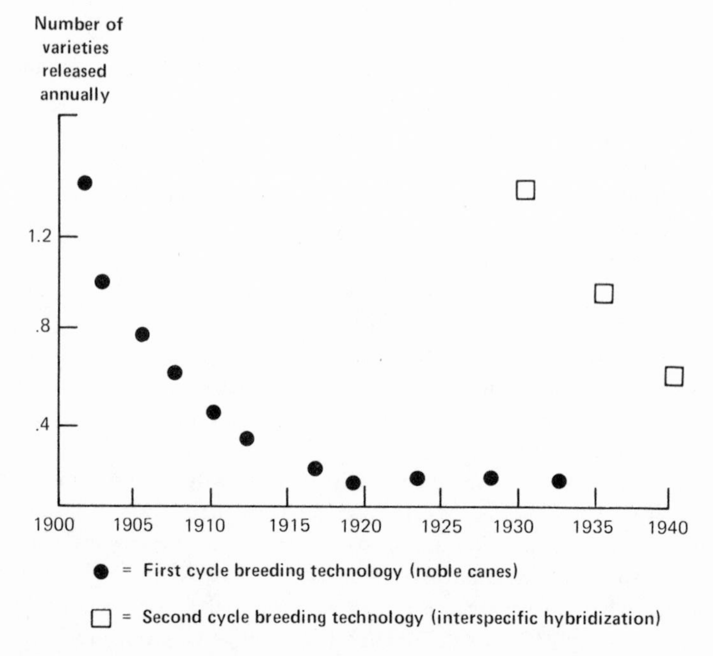

Figure 8-1. Approximate pattern of the release of commercially important sugarcane varieties.

Are there more general inferences about the discovery of technology to be drawn from this evidence?

There is relatively little economic theory which applies directly to these questions. The theory of induced innovation associated with the optimal growth literature and applied to agricultural growth by Hayami and Ruttan does not explain certain shifts in inventive activity which might result in new productivity growth.[16] Some of the crop-breeding literature is at least indirectly relevant. And the economic theory of search and of information is likewise of relevance. These bits and pieces of analytical models can be crudely fashioned to provide at least some insight into the discovery process.

A Simple Search Model

It will be useful to begin to draw upon these concepts by first noting more tangible aspects of the discovery of improved sugarcane varieties in the Barbados Experiment Station. Recall that Barbados scientists shared in the basic discoveries which established the fertility of the sugarcane plant. These discoveries were naturally followed up with a first cycle breeding program where noble canes were crossed and first cycle varieties developed. Later, the station also produced and adapted second cycle varieties. Figure 8-1 portrays the approximate pattern over time of the release of *commercially important* varieties from the station.

If this measure – commercially important varieties – can be regarded as at least an approximation of the output of economically valuable technology by the station, these data show clear cyclical behavior in technology discovery. The cycles are also clearly related to the basic cane-breeding methodology. The first cycle varieties were produced in the noble cane-breeding program utilizing the limited genetic stock of the eighty-chromosome *S. officinarum* species. It is readily apparent that the ratio of discoveries to discovery effort is declining over time, suggesting an exhaustion of potential.

The Barbados station was relatively slow to introduce the interspecific hybridization breeding program. It began work with this breeding methodology about 1929, several years after the stations in Java and India had established the superiority of the method. From 1929 to 1939 both breeding programs were maintained. As the figure indicates, the interspecific hybridization program was clearly superior. The ratio of the commercial testing stage was 1:1800 for the first five interspecific hybrids and 1:2700 for the next nine. It was only 1:13,000 for the exhausted noble program during this period.

This particular sequence has elsewhere been treated as the consequence of search processes.[17] Only the major features of that analysis will be repeated here (see chapter 10 for an extended treatment of this model). The basics of the search model[18] are as follows.

1. That the scientific knowledge of sugarcane breeders during the 1900 to 1920 period and the genetic materials available to these breeders determined a distribution of potential cane varieties with varying economic values. (Later this will be referred to as the "architecture of search.")

2. That breeders were constrained to search for economically superior varieties within this distribution of potential cane varieties. They became subject to the diminishing marginal productivity of search and had effectively exhausted most of the potential within a few years.

3. That the basic parameters of the distribution of potential cane varieties remained quite stable until the introduction of the interspecific hybridization breeding methods. The decision to shift to the improved breeding system was delayed by the resistance of older scientists to change and by the fact that resources were required to make the change.

4. That the new scientific base created a new distribution of potential varieties with a much higher variance and possibly a higher mean. The consequence of this is that researcher productivity increased markedly. (Note that an increase in genetic diversity will increase the variance of the distribution of potential varieties and the rate of discovery of new varieties.)

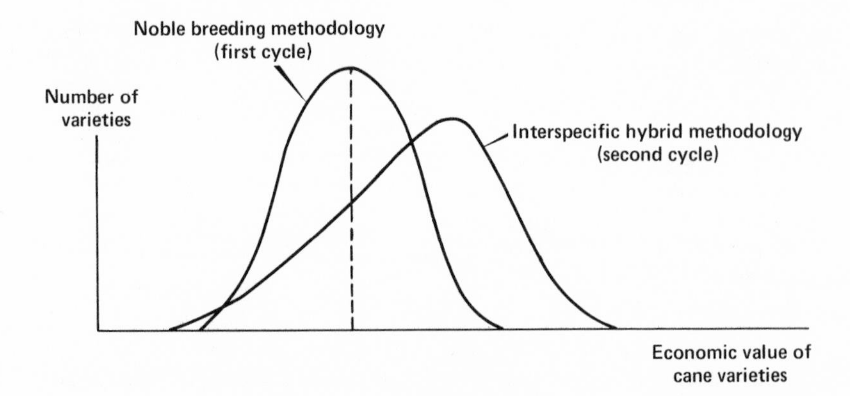

Figure 8-2. Distributions of potential sugarcane varieties.

In other words, search activity undertaken subject to the two potential distributions portrayed in Figures 8-2 and 8-3 is sufficient to generate the data in Figure 8-1.

This simple model of search activity with the somewhat erratic or at least discontinuous shift in the scientific base actually fits most of the cycles described in this chapter rather well. Several relatively straightforward modifications to the model can easily be made.

Productivity of search can be made to be a function of time. That is, the expected discoveries from doubling search activity in one time period will be less than the expected discoveries from the same activities extended over two time periods. Expansion of this concept will lead to optimal search patterns over time.

The search process can be extended to incorporate search over several parameters. In general, a plant variety can be described as a collection of n "traits," each with an economic value. The objective of plant breeders will then be to maximize the change in the total economic value of the traits from one time to the next. Knowledge of the intergenerational heritability matrix, which describes the expected value of the traits in a given generation of plants (or animals) as a function of the selected traits of the parent population, is required for maximization. Incidentally, changes in the economic value of traits will then lead to changes in the mix of traits or characteristics produced as new varietal technology or other technology. It does not follow, however, that the actual rate of change in the trait mix over time will be correlated with initial relative prices of traits.

The search process can be modeled as a more complicated process. Searchers will naturally acquire information about the distribution of potential

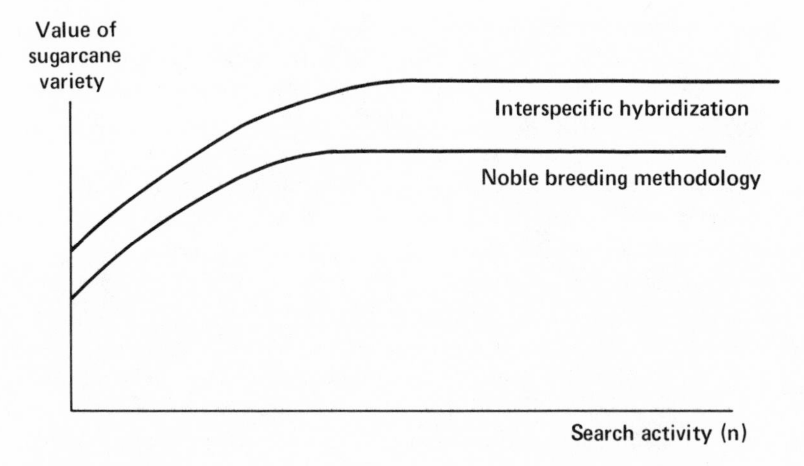

Figure 8-3. Expected maximum value of sugarcane varieties as a function of search.

technologies as they search. At a minimum they will have estimates of the mean and variance of the distribution sufficient to develop "stopping rules." In addition, they will generally be able to sequentially "rule out" unlikely and unpromising directions and areas of search. This sophistication will alter the expected pattern of discovery, but it does not alter the basic exhaustion property.

This type of modeling can be linked to the more conventional growth literature along the lines suggested by Binswanger.[19] Indeed it enriches the induced innovation literature significantly by making it clear that the probability of invention in a particular factor-augmenting direction is a function of the "state of exhaustion." Hence the simple propositions regarding a relationship between factor augmentation bias and factor prices do not hold.

Dynamic aspects of the model can be explored along the lines investigated by Kislev (chapter 10).

Even with such complexities, this approach is still limiting in at least two respects. It does not explain why the scientific base for technology discovery, which originates the cycles, shifts in a discontinuous fashion. And it does not deal with the question of the diffusion of technology to producers facing different environments.

Discontinuities in the Growth of Scientific Knowledge

Searchers for technology, of course, are not indifferent to the state of scientific knowledge. Nor is the inference that research personnel can be conveniently categorized as searchers for technology or as searchers for scientific knowledge justified except in relatively primitive research systems. A primitive system will tend to be organized as two distinct systems, one for technology search and another for scientific knowledge search. The skill mix and administration within each system constrain their output to simple search activity, and they are particularly subject to the exhaustion phenomenon. Such systems produce little new technology or scientific knowledge except as they adapt it when it is introduced from outside the system.

In more advanced systems, one finds a more complex organization. In particular, as the skill mix is increased, new specialities and motivations are developed. In chapter 15 Swanson discusses the role of the "biological architect" in designing and improving search methodology. One could expand this concept to incorporate other types of architects in the physical and social sciences as well. These architects play a key role. Their design activity is fundamental to the implementation of scientific findings. In addition, they serve to direct and orient scientific discovery toward economic objectives.

Within professional and academic circles, there are powerful social incentives to belong to either an elitist scientific organization or a professional society. The biological architect does not fit well in either. These interests tend to be built into graduate programs in the agriculturally related sciences.

The history of most agriculturally related scientific systems clearly shows that the classical organization of scientific discovery was poorly suited to the development of either the technology searcher or the biological architect. Agriculture and engineering research required the creation of separate and competing systems. The land grant system of the United States appears to have been reasonably successful in bringing about the integration of science and technology, but most developing countries have not achieved this.

A basis can be found in this discussion for the high degree of sensitivity to exhaustion which appears to characterize the agricultural research systems of developing countries. The skill mix required for a more advanced organization is such that it is likely to be some time before many systems in the tropics mature to the point where these factors will be less important. It should be expected then that real changes in the scientific base will tend to be discontinuous in research systems without large numbers of skilled and motivated scientific architects.

The Diffusion of Technology and Knowledge

The diffusion of specific items of technology across regions is clearly related to a range of soil, climate, and economic factors. Virtually all forms of technology, and especially those in agriculture, are sensitive to these factors in that minimum average costs (excluding rents to fixed factors) of production with any particular set of techniques will change when production is undertaken under other conditions. A particular variety of rice, for example, performs well on one type of soil and poorly on another.

Consequently, for every possible set of soil, climate, and economic conditions, there exists *in principle* a *unique* set of technologies for most efficient production. Thus, if it did not cost anything to tailor technology to environmental factors, agricultural technology would be immensely rich in detail. An improvement in one particular item of technology would temporarily be diffused over a limited range of environmental factors, but researchers would then adapt and tailor the improvement so that, when a new equilibrium was reached, every environment set would have a unique technology set. In such a world, technology diffusion per se would be virtually nil and would only exist because adaptation took time.

Obviously, in the real world, adaptation is a costly process, and as a result technology is not tailored to every environmental detail. Some degree of tailoring to "aggregated detail" will hold.[20] Nonetheless, the costless tailoring case affords some insight into the diffusion patterns described for sugarcane, rice, and wheat. It is consistent with the temporary character of direct technology diffusion and the inducement given to adaptation.

With costly adaptation, research systems will be designed to take into account both the supply of research skills and environmental conditions. The scarcer the high-level skills of the biological architect, the more efficient it is to design systems purely to undertake a low-level adaptive function based on simple search activities. Such systems have few scale economies to particular experiment stations and consequently are usually designed to achieve a high level of tailoring. Generally one finds a proliferation of small research units under these conditions. These systems are dependent on "mother" institutions for genetic material and search design.

When higher levels of skills are available (and not "wasted" in administrative tasks), the efficient system will have some biological architect capability and will be somewhat independent of other systems. It will sacrifice tailorability to realize the gains to be had from its independent "architect" capability. Because isolation is more costly in such a system, it will tend to have fewer experiment stations. Naturally the system will undertake adaptive research, but it will tend to be much more productive than the purely adaptive systems.

The adaptive research process is, of course, itself a diffusion process. Knowledge of some type is being diffused. Scientific equipment and genetic material can be considered to be forms of embodied knowledge. Knowledge is also sensitive to soil, climate, and economic factors. The physiological relationships of the rice plant in the temperate zone are sufficiently different from these relationships in the tropics. New knowledge for the tropics must be discovered independently of the temperate zone stock of knowledge. In this context, the tropical zone system requires a high level of biological architect capability to achieve this discovery.

Concluding Comments

This chapter offers neither a thorough historical study of technical changes in sugarcane, rice, and wheat nor a strictly formal treatment of the process of technology discovery. It has attempted to draw inferences from certain observable aspects of technical change which have relevance for the eventual formal models of technology discovery. No doubt many points of interest have been missed, and at least some of the inferences and interpretations are subject to improvement. But one must begin somewhere. The failure to understand even the basics of technology discovery and its diffusion has plagued developed programs for years.

In this chapter, observed cyclical behavior in realized productivity gains and in tangible measures of technology discovery has served as a basis for inferences regarding the role of scientific knowledge. This knowledge was interpreted as the architecture of the process of searching for technology.

Development literature generally attributes much less importance than has been given here to the rolė of scientific knowledge. In particular, although there is probably general agreement that little direct diffusion of technology between widely different climate zones takes place, the presumption that scientific knowledge is widely diffused appears to be maintained. The interpretation given to the record of sugarcane, wheat, and rice technology in this chapter is that the diffusion of knowledge is also limited by soil, climate, and economic factors. A full test of the implicit alternative hypotheses put forth in this chapter, however, awaits both better measurement and better theory.

NOTES

1. This section draws heavily on an earlier paper by R. E. Evenson, J. P. Houck, Jr., and V. W. Ruttan, "Technical Change and Agricultural Trade: Three Examples – Sugarcane, Bananas and Rice," *The Technology Factor in International Trade*, ed. R. Vernon (New York: National Bureau of Economic Research, Columbia University Press, 1970), 418-483.

2. See W. R. Akroyd, *The Story of Sugar* (Chicago: University of Chicago Press, 1967); A. C. Barnes, *Agriculture of the Sugar Cane* (London: Leonard Hill, 1967); A. R. Grammer, *A History of the Experiment Station of the Hawaiian Sugar Planters Association, 1895-1945*, Hawaiian Planters Record no. 51 (1947), 177-228.

3. G. C. Stevenson, "Sugar Cane Varieties in Barbados," British West Indies Bulletin no. 39 (Barbados: British West Indies Central Sugar Cane Breeding Station, 1968).

4. A. J. Mangelsdorf, *Sugar Cane Breeding in Hawaii, Part I, 1778-1920*, Hawaiian Planters Record no. 50 (1946), 141-160. *Sugar Cane Breeding in Hawaii, Part II, 1921-1952*, Hawaiian Planters Record no. 54 (1953), 101-137.

5. It was shortly after this that Shull and East in the United States used the same principle to develop the "hybridization" of corn. See Stevenson, "Sugar Cane Varieties."

6. E. P. Alvord and G. Van de Wall, *A Survey of the Land and Feed Resources of the Union of South Africa* (Pretoria: J. L. Saik, 1954); M. H. DeKock, *Economic History of South Africa* (Johannesburg: Juta, 1924).

7. Boris C. Swerling and Vladimer P. Timoshenko, *The World's Sugar* (Stanford: Stanford University Press, 1957).

8. The South African yield-by-age data indicate that a 1 percent increase in plant cane, over and above the normal increase needed to maintain the average age, would result in an increase in average yields of about .2 tons per acre. This is the maximum adjustment which might be made. It would represent about one-third of the Australian yield effect and all of the Caribbean and South African effect. The South African coefficient is approximately what it should be according to the yield comparison data.

9. Yujiro Hayami and S. Yamada, "Agricultural Growth in Japan, 1880-1970," *Agricultural Growth in Japan, Taiwan, Korea and the Philippines*, ed. Yujiro Hayami, Vernon W. Ruttan, and Herman Southworth (Honolulu: University Press of Hawaii, in press).

10. Robert E. Evenson, "The Green Revolution in Recent Development Experience," *American Journal of Agricultural Economics*, 56:2 (May 1974), 387-394; D. D. Evenson, D. J. Martin, and Robert E. Evenson, "Inventive Activity in U.S. Agriculture, 1890-1940: Lessons for Developing Countries," Economic Growth Center, mimeographed (New Haven: Yale University, 1974).

11. Okabe attributes the success of the experiment stations in the second cycle to the result of ecological breeding. In other words, the experiment stations could compete with the veteran farmer only by utilizing their selection procedures. S. Okabe, "Breeding for High Yielding Varieties in Japan," *Rice Breeding* (Los Baños: International Rice Research Institute, 1972), pp. 47-60.

12. Yujiro Hayami and Vernon W. Ruttan, *Agricultural Development: An International Perspective* (Baltimore and London: Johns Hopkins Press, 1971), pp. 198-212.

13. N. Parthasarathy, "Rice Breeding in Tropical Asia up to 1960," *Rice Breeding* (Los Baños: International Rice Research Institute, 1972), pp. 5-30.

14. See chapter 9 in this volume. Also Dana G. Dalrymple, *Development and Spread of High-Yielding Varieties of Wheat and Rice in the Less Developed Countries*, Foreign Agricultural Economic Report no. 95 (Washington, D.C.: Economic Research Service, USDA, 1974).

15. Evenson, "Green Revolution in Recent Development Experience."

16. Yujiro Hayami and Vernon W. Ruttan, *Agricultural Development*.

17. Robert E. Evenson and Yoav Kislev, "Investment in Agricultural Research and Extension: A Survey of International Data," *Economic Development and Cultural Change*, 23 (April 1975), 507-521. The data in this article have been further extended in James K. Boyce and Robert E. Evenson, *National and International Agricultural Research and Extension Programs* (New York: Agricultural Development Council, Inc., 1975).

18. G. J. Stigler, "The Economics of Information," *Journal of Political Economy*, 69 (1961), 213-215.

19. Hans P. Binswanger, "A Microeconomic Approach to Induced Innovation," *Economic Journal*, 84 (December 1974), 940-958.

20. Delane E. Welsch, "Relationships of Regional Agricultural Planning to International Agricultural Research Efforts," Department of Agricultural Economics, Staff Paper no. 15 (Bangkok, Katsetsart University, 1974), discusses economic aspects of tailoring technology to aggregated niches. This paper, which unfortunately is not published, offers many useful insights into this question.

9

Comparative Evidence on Returns to Investment in National and International Research Institutions

Robert E. Evenson

Agricultural research institutions in the developing countries have undergone marked changes in organization and support in the past twenty-five years. In the 1950s, most development programs were designed with the expectation that it was possible to achieve low-cost diffusion of much modern (developed country) technology to the developing economies. Major development aid was given to extension, credit, and community development programs. During this period, investment levels for agricultural research in developing countries were roughly one-half the levels for extension. This was in sharp contrast to investment patterns in the developed countries, where the level of investment in agricultural research was at least double that in extension.[1]

Financial support for buildings, equipment, and project financing from international aid agencies in the late 1950s was on the order of $50 to $60 million annually. This represented more than one-third of the total resources directed toward agricultural research in the developing world at that time. In addition, large numbers of technical advisers, visiting professors, and other personnel were located in developing countries. Hundreds of scholars from developing countries were provided with support for graduate study in universities in developed countries.

During the 1960s these patterns changed markedly. A number of appraisals of the developing-country research and extension programs led policy makers to the conclusion that neither the research nor the extension systems were effective. The problems involved in simple technology transfer were beginning to be recognized, and the failure of the extension systems to produce change were properly attributed to the paucity of real technology to extend. The fledgling research systems in developing countries were seen as lacking in scientific skills, subject to bureaucratic rigidities, and misguided about their objectives. By 1970, international financial support to the national research systems had declined, as had support for international scientists and technical personnel located in the LDC institutions.

While this decrease in support for national programs was taking place, the basis for a significant subsequent development was being laid. This was the establishment, by the Rockefeller and Ford foundations in cooperation with host governments, of two international crop research institutes: the International Rice Research Institute (IRRI) in the Philippines in 1962 and the International Maize and Wheat Improvement Center (CIMMYT) in Mexico in 1966. Several other such institutes followed in the late 1960s. The United States government began to provide financial support in 1969 and 1970. Significant international support for these and other international centers began in 1972 through the newly established Consultative Group on International Agricultural Research.

Many observers have concluded that the shift in emphasis to international agricultural research showed extraordinarily good judgment, perhaps the best in regard to the allocation of development resources since World War II. This feeling was particularly prevalent in the late 1960s and early 1970s, the heyday of the green revolution. The success of the IRRI rices and the Mexican dwarf wheats and the adoption rates and increased production breathed new life into many development programs. The subsequent slowdown in the rate of growth of yields, however, and the failure to realize some unwarranted expectations have resulted in an equally unwarranted state of pessimism regarding future productivity gains.

This recent history tells us a great deal about the process of change in agriculture, about the factors which alter the probabilities of discovery of new, more efficient methods of producing food and fiber, and about the factors that determine the diffusion of such methods across regions. We are still at a primitive stage in our ability to analyze this experience. We have now accumulated what appears to be an impressive collection of "rate-of-return" studies, all of which show extraordinarily high returns to investment in research. Some advances have been made in modeling discovery processes and invention. And the roles of climate, soil type, and prices as determinants of the diffusion of technology have been clarified in recent studies. But in spite of this, we remain some distance from a fully convincing theory of technology discovery.

Rate-of-return estimates have been useful in establishing certain basic propositions regarding the productivity of research. Unfortunately, the available estimates have been abused in policy discussions. The inconsistent citation of extraordinarily high rates of return – especially the oft-quoted 700 percent return on hybrid corn research – has left the impression that the estimates themselves are subject to such a degree of error that only those above 100 percent or so are really significant! This is somewhat similar to the old notion that farmers would not adopt a new technology unless the new technology possessed a degree of superiority of 75 percent to 100 percent over the technology in use. This simply has not been true. Such studies err primarily in their estimation of the real degree of superiority of the new technology.[2]

This is a serious matter because some of the estimates have been derived through the use of systematic econometric formulations, and standard errors have been estimated and reported. Others have simply been computed using, in most cases, supposedly "conservative" assumptions designed to ensure a downward bias in the estimate. Still other computations have been based on costs and benefits of entire research programs.

The computation of rates of return to investment in national and international research without a systematic, if crude, analytic framework provides little useful information to policy makers. Policy issues cannot be effectively addressed with one or two simple estimates. They involve some fundamental questions regarding the externalities attendant upon the diffusion of technology and scientific knowledge between regions. The knowledge that an investment of $10 thousand or $100 thousand in research on sorghum technology will yield an expected 50 percent internal rate of return is not very useful if one cannot say something about the incidence of the research consequences. If a country makes the investment, how much will its producers or consumers appropriate? And how much can they get for nothing, from research undertakings in other countries?

The body of existing economic theory does not provide ready answers to these questions. But enough work on the topic has now been completed to develop partial theoretical and empirical formulations which at least are cognizant of major features of the underlying processes. In the first two sections of this chapter some of those issues are discussed and reviewed. An econometric study of the contribution of research is then reported. The final section develops some of the policy implications.

Elements of a More Systematic Formulation

As a starting point, it will be useful to classify the different skills involved in the discovery of new production technology. This classification will help in understanding the constraints imposed by supply conditions of different types of skills, because there is a close relationship between skill types and efficient organization of research activity. A degree of arbitrariness attends all classification schemes, and the categories of skills proposed here are intended only to capture major features.

A Typology of Researcher Skills

Skills of relevance in research institutions can be classified in the following categories.

Inventive skills. Inventiveness has distinguished all developed societies and is a fundamental human characteristic. Effective application of inventive skills does not always require highly developed skills of other types. A great deal of the mechanical invention relevant to agriculture was undertaken by farmers, mechanics, and blacksmiths without formal training in engineering. Even in some biological technology fields, inventive ability can be sufficient to produce significant new technology, as when the *rōnō* (veteran farmer) selected rice varieties in Japan.

Technical and engineering skills. Professional undergraduate programs in engineering and agriculture seek generally to produce technical competence. Not all programs achieve this, but many succeed in establishing a basic ability to apply certain well-established "textbook" principles to technical problems. It is significant that the supply conditions in many developing countries are such that the research personnel in many research institutions have only the latter capability.

Technical-scientific skills. Technical-scientific disciplines such as agronomy, plant breeding, and agricultural engineering have evolved over the years and form the core of graduate programs in the agricultural sciences. Some of these disciplines bear a close relationship to a classical scientific discipline; others have developed as "interdisciplinary" fields. They vary greatly in their intellectual rigor and degree of closeness to a "scientific frontier of knowledge." Most have legitimate scientific frontiers of their own, but these tend to be closely associated with and form part of more general scientific frontiers.

Most M.S. programs and many Ph.D. programs in the agricultural sciences impart technical-scientific skills. These skills go beyond technical understanding and involve a degree of scientific abstraction. That is, they include the ability to generalize and to conceptualize in abstract terms. This skill category is designed to capture both high-level technical and limited conceptual skills.

Conceptual-scientific skills. Experience at the scientific frontier is required for the development of a high level of conceptual skills. Graduate programs and associated research experiences at the Ph.D. level attempt to develop those skills. Many programs do not accomplish this, however, either because a significant frontier in the discipline is lacking or because a program does not provide the opportunity to acquire the technical or "tool" skills sufficient to permit work at the frontier.

The observed output of scientists with these skills may primarily be in the form of published *scientific studies*. That is, many will not produce *technology* and will not attempt to do so. They will have the depth of understanding necessary, however, to direct their work toward the overall design and other features of technology discovery systems.

Experience at the scientific frontier is extremely important in providing leadership and the design of research activities. Many "successful" agricultural scientists have shifted their work from the scientific to the technology frontier during their careers. Few have managed to remain at the scientific frontier for long periods of time. Some of those who have not opted to move to the technology frontier have continued to be effective teachers to students.

This typology of skills is somewhat general and could be applied, with minor modifications, to systems oriented toward engineering technology or medical technology. A similar typology for the classical science institutions could be developed as well.

Skills and Institutional Organization

The discussion of skills is relevant to the comparison between national systems in developing and developed countries and the international institutes because supply conditions of skills vary greatly between these systems. Inventive skills are generally abundant in most societies, but their exercise in pursuit of new technology depends on a number of factors. Technical and engineering skills are relatively abundant in developing countries, but technical-scientific skills are in very limited supply and conceptual-scientific skills are almost nonexistent in some countries.

A correlation between research system organization and the supply of skills is apparent in both historical and international cross-sectional evidence. Again, at the risk of oversimplification, a stylization of this relationship would be useful.

Low-skill level systems. Typically, in the agricultural research systems that rely primarily on technical and engineering skills, experiment stations are widely diffused and tend to be commodity-oriented. Field trials and experiments tend to be "cookbook" in nature, since little real capacity to design flexible and changing methodology exists. Stations tend to be small and seldom have strong ties with graduate teaching institutions. The system attempts to take advantage of the opportunity to undertake pure adaptation of technology exogenously delivered to it.

Intermediate hierarchical systems. As appreciable technical-scientific and some conceptual-scientific skills become available, hierarchical systems tend to emerge and specific institutions assume leading roles. These stronger institutions respond to the economies of concentration and in many cases to the complementarities between graduate teaching and research. Economies of concentration are based on the capacity to exchange knowledge with other institutions and to create and modify the basic research designs or the "architecture" (as suggested by Swanson in chapter 15), of the research program.

Generally, one observes a move to consolidate and "coordinate" the diverse, isolated station programs. A main station-branch station system tends to emerge with a high degree of centralized control. The central or main stations develop some capability to engage in independent as opposed to primarily adaptive research.

Advanced science-based systems. As the supply of conceptual-scientific skills increases, agricultural research systems are more often organized to take advantage of economies of "scale" based on communication and the exchange of knowledge. Stronger ties to disciplines and to scientific frontiers are maintained, and most institutions exploit the complementarities between graduate teaching and research. Indeed, institutional quality is most often judged on the strength of graduate teaching and the ability to produce original Ph.D. dissertations.

Branch station systems generally are consolidated and as a rule serve very minor functions in the research programs. In such systems, technical-engineering skills are utilized in a purely operative fashion. Low-level technical-scientific skills (masters' level) are of little value if higher level skills are available. The highly regarded stations are leaders in refining theory, and research that does not have direct technological objectives is recognized as a vital and important part of the process.

Technology Diffusion, Knowledge Diffusion, and Organization

The sensitivity of virtually all forms of agricultural technology to soil, climate, and economic factors has influenced the organization of agricultural research systems. Indeed, the scattered experiment station organizations under low-skill-level conditions are designed to adapt technology to relatively minor gradations in soil, climate, and economic conditions. If there were no economies to the concentration of research skills (and there are few with technical and engineering skills), agricultural technology would be extremely rich in variety. Each minor gradation in soil, climate, and economic conditions, each "ecological niche," would have a unique set of technologies. Its adapted technology set would be superior to any other, and no opportunity for diffusion of technology from one niche to another would exist.

Furthermore, when new technologies were introduced from outside, diffusion across niches would be temporary in character. It would always be possible in each niche to discover an adaptation which would be superior to the imported technology. The diffusion of technology in the simplest sense of the term would not take place. In fact, of course, agricultural research systems are neither geared to tailor technology to minor niches nor capable of adapting all technology quickly. An appreciable amount of unrefined direct technology transfer does, necessarily, take place.

The low-skill-level systems with only adaptive capability are dependent on the diffusion to them of technical and engineering knowledge. They have sufficient skills to conduct agronomic trials and to participate in simple plant-breeding and selection programs. If "mother" institutions feed them new genetic material and new fertilizers, chemicals, and machines, they can be productive in adapting and modifying this particular form of knowledge.

The niches to which technology is tailored are based on gradations in the relevant soil, climate, and economic factors. Niches are not necessarily geographically contiguous, though an attempt is made to locate experiment stations where they will produce adaptations suited to modal conditions within the niche, in order to maximize the economic value of these adaptations.

Main stations now engage in a more complex form of knowledge diffusion. They have some capability for feeding technical and engineering knowledge to the branches. This capability depends on their own ability to scan other regions for technical and scientific knowledge. To some extent, scientific knowledge may be fed to the branches (to improve field plot design, for example) but its primary benefit is the facilitation of *independent* technology discovery on the part of the main station. This is a kind of high-level adaptive research.

In a research system with significant conceptual-scientific skills, research organizations develop complex specializations and explicitly seek to produce scientific knowledge. It is in these stations that the stock of knowledge which serves as a foundation for the design of technology discover systems is itself discovered and conceptualized. Much of this design work requires a thorough understanding of scientific facts and generalizations and experience at the frontier. The capacity to be an "architect" thus is developed in graduate schools, though some research experience is also required. Consequently much of the contribution of the science-oriented research and graduate teaching center is in the form of the intellectual capital created in graduate training.

Diffusion of knowledge has been a significant part of all types of research systems. In low-skill systems this has been primarily diffusion of technological and engineering knowledge in which a "disclosure" effect was involved. That is, the simple disclosure of the feature of new technology is sufficient to guide researchers toward further improvements. In more advanced systems, more abstract knowledge is diffused as well (see chapter 8).

The Role of the International Centers

With the exception of CIMMYT and IRRI, most international centers are still in a relatively early stage of development and have yet to establish a place in the research systems of the tropics. Both CIMMYT and IRRI have clearly established themselves as dominant centers as far as the capability to produce both new technology and new technical and engineering knowledge is concerned. The establishment of this dominant position by both institutions was based on two factors: imaginative administration and researcher support, and the intensive application of conceptual-scientific skills in the research program. One or both factors were, by comparison, usually lacking in the national research programs in the tropics and semitropics.

The emergence of the international centers for wheat and rice research has markedly changed the relationship between national research institutions. In a crude sense, the "main" stations in national research programs have become "branches" to either CIMMYT or IRRI. That this should happen in the short run appears to be inevitable. These centers are producing superior varieties and superior genetic material which virtually all national programs find of use in their own breeding programs. In the long run this will change as national experiment stations are able to mature into more independent institutions.

The emerging international hierarchical system in which national research systems are to a considerable extent dependent on the international centers is based only in part on the relative strength of skills in the institutions. In the past few years a new element has entered the picture. It is best seen in CIMMYT's international breeding programs (see chapter 13). With the systematic use of international sites in breeding programs, incentives exist for the development of a simple service-oriented hierarchy. It appears that CIMMYT wheat breeders are convinced that the broad-ranging breeding program is paying off. It is not surprising then that CIMMYT is primarily interested in training researchers who will both service the international programs and be fed from the center (see chapter 15). IRRI appears to be moving in a similar direction (see chapter 12).

Implications for Econometric Investigations

The major implications of the foregoing observations for econometric studies are the following:

1. An attempt to distinguish between the different types of skills should be made. These skills enter the "technology production function" in different ways.

2. A simple geographic correlation of research investment (or a research capital stock) and productivity is not an adequate specification.

3. "Average" products for research investment are generally meaningful only in a historical sense. As with any average product, they have limited relevance to policy. In certain simple models they are easier to estimate than marginal products, but in models of the type that we are dealing with here the more conventional marginal product of "research capital" formulation is the appropriate model.

4. International data are required to identify patterns of knowledge diffusion. This forces the analyst to face up to the problems of utilizing such data and devising models accordingly. It is just not possible to investigate this issue with ideal data.

An approach which is tractable, given available data, is to utilize a basic "determinants-of-productivity-change" model in which several types of research capital variables are included. These are as follows:

1. Technology-oriented research conducted in the country or region on which the observation is based.

2. Science-oriented research conducted in that country. Note that this is agriculturally related science, not classical science as taught in the liberal arts college. It includes plant physiology, plant genetics, experimental design, microbiology, and the like.

3. Technology-oriented research in *other* countries or regions of sufficient ecological similarity that diffusion of technical and engineering knowledge is possible. Some of this may be diffused in the form of direct technology transfer, but the bulk of research should be expected to induce adaptations by disclosure. Ideally, a continuous measure of degree of similarity should be used to define similar regions. As a practical matter at this point, one has to rely on climate classification schemes to define this variable.

4. Science-oriented research in *other* countries of ecological similarity such that scientific knowledge transfer can be expected. The diffusion of scientific knowledge takes place over a much broader range of conditions than is the case for technical and engineering knowledge.

5. The contribution of the international institutes should be treated as a temporary departure from the basic developing-country research process. This is the case only because of the timing of the discovery process. Had the investment in the research leading to the improved wheat and rice varieties been greater and earlier, their development would not have resulted in a revolution but would have been part of the larger pattern of technology discovery.

*Empirical Studies Based on Diffusion and Skill Specifications**

Relatively few studies utilizing the family of specifications suggested in the preceding section have been undertaken. In fact, only three or four studies utilizing international data have been reported. And the utilization of geoclimate data to specify an implicit relationship of knowledge transfer is also confined to a very few studies.

The Early International Studies

The first international study of factors influencing relative efficiency was reported by Hayami and Ruttan. This pioneering study established the feasibility of utilizing an international "meta-production function" as a tool of analysis.[3] At the time of this study, however, a set of reliable international data on investment in research was not available. Hence the inferences regarding the role of research and related investment were quite general.

With the availability of internationally comparable data on research and extension investment, new possibilities have been opened up. Evenson and Kislev report an extension of the basic Hayami-Ruttan results based on updated data and the specific inclusion of a research capital variable.[4] Their study reported estimated marginal internal rates of return of 42 percent to technologically oriented research in developing countries and 21 percent in developed countries. The study also computed an indirect rate of return to related scientific research of 60 percent for the developing countries and 36 percent for the developed countries. These results, while based on quite aggregate data and indirect methods, foreshadowed those obtained in later work (reported below).

Geoclimate Specifications

Perhaps the more important specification improvement, however, was the development of knowledge transfer specifications. The first study to utilize geoclimate regional data was an international study of wheat and maize productivity.[5] The basic model utilized in the study specified that productivity in a country (measured by yields) was causally related to country-specific factors, weather effects, the stock of knowledge created by indigenous wheat (maize) research, *and* the stock of knowledge extant in other countries with similar geoclimate conditions. That is, in this model a country could benefit from research undertaken in other countries. The mechanism was essentially what has been described here as the transfer of technical-engineering knowl-

* This section has been abridged. A number of the author's original technical discussions and tables have been eliminated in order to simplify and shorten the presentation. Further details are provided in Appendix 9-1 and in references cited by the author. – Ed.

edge with a disclosure effect. This study imposed complementarity between the indigenous stock of knowledge and the borrowable stock of knowledge. The results showed that countries without the capability for significant indigenous research realized almost no transfer benefits. High returns to the research activities were computed. Research was shown to be productive by leading directly to improved technology and by facilitating technology transfer.

This basic model has been utilized in two studies of national research programs. The study of most relevance here was based on Indian data.

The Indian National Study

The case of India is an especially relevant national system. It could probably be categorized as an intermediate-skill-level system of a basically hierarchical nature. Relatively few national systems in developing countries are as advanced.[6]

An analysis of crop-related research expenditures by states shows that the government has pursued significantly different research investment programs in the several states in India. The results of the study indicate that an increment to the state *research* capital stock of 1,000 rupees is associated with a direct increase in the value of agricultural product of 6,600 rupees. An additional contribution of 1,300 rupees is forthcoming through interaction when extension activity (at the mean level) is undertaken. An increment of 1,000 rupees to the research capital stock cannot be obtained by increasing investment in research in the most recent years, because of the distributed lag construction. Thus the 7,900 rupees (6,600 + 1,300) can be viewed as the level to which the generated income stream grows after the distributed lag period. Consequently the 7,900-rupee income stream represents a 46 percent internal rate of return. (The "time shape" of the benefits is discussed in a subsequent section.) The same increment to the research capital stock is associated with an additional 800-rupee income stream owing to the acceleration of knowledge transfer.

A 1,000-rupee increment to investment in *extension* capital generates an income stream of 175 to 200 rupees. With a short time-lag between investment and the realized income stream, this represents a relatively normal rate of return of 12 to 15 percent. A related study of Indian productivity, based on district data designed to measure the contribution of the Intensive Agricultural Districts Program, indicated that the return realized on the resources devoted to extension and general development activities under the program was also approximately 15 percent.[7]

The Indian study serves to demonstrate that knowledge transfer between regions can be identified with the geoclimate data. It also provides hard evidence that national research systems like those in India are, indeed, productive. (However, the fact that they are productive does not necessarily mean that they are efficient.) Data from one country, even one as large as India, are not entirely adequate to investigate the larger question of knowledge transfer. Nor was it possible with the Indian data to explore the issue of the relative contribution of different types of knowledge procured by different types of skills.

Productivity in Cereal Grains

An international study of cereal grain productivity reported by Evenson attempted to explore further both the issue of international knowledge transfer and the question of the relationship between different types of knowledge.[8] The study is of particular interest in the context of this chapter in that the contributions of the high-yielding varieties (HYV's) produced in part by CIMMYT and IRRI are incorporated into the model. A rough measure of the relative marginal contributions of additional investment in national versus international research programs and in technology-oriented versus science-oriented research activities is possible. In addition, it is possible to distinguish gains appropriated by the investing country from those accruing through knowledge transfer to ecological neighbors of the investing country.

This study does have data limitations. It is not possible to obtain measures of several conventional factors of production. One cannot know how much labor or machine service was devoted to the production of specific cereal grains. Many studies of agricultural productivity have utilized the simple measure of partial productivity – production per unit of land. It is, on the whole, not subject to serious error as an index of efficiency gains over time, but it is not an ideal measure. Fortunately, it was possible in this study to add a second factor, fertilizer use, to the analysis.

The methodology and findings of the study are presented in Appendix 9-1; the implications will be outlined below.

Economic Implications for Research Investment

The several studies reviewed here have considerable significance for the productivity of research. They may be summarized in the form of estimated marginal benefit streams associated with an increment of $1,000 to the research capital stock. These are not the usual calculations of average product type. The reader can convert the streams into rates of return or cost-benefit ratios as desired. If the increment of $1,000 to the capital stock is viewed as being distributed over time according to the construction of the stock, the time shape of the return is as shown in Figure 9-1. Investment in time t pro-

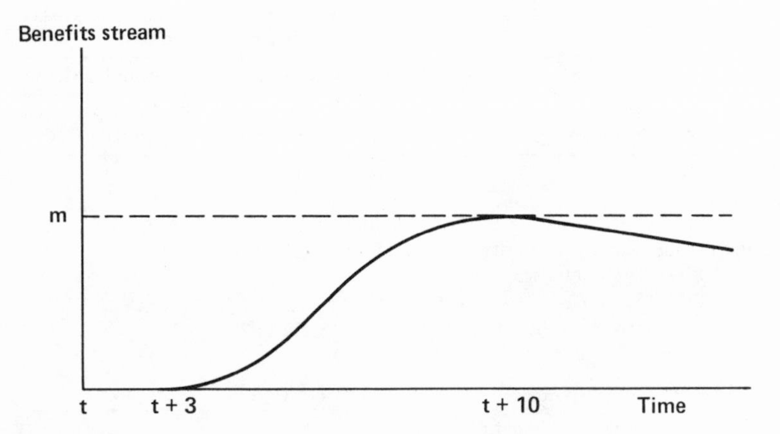

Figure 9-1. The timing of research benefits.

duces an expected stream beginning in year t + 3 and rising to the level m by year t + 10. Since technology is subject to depreciation, it is quite possible that gains once realized will be lost. The decline after year t + 10 reflects this possibility.

Table 9-1 summarizes the economic implications of the cereal grains study. In the first part of the table, computations based on the cereal grains regressions in Appendix 9-1 are presented. These computations are relevant to investments in national systems. The developing-country computations are based on a regression excluding the HYV variables (see Table 9-6 in the appendix) on the grounds that it measures a more stable long-run relationship. If they were to be based on the HYV regression they would be slightly lower.

One of the advantages of the particular functional form utilized is that one can estimate how much of the expected generated income stream is appropriated by a typical investing country and how much flows to ecological neighbors. This, of course, depends on how much research the neighbors do. The computations in part 2 of Table 9-1 are based on mean level of research capital in neighboring countries.

The relative size of the estimated income streams in developed and developing countries should not be surprising. It does not indicate that researchers in developing countries are more efficient than those in developed countries. It reflects two factors. First, developing-country research skills are relatively low-priced; they cost less than half what they cost in developed countries. The research measure utilized partly corrects for this. Second, and more importantly, the size of the streams reflects that the ratio of research capital to conventional factors of production is only one-fifth as high in the develop-

Table 9-1. Estimated Marginal Benefit Streams Associated with National Research Investment of $1,000[a]

Benefit Streams	Research Investment: Developed Countries: Technology-Oriented	Developed Countries: Science-Oriented	Developing Countries: Technology-Oriented	Developing Countries: Science-Oriented
Part 1				
Appropriated by investing country				
(a) Direct contribution	$ 630	$12,300	$ 3,710	$35,600
(b) Through complementarity with research in other countries.	1,620	1,620	7,200	7,200
Total appropriable benefits	$2,250	$13,920	$10,910	$42,800
Part 2				
Contributed to other countries[b] . . .	5,150	17,000	49,000	37,300
Total international benefit stream (Part 1 + Part 2)	$7,400	$30,920	$59,910	$80,100
Part 3				
Realized by a typical country from research investment by other countries in similar climate zones (or regions)[c]				
(a) With average indigenous research capability	$8,580	$ 520	$55,000	$ 1,700
(b) With no indigenous research capability	$4,560	$ 520	$ 1,700	$ 1,700

Source: Computations were based on cereal grains research regressions (1) and (2) in Table 9-6 (Appendix 9-1).

[a] Estimated levels, in 1973 United States dollars, to which benefit streams associated with a research investment of $1,000 will rise eight to ten years after initial investment. If a normal 12 percent rate of return is realized, the level will be approximately $1,000. A 20 percent compound internal rate requires roughly a $3,000 stream and a 50 percent rate requires $16,000. It is useful to think of this investment as a purchase of income streams or as a purchase of economic growth. The price of the streams is the inverse of the internal rate of return.

Computations were based on mean values of variables in the derivatives from the two data sets. The derivatives are in terms of the effect on production of a change in the knowledge stock. The knowledge stock is converted from publications to dollars, based on the data in Robert E. Evenson and Yoav Kislev, *Agricultural Research and Productivity* (New Haven: Yale University Press, 1975), chapter 2, Table 2.4. Arithmetic rather than geometric means were utilized in the computations on the grounds that they are more representative of typical countries. Cereal grain product is valued at $80 per metric ton (approximate 1971 prices).

[b] Contributions to other countries were based on the average number of other countries in similar regions for T (0.6 in developed countries, 0.9 in developing countries), and zones for S (33 for developed countries, 23 for developing countries) (see Table 9-8).

[c] Benefits realized from other countries were computed as the marginal products of regional research. 3(b) was computed setting indigenous research equal to zero. Note that the construction of the model is such that the contribution of zonal science-oriented research depends only on regional research and not on indigenous research.

ing countries, so that marginal products of research will be high for this reason if standard production relationships hold (see chapters 8 and 10).

The fact that the science-oriented research should be more productive than the technology-oriented research is also not surprising. It reflects that conceptual-scientific skills in the developing countries are scarce and have high value. Actually, the special nature of the research capital stock construction enables it to measure relatively high-level skills in the technology-oriented research as well. The inference that engineering-technical skills (B.S. level) are of high value is not warranted. These results are complementary with the Indian study results referred to earlier. Though based on different data, the implied appropriated streams in this study are close to those estimated for India. In the developing countries a considerable amount of transfer of benefits based on knowledge diffusion takes place. Part 3 of the calculation shows, however, that such transfer depends heavily on the capacity of indigenous research. The strategy of waiting for the neighbor's technology to "spill in" just doesn't work. The country without an indigenous research capability benefits very little from its neighbor, even when the neighbor is considerate enough to invest in research.

The levels of the income streams are extraordinarily high and imply returns to research investment several times higher than those realized on normal investment. The research dollar buys income streams many times as large as the average development dollar, though given the time-lags the price of growth through research investment is probably one-fifth to one-eighth as high as it is in most development projects.

An Ad Hoc Computation of Returns to International Center Research in Wheat and Rice

The computation of returns to investment in international center research cannot be made without first making some rather arbitrary assumptions. Were it not for the tremendous importance for policy of such a calculation, it probably should be avoided. The procedure utilized here starts with the reasonably solid information summarized in Table 9-2. These estimates of the economic value of wheat and rice which would *not* have been produced if none of the HYV's had been discovered is based on the regression analysis in Table 9-6. This, of course, is most certainly *not* an estimate of the contribution of CIMMYT and IRRI. In fact, the crude allocation of varieties by center, joint center-national, and independently produced varieties suggests that in the absence of CIMMYT and IRRI some HYV's would have been produced.[9] It is even likely that a large number of HYV's might have been produced eventually, since the independently developed varieties have been adopted at a significant rate in comparison with IRRI and CIMMYT varieties.

Table 9-2. Benefit Streams Associated with High-Yielding Varieties in Asia and North Africa

Benefit Streams	1965-66	1966-67	1967-68	1968-69	1969-70	1970-71	1971-72	1972-73
Wheat								
Aggregate adoption level of all HYV's (%)	.001	1.5	9.7	18.9	20.5	24.0	28.3	34.1
Proportions owing to								
CIMMYT-bred varieties.	.001	1.5	8.8	16.6	17.0	18.7	20.4	23.0
Joint CIMMYT-national	0	0	.8	1.9	2.5	3.6	5.6	7.5
Other independent	0	0	.1	.5	1.0	1.7	2.3	3.6
Value[a]								
at $75 per metric ton (million dollars)	.03	30	170	325	340	403	445	523
at $130 per metric ton (million dollars)	.04	52	293	563	590	697	772	906
Rice								
Aggregate adoption level of all HYV's (%)	.001	1.1	3.4	6.0	9.9	13.1	17.1	20.9
Proportions owing to								
IRRI varieties	.001	1.0	2.9	4.6	7.0	9.1	10.5	13.3
Joint IRRI-national.	0	0	0	.4	.9	1.4	3.2	3.4
Other independent	0	.1	.1	1.0	2.0	2.6	3.4	4.2
Value[a]								
at $100 per metric ton (million dollars)	.07	76	233	420	695	905	1155	1359
at $175 per metric ton (million dollars)	.11	133	407	736	1216	1584	2022	2379

[a] In computing these benefit streams, the following geographic groupings were utilized: *Wheat*. Production from all developing countries plus the Southern European countries. *Rice*. Production of rice from all tropical and semitropical developing countries, excluding South Korea and Taiwan.

The lower prices are representative of those in the 1970-72 period; the higher prices are more representative of 1975.

These figures differ somewhat from the author's earlier computations reported in "Consequences of the Green Revolution," Economic Growth Center, Yale University, 1974 (mimeographed). The major differences are attributable to the use of geometric rather than arithmetic means in the computations, different geographic groupings, and somewhat more accurate production data. The results reported here are more consistent with Dalrymple's estimates than were the earlier figures; for comparisons see chapter 7.

Table 9-3. Income Stream and Cost Calculation for International Center Research

Item	First Generation Varieties (1966/67-1969/70)		Second Generation Varieties (1970/71-1972/73)	
	Wheat	Rice	Wheat	Rice
Annual increment to income stream (1973 million dollars)[a]	182	142	74	301
Associated cost on annual basis (1973 million dollars)[b]	.6	1.0	1.2	2.8
Income stream per $1,000 investment.	$303,000	$142,000	$62,000	$108,000

[a] Computed from Table 9-2 utilizing prices of $130 per metric ton for wheat and $175 per metric ton for rice.

[b] Computed from Dalrymple, chapter 7. The second generation costs are based on IRRI and CIMMYT Annual Budgets. That is, it is supposed that the research program during these years was primarily responsible for the production gained during the second generation period. First generation costs are all prior costs at IRRI (capital expenditures are amortized), and a capital adjustment is made for CIMMYT costs to make them roughly comparable with IRRI costs.

One could attempt to estimate this alternative pattern of varietal development to determine how much should be attributed to the centers, but this would be very difficult. It would also bias the result against the centers, since the development of the independent varieties is not truly independent. Instead, a computation which is something of an overestimate will be made. First we compute income streams to each center based on the center varieties plus one-half the center-national joint varieties. These are averaged for the first five years of the adoption pattern and for the last three years. These can be taken to represent the income streams associated with the first and second generations of center varieties. Then, if the cost data can be matched up with these benefit streams, an estimate of the income stream associated with an investment of $1,000 is possible. Table 9-3 summarizes this calculation.

The results of this calculation show, not surprisingly, that the income streams purchased by investing in both the CIMMYT and the IRRI programs have been truly extraordinary. The Mexican wheats, when introduced in India and Pakistan, were quickly adopted in the regions to which they were suited. The IRRI-type rice, on the other hand, was adopted at a slower pace and required more systematic screening and the adaptation of complementary technologies. This is reflected in the calculated income streams. The distinction between first and second generation varieties is somewhat arbitrary and is designed primarily to show marginal gains to the extent possible. The CIMMYT income stream in the first generations of varieties going into India is extremely high but drops considerably in the second generation time period because new adoption has slowed down. The IRRI-generated stream declines only moderately, as adoption of the varieties in the second period was almost as great as in the first period. The second generation income streams, while still far above income streams consistent with efficient investment, do not greatly exceed those indicated for national system investment.

It should be noted, however, that both the national system and the international center computations are subject to error and are not strictly comparable in terms of the timing of benefits. In addition, these computations are based on the results of relatively new research programs. As these programs mature and expand it is highly improbable that benefit streams of the order measured here could be attained. Thus it would not be realistic to expect these high returns to continue to be realized as the international system expands and as national system investment is increased.

Policy Implications of Agricultural Research Productivity

The estimated income streams associated with investment in virtually all types of research programs from a large number of studies must be taken as evidence that investment levels are less than optimal. The computed rates of returns are well above the social opportunity cost of investable funds. If one considers the available evidence regarding the payoff to investment in the full range of development projects undertaken in the low-income countries, it simply is not possible to match the returns realized to research investment. Still we find that research system development does not have high priority in the programs of many developing countries. And international funding has given it low priority as well.

A recent study by Boyce and Evenson[10] provides new data on the willingness of countries to invest in agricultural research and extension. Table 9-4 summarizes investment in agricultural research and extension in constant (1971) United States dollars by major regions of the world. The research data include private industrial sector research and agriculturally related scientific research. The extension data do not include private sector extension activity.

The research data also include international center investment which in 1974 totaled some $30 million, less than 5 percent of low-income country investment. The study in question estimated that total international aid agency funding of low-income country research in the 1950s was on the order of $50 million per year, representing 40 to 50 percent of total investment. By 1965, aid funding had risen to roughly $100 million per year and still accounted for roughly one-third of national system investment in the low-income coun-

Table 9-4. Expenditures for Agricultural Research and Extension by Major World Regions, 1951-74

Region	1951	1959	1965	1971	1974
	Total Annual Expenditures for Research (millions of 1971 constant U.S. dollars)				
Western Europe	130.0	172.3	407.4	671.0	733.4
Eastern Europe and USSR	132.2	365.2	626.8	818.0	860.5
North America and Oceania	365.7	540.0	805.9	1203.4	1289.4
Latin America	29.7	39.2	73.0	146.4	170.3
Africa	41.3	58.0	113.5	138.5	141.1
Asia	70.0	131.0	356.4	610.2	646.0
World Total	768.9	1305.7	2383.0	3587.5	3840.7
	Proportion of Total Annual Expenditures for Research Accounted for by Industrial Sector Research (%)				
Western Europe	12.6	12.4	11.7	10.8	10.8
Eastern Europe and USSR	7.5	7.4	8.1	8.3	8.3
North America and Oceania	28.0	28.3	26.9	24.9	25.4
Latin America	3.3	3.6	3.6	3.2	5.1
Africa	2.9	3.5	3.5	2.9	2.9
Asia	2.8	2.5	2.4	2.2	2.2
World Total	17.4	15.9	13.9	12.9	13.1
	Proportion of Total Annual Expenditures for Research Accounted for by "Agriculturally Related" Scientific Research (%)				
Western Europe	19.8	19.5	24.8	27.6	27.6
Eastern Europe and USSR	27.0	26.4	19.0	17.2	17.2
North America and Oceania	11.7	11.7	12.2	16.3	16.4
Latin America	9.2	9.2	11.5	14.1	14.0
Africa	6.7	5.8	6.9	9.2	9.2
Asia	19.8	18.9	23.3	25.9	25.9
World Total	11.3	17.2	13.3	19.9	20.5
	Total Annual Expenditures for Extension (millions of 1971 constant U.S. dollars)				
Western Europe	n.a.	99.4	169.5	196.6	183.3
Eastern Europe and USSR	n.a.	128.0	90.0	230.0	250.0
North America and Oceania	n.a.	163.1	198.4	263.5	287.6
Latin America	n.a.	32.4	51.1	102.8	121.9
Africa	n.a.	90.7	161.0	217.0	224.5
Asia	n.a.	73.2	160.0	249.5	258.5
World Total	n.a.	586.8	930.0	1259.4	1325.8

Source: James K. Boyce and Robert E. Evenson, *National and International Agricultural Research and Extension Programs* (New York: Agricultural Development Council, Inc., 1975).

Table 9-5. World Expenditures on Research and Extension as Percentages of the Value of Total Agricultural Product by Annual Per Capita Income Group, 1951-74

Income Group (1971 U.S. dollars)	1951	1959	1965	1971	1974
			Agricultural Research		
Over $1,750	1.21	1.26	1.80	2.48	2.55
$1,000-1,750	.83	1.19	1.95	2.34	2.34
$400-1,000	.40	.57	.85	1.13	1.16
$150-400	.36	.37	.62	.84	1.01
Under $150	.22	.28	.47	.70	.67
			Agricultural Extension		
Over $1,750	n.a.	.45	.52	.61	.60
$1,000-1,750[a]	n.a.	.17	.22	.33	.31
$400-1,000[a]	n.a.	.26	.40	.46	.40
$150-400	n.a.	.67	.99	1.44	1.59
Under $150	n.a.	.57	1.04	1.76	1.82

Source: James K. Boyce and Robert E. Evenson, *National and International Agricultural Research Extension Programs*. New York: Agricultural Development Council Inc., 1975).

[a] Excluding Eastern European countries.

tries. After 1965, international aid to national research systems declined to the $60 to $70 million level by 1971. Some increase may have taken place since 1971.

Table 9-5 summarizes the data in Table 9-4 by grouping countries by per capita income level (as of 1971). Here we can see a rather extraordinary correlation between level of development and the propensity to invest in research and extension. The low-income countries have clearly opted to expand extension systems. (It should be noted, however, that the lack of private sector extension activity, which would be more heavily concentrated in the high-income countries, biases this picture.) The research data indicate that from 1965 to 1971 low-income countries did expand research system investment even though international funding was declining during this period. The 1974 data, on the other hand, show little further increase except in the $150-400 per capita income group. This lack of progress is very likely to have serious future consequences.

There are several economic explanations for the relatively high estimated rates of return to research investment. The supply of research skills in the developing countries may be such that the real marginal cost of skills from a social perspective is above the average cost of the skills. (In fact, most of the high-level skills have been created with fellowship support from international agencies. The expected returns to fellowship investment are presumed not to be fully captured by the fellowship recipient. The computed rates of return then do not fully account for the actual investment.)

Externalities associated with the diffusion of technology are probably an important factor in holding down investment levels. The proposition that technology is relatively easily transferred through investment in extension activities has been stressed in a great deal of the development literature. Countries have been led to believe that much of their own investment will produce benefits which they will not be able to appropriate. Conversely, with a little investment in extension they have expected to appropriate the results of other research programs. The studies reported in this chapter indicate that this policy strategy has not paid off. The degree of diffusibility of technology has been much lower than supposed. Countries without an indigenous research capability have benefited little from research in other countries. Had these countries' judgment been correct, the expected rates of return to the appropriated income streams from research would have been normal, but the returns computed from the full income streams would be above normal.

National systems may be discounting the expected income streams to take into account possible social costs associated with factor adjustment. These have not been considered in the studies reported here. Two recent studies of the determinants of research investment provide evidence that countries which produce higher proportions of both exported and imported commodities do invest more in agricultural research.[11] Since these commodities have relatively high demand elasticities, realized technology change will create fewer adjustment pressures.

Finally, *ex-post* measures cannot automatically be taken to reflect *ex-ante* expectations. This is especially serious when research programs are obvious successes in an *ex-post* sense. The extraordinarily high rates of returns to investment in CIMMYT and IRRI probably could not have been expected in advance. On the other hand, those studies which include the entire spectrum of research on a commodity or commodity group, such as the cereal grains study, are not subject to serious bias on this score.

Several additional factors have probably also been important in guiding research investment policy in both national and international systems. Not the least of these is the high level of administrative skill demanded in the stronger research institutions. Extension and rural development projects are much easier to organize and administer than are effective research programs. A second major factor is the persistent preference for quick results in development projects.

The prognosis for future productivity advances in the developing countries is mixed. On the one hand, while national system investment is well below optimum, significant progress has been made, at least until 1970. The improvement in many national systems in recent years should lead to improved agricultural performance in the late 1970s. But only a few countries have really given research system development a major place in development plans. For much of the developing world, investment in the expansion of research capacity is so low that the prospects for rapid technology improvement are dim.

The further development and expansion of the international centers system, although probably having a high expected payoff, do not really hold promise for many of these regions. It simply is not practically possible for a single research organization to produce technology which will be relevant to more than a small fraction of the world's producers. Consequently, the impact of the international centers ultimately depends on the existence of strong national systems. The international centers have demonstrated that they can be tremendously productive in a setting of weak national systems. As strong national systems are developed, their role will change and they will have to move toward a greater emphasis on their comparative advantage in more basic science research.

APPENDIX

Appendix 9-1. Cereal Grains Productivity Analysis

The basic specification actually utilized in the cereal grains productivity analysis was

$$P = CL^{a_1}F^{a_2}TR^{(a_3 + a_4 SR)}RTR^{(a_5 + a_6 ZSR + a_7(TR + SR) + a_8(TR + SR)^2)}RP^{(a_9 + a_{10}TR)}$$

where

P, L, and F are measures of production, land, and fertilizer use in quantity units for each country. Each is expressed relative to the average 1948-50 base level of the variable. Thus the analysis is of changes in productivity relative to the base period.

TR and SR are knowledge capital stocks in each country. TR measures technology-oriented research capital, and SR measures science-oriented research capital. Again, a price must be paid for this specification in that the only possible measure of research activity by orientation in international data is by publications. A direct measure of scientist man-years or expenditures would be preferable. Fortunately, we have available internationally consistent data on publications oriented to specific cereal grains from *Plant Breeding*

Abstracts. The cumulated sum of the publications for each cereal grain with a distributed lag forms the TR variable.

$$TR(t) = \sum_{t=1942}^{t-5} P_t + .8P_{t-4} + .6P_{t-3} + P_{t-2} + .2P_{t-1}.$$

With a three-year lag from investment to publication the formulation implies an eight-year lag until research becomes fully productive.[12] The SR variable is constructed in a similar manner from publications data in scientifically oriented publications in the fields of plant physiology, phytopathology, and soil science abstracted in *Biological Abstracts.*

RTR and ZSR are the counterpart "borrowable" research capital stocks available from ecological neighbors. Ecological neighbors are defined as being in the same geoclimate *region* in the case of RTR or the same geoclimate *zone* in the case of ZSR. The geoclimate regions and zones are adapted from the work of Papadakis.[13] The specification is designed such that the science stock does not directly contribute to productivity. Its contribution comes through improving the productivity of the technology-oriented knowledge stock. The terms (TR + SR) and $(TR + SR)^2$ in the exponent of RTR measure the degree of complementarity or substitutability of indigenous research-based knowledge to the borrowable stocks.

The term RP is an index of productivity change in cereal grain production in ecologically neighboring countries. It is, of course, itself a function of research capital stocks, but it measures actual realized technical change. It reflects some random elements as well as development and adoption investment and is designed to distinguish between direct technology transfer and knowledge transfer. When a country is in more than one geoclimate zone (as most are), the allocation to each zone for purposes of the definition of RTR, RP, and ZSR is proportional to the crop production area. In addition, research variables are "deflated" by the number of geoclimate regions in the country (adjusted for the size of regions).[14]

The results of regression analysis of this model were carried out for three alternative sets of cereal grains data reported in Table 9-6.

The reported regressions were estimates which utilized the Nerlove-Baelestra procedure for modified generalized least-squares estimates with combined time-series cross-sectional data. The principal features of the first two regressions are the following:

1. The land coefficients are approximately what would be expected if land serves as a proxy for the "left-out" variables, labor and power. This is not the only interpretation possible, but it suggests that the problem of missing data is not too serious. The fertilizer coefficients are reasonable.

Table 9-6. Sources of Variation in the Index of Cereals Grain Production[a]

	Independent Variable	Developed Countries[b] Regression (1)	Developing Countries[c] Regression (2)	Developing Countries[c] Regression (3)
a_1	Land	.965	1.011	1.083
	[LN(Land)]	(199.8)	(288.7)	(222.9)
a_2	Fertilizer	.0333	.0318	.0273
	[LN(Fert)]	(8.67)	(6.26)	(5.38)
a_3	Country T research	.00707	.00231	.0021
	[LN(TR)]	(2.09)	(.75)	(.70)
a_4	Country T x country S research	.00000404	.0000684	.0000524
	[LN(TR)*SR]	(1.64)	(7.33)	(5.44)
a_5	Regional T research	.01611	− .00014	− .00231
	[LN(RTR)]	(2.46)	(.05)	(.71)
a_6	Regional T x zonal S research	.0000639	.000147	.000157
	[LN(RTR)*ZSR]	(12.09)	(10.56)	(11.40)
a_7	Regional T x country T + S research	.0000093	.000095	.00010
	[LN(RTR)*(TR + SR)]	(2.81)	(5.17)	(5.18)
a_8	Regional T x country T + S research squared	.00000000023	− .000000045	− .000000065
	[LN(RTR)*(TR + SR)^2]	(1.46)	(16.06)	(7.49)
a_9	Regional yield index	.1753	.0627	.0026
	[LN(RY)]	(5.38)	(2.26)	(.09)
a_{10}	Regional yield x country T research	− .000215	− .00061	− .00036
	[LN (RY)*TR]	(10.12)	(8.94)	(5.07)
a_{11}	Percentage area planted to high-yielding varieties			.00574
	[HYV]			(2.93)
a_{12}	HYV percentage squared			− .00154
	[(HYV)^2]			(3.67)
a_{13}	HYV percentage x country T research			.0000144
	[HYV*TR]			(5.81)
a_{14}	Dummy for wheat	− .2233	− .018	− .060
		(10.47)	(1.53)	(4.67)
a_{15}	Dummy for barley	− .2777	− .081	− .094
		(14.29)	(4.69)	(5.46)
a_{16}	Dummy for rice	− .3455	− .097	− .1164
		(12.57)	(7.29)	(8.81)
Constant		.565	.026	.087
		(10.55)	(.51)	(2.06)
R^2 (Adj.)		.981	.986	.987

[a] Regressions weighted by area and estimated using Nerlove-Baelestra techniques; t ratios in parentheses.
[b] Eighty-seven crop-country combinations, 1948 to 1971 (2,088 observations).
[c] Seventy-eight crop-country combinations, 1948 to 1971 (1,872 observations).

Table 9-7. Wheat and Rice Production: A Simplistic Regression Model

Independent Variable[a]	Wheat Production[b]	Rice Production[c]
Land [LN(Land)][d]	.9836 (.0030)	1.0374 (.0030)
Fertilizer [LN(Fert)][d]	.0411 (.0058)	.0477 (.0021)
Percentage area planted to high-yielding varieties [HYV][e]	.0430 (.0018)	.0052 (.0010)
HYV percentage squared $[(HYV)^2]$	−.00085 (.00004)	−.00005 (.00002)
Constant	.1758	−.2208
R^2	.9859	.9965

[a] Dependent variable is LN(Production). Production is scaled relative to average levels in 1948-1950.

[b] In thirteen Asian and Middle Eastern countries, 1948 to 1971 (307 observations); regressions weighted by area harvested; standard errors in parentheses.

[c] In twelve Asian and Middle Eastern countries, 1948 to 1971 (282 observations); regressions weighted by area harvested; standard errors in parentheses.

[d] Land and fertilizer are scaled relative to average levels in 1948-1950.

[e] HYV is the percentage of the acreage of wheat or rice planted to high-yielding varieties as defined by Dana G. Dalrymple, *Development and Spread of High-Yielding Varieties in the Less Developed Countries*, Foreign Agricultural Economic Research Report no. 95 (Washington, D.C.: Economic Research Service, USDA, 1974).

2. Indigenous technological research (TR) is primarily productive when interacting with scientific research (SR). The a_4 coefficient is highly significant in all but the developed-country regions. Note that the productivity of TR depends on both a_3 and a_5. A negative a_3 (as in the rice regression) does not mean that TR is unproductive. Its net contribution is positive.

3. Research by ecological neighbors is of value as indicated by the a_6 and a_7 coefficients. Again the productivity is primarily in the form of interactions. The (TR + SR) terms indicate that indigenous research *complements* the borrowable research at low levels but *substitutes* for it at high levels.

4. The RP variable indicates that indigenous research is less productive when the rate of productivity in ecologically neighboring countries is higher. Technically this is the case, holding constant the research capital stocks in the neighboring countries. Thus it bears the interpretation that the more efficient neighboring countries are in converting research capital into productivity the more likely it is that the resultant technology or technical knowledge will be transferred.

Regression (3) in Table 9-6 is based on an extension of the model to incorporate the extraordinary productivity gains associated with the international institutes. The variable HYV is measured as the percentage of wheat (or rice) area planted to high-yielding varieties as defined by Dalrymple.[15] The results

Table 9-8. Regression Analysis: Wheat and Rice Production in Asian and Middle Eastern Countries, 1948-71[a]

Independent Variable	Wheat Production[b]	Rice Production[c]
Land	1.0050	1.0217
[LN(Land)]	(63.5)	(107.4)
Fertilizer	.0693	.0409
[LN(Fert)]	(4.05)	(2.91)
Country T research	.0112	– .0144
[LN(TR)]	(1.10)	(2.54)
Country T x country S research	.00067	.00024
[LN(TR)*SR]	(4.36)	(2.00)
Regional T research	.1656	– .0179
[LN(RTR)]	(6.16)	(2.08)
Regional T x zonal S research	.000046	.000010
[LN(RTR)*ZSR]	(2.29)	(1.52)
Regional T x country T + S research	– .00068	.00002
[LN(RTR)*(TR + SR)]	(2.39)	(.18)
Regional T x country T + S research squared	– .00000074	– .00000018
[LN(RTR)*$(TR + SR)^2$]	(4.35)	(6.00)
Percentage area planted to high-yielding varieties	– .002569	– .0097
[HYV]	(.61)	(1.96)
HYV percentage squared	– .0000716	– .000018
[$(HYV)^2$]	(.86)	(.27)
HYV percentage x country T research	.0089	.000039
[HYV*TR]	(3.63)	(5.70)
Time	.296	– .0181
[LN(Time)]	(2.33)	(3.63)
Constant	– 2.47	1.22
R^2 (Adj.)	.987	.998

[a] Regressions weighted by area and estimated utilizing Nerlove-Baelestra techniques; t ratios in parentheses.

[b] In ten Asian and Middle Eastern countries (240 observations).

[c] In twelve Asian and Middle Eastern countries (120 observations).

in regression (3), Table 9-6, show that as the percentage planted to HYV's increases the contribution to production decreases (the negative HYV^2 term). They also clearly show that the contribution of HYV's depends on the indigenous research capital stock. Table 9-7 is basically intended to show how misleading a simplistic approach to the green revolution can be when the HYV variables are postulated to be the only determinants of productivity change in the regressions reported there. The HYV category includes varieties developed directly by CIMMYT or IRRI, varieties developed by national systems utilizing CIMMYT or IRRI genetic material, and varieties developed independently of the international centers (e.g., the ADT-27, Mahsuri, TN-1, C4-63, and related rice varieties, Bezostaya [a Russian wheat variety], and other Iranian, Tunisian, and Italian varieties).

These regressions are based only on the wheat and rice data. They indicate that the productive contribution declines to zero at 26 percent adoption for wheat and 52 percent adoption for rice. This of course is to be expected if the HYV mix is a fixed set of varieties. The productive effect will be highest in the areas of earliest adoption and will be exhausted as the acreage expands in geoclimate conditions less suited to the varieties. But Dalrymple's HYV measure is a changing indicator of technology. In particular, it includes varieties produced by national systems.

Table 9-8 presents regression results based on wheat and rice data when research capital is incorporated into the specifications. Note that the interaction with indigenous research, TR, is the important variable, as it was in the regressions based on data from all cereal grains. The diminution of the productivity of HYV's is maintained in the Table 9-8 regressions, if TR is held constant. At high levels of TR, however, the scope of the coverage of HYV's will be much greater than 25 to 50 percent before the productive effect is exhausted.

NOTES

1. Robert E. Evenson and Yoav Kislev, "Investment in Agricultural Research and Extension: A Survey of International Data," *Economic Development and Cultural Change*, 23 (April 1975), 507-521.

2. Comments by Webster and Ulbricht at the Airlie House conference reflected appropriate skepticism toward some of the cruder estimates.

3. Yujiro Hayami and Vernon W. Ruttan, *Agricultural Development: An International Perspective* (Baltimore and London: Johns Hopkins Press, 1971).

4. Robert E. Evenson and Yoav Kislev, *Agricultural Research and Productivity* (New Haven: Yale University Press, 1975), chapter 5.

5. Robert E. Evenson and Yoav Kislev, "Research and Productivity in Wheat and Maize," *Journal of Political Economy*, 81 (September-October 1973), 1309-1329.

6. For a more complete discussion of this study see Robert E. Evenson and Dayanath Jha, "The Contribution of Agricultural Research Systems to Agricultural Production in India," *Indian Journal of Agricultural Economics*, 28 (October-December 1973), 212-230.

7. Rakesh Mohan and Robert E. Evenson, "The Intensive Agricultural Districts Program in India: A New Evaluation," *Journal of Development Studies*, 11 (April 1975), 150.

8. Robert E. Evenson, "The Green Revolution in Recent Development Experience," *American Journal of Agricultural Economics*, 56:2 (May 1974), 387-394.

9. Based on footnotes from Dana G. Dalrymple, *Development and Spread of High-Yielding Varieties in the Less Developed Countries*, Foreign Agricultural Economic Research Report no. 95 (Washington, D.C.: Economic Research Service, USDA, 1974).

10. James K. Boyce and Robert E. Evenson, *National and International Agricultural Research and Extension Systems* (New York: Agricultural Development Council, Inc., 1975).

11. Evenson and Kislev, "Research and Productivity in Wheat and Maize," "Investment in Agricultural Research."

12. The eight- to ten-year time-lag is approximately the lag estimated for the United States. Robert E. Evenson, "Economic Aspects of the Organization of Agricultural Research," *Resource Allocation in Agricultural Research*, ed. W. L. Fishel (Minneapolis: University of Minnesota Press, 1971).

13. Juan Papadakis, *Climates of the World and their Agricultural Potentialities* (Buenos Aires: privately printed, 1966).

14. See Evenson and Kislev, *Agricultural Research and Productivity*, for the adjustment rationale.

15. Dalrymple, *High-Yielding Varieties*.

10

A Model of Agricultural Research[1]

Yoav Kislev

There is substantial evidence to support the notion that the rate and direction of agricultural research are influenced by economic circumstances.[2] The issues involved in this proposition are usually discussed in terms of "demand-oriented research" or "biased technological change" and focus on the payoffs to potential innovations.[3] However, as is so often the case, equilibrium is determined by both demand and supply. The basic building block of the theory of supply is the theory of production. The purpose of this chapter is to suggest a formal model of research, particularly of applied research, in the hope that this model will contribute a theory of the production of knowledge and help in understanding the supply side of the process of technical change.

The most important difference between the production of knowledge and the production of tangible goods is the strong element of uncertainty associated with the outcome of research work, and it is on this element that the present model focuses. The model is inspired by Stigler's work on the economics of information and is similar in some aspects to Nelson's treatment of research and development.[4]

The discussion starts with a presentation of an actual example. Applied research is modeled as a search in a distribution of unknown outcomes. One section discusses the properties of the model, and another is devoted to the effect of advancement in basic knowledge on applied research.[5] The last section draws some implications for the problems of the international agricultural research centers and highlights a severe shortcoming in our understanding of the operation of research systems, particularly publicly supported systems.

The Development of Modern Sugarcane Varieties

To create the model, it will be useful to open our discussion with a short history of sugarcane variety development (see also chapter 8).

Four stages can be identified in the history of sugarcane breeding. First, sexual reproduction of the cane plant was not known, and as a result, improvements in the plant were very slow to come, based on occasional and rare cases of "natural" sexual reproduction. The second stage can be dated from 1887, when it was discovered that proper light and temperature conditions can induce flowering and thus sexual reproduction. Crossings were then made and offspring observed, and seedlings with superior potentials were selected and propagated vegetatively. At this stage, crossing was random. The major innovation of the third stage was the directed crossing of selected parents, aimed at specifically influencing characteristics of offspring. At this stage, wild, disease-resistant cane varieties were introduced into the breeding program. The fourth stage marks the modern period in sugarcane breeding wherein most of the effort has gone into the development of species suited to specific local climate and technology conditions.

Starting with the second, each stage was characterized by an innovation in the method of research. This innovation augmented significantly the productivity of research, which in turn encouraged the establishment of new experiment stations and the development of new varieties. After a while, however, returns to research began visibly to diminish, as the potentials revealed by the innovation were gradually exhausted.

In statistical terms, each crossing is a random drawing from the population of all genetically possible types. This selection is a process of search, by drawing, in a population of outcomes. Figure 10-1 depicts the four stages of sugarcane development. In the first stage virtually all the observations – cane plants in the field – were concentrated around one value with very little probability of discovering different types. In Stage 2 the *sample variance* was increased tremendously; both inferior and superior types were observed among the seedlings, and the best were selected. By identifying parents, search in the third stage was limited mostly to one portion of the population. That portion is in itself a new population centered around a higher mean than those of the original Stage 1 and Stage 2 populations. Similarly, Stage 4 marks a shift to a new, different population.

The selection depicted in Figure 10-1 is unidirectional, aimed, say, at achiev-

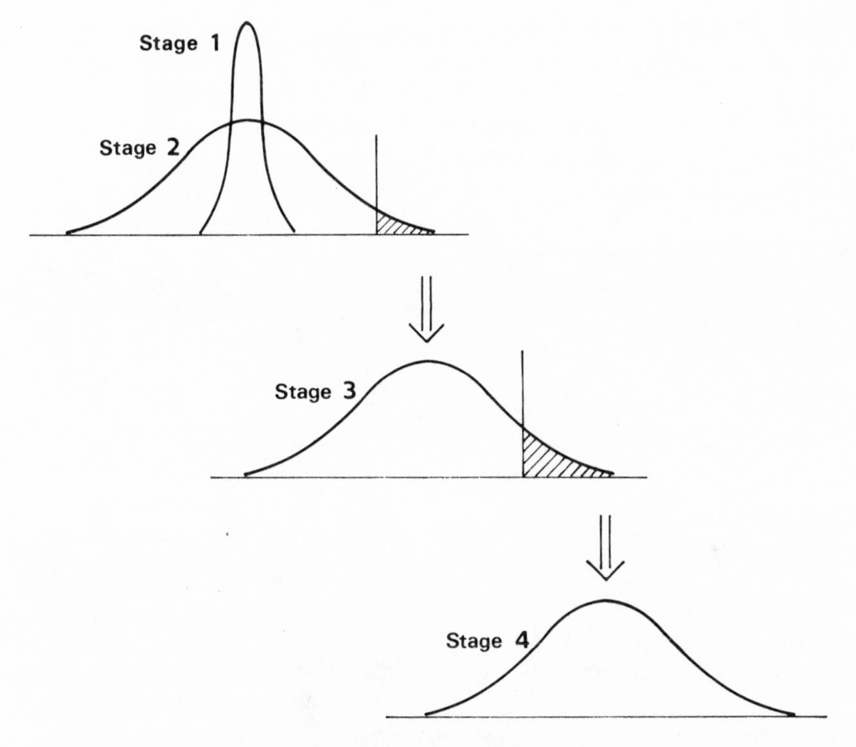

Figure 10-1. Stages in the development of sugarcane varieties.

ing higher yields. In reality, selection processes usually cover more than one characteristic of the organism. For simplicity, at this stage, the model to be developed below is also limited to unidirectional selection processes.

For concreteness, our discussion will be couched in terms of genetic-selection application and yields, but the model is of wider application. Research in other areas, such as organic chemistry, pharmacology, and the development of plant protection compounds, is technically a search and selection process, similar to breeding processes.

The Model

Imagine a scientist (or a scientific team) working to improve technology to increase the yield of a crop. To simplify, let us assume that income is proportional to yield and that the objective of the scientist is to maximize income of the whole system – research and production. Here, of course, is hidden one of the crucial issues of applied research: What signals does the research system get from its "customers," and what dictates its response? We simply assume that the scientist is aware of the economic implications of his work and acts to maximize net benefits. The scientist faces a given distribution of genetic types, which he searches to find the best. An experiment is, in our formulation, a sample drawn from the distribution of unknown outcomes. In each period the scientist conducts such an experiment, that is, he draws one sample. The genetic types obtained in the sample are ordered by their yield level, the highest yield regarded as the outcome of that experiment. If the outcome of the experiment is better than the currently practiced technology (variety), the newly discovered technology will replace the old and yield (income) will rise. If not, the current variety is maintained. Experimentation may then continue.

The process is illustrated with the exponential distribution which can be taken to approximate the right-hand tail of the normal distribution. The exponential distribution is $f(x) = \lambda e^{-\lambda(x-\theta)}$ and it can be shown that for large samples the distributions of the maxima (which is what we are interested in here) converge to the same distribution for a large class of original distributions including the normal and the exponential.[6] In Figure 10-2 yield is measured along the horizontal axis, cumulative probability and probability density along the vertical axes. The experiment depicted in the figure is composed of a sample of three drawings from the random exponential distributions – with yields x_1, x_2, x_3. In the figure x_3 exceeds the current yield level, marked y; therefore a variety replacement will take place, yield increment being Δy – the difference $x_3 - y$. Had the yield of the best variety in the experiment been smaller than y, the use of the current variety (with yield y) would have continued.

Since the actual outcome of the experiment is not known in advance, the value of the *expected* contribution, $E(\Delta y)$, for a given distribution of potential outcomes depends on the following two magnitudes.

1. The extent of experimentation, that is, the number of observations. The larger this number, the higher the probability of finding x values of larger magnitude. The probabilities, the density functions for the highest value in a sample, are depicted for five values of n in Figure 10-3.

2. The value of y – the current yield level. The higher the current level of technology, the less probable it is that an improvement will be discovered.

The economic value of the expected technological improvement is the present value of its future contributions. For simplicity of exposition we shall regard it here as $1/r\,[E(\Delta y)]$. This will be the correct contribution if experimentation is to take place only once; if it continues in later periods, an improvement today "spoils" the chances of improving technology tomorrow. This

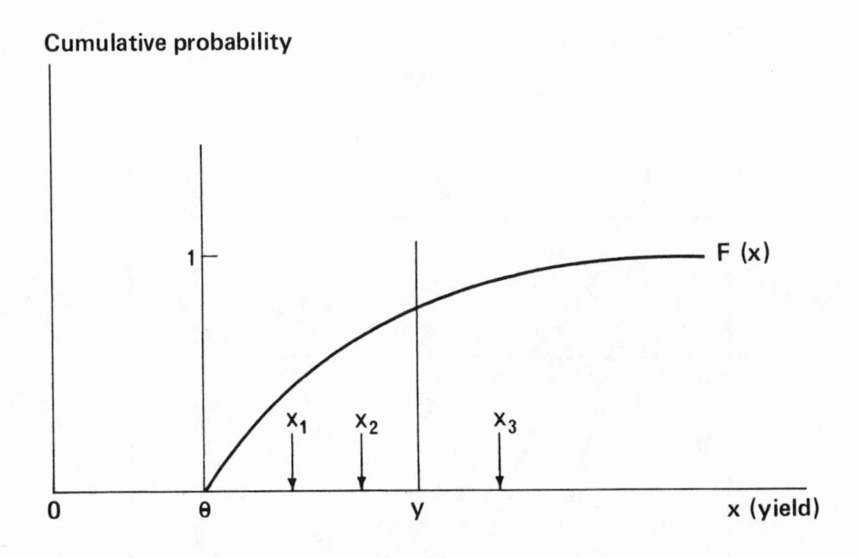

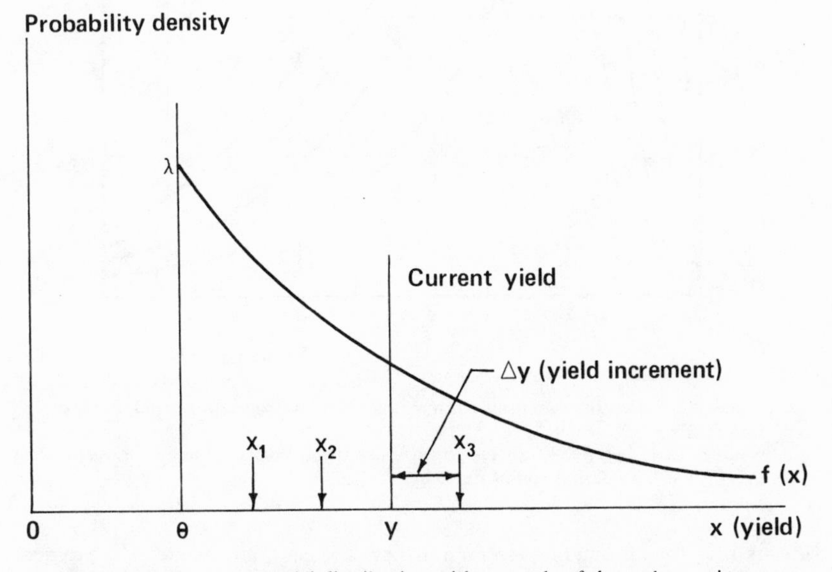

Figure 10-2. The exponential distribution with a sample of three observations.

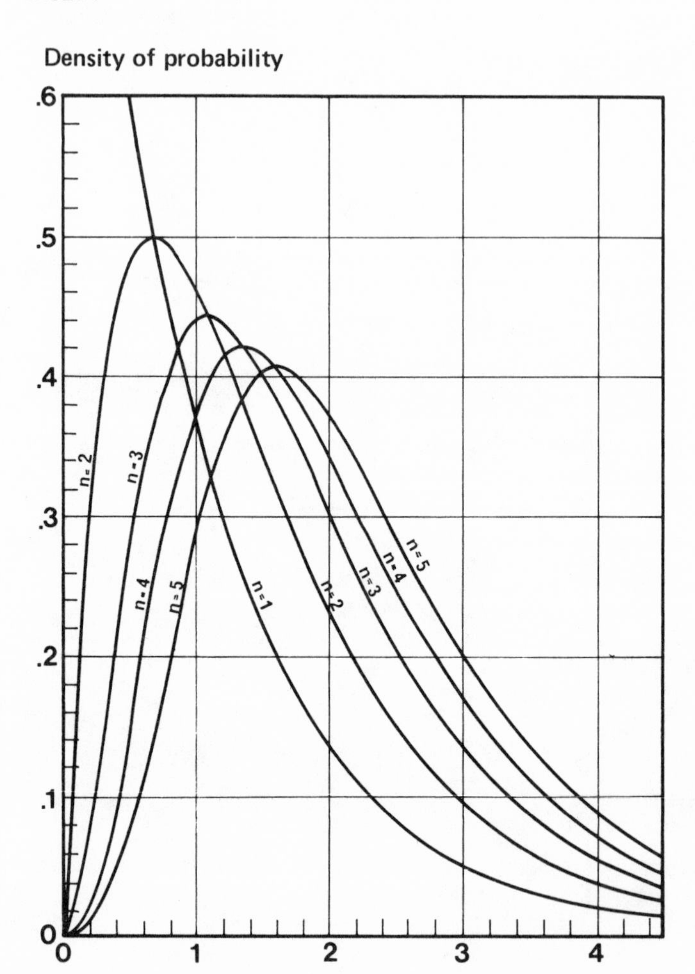

Figure 10-3. Density functions for maximal values of the exponential distribution $f(x) = e^{-x}$ for $n - 1, \ldots, 5$.
Source: E. J. Gumbel, *Statistics of Extremes* (New York: Columbia University Press, 1958), by permission of publisher.

means that for an ongoing research project the economic value of the expected technological improvement is lower than $1/r\,[E(\triangle y)]$, but this does not affect the relevant properties of the research system to be analyzed below.

The economic problem of the research system is to decide on the optimal amount of experimentation to conduct in each period – on the optimal num-

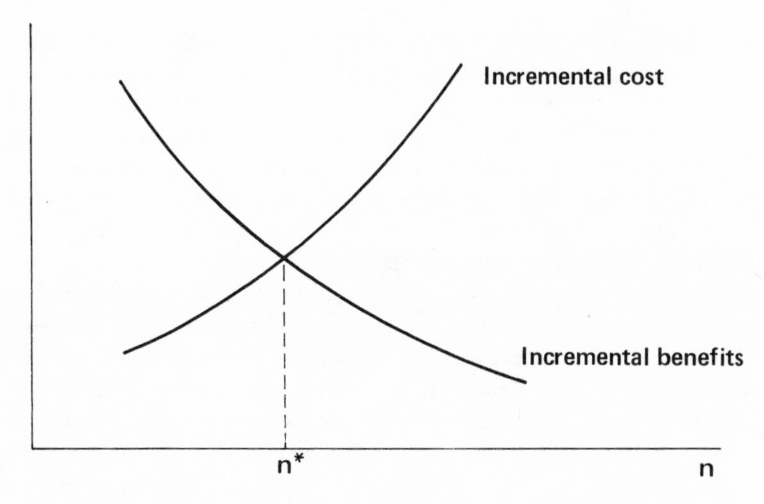

Figure 10-4. Optimal number of observations, n* (the smooth graphs in the diagram approximate step functions).

ber of observations, n, to draw. A reasonable assumption is that cost of research is an increasing function of n. Average costs may decrease for small values of n. This introduces nonconvexities which are assumed away at this stage. As an examination of Figure 10-3 will reveal, incremental technology improvement is a decreasing function of the number of observations; that is, each additional observation adds something to the expected technology increment, E(△y), but these additions get smaller and smaller as the number of observations increases. The magnitude $E_n(\triangle y) - E_{n-1}(\triangle y)$ is positive and decreasing with n. We get, therefore, the familiar equilibrium position determined by the intersection of marginal (incremental) cost and returns (see Figure 10-4).

Implications and Extensions

Technological research is more fruitful the wider the divergency between the level of theoretical scientific knowledge and the level of technology in practice. Nelson and Phelps made this assumption in their model of technological diffusion and schooling.[7] Evenson and I estimated the increase in the productivity of research in one country owing to the availability of relevant knowledge in other countries.[8] Technological gaps also explain rates of international technology diffusion when adaptive-type applied research is needed to facilitate the transfer of knowledge (see chapter 8).

In our framework a formal definition of the *technological gap* is the difference $y - \theta$ (see Figure 10-2). Optically, this may seem inappropriate, as one is inclined to define the technological gap as the difference between some potential technological ceiling and the current level of technology. But such a ceiling can probably never be defined, and our measure, the magnitude $y - \theta$, measures the ease at which new technological improvements can be achieved. On more formal terms, the *smaller* the difference $y - \theta$ the higher the probability of finding technologies superior to the currect practice for a given rate of experimentation – a given n.

With our assumptions, $\triangle y$, the technological change, can be only positive – technology can only improve with time, closing the technological gap as it improves. Even if experimentation continues at the same rate (n = const.), technological change will decrease with time. Moreover, since the marginal expected returns to research will decrease with time (as the technological gap closes), optimal n will decrease with time, reducing even further the acceleration rate of technology. Eventually, a point will be reached at which expected returns will be smaller than the cost of a single observation, n = 1, research will stop, and technology will stagnate forever.

It is worthwhile to recall at this point the assumptions that led to our conclusions: (a) technology will stagnate so long as basic knowledge is constant (θ = const.); and (b) research will stop as its payoff diminishes. Even if θ = const., research will continue if technology deteriorates or is subject to obsolescence. This might be termed maintenance research – the research necessary to maintain current productivity levels.

The optimal rate of experimentation depends on the present value of future probable benefits from research – $1/r\,[E(\triangle y)]$. The *lower* the rate of interest, the higher the present value of future benefits and the higher the optimal level of technological research. Like any investment, research (investment in knowledge) is a decreasing function of the rate of interest.

Scale can enter the economics of research in several ways. Evenson discusses the effect of the scale of the experiment station.[9] Here economies of scale in the creation of knowledge stem mainly from the interaction of scientists working in different disciplines and from the integration of research and graduate university training, factors which cannot be introduced into our model in its present, simplified one-product stage.

Another, different aspect of scale effect is the size of the industry which is affected by the new technology. The larger the industry, the higher the benefits to research. But, particularly in agriculture, size of industry as measured by acreage, for example, is also associated with diversity in conditions of production. In this case benefits will not be directly proportional to scale.[10]

Recall that the first stage in the development of the modern sugarcane va-

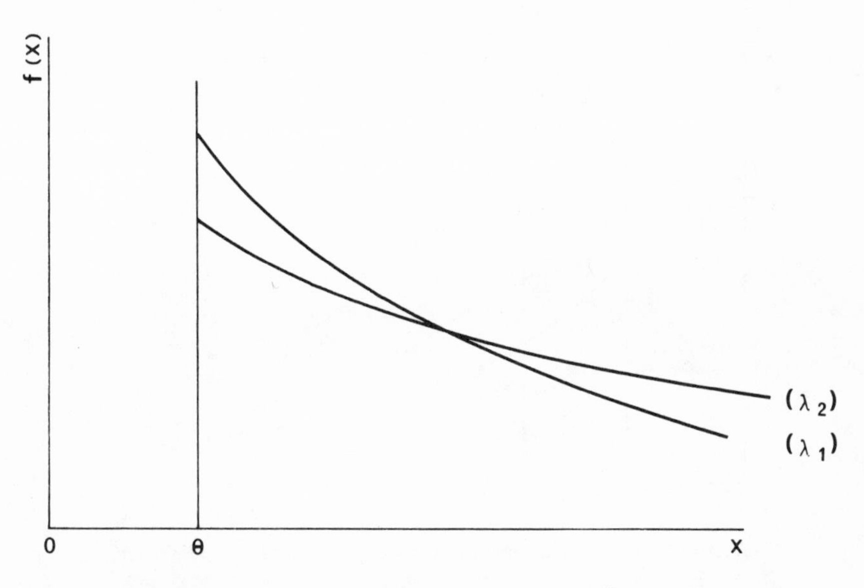

Figure 10-5. The effect of an increased variance [$\text{Var}(x) = \frac{1}{\lambda^2}$, $\lambda_2 > \lambda_1$].

rieties consisted, in fact, of the discovery of a variance-increasing technique. The importance of research designed to develop a genetic pool that incorporates greater diversity (variance) is not always properly appreciated. The collection of varieties, radioactive radiation, methods for creating new chemical compounds – all these are variance-increasing techniques which are followed by search and selection types of research work.

In our model the variance parameter is λ. (In the exponential distribution it is also the mean parameter [$E(x) = \theta + 1/\lambda$].) Reducing λ will increase the probability of finding higher x values (see Figure 10-5) of improving technology. Often applied research will not be economically justified until ways to increase the variance of the samples observed are developed.

Basic Research

Basic research widens the technological gap and increases the probability of finding superior technologies. In our model, basic research will shift the parameter θ to the right (see Figure 10-6) and will increase optimal experimentation.

It is interesting to note here a *steady-state property*. Assume that basic research and applied research proceed at constant rates (per period $\Delta\theta$ = const., n = const.). Then the rate of advancement of technology in practice will ul-

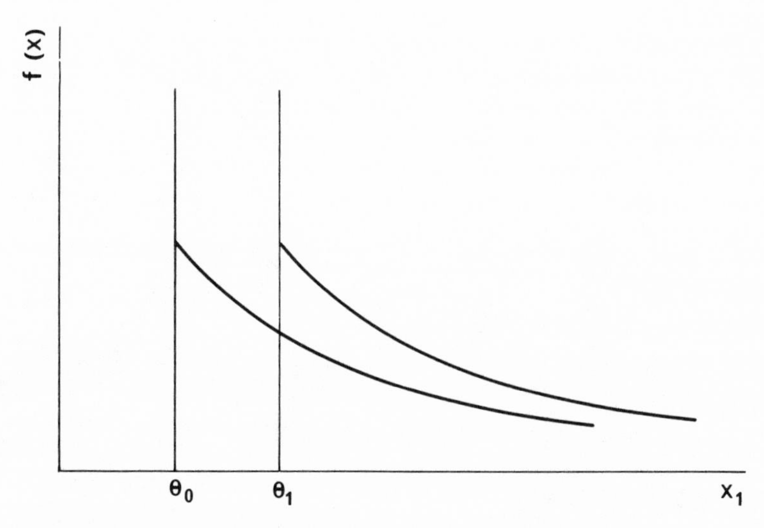

Figure 10-6. Basic research shifts the population searched to the right.

timately converge to the rate of advancement of basic knowledge. (A special case has already been encountered: in the absence of basic research, technology will eventually stagnate at a constant level.) The level of technological research in the steady state does not affect the *rate* of technological advancement, but only the *level* of technology in practice (see Figure 10-7).

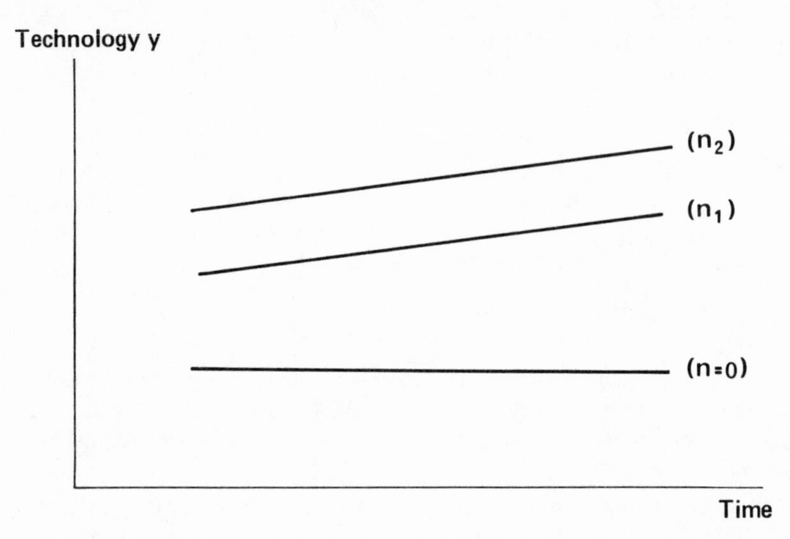

Figure 10-7. Time path of technology in the steady state at different levels of experimentation ($n_2 > n_1$).

The proof of the steady-state property is based on the fact that technological improvements at a rate faster than the advancement of basic knowledge will close the technological gap and decrease the probability of further technological change. Slower technological improvements will increase that probability. The endogenous variable in the system is the technological gap – it will be adjusted to maintain a steady-state equilibrium, with technological change fluctuating stochastically around a constant rate of change.

It can also be shown that in the long run, if basic knowledge proceeds at a constant rate, constant rate of experimentation is optimal. Thus in the long run the steady state is the optimal state of a research system fueled by a constant rate of new basic knowledge.

Concluding Remarks

Perhaps the most dramatic recent development in the field of agricultural research is the establishment of the international research centers. With their size and resources they can draw the best scientists, and they are well equipped to perform their task. Two elements contribute substantially to the success of these centers: (1) a wide technological gap – an example which comes immediately to mind is the case of the *Norin* 10 Japanese dwarf wheat variety which served as the basis for the development of the Mexican wheats and later provided principles for the breeding of the "miracle rice" varieties; (2) the large variance of the populations they search – the International Rice Research Institute (IRRI) alone has a collection of 30,000 rice varieties, and the International Center for the Improvement of Maize and Wheat (CIMMYT) performs tests all over the world. The subtropics and the tropics have been comparatively neglected in agricultural research, and the potential exists there for successful international efforts. (On these issues see chapters 7 and 9.)

Yet eventually, once the gap has been closed and the large variance exploited, the rate of progress of technology will be limited by the advancement of basic knowledge. Thus, in principle, the contribution of the international centers can be expected to decline in time (after an initial stage of acceleration).

The initial focus by the centers on wide-base technologies was clearly consistent with the objective of efficient use of research resources. Mexican wheats are now grown over large areas and under diversified geoclimate conditions. In following this procedure, the centers opened up technological gaps in the countries that were recipients of the new technologies and dramatically increased the payoffs to local adaptive research. Such explicit signals are being recognized by public authorities. As the potential gains are recognized, the effect is to expand and strengthen research institutions in many developing countries. The more successful the international centers, the more productive will local research work be. This will have a significant impact on the future role of the international centers. They cannot be expected to remain the sole producers of modern technology. Rather, additional emphasis will be given to their function as clearing houses for knowledge and genetic material. An excellent example is IRRI's recent International Rice Testing Program in which "scientists are cooperatively developing a 'critical mass' of genetic technology that will feed improved genetic materials and breeding information to scientists across the rice growing world." The main feature of the program is that "each nation contribute[s] . . . genetic materials, testing results and breeding information to help develop the critical mass effect."[11] Such developments will reduce the *relative* contribution of the international centers to global agricultural research. Furthermore, more effective scientific communication systems can be expected to reduce the importance of the geographic concentration of scientific work in one locality. These effects should not be expected to materialize fully in the near future, but the centers should be ready for them, as they should be welcome. The success of the international research efforts will induce competition among national systems. This result should be welcomed as an indication of the successful contribution of the institutes to global research capacity.

I have tried to point out clearly and explicitly the simplifying assumptions of the model presented in this chapter. I hope that despite its limitations the model can serve as a starting point for a theory of research and as a stepping-stone for further analysis. Two reservations are worth mentioning in conclusion.

In the present formulation, basic scientific knowledge is exogenous to the applied research system. Ruttan has pointed out the reverse link whereby advances in technology – in instrumentation, for example – contributed to further advances in basic science.[12] Changes in factor supplies or in demand may also exert an independent impact on the productivity of technological research. This observation sheds strong light on the "system" nature of the science-technology complex which stretches even further. For example, the level and quality of university education is to a large extent a function of the quality of the research conducted at the institutes of higher learning. Thus a high-quality scientific community breeds the technical personnel that later conducts the applied and adaptive research. On the other hand, the demand for technical skills, if properly channeled, strengthens institutions of higher learning and "pure," basic science. (I am, of course, abstracting from a host of complicated and important issues which are outside the scope of our present discussion.) Against the background of this broad view, ours is a partial analysis – one building block of the system.

There remains one shortcoming in our understanding of the operation of public research systems which I should like to reemphasize here, and this is the nature of the connection between the demand, or the payoffs to research, on the one hand, and the policy maker and the scientist on the other. What is the "market" in which such a system operates? What signals does it follow? What "profits," if any, are maximized? Hayami and Ruttan showed that research in Japan and the United States reacted efficiently to economic incentives.[13] Why did it not react in a similar manner in many other places? This gap in our knowledge and understanding is part of a larger gap: the lack of an economic theory of bureaucracy and our meager understanding of the process of economic development.

NOTES

1. This study was supported in part by the United States-Israel Binational Science Foundation.

2. Yujiro Hayami and Vernon W. Ruttan, *Agricultural Development: An International Perspective* (Baltimore and London: Johns Hopkins Press, 1971).

3. An exception is the "innovation possibility frontier" suggested by Charles Kennedy, "Induced Bias in Innovation and the Theory of Distribution," *Economic Journal*, 74 (1961), 541-547. For a critique of Kennedy's model see H. P. Binswanger, "A Microeconomic Approach to Induced Innovation," *Economic Journal*, 84 (1974), 940-958. See Hayami and Ruttan, *Agricultural Development*, for references to other sources on these issues.

4. G. J. Stigler, "The Economics of Information," *Journal of Political Economy*, 69 (1961), 213-215. R. R. Nelson, "Uncertainty, Learning and the Economics of Parallel Research and Development Efforts," *Review of Economics and Statistics*, 43 (1961), 351-364.

5. This is a nonmathematical exposition. For rigorous formulations and proofs, see Robert Evenson and Yoav Kislev, *Agricultural Research and Productivity* (New Haven: Yale University Press, 1975), chapter 8.

6. B. Epstein, "Elements of the Theory of Extreme Values," *Technometrics*, 2 (1960), 27-41.

7. R. R. Nelson and Edmund S. Phelps, "Investment in Humans, Technological Diffusion, and Economic Growth," *American Economic Review*, 56 (1966), 69-75.

8. Robert E. Evenson and Yoav Kislev, "Research and Productivity in Wheat and Maize," *Journal of Political Economy*, 81 (September-October 1973), 1309-1329.

9. Robert E. Evenson, "Economic Aspects of the Organization of Agricultural Research," *Resource Allocation in Agricultural Research*, ed. Walter L. Fishel (Minneapolis: University of Minnesota Press, 1971).

10. For an empirical attempt to cope with some of these problems see Evenson and Kislev, "Research and Productivity in Wheat and Maize."

11. *IRRI Reporter* (Los Baños: International Rice Research Institute, February 1975).

12. Vernon W. Ruttan, 1975, personal communication.

13. Hayami and Ruttan, *Agricultural Development*.

Organization and Development
of the International Institute System

11

Development of the International Agricultural Research System[1]

J. G. Crawford

This chapter is about international agricultural research and, more especially, developments since May 1971. In that month the first meeting of the Consultative Group on International Agricultural Research (CGIAR) was held. The membership comprised the sponsors, which included the World Bank (IBRD), the Food and Agriculture Organization (FAO), the United Nations Development Program (UNDP), and fifteen additional members – nine national governments, two regional banks, three foundations (Rockefeller, Ford, and Kellogg) and the very young International Development Research Centre (IDRC) of Canada. (See Appendix 11-1 for a complete listing of the membership.)

The Consultative Group had met informally in January 1971 following continuing talks between the Ford and Rockefeller foundations, IBRD, FAO, and UNDP since October 1969. The leadership was provided by the World Bank, which now provides the chairman and secretariat of the group. The May meeting adopted a number of objectives (see Appendix 11-2) designed to strengthen existing international research in Mexico (CIMMYT), the Philippines (IRRI), Latin America (CIAT), and West Africa (IITA), and to develop new activities to meet priority needs as determined by the Consultative Group.[2]

The impetus for this action was the desire to encourage more research to assist developing nations increase the quantity and improve the quality of

Table 11-1. International Agricultural Research System in 1975

Center	Location	Research	Coverage	Date of initiation	Proposed budget for 1975 ($000)
IRRI (International Rice Research Institute)	Los Baños, Philippines	Rice under irrigation; multiple cropping systems; upland rice	Worldwide, special emphasis in Asia	1959	$8,520
CIMMYT (International Center for the Improvement of Maize and Wheat)	El Batan, Mexico	Wheat (also triticale, barley); maize	Worldwide	1964	6,834
CIAT (International Center for Tropical Agriculture)	Palmira, Colombia	Beef; cassava; field beans; farming systems; swine (minor); maize and rice (regional relay stations to CIMMYT and IRRI)	Worldwide in lowland tropics, special emphasis in Latin America	1968	5,828
IITA (International Institute of Tropical Agriculture)	Ibadan, Nigeria	Farming systems; cereals (rice and maize as regional relay stations for IRRI and CIMMYT); grain legume (cowpeas, soybeans, lima beans, pigeon peas); root and tuber crops (cassava, sweet potatoes, yams) Maintaining fertility in humid tropics	Worldwide in lowland tropics, special emphasis in Africa	1965	7,746
CIP (International Potato Center)	Lima, Peru	Potatoes (for both tropics and temperate regions)	Worldwide including linkages with developed countries	1972	2,403

Table 11-1 – continued

Center	Location	Research	Coverage	Date of initiation	Proposed budget for 1975 ($000)
ICRISAT (International Crops Research Institute for the Semi-Arid Tropics)	Hyderabad, India	Sorghum; pearl millet; pigeon peas; chick-peas; farming systems; groundnuts	Worldwide, special emphasis on dry semiarid tropics, nonirrigated farming. Special relay stations in Africa under negotiation	1972	10,250
ILRAD (International Laboratory for Research on Animal Diseases)	Nairobi, Kenya	Trypanosomiasis; theileriasis (mainly east coast fever)	Africa	1974	2,170
ILCA (International Livestock Center for Africa)	Addis Ababa, Ethiopia	Livestock production systems	Major ecological regions in tropical zones of Africa	1974	1,885
IBPGR (International Board for Plant Genetic Resources)	FAO, Rome, Italy	Conservation of plant genetic material with special reference to cereals	Worldwide	1973	555
WARDA (West African Rice Development Association)	Monrovia, Liberia	Regional cooperative effort in adaptive rice research among 13 nations with IITA and IRRI support	West Africa	1971	575
ICARDA (International Center for Agricultural Research in Dry Areas)	Lebanon	Probably a center or centers for crop and mixed farming systems research, with a focus on sheep, barley, wheat, and lentils	Worldwide, emphasis on the semiarid winter rainfall zone		

their agricultural output and thus to raise standards of living. All parties realized that increased agricultural productivity was essential to economic and social development in the great majority of these countries.

Additionally, the two foundations which had initiated the research centers had indicated that the future needs of these and other new centers would be beyond their financial capacity. They indicated continuing strong support but nevertheless felt the need, in the unofficial words of one senior official, of "going public." I believe the step the foundations took will be given a high and honorable place in the history and achievements of international cooperation.

From this beginning, the international research system has grown in breadth and complexity. It now comprises eleven centers, located in Asia, Africa, Latin America, and Europe, which are involved in varied research programs, as can been seen in Table 11-1. An additional center in the Middle East is being planned.

From a cost of about U.S. $12 to 14 million in 1972, the program financed by the Consultative Group was close to $34 million in 1974 (capital, core budget, and outreach programs) with a commitment of about $45 million for 1975. While an element of this increase is attributable to inflation, it does represent a growth from four centers in 1971 to nine in 1975. In a short period of three years, the CGIAR system has given solid evidence of its willingness and ability to back its judgment that international research has a vital role to play.

The Technical Advisory Committee

To assist it in its work, the Consultative Group established in May 1971 the Technical Advisory Committee (TAC) of which I have the honor to be chairman. It comprises twelve scientists and me. (See Appendix 11-3 for a list of members.) FAO provides the secretariat. Put in a sentence, the task of TAC is to define priorities for research and to recommend action.

TAC may either act on its own initiative or consider proposals submitted from the Consultative Group through its sponsors. (See Appendix 11-4 for TAC's full terms of reference.) It has to be remembered that the established four centers – IRRI, CIMMYT, CIAT, and IITA – had virtually preempted judgment on the matter of research priorities. While TAC had little difficulty in approving the main work of these bodies, it has begun to encourage some new thrusts, such as the move into rainfed rice production by IRRI.

The work of TAC is explicitly related to the problems of developing countries both in technical (agricultural) and socioeconomic fields. National research in developed countries is often highly relevant but is of formal concern to TAC only when it is or can be linked with the problems of developing countries.

TAC has defined "international research" broadly to mean "research which, although based in one country, is of wider concern, regionally or globally; is independent of national interest and government control; and retains appropriate links with national and other regional or international research systems to ensure the necessary testing of results and feedback of both results and needs."

TAC is advisory only: the Consultative Group will not act on important research proposals without prior advice from TAC, but it remains free to reject or modify the advice it receives. It follows that the Consultative Group has to be persuaded by TAC in its development of ideas on priorities.

Priorities and Programs

TAC, confirmed by the Consultative Group, places the highest importance on research directed toward increasing the amount and quality of food produced. This is hardly surprising in the Malthusian situation in which much of the world finds itself. At best, TAC believes, research will buy time while population growth is brought under control, but it is vital even for this purpose. This is reflected in TAC's statement on the cereals:

> In the first place cereals provide the mainstay of the diet in most developing countries, especially for the poorer people, supplying an average 52 percent of the calories and nearly half the total protein. It has been shown that if there is a serious deficit in calories in the diet the body consumes protein for energy. Since cereals generally make the largest single contribution of any commodity to both energy and protein, research to increase their yield and protein content is of crucial nutritional importance. Upgrading their amino-acid composition could, at no extra cost to consumers, make a further improvement in the quality of the diet. Secondly, despite the real successes in increasing wheat and rice output, cereal production in developing countries has barely kept pace with population *and* income growth during recent years, and experience in Asia in the last two years shows how fragile is the base on which these critical supplies rests. Income elasticity of demand for cereals is still high in the poorer countries, quite unlike the situation for food grains in the developed economies, and an important indication that food consumption levels are inadequate. In a number of countries failure to increase production rapidly enough to meet domestic demand has led to increasing imports, draining foreign exchange required for social and economic development. Third, cereals are the lynchpin of the cropping system in many developing countries and contribute signifi-

> cantly to income and employment. Finally, a faster growth of grain production will be necessary if feed supplies are to become available in sufficient quantity at prices which will permit their economic use in livestock rations. It is relevant to note the rapidly rising demand for feedgrains in the more affluent nations, which has been one factor contributing to the recent stringency and high prices of cereals and soya beans. This has in turn affected the availability of grains on concessional or normal trade terms to developing countries and further emphasizes the need to increase output in these countries as rapidly as possible.

The TAC report goes on to observe the growing necessity to raise yields per hectare as new arable land becomes more and more limited in relation to population. Its report comments:

> It thus becomes increasingly necessary to turn towards raising yields and crop intensities per acre as the major source of future growth, and since cereals occupy the largest share of the arable area in a wide range of environments, they hold the key to the more effective use of land and water resources. *Unless their yields can be increased or their time to maturity reduced, it will be correspondingly more difficult to make significant progress with other crops and livestock since more and more land will have to be devoted to satisfying basic calorie requirements.* The alternative – increasing imports – is open only to a few countries.

The main cereals supported are rice, wheat, barley, triticale, sorghum, and millet.

Turning from cereals to other key commodities, the TAC has accorded high priority to those which will improve the quality of the diet, especially in respect to protein. In particular, it has focused atttention on the food legumes and on ruminant livestock. (It recognizes fully the place of pigs and poultry, especially in developing their production by labor-intensive methods. Most members have felt that this would not require extensive research, but that the opportunities open for such development could be seized by the application of known methods of disease control, feeding, and management.) TAC is also supporting research in starchy foods including cassava, potatoes, yams, and sweet potatoes. The importance of these crops in many developing areas with poor resources in relation to population – as in tropical Africa – is very great indeed. TAC has yet to determine its position in respect of aquaculture but there is evidence of scope for research and training with definite promise of breakthrough.

TAC does recognize a second-level priority for food research. To quote its statement again:

> Having taken a firm position on its priorities for cereals, food legumes, roots and tubers, and ruminant livestock (especially cattle), and placed

a temporary questionmark against aquaculture, the TAC has been less decisive on some other foods, in particular oilseeds, vegetables and tropical fruits. This is partly the result of pressure of work related to the commodities listed above, which, it decided in its earliest session, were of highest priority, and where some good projects were already in the pipeline, but it also reflects a lack of sound proposals for research in other food commodities.

This brings me to the important question of "nonfood," "industrial" or "agricultural raw materials" which have considerable importance in the economies of many developing countries. Cotton, jute, rubber, and forests are examples. Given the probable order of financial constraints, which I discuss later, TAC has been firm that it would be unhappy to give preference to research in these crops (to be supported by the Consultative Group) if this was likely to impair necessary programs of food research. TAC is willing to consider proposals for nonfood crops referred to it provided that "the overriding need to secure the staple food supplies of the mass of the people was first covered by existing or new international and regional research programmes."

Factor-Oriented Research and Systems Research

I have outlined our priorities in commodity terms. This at least has the merit of clarity and easy definition. Nevertheless, TAC has also had before it certain proposals for research relating to what might best be defined as factors of production – water use and management, fertilizers, integrated pest control, pesticide residues, etc. – which have caused it some difficult moments. TAC stated its position as follows:

> In general, members have taken the view that such problems are most meaningfully studied in relation to specific commodities rather than as ends in themselves. They have argued that one of the reasons for the success of the rice and wheat programmes has been the realisation by IRRI and CIMMYT of the need to develop and present to the farmer an integrated "package" of technology appropriate to their new varieties, and not just the latter in isolation unsupported by other essential inputs.
>
> While there is much merit in this argument, there are nevertheless instances which can be identified where it may be an inadequate approach and where it is essential to move from the study of the commodity or package of technology to that of the system. Except in monocultures, water use and management has to be related to the crop-mix rather than to the individual crop; fertiliser and pesticide residues contributing to environmental pollution again come from the totality of the farm

> and not just one enterprise. The introduction of small-scale livestock often implies a major revision of an established system. Multiple cropping depending on high output per annum involves radically different management *and* plant breeding and cultural concepts than systems which depend principally on high yield per individual crop. Inadequate survey and exploration of surface and sub-surface water resources combined with insufficient research on soil/plant/water relationships is frequently a serious obstacle to sound design, good water management and the development of optimum production systems in irrigated areas. Storage, and control of certain causes of crop loss, e.g., rodents, may present problems of a broader nature than a single crop.

It is with these thoughts in mind that TAC has, of late, been stressing the scope for systems work which leads into socioeconomic research also. It is naturally concerned, within this framework, with devising means of intensifying agriculture as a means of raising total productivity (of two or more crops) per hectare through better resource utilization. In doing so, it may at times be forced to recognize a degree of location specificity not normally a constraint on commodity-oriented research. This in turn gives emphasis to regional and national research of the kind being considered in Africa (livestock management) and in the Middle East.

Despite the greater difficulty confronting TAC in looking at research in noncommodity-oriented terms, TAC is prepared to do so.[3] Its general position is summed as follows:

> *But although increasing yields and production of basic staple foods must remain a priority goal, the ultimate objective of agricultural research is development and the economic well-being of people. We must not be so bewitched with the hopes of further spectacular successes with single crops that we fail to recognize that other pathways to growth may exist. In some regions, for ecological, social, or economic reasons, research of a broader nature – even if it appears more complex, may offer the better hope of a solution. Where such an approach seems desirable the TAC and the Consultative Group must grasp the nettle boldly.*

Socioeconomic Research

The very real problems of, and opportunities for success in, the green revolution have aroused widespread demands for a single international center in socioeconomic research. At the other extreme, many expect IRRI and CIMMYT to carry the whole burden of socioeconomic research associated with the national application of the rice and wheat technologies emerging from their work. Neither approach alone makes sense.

TAC has recognized three levels of action:

(i) *research at the micro-level* (farm or village community), to identify the socio-economic constraints to the successful adoption of new technology, and to guide scientists at the International Centres and elsewhere as to the types of technology most likely to be acceptable to farmers.

(ii) *research at the level of public policy*, e.g., to determine the measures and incentives needed to accelerate the use by farmers of technical innovations, to give early warnings of possible "second generation" effects of such innovations, e.g., on employment or prices, and to illuminate the choice of alternatives.

(iii) *research at the macro-level* on broad issues affecting more than one country, or the economy of a country as a whole, e.g., on commodities and trade, some aspects of nutrition, sectoral analysis, etc.

It is clear that the international research centers can do much under (i); and TAC has recommended accordingly. This first category leads into (ii) – *public policy* social and economic issues. While external assistance and investment support can be given in these areas, the identification of problems and plans for their solution is very much a problem for research, planning, and governments in the nations affected. They can be helped, but no more, by training and seminars conducted by the international centers.

Where issues affect more than one country, there is scope for more international action. This will become increasingly apparent in fertilizer supplies, commodity trade, pricing problems, and investment aid for development. In some of these matters, e.g., providing an early warning system for cereal production forecasts or for major issues of world food policy, there is undoubtedly room for concentrated international effort – governmental and nongovernmental – but these are beyond my terms of reference.

The real need, therefore, is strengthened socioeconomic work associated with the development of new technologies at research centers: international and national (including universities); national work on public policy implications of new technologies; and international effort especially on economic and environmental efforts beyond the scope of national governments alone.

Basic and Applied Research: Flexibility in Research Organization

The TAC report does touch on this question. My comment must be extremely brief and avoids the problems of theoretical delineation of the two terms. Indeed I shall make only one point: it may not be wise or necessary to tackle "basic" problems through new institutions. Thus *triticale* is the product of basic work in Canada and elsewhere; *soybean* problems are probably better handled in such places as the University of Illinois, and some of the worrying, relatively low-yield characteristics of legumes which may have physiological and morphological explanations could be dealt with by contracts to university centers for research. TAC has started to consider this last possibility. The contractual approach is an illustration also of needed flexibility in our approach to international research. Not everything calls for an IRRI or a CIMMYT in organization terms.

Strengthening National Institutions

TAC could not possibly handle requests for strengthening national research systems, nor is the CGIAR established for this purpose. Nevertheless, TAC is clear that unless national research capacities are strengthened to an extent that enables them to take advantage of the results of international research the dividend from international research will be limited. Moreover, I do not hesitate to stress again the importance to the international centers of feedback from adaptive research within national boundaries. Sometimes this is effected by outreach programs; but Indian and even Indonesian research capacities are not typical of the many very poor countries in Africa and Asia. Accordingly TAC has strongly urged more financial and organizational support from FAO, UNDP, the World Bank, and bilateral donors for national research efforts.

Financial Constraints

As I have pointed out above, the total of core, capital, and outreach programs has risen from $12 to 14 million in 1972 to an estimated $45 million in 1975. At constant prices this could – given continued support by the group – reach $57 million in the late seventies and perhaps $64 million in the early eighties. These two figures could be $65 million and $87 million if inflation continues at recent rates. Of course, these figures would quickly increase if the group were to invite TAC to cross the borderline between regional and national research.

We have here a dilemma of concern both to TAC and to the Consultative Group. On the one hand, TAC cannot assume unlimited support; on the other, once the group offers support it must also assume reasonable continuity in that support. The group has given TAC very strong backing thus far. For its part, I believe TAC has, in its priorities, acted with care and financial responsibility. It knows very well that research programs must be reviewed from time to time and unnecessary or unpromising work deleted. It is about to establish working relations with the centers to this end. On the other hand, it will continue to press for support for new work which it considers to be a vital contribution to the solution of the world's Malthusian situation.

I think I should finish on a note of confidence. The CGIAR/TAC system is a unique venture in international collaboration. It has succeeded beyond anyone's real expectations in 1971. We confidently await an increasing and usable output from the rising research investment. Much now depends on international and national economic and social policies to apply the actual and anticipated research results. This topic, which embraces the whole meaning of, and prospect for, development in the poorest areas of the world is beyond my brief in this chapter. However, I conclude simply by saying I do not belong to the band of hopeless pessimists. The CGIAR/TAC experiment has buoyed my hopes and expectations that good sense will yet prevail.

APPENDIXES

Appendix 11-1. Membership of the Consultative Group on International Agricultural Research

Membership as of November 1, 1974 comprised the following: Australia, Belgium, Canada, Denmark, France, Germany, Japan, Netherlands, Nigeria, Norway, Sweden, Switzerland, United Kingdom, United States, United Nations Environment Program, three regional development banks (African, Asian, and Inter-American Development Banks), the Commission of the European Communities, three private foundations (Ford, Rockefeller, and Kellogg foundations), and the International Development Research Centre, an independent Canadian organization.

The five major developing regions of the world participate in the Consultative Group through representatives designated for a two-year term by the membership of FAO. Each region has designated two countries which alternate as members at their discretion. Representing Latin America are Argentina and Brazil; representing Africa: Morocco and Nigeria; representing Asia and the Far East: Malaysia and Thailand; representing the Middle East: Egypt and Pakistan; representing southern and eastern Europe: Israel and Rumania.

The World Bank serves as chairman of the Consultative Group, FAO and UNDP as cosponsors.

Appendix 11-2. Objectives of the Consultative Group on International Agricultural Research

The main objectives of the Consultative Group (assisted as necessary by its Technical Advisory Committee, or TAC) are as follows:

(i) On the basis of a review of existing national, regional and international research activities, to examine the needs of developing countries for special effort in agricultural research at the international and regional levels in critical subject sectors unlikely otherwise to be adequately covered by existing research facilities, and to consider how these needs could be met.[4]

(ii) To attempt to ensure maximum complementarity of international and regional efforts with national efforts in financing and undertaking agricultural research in the future and to encourage full exchange of information among national, regional and international agricultural research centers.

(iii) To review the financial and other requirements of those international and regional research activities which the Group considers of high priority, and to consider the provision of finance for those activities, taking into account the need to ensure continuity of research over a substantial period.[5]

(iv) To undertake a continuing review of priorities and research networks related to the needs of developing countries, to enable the Group to adjust its support policies to changing needs, and to achieve economy of effort.

(v) To suggest feasibility studies of specific proposals to reach mutual agreement on how these studies should be undertaken and financed, and to exchange information on the results.

In all of the deliberations of the Consultative Group and the Technical Advisory Committee, account will be taken not only of technical but also of ecological, economic, and social factors.

Appendix 11-3. Original Members of the Technical Advisory Committee on International Agricultural Research[a]

1. *Sir John Crawford* (economist), Australian National University, Canberra. Chairman. — Australia
2. *Ing. Manuel Elgueta* (agronomist), Ex-director, Chilean Agricultural Research Institute; now working with IICA as director of proposed Turrialba Research Corporation — Chile
3. *Professor Dr. Hassan Ali El-Tobgy* (geneticist), Undersecretary of Agriculture and chairman of the Research Committee — UAR

[a] *Professor D. Bommer*, Head, Institute for Plant Cultivation and Seed Research, Agricultural Research Centre, was added as the thirteenth member in June 1972. — Fed. Republic of Germany

4. *Professor H. Fukuda* (irrigation specialist), Vice president, International Commission for Irrigation and Drainage, Tokyo University	Japan[b]
5. *Dr. G. Harrar* (plant pathologist), President , Rockefeller Foundation	USA[c]
6. *Dr. W. D. Hopper* (economist), President, International Development Research Centre	Canada
7. *Dr. Luis Marcano* (agronomist), President, Shell Foundation	Venezuela
8. *Dr. T. Muriithi* (animal health), Director, Veterinary Services	Kenya
9. *Dr. J. Pagot* (animal production), Directeur general, Institut d'Élevage et de Médecine Vétérinaire des Pays Tropicaux	France[d]
10. *Dr. H. C. Pereira* (physicist), Director, East Malling Research Station, Kent (Previously director, Central African Research Organization)	UK
11. *Dr. L. Sauger* (agronomist), Directeur, Centre de Recherche Agronomique du Bambey	Senegal
12. *Dr. M. S. Swaminathan* (geneticist), Director, Indian Agricultural Research Institute, New Delhi	India

Appendix 11-4. Terms of Reference of the Technical Advisory Committee

TAC will, acting either upon reference from the Consultative Group or on its own initiative

(i) advise the Consultative Group on the main gaps and priorities in agricultural research related to the problems of the developing countries, in both the technical and socio-economic fields, based on a continuing review of existing national, regional, and international research activities;

(ii) recommend to the Consultative Group feasibility studies designed to explore in depth how best to organize and conduct agricultural research on priority problems, particularly those calling for international or regional effort;

[b] Since succeeded by *Dr. N. Yamada*, Director, Tropical Agricultural Research Center, Ministry of Agriculture and Forestry — Japan

[c] Since succeeded by *Dr. V. W. Ruttan*, Agricultural Development Council, New York — USA

[d] Since succeeded by *Dr. Guy Ch. Camus*, Directeur general, Office de la Recherche Scientifique et Technique Outre-Mer — France

(iii) examine the results of these or other feasibility studies and present its views and recommendations for action for the guidance of the Consultative Group;

(iv) advise the Consultative Group on the effectiveness of specific existing international research programs; and

(v) in other ways encourage the creation of an international network of research institutions and the effective interchange of information among them.

These terms of reference may be amended from time to time by the Consultative Group.

NOTES

1. The present paper is an edited and updated version of my Hannaford Lecture entitled "International Agricultural Research: An Encouraging Venture in International Collaboration," given at Adelaide University, Adelaide, Australia, on November 26, 1973.

2. For background on the early development of the international system, see E. C. Stakman, Richard Bradfield, and Paul C. Mangelsdorf, *Campaigns against Hunger* (Cambridge, Mass.: Belknap Press of Harvard University Press, 1967); also, Sterling Wortman, "Extending the Green Revolution," *World Development*, 1:12 (December 1975), 45-51.

3. An example is plant nutrition which is currently under consideration by TAC.

4. Research is used in this document in a broad sense to include not only the development and testing of improved production technology, but also training and other activities designed to facilitate and speed effective and widespread use of improved technology.

5. Final decisions on funding remain a responsibility of each member in connection with specific proposals.

12

The International Rice Research Institute (IRRI) Outreach Program

Nyle C. Brady

International agricultural research centers are an innovative approach to the use of science in solving the world's food problem. They are sharply focused scientific establishments which, though located in developing countries, are as well equipped and manned as any in the developed world.

There are two primary criteria for the success of international agricultural research centers. First, they must be *centers of excellence*, applying the world's best scientific talent to the practical problems they were established to solve. They must develop superior varieties, strains, cultural practices, and farming systems on which improved technology for the developing world can be based.

Second, the international centers must serve as *stimulating and collaborating forces to improve the quality and output of national research programs*. They must do more than merely make their products available for use by other countries. They must work collaboratively to improve the scientific expertise, operational efficiency, and output of the national research programs.

IRRI's Traditional International Involvement

From its inception, IRRI has fulfilled this dual role – as an emerging center of excellence and as a collaborator with rice-production countries.[1] Although it was necessary in the early days of IRRI's history to emphasize the development of excellence, the international role was not neglected nor could it have been.

The Collection of Germ Plasm

The original seeds collected for IRRI's germ plasm bank came from existing national stocks or from samples collected by collaborators in cooperating countries. The practice of splitting each sample collected, retaining one portion in the country of its origin, and sending the other to IRRI set the stage for this mutually beneficial program. More than 33,000 samples have been collected from cooperating countries.[2]

As the seed bank has grown, the return flow of samples to national programs has expanded. In 1973, nearly 8,000 samples were sent to scientists working in national programs. Similarly, seeds of lines from IRRI's breeding programs are furnished upon request to country scientists. About 8,000 samples of these lines were sent in 1973.

Publications

In 1963, IRRI published a bibliography of the world's literature on rice. This publication, which is supplemented annually, is very helpful to researchers in developing countries who cannot easily translate Japanese, the language in which much of the rice literature is published. Photocopies are made available upon request, a significant service to national research agencies.

IRRI scientists have written a number of books and special publications on rice and its enemies. The institute also publishes the proceedings of important conferences and symposia and makes them available to scientists in cooperating countries.

The quarterly *IRRI Reporter* provides brief summaries not only of research findings at IRRI but of research done in cooperation with scientists in other countries. The IRRI annual report contains more detailed information on research accomplishments.

Conferences, Symposia, and Workshops

Conferences, symposia, and workshops which provide opportunities for communication among rice workers have been held regularly since the institute was established. The annual international rice conference, initiated in 1969, has traditionally provided opportunities for scientists and research administrators to review research results from all important rice-growing areas. This annual conference is now being used also for making cooperative plans for future programs.

A series of special symposia has permitted rice scientists to explore in depth subjects of broad interest to them. Scientists from both the developing and the more developed world are invited. A symposium was recently held on "Climate and Rice," the eighth major symposium held since the institute was founded.

Training

IRRI's training program seeks to upgrade the expertise of rice scientists and educators in cooperating countries. Selected in consultation with officials in their home countries, the trainees participate in one of two types of training. Those with extension and applied research orientation are involved in one of two six-month production research training courses. In the first course, they gain practical experience in all phases of rice production; in the second, they study rice-cropping systems. After completing these courses, participants are prepared to give similar training to extension workers in their own countries. Such training imparts needed knowledge and skills, but perhaps more importantly it is a source of pride to those who have worked together as a team using science to help farmers produce more food.

Those trainees who are research-oriented can participate in short on-the-job research training programs or, if their needs dictate, they can take course work at the University of the Philippines at Los Baños to fulfill the requirement for the M.S. or the Ph.D. degree. Their thesis research is done at IRRI, giving them an opportunity to gear this research to the practical problems they will face when they return home.

In some cases it is desirable for the trainee to take some graduate courses at a university in the United States or in Europe. However, the scholar must return to IRRI to do his research work on a problem of some relevance to his own country. Whether the IRRI scholars and fellows obtain their academic training at Los Baños or in the Western Hemisphere, they have the advantage of orienting their research to the solution of practical problems. Furthermore, they do their research in a cultural environment more similar to their own than are those of the United States or Europe.

Postdoctoral research is provided for a select group of young scientists from rice-growing countries. Some are scientists working in national programs. Others have recently completed their Ph.D.'s and need to focus on rice-production problems before returning home.

Since IRRI began its training program in 1962, about eight hundred man-years of training have been given to scientists and educators from forty-five countries. Most of the students have come from South and Southeast Asia although a number have come from Africa and Latin America and a few from Europe and North America. Currently, IRRI provides about ninety man-years of training each year, twenty-five of which are for nondegree scholars and fellows, fifty-five for those registered for M.S. and Ph.D. degree training, and ten for postdoctoral fellows.

Four Cooperative Approaches

IRRI's cooperative programs with national research organizations have four primary objectives: to do location-specific research which cannot be done effectively at IRRI headquarters in the Philippines (this is called collaborative research); to develop international research networks on problems of common interest (international testing); to strengthen national research capabilities (outreach services); and to strengthen the countries' capacities to utilize research findings in rice production programs. Any given cooperative project may have more than one of these objectives. In some cases, all may be involved.

Collaborative Research

IRRI scientists often find that major problems cannot be attacked conveniently at our headquarters in Los Baños. For example, some serious insect and disease pests are not found at that location, and yet it is essential that IRRI's varieties and breeding lines be thoroughly tested for resistance to the pests in question. Examples are research on the gall midge insect in India, on the tungro virus disease in Indonesia, and on a suspected biotype of the brown plant hopper in India.

The tolerance of different rices to toxic soil conditions (such as those brought about by excessive salt, acidity, alkalinity, or iron) can also best be ascertained if tests are run where the problem exists. Arrangements are being made with scientists in India and Sri Lanka to carry out field screening trials to identify varieties with greater tolerances to toxic conditions.

In some cases, collaborative research can be done overseas with little direct input from IRRI other than in the planning stages. The research is sufficiently important to the cooperating countries to justify additional national financial and personnel inputs without outside assistance. Testing IRRI lines in parts of Indonesia infected by tungro virus and in areas in India with brown hopper pressures are examples. In other cases, IRRI scientists are involved not only to help with the overseas research but to carry out supplementary and complementary experiments at IRRI headquarters. Research on flood-tolerant lines and varieties in Thailand and at IRRI headquarters is an example.

Collaborative research may also be undertaken with more than one other country. An example is research being planned on deep-water rice, a type of culture found in approximately 10 percent of the rice area of Asia. The area of deep-water rice in the Philippines is insignificant, but large areas are found in Bangladesh, Thailand, India, and Indonesia. We are developing a cooperative research program with the Ministry of Agriculture in Thailand to work on deep-water rice. Two IRRI scientists located in Thailand will collaborate with their Thai counterparts in expanding and strengthening an ongoing deep-water research program there. They will also collaborate with scientists in other deep-water rice countries.

There are many research areas of mutual interest to IRRI scientists and their associates in India. We have signed a memorandum of agreement with the Indian Council of Agricultural Research which provides for collaborative research planning and implementation as well as for the exchange of scientific personnel. Each year a work plan is developed. This plan clearly identifies the areas of mutual interest, the division of responsibility in carrying out the needed research, and the areas in which personnel will be exchanged. This procedure has many advantages and is being pursued in a modified form in the Philippines and in Thailand.

Although the primary objective of collaborative research is to find answers to specific problems, a secondary objective is the strengthening of national research capabilities. In implementing the research, both IRRI scientists and their counterparts improve their capabilities. New techniques are developed and utilized. Interdisciplinary approaches are fostered. This procedure provides training in a framework that is satisfying to scientists from IRRI and the developing countries alike.

International Networks

The second phase of IRRI's outreach program is the development of international research networks. These networks permit scientists from different countries to plan and implement research projects in several countries at once using common objectives and procedures. At present, three networks are in existence or are being set up: the international testing program for Genetic Evaluation and Utilization (GEU); the International Rice Agro-Economic Network (IRAEN); and the International Cropping System Network (ICSN). Other informal networks exist for research on herbicides and fertilizers.

IRRI's initial objective is to serve as a catalyst for these networks. Once a general area of mutual interest is identified, scientists from cooperating countries are brought together to set up the general framework for the network and to determine the specific experiments, surveys, or studies to be done.

In some cases, IRRI or one or more of the cooperating countries may have already run some pilot experiments in the research area which serve as a guide. For other projects, a loose cooperative framework may already exist which needs only formalization for the specific experiments in question.

National research personnel must be fully involved in the planning and the implementation of the network. In no case can the impression be given that the network is merely an extension of IRRI's program. To be successful, the network research must be planned and carried out by the country scientists.

In each of the three projects currently under way we have assigned an IRRI scientist to serve as network coordinator. In the genetic (GEU) trials, for example, the coordinator is responsible for collecting seeds of lines and varieties to be tested and for disseminating them among the cooperators. He facilitates communications among cooperators and coordinates the exchange of biological materials, the collection and collation of data, and the planning of workshops or conferences. The coordinator visits the countries involved and reviews ongoing experiments and program plans with cooperators.

If funds permit, scientists from one country are encouraged to visit experiments in other countries so they can see more clearly how the international network can be useful to them. These visits also have some training value, because new techniques are demonstrated and innovations evaluated.

The international testing network of the Genetic Evaluation and Utilization (GEU) program has been in progress since the early sixties. In cooperation with national research centers, nurseries have been set up to screen rice lines and varieties for insect and disease resistance. For example, more than 300 international blast nurseries have been conducted in twenty-five countries since 1963. Similar tests for bacterial leaf blight were initiated in 1972 and for sheath blight in 1973. International yield trials were initiated in 1973, and observational nurseries are being established for the general evaluation of several hundred of the best selections from both national and international sources.

Recently, the international testing program has been expanded to include international nurseries for adverse soil and weather conditions as well as for the major insects and diseases. Observational nurseries and yield nurseries are also included. The tests involve upland rice as well as paddy rice nurseries.

The Agro-Economic Network (IRAEN) is based upon the success of a preliminary cooperative study among economists and agronomists from several countries who were concerned with rice yields in relation to different types of farming. The IRAEN has been concerned initially with international study of constraints on yields in farmers' fields. It is innovative in that its success depends upon the close collaboration of economists and agronomists who will attempt to measure the relative importance of different factors constraining rice yields under different environmental conditions in South and Southeast Asia.

Planning and implementation of the Cropping System Network are accomplished using procedures similar to those used by the GEU and IRAEN. Potential research locations have been chosen on the basis of broad agroclimatic regions. A network coordinator has visited most of the countries of Southeast Asia. In consultation with local scientists, potential sites for cropping systems trials have been identified in Indonesia and the Philippines, and discussions are under way with scientists from other countries.

IRRI scientists have great expectations for the international networks system. It gives country scientists experience in conducting research. It illustrates the international nature of science. It provides IRRI scientists with a better understanding of the problems faced by farmers and by researchers in the cooperating countries.

Strengthening the Capacity of Country Research Programs (Outreach Services)

Most of IRRI's activities – collaborative research, international networks training, distribution of seed samples – are aimed at strengthening national research capabilities. They are complemented by formal country-assistance projects.

The prime objective of these projects is to enable local scientists to improve their skills and the national agency to develop a workable research system. In these projects, IRRI scientists are located in the cooperating country. These scientists function as members of the local staff, not merely as IRRI overseas employees. Their operational support comes largely from local sources.

The three major functions of IRRI scientists working in cooperative country programs are the following: to provide temporary research expertise which permits national research programs to begin while local staff are receiving formal training outside the country; to offer on-the-job training for their counterparts in national programs; and to assist in the development of a viable system of rice research and of a managerial framework within which that research can be implemented.

Since 1960, IRRI has undertaken twelve country projects in Pakistan, Bangladesh, India, Sri Lanka, Indonesia, Vietnam, Egypt, and the Philippines. Although not all of these projects have been successful, together they have played a major role in the general steady improvement of the research and institutional capabilities of the countries involved.

Strengthening the National Capacity to Utilize Research Findings

For several years, IRRI has been involved in a pilot project aimed at putting research results into production more rapidly. This project involves a series of applied research trials planned and implemented cooperatively with the national research and extension organizations. These in turn become the basis for a pilot action program operated by the cooperating country and, if this is successful, a nationwide production effort based on the applied research findings.

This procedure has been tried in the Philippines with some success. The applied research trials were concerned first with a "package" technology approach and then with a direct-seeding two-crop management system which will permit two crops to be grown in rainfed areas where only one grew in the past using the conventional cropping systems. The trials were run on farmers' fields with the aid of IRRI-trained technician-employees of the extension service. The Philippine government utilized the technology demonstrated in the fields to set up the "Masagana-99" national rice production program in the Philippines. This program appears to have been reasonably successful in spite of shortages and high costs of inputs as well as typhoon and flood damage during the past two years.

The Philippine experience is still considered a pilot operation. Depending on a final assessment of this preliminary work, similar programs in other countries may be initiated, starting in 1976.

Operational Constraints

IRRI scientists and administrators have given considerable thought to steps which might improve IRRI's international programs, particularly the cooperating-country projects.

We have identified at least three major problem areas which tend to limit our success. These are limitations on IRRI's ability to provide the needed assistance; limitations on the ability of the national agency to use the assistance effectively; and limitations stemming from the fragmentation and lack of continuity of donor inputs into the national research programs.

IRRI faces several problems in working effectively to improve national research capabilities. First, there is the difficulty arising from the dual role played by IRRI: an aggressive "doer" of research on the one hand and a less aggressive "tutor" for research on the other. Some difficulty is experienced in taking steps to strengthen the training and the organizational and managerial capabilities of country programs without giving the impression that the external organization is dominating the local scene. Fortunately, in most cases the selection of IRRI staff for the overseas assignments has been such that the "dual role" has been minimized. Working relationships at the country level are good.

A second limitation on IRRI's ability to be more helpful relates to the lack of research coordination and management expertise in national agencies. Frequently, the greatest need does not stem from the inadequacy of the local scientist but rather from the inadequacy of the research system of which he is a part. This need for managerial assistance is attributable to both the constraints on the national researchers and the complexities resulting from the fragmentation of donor assistance.

The limitations experienced by the national research agencies are well known. In some cases government restrictions and inflexibilities set up almost insurmountable roadblocks to the development of viable agricultural research programs regardless of the external support provided. In others, low staff salaries and inadequate operational support funds give little opportunity for local staff to innovate and reorient programs. In still others, the rate of change needed in the organizational and operational frameworks is more rapid than any but the most innovative administrators and political decision-makers will permit.

The problems facing national research agencies are attributable as much to organizational and managerial weaknesses as to the inadequate training of scientific personnel and the low quality of the research being conducted. At the same time, research achievement can solve the problem in some cases. It has been said, for example, that IR-8 and other high-yielding varieties brought about more change in the organizational and managerial frameworks of national rice research programs than all the research coordinators, administrators, and other nonresearch advisers combined. Although this may be an overstatement, IRRI is working on the assumption that research accomplishments can influence decision-makers. We will do all we can to help the country researchers achieve these accomplishments. At the same time, other efforts will be continued to help improve the rice research systems of our cooperators.

The fragmentation and lack of coordination of donor assistance to national rice research programs are serious obstacles in some cases. Each donor provides assistance to alleviate the constraints on research as perceived from his institutional point of view. In some instances this viewpoint has a political flavor which relates as much to the objectives of the donor as to those of the cooperating country. While this situation appears to be inevitable where donor assistance is given by outside national aid agencies, it does not make research coordination easy to achieve.

Fortunately, donor agencies do at times coordinate their activities well, making IRRI's job of assistance easier. Plans are made jointly by representatives of the national agency, the donors, and IRRI. Support from each agency is agreed upon, and the program is implemented.

A second donor-related constraint is the lack of continuity of funding for the cooperating country projects. These projects are often of two to three years' duration only. Although such time limits accord with justifiable donor policies, they constitute a serious constraint on the employment of competent personnel. Also, the seemingly unavoidable delays in approval of project extensions is bad for staff morale. Furthermore, with the fate of IRRI international staff at stake, it is very difficult to bring about needed changes in projects as they are renewed. Moral commitments to existing staff limit our bargaining power to obtain needed changes in contract provisions.

What of the Future?

We are optimistic about future cooperative relationships between international research centers and country agencies. These relationships are helping to move research results from the experimenter's plots to the farmer's field. Furthermore, they are identifying which of the experimenter's results will be most useful to the farmer.

Our optimism is based on the assumption that certain clear guidelines will be followed in center-country program relationships. These guidelines include the following.

1. There must be a clear delineation of the responsibilities of each center in respect to both the coverage of science subject matter and the methodology to be used. This delineation must be made by the administration and governing boards of the centers and must be clearly understood by donors and national organizations alike. Centers should not be called upon to perform activities for which they do not have a comparative advantage. For example, they should not be used as substitutes for the type of general-assistance programs formerly carried out by donors.

2. Each special project involving a center and a national program should be directly related to the long-term goals of both the center and the country program. To assure this, a joint analysis by both organizations of the country's long-term agricultural research and training goals is desirable.

3. The quality of center personnel in country projects should, to the extent feasible, be of the same caliber as that found in the center's core program. The scientist's training must fit his responsibilities in the country, but scientists assigned to country programs should not be second rate. If we are to have competent scientists in country programs, longer term commitments must be made to these scientists, and they must be given perquisites comparable to those enjoyed by scientists located at center headquarters.

4. A fiscal and personnel management system must be developed at each center to permit the full exploitation of opportunities for collaboration and cooperation with country programs. This means that core program donors should realize that some scientist and administrator time will be involved in assisting country programs. This time should be recognized as being as legiti-

mate in carrying out the goals of the center as is any other phase of the core program.

To accommodate expanding country-center programs additional scientific and managerial talent must be recruited. This additional help is needed to prevent the erosion of core programs, which in many instances are thinly staffed.

A sounder financial base must be provided for centers to carry out their obligations and opportunities with cooperating countries. Funds are also needed to support exploratory studies and the staff time required for the development of projects before they are funded by an outside donor.

5. The role of the international centers in providing assistance to national programs must be subject to continuing scrutiny, not only by the center governing bodies but by the consultative group as well. Projects dealing with subjects central to the missions of the centers should be developed and initiated by or in cooperation with the centers.

The challenge to both the centers and the national programs is to carry out their symbiotic relationships without endangering their common primary function – to bring science to bear effectively on the solution of the world's food problems. The challenge for donors is to provide long-term funding for quality center-country programs which have as one of their prime objectives the enhancement of research capabilities within countries.

NOTES

1. A. Colin McClung, "IRRI's Role in Institutional Cooperation in Asia," *Rice, Science and Man* (Los Baños: International Rice Research Institute, 1972), pp. 19-40.

2. See T. T. Chang et al., "The Genetic Conservation Program of IRRI," report prepared by a committee of IRRI scientists (Los Baños: International Rice Research Institute, December 1974).

13

The International Maize and Wheat Improvement Center (CIMMYT) Outreach Program

Haldore Hanson

Background on CIMMYT

CIMMYT's mandate is to improve the quantity and quality of maize and wheat wherever they can be grown efficiently, but especially in developing countries.

The maize program includes cool-tolerant sorghum, a possible substitute crop for maize at high elevations, and wide crosses attempted between maize and other species of plants (for example, maize x tripsacum, maize x sorghum).

The wheat program includes bread wheat, durum wheat, barley, triticale (a cross of wheat x rye), and other wide crosses attempted between small grains (for example wheat x barley, wheat x oats).

The reference in the mandate to the "quality" of crops applies particularly to the protein content of cereals.

Wheat or maize is the basic foodstuff of a majority of the developing countries. These two crops are eaten by more than one billion people. Together with rice, which stands first in production in the developing countries, they constitute over 80 percent of the cereals eaten by poorer nations.

If we arbitrarily decide that 100,000 hectares planted to a given cereal crop makes that crop an important national food, then there are seventy maize-growing countries and sixty wheat-growing countries in the world.

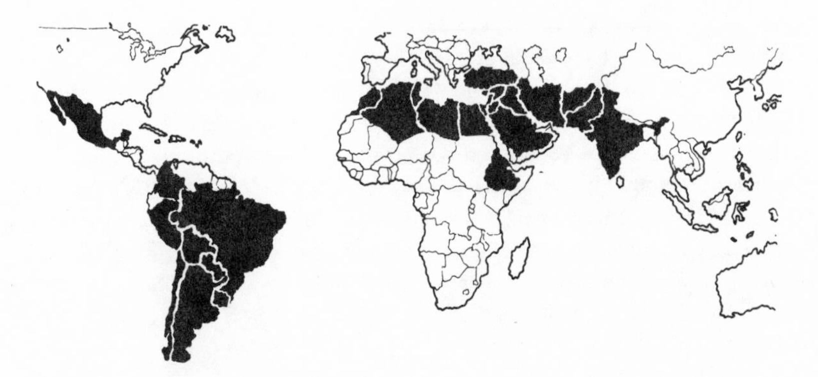

Figure 13-1. Wheat: Developing countries in which wheat is an important crop (over 100,000 hectares of production per country or over 25 percent of national calories derived from the crop).

Fifty-three of the maize growers and thirty of the wheat growers are developing countries.

If we add to the list of developing countries those which produce less than 100,000 hectares of maize or wheat but derive more than 25 percent of their total calories from one of these crops, we find a total of more than sixty national maize programs and forty national wheat programs which deserve attention from CIMMYT (see Figures 13-1 and 13-2).

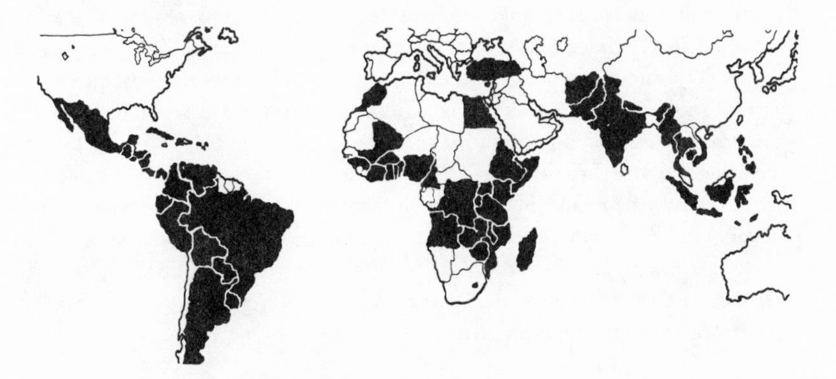

Figure 13-2. Maize: Developing countries in which maize is an important crop (over 100,000 hectares of production per country or over 25 percent of national calories derived from the crop).

When we compare this potential workload with the size of CIMMYT's headquarters staff, which includes thirteen scientists for wheat and thirteen for maize, it is clear that only a remarkably efficient use of resources will permit so small a staff to have an impact on production in so many countries.

One fact that influences CIMMYT planning is revealed in the statistics cited above: although the wheat crop in developing countries is somewhat larger than the maize crop – 79 million tons compared with 62 million tons – more countries are staple maize-eaters than wheat-eaters. Consequently, the maize staff at CIMMYT must deal with more governments than must the wheat staff. Later in this chapter, when we discuss proposals for regional programs involving groups of producing nations, it will be seen that CIMMYT recommends six maize-producing regions compared with four wheat-producing regions.

Services for National Programs

The requests for assistance which CIMMYT receives suggest that the primary needs of many countries are (1) better germ plasm (seed) and improved production technology; (2) the training of institution/agency staff who will test and introduce better technological methods into their countries; and (3) advice on all aspects of food production for such decision-makers as the president, the staffs of various agricultural service agencies, and the farmer himself. An international center must be prepared to participate in most of these fields if its effectiveness is to be measured by rising national crop yields.

There are many aspects of national crop improvement in which the international center can offer only marginal expertise. Salient among these are investments in fertilizer factories and social reforms affecting land tenure and political-economic decisions, such as whether to grow more food at home or to import more food at concessional prices. (The latter issue held back wheat production in the Andean region for more than two decades.)

A successful breakthrough in crop technology, such as the packaging of Mexican dwarf wheat and its production practices, has conferred a creditability upon CIMMYT which carries far beyond the agricultural sciences and causes governments to ask advice on a wide range of production factors. In this situation CIMMYT must exercise caution and refer some requests for advice to more appropriate specialists.

CIMMYT's Predecessor

CIMMYT became a legal entity with an international mandate in 1966, but during the preceding two decades, starting in 1943, CIMMYT's predecessor agency was a cooperative research and production program for basic foodstuffs in Mexico, cosponsored by the Mexican government and the

Rockefeller Foundation. During these two decades, CIMMYT's research, training, and outreach activities took shape.

Norman Borlaug, director of the wheat program since 1944, conducted his wheat breeding in Mexico in two cycles a year, one winter crop under irrigation in Sonora State at latitude 29° N (sea level) and one summer crop in rainfed conditions on the Mexican high plateau near Toluca at 19° N (elevation 8,700 feet). The result of this biannual movement over many cycles was a wheat type which was adaptable to the day length, temperature range, moisture conditions, and tolerance for diseases of the two climates. This adaptation became a major factor in the success of Mexican wheats when they were moved half way around the world to the Punjab of India-Pakistan in the 1960s and were found to be well adapted there.

Another development of this early period was the Mexican training program. More than 400 Mexicans were given in-service training for research on wheat, maize, and beans. More than 200 did work toward the M.S. degree, and more than 80 received their doctorates. This group of agricultural scientists trained in the 1940s, 1950s, and 1960s now provides leadership for most of the agricultural agencies in Mexico, from the minister of agriculture on down. The lessons learned from this training program still offer insights for the needs of many other countries.

Finally, the Mexican program in the 1950s and 1960s served as a stepping stone for Rockefeller Foundation scientists who first gained experience in Mexico and then helped other Latin American countries – Colombia, Ecuador, and Chile – establish similar crop improvement programs. These foundation men carried with them the experimental maize and wheat from the Mexican program, and soon there were inter-American nursery trials, which set the pattern for the later international workshops among this group of nations.

Thus many of the activities that CIMMYT now calls "outreach" were tested and improved in those early years before CIMMYT became an international center.

Services from CIMMYT Headquarters

Strengthening national production programs is the basic objective of CIMMYT and the other international centers. Our headquarters program is organized to achieve this end. CIMMYT distributes seed, production technology, and training services to national production programs in a variety of ways.

Consulting with Governments

CIMMYT's senior staff spend between 15 and 20 percent of their working time traveling outside Mexico, consulting with governments of wheat- and

Table 13-1. Locations of CIMMYT Nursery Trials in 1974

Region	Countries Conducting Trials[a]		Trials per Region	
	Wheat	Maize	Wheat	Maize
Latin America	14	12	297	127
Asia and Pacific	12	7	196	60
North Africa and Near East	17	4	273	13
Africa south of the Sahara	17	13	122	36
Europe, Canada, U.S.	26	3	316	1
Total	86	39	1,204	237

[a] A total of ninety-three different countries.

maize-growing countries, or exchanging information with research institutions. This international travel required 1,823 man-days (about five man-years) in 1972 and 2,500 man-days (about seven man-years) in 1974.

Consultation by CIMMYT staff has been increasing steadily over the past five years and has reached an approximate limit for the present size of the headquarters staff. Future increases in consulting will be accomplished by stationing CIMMYT staff members in the various producing regions, as will be discussed later.

International Nursery Trials

International nursery trials distributed from CIMMYT were grown in ninety-three countries in 1974. These trials represent CIMMYT's principal method for distributing improved germ plasm and outstanding breeding materials to developing countries.

An "international trial" consists of identical packages of experimental seed sent to a network of collaborating scientists throughout the world. These scientists are asked to grow the seed under a standard set of procedures, using their best local varieties as checks, and to return the data to CIMMYT. CIMMYT then analyzes and publishes the results.

CIMMYT began its international nurseries for wheat in 1960 and for maize in 1971. In 1974, wheat trials were grown at 1,204 sites in eighty-six countries and maize trials were grown at 237 sites in thirty-nine countries. The objectives are (1) to test new lines of wheat and maize under widely differing conditions of day length, temperatures, moisture, diseases, and insects; (2) to obtain yield data which can guide the breeding work of the entire network, including CIMMYT; (3) to train a network of cooperating scientists; and (4) to obtain from these scientists, in exchange, their best experimental germ plasm for inclusion in future trials and in CIMMYT's crossing program.

Scientists estimate it would take any one collaborator fifty years of repeat-

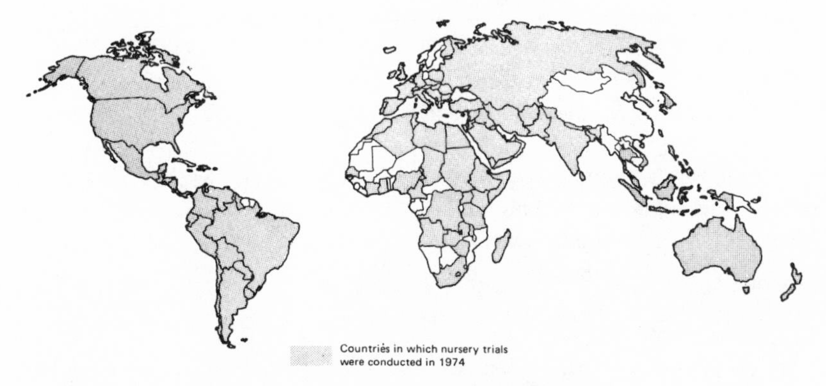

Figure 13-3. Location of CIMMYT nursery trials, by region, 1974.

ed trials in his own stations to obtain data as comprehensive as that from one year's trial on a worldwide basis.

Locations of the CIMMYT nursery trials of 1974 are shown in Table 13-1 and Figure 13-3.

The Germ Plasm Bank

CIMMYT maintains a germ plasm bank for maize seed – the largest maize collection in the world – from which breeders may request samples of seed carrying particular genetic characteristics. For example, a breeder in Argentina seeking genetic resistance to a pest called fall army worm can request seed samples for all bank entries which CIMMYT believes will provide this characteristic. The maize bank contains about 11,000 entries for the maize species and its close relatives. In 1973 the bank made forty-four seed shipments to breeders in nineteen countries.

CIMMYT's wheat staff does not maintain a world collection for bread wheat, durum wheat, and barley, but only working collections. Therefore, CIMMYT sometimes refers wheat requests to the USDA, which maintains a world collection of small grain germ plasm.

In-Service Training in Mexico

From 1966, when CIMMYT began, through 1974, 500 young scientists from developing countries have come to CIMMYT for practical experience in research and production. Courses last six to nine months and include one complete cycle of crop research plus some lectures in basic agricultural sciences. No degree is awarded. CIMMYT now receives about 100 fellows a year – of whom a small percentage are women – for this type of training.

The fellows who come from research services participate at CIMMYT in data analysis and in all research activities from planting to harvest. Those employed as production agronomists receive some research experience at CIMMYT but also help to lay out field trials on private farmers' land. This is an activity which the agronomist will be expected to perform upon returning home.

All CIMMYT training courses stress "learning by doing," "dirty hands," and the discipline of working under heat, humidity, heavy rains, and long hours. We do not know how to teach motivation, but we find that many training fellows leave Mexico with a new drive.

Visiting Scientists in Mexico

A program for visiting scientists and administrators brings over 100 persons each year from developing countries to CIMMYT. This activity reaches a different age group and serves a different purpose than does the in-service training program.

In developing the visiting scientist program, CIMMYT reasoned that, if it gives in-service training to a group of young scientists from a country like Tanzania, it is useful for the research director of that country to spend one or two months at CIMMYT during a harvest season observing how research decisions are made. Moreover, CIMMYT felt that such a country's vice minister or minister of agriculture could spend a profitable week at the institute.

Several benefits can be observed after these visitors return home. First, those who are practicing scientists make greater use of international germ plasm and become key members of the international network which grows the international nurseries. Second, the visitors give active support to the CIMMYT training programs, nominating the best candidates, helping to arrange study leaves, and showing interest in the work of the trainees after they return to their posts. Finally, administrators among the visitors take an active role in the food production problems of their countries.

Doctoral Fellows in Mexico

Candidates for doctoral degrees at North American or European universities come to CIMMYT for twelve to eighteen months to do their thesis research under supervision of CIMMYT scientists and then return to their universities to qualify for the degree. Under another option, postdoctoral fellows are invited by CIMMYT to serve two years as junior members of the CIMMYT staff.

Since 1966, seventeen predoctoral and twenty-five postdoctoral fellows have received grants to spend one or two years in Mexico. Of these, twenty-two have completed their work and – without exception – each is now employed in crops research work, either by his own government or by one of the international centers. There were twenty fellows holding pre or postdoctoral grants in 1974.

Assistance with Economic Studies

CIMMYT established an economics program in 1971, and these services to developing countries are still evolving.

We believe that CIMMYT can best conduct its economics work through indigenous social scientists in national programs, helping them gather better farm and market data, which is needed by local policy officials and local biological researchers.

In 1974 CIMMYT was working with twelve countries in Asia, Africa, and Latin America on three types of problems.

1. What characteristics of farms, farmers, and agricultural policy have influenced the adoption of new technology for maize and wheat? Eight studies, covering the following crops and countries, were completed in early 1975: *Wheat* – India, Iran, Turkey, and Tunisia; *maize* – Colombia, El Salvador, Mexico (Plan Puebla), and Kenya. Most of these individual studies will result in a Ph.D. thesis by an indigenous researcher, reflecting CIMMYT's concern for augmenting national capacity to do micro-research.

2. What information is needed by policy makers to promote new technology, and how can these data be assembled? On this topic as well, CIMMYT is collaborating with indigenous researchers.

3. How can economists and agronomists better work together in national programs? This research is motivated by the question: what is the minimum amount of information necessary to make useful recommendations to farmers, given that researchers have better control over the crop's environment than do farmers and assuming that farmers are risk-averting income seekers.

The economics staff also works with doctoral fellows from developing countries who do their thesis research in agricultural economics at CIMMYT. Each thesis topic is focused on a theme of interest to CIMMYT.

Beyond this the economists are collaborating with CIMMYT plant physiologists in identifying agroclimatic regions of the world, structured in terms that are significant for maize and wheat research. This information, coupled with socioeconomic data, will help define research priorities at CIMMYT geared to the needs of the producing countries.

Assistance in Laboratory Management

Two of CIMMYT's laboratories in Mexico – for protein analysis and wheat industrial quality – have assisted a number of developing countries in setting up similar laboratories.

These CIMMYT laboratories have trained more than twenty technicians from national programs during 1972-74. They have also produced "cookbook" bulletins on laboratory procedures for use in developing countries.

With funds made available by the UNDP and others, CIMMYT has assisted in the purchase of equipment and the establishment of protein-quality laboratories in twelve countries (Algeria, Colombia, Egypt, El Salvador, India, Mexico, Nepal, Pakistan, Peru, Philippines, Thailand, and Tunisia).

As the problem of protein deficiency in national food supply becomes more acute, CIMMYT's assistance for protein laboratories is expected to rise. The demand is immediate for those countries introducing high lysine maize; these must maintain constant surveillance over the amino acid content of the experimental maize varieties.

Assistance for Research Station Management

The manager of the eight experiment stations used by CIMMYT in Mexico — Mr. John Stewart — consulted with six countries on experiment station management in 1973-74. These countries were Algeria, Nepal, Pakistan, Tanzania, Turkey, and Zaire.

Another six countries have sent experiment station managers to Mexico for training under Mr. Stewart at the CIMMYT stations: Brazil, Bolivia, Egypt, El Salvador, Ivory Coast, and Nigeria.

During a staff review of CIMMYT programs in 1974, the maize and wheat staffs said that the consulting work by Mr. Stewart, outside Mexico, had made the difference between research data which were useless and data which were highly significant. The changes were brought about by land leveling, fencing, better and more timely seedbed preparation, more accurate fertilizer placement, more timely plant protection, and better maintenance of equipment.

International Symposia

During 1971-74 CIMMYT held six international symposia in Mexico, bringing together a part of the world network of scientists on wheat or maize with a group of scientists from advanced countries, to review past research and make future plans. Most symposia last five days. About half the participants are from developing countries.

A record of proceedings for each symposium becomes a major publication for distribution to the network of scientists. In addition, the presence of so many visitors at CIMMYT leads to side meetings at which scientists from developing countries are able to negotiate for financial grants with donors, select germ plasm from CIMMYT's research fields, interview their trainees, or hold group meetings for neighboring countries of Southeast Asia, tropical Africa, and so on.

Table 13-2. CIMMYT Bilateral Assistance to National Programs in 1974 (Financed by Special Grants)

Country	Crop[b]	Number of CIMMYT Staff 1974	Donor	Starting Year	1974 Budget through CIMMYT (rounded)
Algeria-Tunisia[a]	Wheat	8	Ford and Rockefeller foundations, USAID	1968	$480,000
Argentina	Maize, wheat	0	Ford Foundation	1968	8,000
Egypt	Maize	1	Ford Foundation	1968	40,000
Lebanon	Wheat	2	Ford Foundation	1973	60,000
Nepal	Maize	1	USAID	1972	60,000
Pakistan	Maize, wheat	3	Ford Foundation	1965	155,000
Tanzania	Maize	2	Ford Foundation, USAID	1973	46,000
Turkey	Wheat	2	Rockefeller Foundation	1970	118,000
Zaire	Maize	4	Government of Zaire	1972	280,000
Total		23			$1,247,000

[a] Combined grant.
[b] Wheat assistance includes bread wheat, durum wheat, barley, and triticale.

Publications by CIMMYT

CIMMYT publishes an annual report, a scientific newsletter, and technical bulletins in three languages: English, Spanish, and French. The mailing list for English totals 4,000, that for Spanish 4,000, and that for French 1,000.

For each publication and each language there is an inner core of significant readers, made up of scientific collaborators, donors, and staff members of other international centers. This core is no larger than 1,000 people. The balance of each mailing list is made up of libraries, universities, government agencies, and requesting individuals and institutions, over half of them in countries where CIMMYT collaborates with maize and wheat programs.

The French language distribution goes largely to nineteen former French or Belgian countries in Africa.

Outreach by CIMMYT Staff Posted outside Mexico

Since 1966 CIMMYT has posted a growing number of staff in national programs to assist their national research and production activities. In 1974 there were twenty-three CIMMYT staff on residential assignment. The bilateral assistance projects are summarized in Table 13-2.

Before being posted abroad, most of this staff served as scientists in CIMMYT headquarters or held postdoctoral fellowships in Mexico.

The majority of CIMMYT's bilateral projects were initiated by donors who were seeking technical leadership for agricultural assistance which they were prepared to finance. In a few instances, CIMMYT initiated conversations directly with the host government and then solicited donor funds.

Bilateral assistance projects involve a number of activities which differ from the work of CIMMYT headquarters. Such activities are the following:

1. Research on farming systems for wheat in Algeria and Tunisia, including the introduction of medicago (forage legumes) as part of the wheat rotation and the use of biennial fallow in North Africa.

2. Selection of winter wheat in Turkey. (Mexico lacks the necessary climate.)

3. Breeding and selection for diseases and insects not prevalent in Mexico.

4. Agronomic testing of maize-legume rotation in eastern Zaire, where chemical fertilizer has always been in low supply.

5. Developing agronomic packages of practices adapted to each national program and formulating local recommendations to farmers.

6. Training local staff within national programs, especially production agronomists.

CIMMYT itself benefits from the bilateral projects. Among the hundreds of international nursery trials, those grown by CIMMYT staff in the countries listed in Table 13-2 are considered to produce more reliable data than the average; hence they serve as checks in reviewing the total returns.

CIMMYT also has its staff from bilateral projects attend each symposium at CIMMYT. The contribution of these scientists is especially useful because they are fully trained in CIMMYT methods, and they observe CIMMYT breeding materials under different environments.

CIMMYT outreach staff residing in countries where projects are under way are able to provide more continuous advice, based on fuller knowledge of the local situation, than are the consultants traveling from Mexico. In addition, CIMMYT outreach staff help select candidates for training in Mexico.

Collaborative Research in Association with National Programs

"Collaborative research" is a new activity in CIMMYT's core budget for 1975. CIMMYT and other international centers have found that a part of their assigned responsibility cannot be carried out at headquarters because the local environment does not provide the needed range of temperatures, moisture conditions, problem soils, or disease and insect conditions which affect the world crop. Therefore, some "core" research must be conducted abroad.

For example, the Potato Center (CIP) has found it cannot study late blight of the potato in Peru and has made arrangements for such research at the CIMMYT station near Toluca, Mexico, although it is charged to the CIP budget in Peru. Similarly, IRRI has found it cannot experiment with deep-water rice in the Philippines, and it has arranged that such research be carried out in association with the Thai national rice program. Again, this is charged to the core budget in the Philippines and not considered as a service to Thailand.

CIMMYT has initiated two projects in "collaborative research" in 1975, one for the surveillance of wind-borne diseases of wheat-barley in the Mediterranean-Near East region and the other for the special testing of CIMMYT maize materials, in several areas of the world, for resistance to three diseases of maize not prevalent in Mexico. These diseases are maize downy mildew in Southeast Asia, maize streak virus in tropical Africa, and corn stunt virus in Central America.

One additional proposal for collaborative research, now under negotiation, is a summer season wheat nursery in cooperation with the Kenyan government at Njoro Station. A summer nursery could test CIMMYT experimental lines under the disease pathogens of the East African highlands together with some nursery materials from national programs in the Mediterranean region. This would help guide the breeding of CIMMYT and the national programs. CIMMYT staff posted in East Africa would also consult with wheat-growing countries in that area.

We anticipate that such activities will expand gradually over a five-year period to a magnitude of possibly $500 thousand a year. All will be conducted outside Mexico, generally in association with a strong national program.

Regional Programs for Groups of Producing Countries

During the remainder of the 1970s CIMMYT proposes to increase gradually its activities with national programs by stationing a few CIMMYT staff members in major producing regions to supplement the activities now conducted from Mexico. This expansion would be the most important change in the CIMMYT outreach program in this decade.

We have in mind six producing regions for maize and four regions for wheat, where staff may be stationed. Discussions have already begun with both governments and donors for the regions listed in Table 13-3.

No two of these regions are alike in population, internal transportation, quality of agricultural services, or production problems for wheat and maize. Nevertheless, CIMMYT calculates that, on average, two CIMMYT staff members placed in each of these regions (a total staff of twenty) could significantly help in the following activities: (1) serve as regional consultants to national programs; (2) maintain surveillance for wind-borne diseases in the region; (3) organize training within the region for national programs; (4) circulate nurseries composed of germ plasm gathered within the region and exotic materials; (5) organize regional workshops; (6) administer a travel fund for sci-

Table 13-3. Maize- and Wheat-Producing Regions Which CIMMYT Believes Will Need Additional Outreach Staff and Services in the 1970s

Region	Number of Countries	Present Production[a] (in tons)
Maize		
South and Southeast Asia	12	15,400,000
Tropical East Africa	14	6,400,000
Tropical West Africa	14	3,000,000
Central America and Caribbean (excluding Mexico)	9	2,300,000
Andean zone and tropical Brazil	6	7,400,000
Brazil	6	20,000,000
Wheat		
Mediterranean and Near East	15	23,900,000
East Africa.	5	1,230,000
Andean region.	5	337,000
Southern cone of South America	5	9,500,000

[a] 1973 figures.

entists in national programs to visit outstanding research in the region; and (7) circulate a regional newsletter.

Each regional program would go through three steps of approval before it began: first, CIMMYT would discuss with producing countries in each region the kinds of service required; second, CIMMYT trustees would be asked to approve the program; and third, financial support would be sought from a donor.

Problems and Issues in CIMMYT Outreach

Like all organizations that have experienced rapid expansion, CIMMYT encounters problems for which answers are still evolving. Several of these problems will be described below.

Regional and National Training for Production Agronomists

Centers with leadership responsibilities for one crop – for example, wheat at CIMMYT or rice at IRRI – have found it possible to provide the in-service training required for breeders, laboratory technicians, experiment station managers, and economists from national programs. The number of candidates in these fields is manageable.

But training for production agronomists (extension workers) involves hundreds of candidates in most countries and thousands in the larger programs such as those for India, Pakistan, Egypt, and the Philippines. Clearly such training can best be organized within producing regions, or preferably within national programs, with some outside assistance provided by international centers. The cost factor alone would argue that this training be done at home, and the existence of various farming systems and various language problems also favors localization of training for agronomists.

Some experience is accumulating for regional and in-country training, but no reproducible system has yet been developed and the volume of training is still far from adequate. Some exploratory experiences are described in the paragraphs below.

1. When IRRI helped national programs introduce the first semidwarf rice varieties in the mid 1960s, IRRI organized training for extension workers within national programs on a three-step basis: first, IRRI brought to the Philippines a few extension leaders who would serve as trainers in a national program and gave them a course in rice production and in training methods; second, some IRRI training staff moved to the country where the training course was to be given and spent several months with the local staff developing the syllabus for the course and the demonstration crops; third, the local trainers and IRRI trainers jointly supervised the first course within the national program. The IRRI staff then withdrew, and further courses were wholly the responsibility of the local trainers.

This system was repeated in a number of national programs, generally just before the planting season for the crop.

2. The Rockefeller Foundation established regional training centers for wheat in Turkey and maize in Thailand, headed by Rockefeller scientists but also using trainers from within the host country. A considerable volume of training has been accomplished, especially by the Thailand center.

3. The Ford Foundation has financed a regional training program for a number of crops through the Arid Lands Agricultural Development Center (ALAD) at Beirut. One advantage of regional training, particularly for the Mediterranean region, has been that courses can be offered in the English, Arabic, or French languages for different groups of trainees.

4. Various bilateral assistance programs such as the United Kingdom's Ministry of Overseas Development (ODM), the Swedish International Development Agency (SIDA), Canada's International Development Research Centre (IDRC), and USAID have financed training programs for extension workers of various crops within national programs of the Mediterranean-Near East region.

5. CIAT has "trained the trainers" for several national programs of maize, rice, cassava, and beef in South America, then sent CIAT trainers to assist the local trainers in conducting the first in-country course. Such a course was co-

sponsored by the government of Ecuador and CIAT in late 1974 for the lowland agricultural area featuring rice, maize, beans, swine, and beef (all crops and animals for which CIAT holds responsibility).

The government of Ecuador has asked CIMMYT and CIP to repeat the same procedure for its highland crops, which include maize, wheat, barley, and potatoes. "Training the trainers" for this course has been completed at CIMMYT.

6. A totally different approach is represented by a Philippine-CIMMYT experiment in 1974. A top training officer for maize in the Philippines has spent a half year serving as the assistant maize training officer at CIMMYT. He has now returned home, carrying the CIMMYT work experience, but he gave us as much as he received.

Other formulas whereby the international centers can help organize more training for extension workers, in-country and in-region, will no doubt evolve.

The need is not yet adequately filled. That is a major reason why CIMMYT is proposing the assignment of trainer-agronomists to six maize-producing regions and four wheat-producing regions of the world.

Establishing Priorities: Need for Data

The international centers operated in the 1960s like horse-and-buggy doctors – diagnosing all problems with general information and common sense, and with surprising success.

Now, after the world crises in food, fertilizer, and energy have gained wide recognition; after more finances have been assigned to centers; after more national production programs have been organized – there is need for more quantitative information in making plans.

"Eyeballing" was a favorite term in the 1960s used to describe the methods for selecting new lines of breeding materials for crop improvement. Now more sophisticated methods of the statistician and the protein laboratory are supplementing the eyeball.

Our knowledge about national programs, their climates, and their governments was also relatively simple in the 1960s. "We worked on the problems that any cow could see from the side of the road" was a common expression. And at that stage of world cereal improvement the problems given the highest priority were also the most obvious.

Now, as the centers reach beyond the conspicuous problems and focus attention on constraints which were considered secondary ten years ago, more detailed information is needed about producing countries. However, the information needed for planning is not necessarily the same information which has traditionally been gathered and published by the FAO, the IBRD, and the USDA.

CIMMYT has begun, through the travel of its staff, to gather some of this information directly from national programs, and answers can be speeded by assigning postdoctoral fellows to some of this work.

We need information about cereal consumption: How many people eat bread wheat, durum, barley, or maize? What is the size of the demand for the different kernel types (flint, dent, floury) and colors of maize, and the different gluten strengths and colors of wheat? How large and populous are the areas where the total intake of protein is deficient and which could profit from hi-lysine maize and barley? What proportion of the LDC barley crop is used for human food, animal feed, and malting?

We also need a range of information about the extent not only of various climatic zones but of cropping areas affected by diseases and insects. Similarly, more information is needed on the extent and frequency of drought and on problem soils.

Answers to all these questions will become steadily more important in CIMMYT's program planning.

Problems in Management in National Programs

Deficiencies in management of national programs, especially in research management, constitute a frequent obstacle to improved food production. These management problems are common to all crops and all services, not just to wheat and maize.

Scores of crop scientists in developing countries, many with Ph.Ds, have left their home governments to join FAO or other international organizations. Their complaints against their home research service include low salaries, staff promotion on a political rather than a merit basis, lack of budget support from policy makers, corruption, and many other grievances.

Beyond staff losses other shortcomings in research management widely observed by CIMMYT consultants include the following:

1. Research managers try to handle too many crops and too many plant materials for each crop, resulting in unreliable data.

2. Research managers fail to test experimental plant materials off the station on private farmers' land, which would give a better understanding of what is impeding the farmers' yields and thus better guidance to research planning.

3. Research managers fail to identify elite plant materials at an early stage and to move these materials expeditiously into seed increase and varietal release for national use.

4. Research managers fail to arrange that the research organization perform a continuing training function for the extension service, by annual workshops at the research station, preceding each crop planting time.

5. Research managers fail to issue an annual report which is timely and objective, describing the status of national production each year and the constraints which are researchable, and discussing the relationship between the research service, extension service, seed increase agency, credit service, and other inputs needed to bring about higher yields.

Successful remedies to these management defects have been few, and the experience to date with international assistance for management has set no reproducible pattern.

Some countries with large food production programs like India, Pakistan, and Turkey appointed commissions to review the management of research in the 1960s. The success of such commissions has varied with the willingness of the Ministry of Agriculture to endorse the findings and support the director of research in making changes.

In 1973 the IBRD financed a mission to review the national research programs of Spain. This mission included a number of foreign consultants, including staff from two of the international centers. An earlier IBRD mission on agricultural education in the Philippines obtained a substantial improvement in the salary scale for those employed in agricultural research.

CIMMYT staff are often asked to comment on management problems in agricultural research and on the relationships between agricultural services in a national program. We have made no careful review of the actions taken after such consultation.

To date there has been no organized effort by the international centers to arrive at a judgment on the management problems in national programs, to list available consultants who could advise on management problems, or to incorporate a management component into the research training provided by the international centers.

14

Impact of International Research on the Performance and Objectives of National Systems

Sterling Wortman

The impact of international research on national systems may occur in a dozen or more ways, some direct and others indirect.

If we are interested – as we must be – in the impact of research on *entire* national agricultural systems, and hence on national agricultural productivity, we must also recognize that any effects of international research are confounded with those of many other technological, economic, and social factors. What portion of the overall effect is the result of research thus becomes a matter of judgment rather than measurement. And such judgments, to be valid, ought to be made by authorities of national systems who know first-hand the effects of research on their efforts. But, judgments by what authorities of what nations?

There are over 125 national agricultural systems: some are large and some small; some are centrally planned and others not; some are well supplied with trained people but most have few specialists; some operate with sizable or even liberal budgets but most have meager funds and facilities; some have a long tradition of distinguished agricultural research while others are just getting organized; and, finally, some belong to countries in which yields have been steadily advancing while others function where stagnation is characteristic. Assessing the impact of international research on such a complex of national systems, even if that assessment were made by a sizable panel of national authorities representative of these diverse interests, calls for considerable humility. Moreover, since the international system is so new (many programs are only two to five years old) and a lag of several years is to be expected in the application of research results, any current assessments must be accepted as tentative at best, with large error terms built in.

It is possible, however, to recall the ways in which the international research system was *intended* to affect national systems and even to describe in a general way what impact there appears to have been. Before proceeding to do this, it will be useful to consider definitions of some of the terms used in this chapter.

Definitions

By "national system" is meant the entire apparatus – public and private, from national governing bodies to the smallest farmer – which must be involved in the process of agricultural change. Included is the "national agricultural research system" with all its components, from the analyses involved in setting production and other goals (planning offices), through central and regional experiment stations and experimental programs of colleges or schools of agriculture, to on-farm experimentation.

Even the phrase "agricultural research" deserves definition. As used here it means the systematic effort to develop new ways to change agricultural productivity, or efficiency, at any level. Importantly, it includes the greatly underappreciated level of on-farm experimentation – assistance to individual farmers in identifying superior crop or animal production systems – a level of research which for too long has been miscategorized as "extension." It includes studies of ways to improve national planning and to identify rational agricultural goals, as well as the conventional field and laboratory work of experiment stations and colleges.

"International research" is here used to describe the total world experimental effort including that of (a) universities and agencies of developed nations, (b) international institutes and agencies, and (c) institutions of developing countries. From the standpoint of any particular nation, "international" research is all that is done elsewhere, plus the activities of any international organizations within its borders.

International Institutes

A substantial number of large and small, public and private efforts combine to affect in widely varying degrees the performance and objectives of national systems. As it is quite impossible in one chapter to examine the impact of such efforts, I shall limit my discussion to the impact of that relative newcomer to the scene of international research: the autonomous, international agricultural research institute.

Research undertaken by the private but international institutes requires the special type of funding provided by the Consultative Group for International Agricultural Research. And, several unique characteristics distinguish research as it is conducted at the institutes:[1]

1. It addresses those complex problems that require long-term attention by teams of scientists backed by expensive facilities and that are not or cannot be handled quickly and effectively by other institutions.

2. It serves to fill important voids in international research and contributes to the development of a cohesive, collaborative world effort; hence, work at the centers gets its impetus and orientation in large part from others whose work it complements or backstops.

3. It provides opportunities for training at several levels of sophistication for the staff of national programs or from other institutions in the world system.

Confounding of Effects

Any impact on national systems by research at the international institutes is – and should be – confounded with contributions of research efforts elsewhere, since the institutes are simply hubs in an international network of cooperative, collaborative activity. Much of the centers' success depends upon the great array of advances made over past decades, upon the scientific and scholarly capital available to the centers in the form of advanced biological materials, equipment, or chemicals, and upon their store of knowledge and understanding as embodied in their staff and consultants or as given expression in their publications.

Success of the program of any international center is enhanced by the continuing flow into that center of the results of advances anywhere in the world. Indeed, one measure of success of a center is the degree to which it facilitates active exchange among nations of information and materials. The stronger the world effort a center helps to generate, the more successful we may judge that center to be.

The International "Network"

Probably the need for an international network, or world system, of agricultural research has been recognized and expressed many times in the past. Those working at the International Rice Research Institute in the early 1960s certainly had in mind the creation of an international *rice* research network;

in fact, the institute staff realized at the outset that its mandate – to increase national average rice yields and total production in Asia – would require the combined efforts of individuals and institutions of many countries. Without such a major cooperative effort – without the increase in yields and total production – the institute would be judged a failure. It was as simple as that.

Creation of a similar network had begun even earlier for spring-type bread wheats. The Oficina de Estudios Especiales of Mexico's Ministry of Agriculture, manned jointly by Mexican and Rockefeller Foundation scientists, was at work on this in the late 1940s, when young technicians and scientists of other countries were brought to Mexico for training, and a system of international cooperative field trials was organized. At about the same time similar efforts were initiated with maize, and in 1954 the Central American Cooperative Corn Program was launched, with agencies of all nations of Central America and Panama participating.

There were early international efforts with some crops sponsored by FAO through FAO-Rome and the International Rice Commission, by USDA, and probably by others.

By 1969, the need for a comprehensive international research network had been recognized by the Rockefeller Foundation, and the concept was presented to the heads of the world's assistance agencies at an April 1969 meeting in Italy (now referred to as "Bellagio I" in agricultural circles):

> Action Recommended
>
> Acceleration of world agricultural output can be fostered by the formation of a worldwide, interlocking complex of national and international scientific institutions, programs, and projects designed to produce scientific information, materials, and manpower required to intensify agricultural production wherever needed. Provision must be made for immediate attention to all areas where agricultural productivity is still low and static and where man:land ratios are most unfavorable, and to control of many internationally serious diseases and pests. The underexploited tropics and certain arid areas can and should be brought into use as required.
>
> Development of such a network of institutions and activities will require that national and international efforts be cooperative and coordinated to the extent possible. Toward this end, increased and periodic dialogue among appropriate leaders should be established, to identify neglected, high-priority needs and to foster cooperation wherever indicated.[2]

The foundation's paper went on to list and discuss some of the necessary components of an international network, including (a) international centers (such as IRRI, CIMMYT, IITA, CIAT), (b) regional research centers, (c) international programs and projects, (d) national research, training, and production systems, (e) colleges of agriculture, and (f) centers of specialization in developed nations.

In a summary statement in the above-mentioned proceedings, prepared by the late Dr. W. M. Myers, the participants recognized the need for support of an international "hierarchy of institutions" and for mobilization of large sums of money for investment in high-technology agriculture.

The elaboration of such a system, with favorable impact on yields, quality, and total production of major food crops and animal species in the developing countries, is the central concern of the Consultative Group for International Agricultural Research.[3] An impressive beginning has been made.

Effects and Interactions

The newly developing international research effort has had a multiplicity of effects on national agricultural systems, on national research programs, on production of certain crops in some regions of the world, and on foundations, national assistance agencies, and international institutions. As we have already noted, there is a high degree of interdependence (the confounding effect mentioned earlier) among these varied organizations, and the fact that they are discussed separately below should not cause us to lose sight of that. Perhaps a few examples of the simpler interactions will be helpful. First, the clear demonstration that new opportunities exist to raise yields and the profitability of farming can cause agencies to invest in new initiatives. Similarly, the decisions of national authorities to push the production of particular crops or animals in particular regions influence the orientation of research.

Second, experimental evidence that large numbers of small farmers can be benefited through promotion of high-yielding cropping systems probably has had an impact on international agencies such as the World Bank. Surely the World Bank's decision to promote increased productivity of small farms will affect research and production efforts of many nations and most international centers.

Third, success or lack of success of national production programs utilizing technology developed locally or internationally will in turn have an effect on the orientation of efforts at international centers.

Fourth, breakthroughs at centers of specialization, including universities, can have far-reaching consequences. For example, the identification at Purdue of the opaque-2 gene effects in corn and Purdue's discovery of lines of sorghum with high nutritive value are having an obvious impact on the research programs at CIMMYT and ICRISAT, as well as on national programs.

Fifth, determinations of the payoff from investment in agricultural research and of the factors which influence such payoff will surely have an effect on investment attitudes of international banks and assistance agencies and, it is hoped, on the leaders who must make national decisions regarding investment in agriculture.

There are many other interactions which might be mentioned, but the foregoing may suffice to demonstrate that interdependencies exist. Studies of factors conditioning the effectiveness of agricultural research can, if constructive, lead to improvements in such research efforts.

Impact on National Systems

In their studies of the impact of international research on the performance and objectives of national systems, some investigators have attempted to identify specific causes and effects as well as potential improvements in the direction or operation of the systems. Almost invariably, such studies require not only narrowing of the issues and the employment of assumptions which may or may not be correct but also the exclusion of some considerations which are so elusive that they cannot be adequately considered. Let me mention a few of these elusive problems.

First, a substantial number of leaders from the developing nations serve on the boards of the international institutes, on the Technical Advisory Committee of the Consultative Group, or as representatives to the Consultative Group. Others are on boards or advisory committees of international banks or international or national assistance agencies. In such roles they participate in a continuing debate on approaches to the development of agriculture and become acquainted with advances in materials or techniques as they occur. Undoubtedly this experience has an impact on the performance and objectives of these leaders' respective national research systems, but how does one measure such impact?

Second, through participation in the various Bellagio conferences which led to the formation of the Consultative Group, through service on the boards of institutes, and through their interactions with leaders of developing countries, the heads of assistance agencies have become increasingly knowledgeable, and some have made major changes in their approaches to lending and/or to providing technical assistance to developing nations. These changes can have an impact on the nations themselves. But it is impossible to quantify the magnitude and effect of such changes, and consequently it is difficult to attribute them to any set of institutions.

Third, the construction of modern agricultural research centers in Africa, Asia, and Latin America has drawn the attention of national political leaders, of both developed and developing countries, to the importance of agricultural research. That such installations were placed in the developing countries certainly has demonstrated in a dramatic way that many of the biological components of agricultural production systems must be developed in and tailored to the ecological and economic requirements of the regions where they are to be utilized. An impressive number of heads of state and ministers of the developing countries have been drawn to these modern centers. That this has had an effect on the quality of facilities provided in some nations, for both research agencies and universities, is without doubt. To trace such effects would be difficult; to suggest probable causes would be risky at best.

Fourth, we have witnessed the strengthening of national agricultural institutions, the initiation of production campaigns, the implementation of decisions to work with small farmers, and the reorientation of activities of universities and national agencies. But the degree to which such decisions have been affected by the international research system is and probably will remain unknown.

Fifth, during the past several years, the Ford Foundation has greatly increased its intellectual and financial involvement in agricultural research. The International Development Research Centre of Canada has been formed and has become an important source of research support and innovation. These and other organizations with a longer history of involvement, such as FAO, USAID, the United Kingdom's ODM (Ministry of Overseas Development), France's ORSTOM (Office de la Recherche Scientifique et Technique d'Outre-Mer) and IRAT (Institut de Recherches Agronomiques Tropicales et des Cultures Vivrières) and the ADC, have participated in a growing international effort to improve understanding of systems, to add to the knowledge base, and to foster the exchange of information among interested individuals and institutions. It all adds up to a growing world capability, with important but immeasurable effects on national programs.

Given these and other influences of unknown magnitude on national programs, we can only point out – we cannot measure – some of the probably important ways in which the international research system has had an impact on the performance and objectives of national efforts.

Impact on National Agricultural Systems

Clearly the international research system has drawn attention worldwide to the importance of agriculture. Some of the early successes, notably those with wheat and rice which were popularly termed the "green revolution," attracted the interest of national authorities. Visits to the centers by national leaders probably created in some the desire to promote scientific agriculture at home, to strengthen national research agencies and universities, and to marshal the necessary funds for research and production efforts.

In the late 1960s, the long-neglected basic food crops (cereal grains, food legumes, root crops, vegetables, animal species) began to receive serious consideration by many nations. It was recognized that work on these crop and animal species would have to be done in the public sector; there simply was no way for private companies to realize returns on investments in such activities. Clearly the present focus on production of the basic food crops in the developing countries is in part attributable to such an emphasis at the international institutes.

There has been a recent trend toward production-oriented, commodity-oriented research and production programs in a number of nations. While not replacing discipline-based research, the new trend reflects the more purposeful nature of new national activities.

There has been a growing realization of the need for more sophisticated agricultural research efforts at the farm level.

The institutes offer new sources of training for research and production personnel from national institutes. More and more often the much sought-after, high-quality educational programs offered by universities in the United States, Europe, and elsewhere are being combined with graduate thesis research at international centers or in the developing countries. This has permitted graduate students to contribute to the solution of problems of the regions in which they live. Specialists or advisers at such universities have become engaged in this work on developing nations' problems, and many lasting friendships have been formed among institutions and individuals—relationships which will be of growing importance. Surely the new forms of training have had and will continue to have a major impact on the strength and the orientation of developing countries' institutions. But, again, effects of training would be hard to measure.

Finally, the successes of new production programs, together with the continuing need to increase agricultural output, have contributed to the growing willingness of some nations to increase their investments in scientific agriculture.

Apparently there is a growing conviction among national leaders that science-based agriculture (plus reduction of population growth rates) offers the only reasonable hope of meeting food needs and raising the standards of living of the rural poor; this must to some degree have been stimulated by the international research system.

Impact on National Research Systems

International research efforts clearly have affected national research systems in ways other than the production of useful technology and trained people.

Priorities established for the international system by the Technical Advisory Committee of the Consultative Group control in a major way the priorities of the international institutes and of related programs. Advanced technology available through the institutes and through the types of training they offer in turn affects the priorities of national organizations.

The production orientation of much of the international research effort, including its emphasis on commodity-oriented research, evidently is gaining acceptance as a logical approach by increasing numbers of nations.

The interdisciplinary team approach to problem solving is spreading. The larger nations can and must undergird their production-oriented research efforts with more basic investigations. The greater number of the smaller nations, however, cannot expect to organize truly comprehensive scientific efforts right now. In the immediate future, the in-depth, interdisciplinary research at the international centers will serve as the more basic research effort for these nations, allowing them to concentrate their limited human and financial resources, as they should, on the tailoring of technology to their own needs and in getting it applied at the farm level. At the same time, we must note a growing sophistication of the agricultural research effort of some nations, which undoubtedly is due to the influence of the international cooperative activities.

New standards of performance are being set. Everywhere there is growing dissatisfaction with the low and static yields so generally being obtained, and scientists are being pressed to do something to raise these yields; this is an issue that no longer is easy to duck. Scientists now freely discuss and compare yields per hectare per crop, per hectare per year, and even per hectare per day. The new higher yields are being demanded by people outside the scientific fraternity, and this is a healthy development.

Research programs are being speeded up — more problems solved, more new materials produced. Two or sometimes three experimental crop generations are grown per year.

More frequent opportunities now exist for national scientists to meet with colleagues from other institutions in their home countries, from institutions in other countries, or from international institutes. This helps to prevent the repetition of experimental work already accomplished elsewhere. Through international conferences innovation is fostered. It is likely that such opportunities for exchange of ideas and information will help to attract and hold competent national researchers on the job.

International cooperative experiments and field tests draw the community of scientists together, allowing them to base conclusions on much more substantial experimentation than would otherwise be possible. By testing new materials and techniques over many locations in a single year, the time required to generate necessary data is shortened. For these and other reasons, one can sense a dramatic acceleration in the world drive toward an improved science-based agriculture.

The international institutes have introduced an element of stability into the world research effort. None of the major assistance agencies, with the possible exception of the World Bank, could arrange with any degree of certainty for the long-term maintenance of an international center. But multilateral support of such efforts provides reasonable security for such operations. The presence of substantial numbers of career scientists at the international institutes ensures that continuing contributions to the world knowledge-base will be centrally preserved, not lost as so often occurs when research efforts are of short duration and participants scatter. Even when national research operations are disrupted for political or other reasons, momentum can be regained more easily if the national agencies have been cooperating with an international center. One reported example of this was the resumption of activity of the Bangladesh Rice Research Institute following the civil war. The fact that IRRI had been involved, along with others, allowed BRRI to pick up its program and move ahead when conditions again permitted it. Stability is an important factor in the world effort, but how does one measure its impact?

Impact on National Production

Clearly the international research effort has had a positive impact on national food production. National and international agencies are attempting to measure it. Dalrymple has provided extraordinarily valuable assistance in monitoring the spread of high-yielding varieties,[4] and a number of investigators have studied the effects of research on levels of production. Indeed, the presence at Airlie House of so many capable people reporting on significant work was testimony to the growing impact of research. Yet we are all aware of the difficulties of isolating the effects of research on production, given the many other factors which obviously are involved.

Impact on the International Community

There is a growing understanding of the requirements for increased agricultural output per unit area per unit of time. That understanding is spreading, albeit too slowly, through the staffs of national and international organizations. Gaps in technology are being identified and, rather systematically, closed.

We seem to be moving toward the identification of strategies of national agricultural development that are much more likely to succeed than were the earlier ones. Moreover, the newly developing acquaintances among individuals and institutions have added an element of mobility to world scientific efforts. The impact on the developing nations should be favorable.

Schultz pointed out several years ago, in his excellent book *Transforming Traditional Agriculture*, that the small farmer is a rational economic being, that he is efficient in the management of scarce resources, and that he will in many cases improve his farming practices if it is feasible and profitable for him to do so. And now at last interest in the small farmer is gaining momentum.[5] In part this is the result of work carried out with some support from international centers: the Puebla Project in Mexico; IRRI's research program in central Luzon, which reportedly has contributed to the government's "Masagana 99" program. The impact of such activities on the decision of the World Bank to press for small farmer programs is unknown to me. But, out of the growing debate on the feasibility and desirability of assisting small farmers has come the realization that they *must* become involved. They can contribute significantly to increases in needed food output. If improvements in their own incomes occur, they can purchase the products of urban industry, thereby increasing the domestic market for those products. In short, the improvement of the productivity and incomes of small farmers in most countries should provide new momentum to the economic development of those nations.

Certainly those involved in the international research network have made a significant contribution to awareness on an international scale of small-farmer problems and potentials. But the task of measuring the magnitude of any such contributions is difficult, if not impossible.

Measuring Magnitude of Impact

In the preceding sections I have attempted to identify some of the contributions which have been made by the international research system.[6]

The next logical question is, What has been the *magnitude* of the impact? The answer is easy: I don't know. The Airlie House conference did not fully answer the question, though it surely has moved us closer to a sound judgment.

Many of the probable effects of the international system are elusive (but not phantom) and are cumulative in their effects, hopefully in a positive direction. My guess is that the effects of the international system on national systems generally will be greatly underestimated, especially in studies in which scientific methodology is rigorously and correctly applied. Error will occur because of the difficulty of sorting out interdependencies required to arrive at net effects. In all likelihood, calculations will be off by several orders of magnitude. However, this should not discourage us from attempting to reach ever better approximations.

Meanwhile, the elaboration of the international system and of most national systems will continue to be based on judgments. There is no alternative, for decisions must be made. People are hungry, populations are growing, and time for effective action is passing. Fortunately, new mechanisms have emerged for arriving at considered judgments, and in the following paragraphs a few of these will be mentioned.

There is a growing number of national authorities, including scientists, who have been watching national development long enough to have witnessed change and who are aware of major factors causing such change. In 1972 the International Rice Research Institute, after ten years of operation, arranged for several people to review the impact of the institute and its work.[7] Three of the participants in that conference had particular competence to evaluate IRRI's contributions. These were Dr. D. L. Umali, now assistant director general for Asian and Eastern Affairs, FAO-Bangkok, and formerly dean of agriculture of the University of the Philippines; Dr. B. P. Pal, who served as director general of the Indian Council of Agricultural Research through the 1960s; and Dr. Gelia T. Castillo, associate professor of rural sociology, College of Agriculture, University of the Philippines. Their papers are very useful and should be read widely by people outside the institute.

A more formal assessment of impact is made by the board of directors of each independent institute. Institute directors are aware of national needs and know how importantly international research efforts may influence, or fail to influence, the efforts of their nations.

The impact of the international system is under constant review by the Technical Advisory Committee of the Consultative Group either through the efforts of its own members or through the review missions which it organizes.

The effect of the international system on national programs is under continuing review by the donor organizations which compose the Consultative Group. These organizations are investing substantial sums of money in the system and are under presssure to provide justification for those investments.

Finally, there are the studies of the impact of the international system undertaken by investigators such as the Airlie House conference participants. These provide a valuable basis for the judgments which must and will be made by individuals whose responsibility it is to make decisions.

Quantitative assessments of the impact of the international system on national research and production efforts clearly cannot be precise. However, judgments on the organization of and investment in research are, one hopes, increasingly based on the more solid information being generated and certainly can be based on a growing body of experience. There is comfort in that.

NOTES

1. For an elaboration of this subject, see "The International Agricultural Research Institutes: Their Unique Capabilities," *Rice, Science and Man* (Los Baños: International Rice Research Institute, 1972), pp. 65-75.

2. See "The Technological Basis for Intensified Agriculture," *Agricultural Development: Proceedings of a Conference Sponsored by the Rockefeller Foundation, April 23-25, 1969, at the Villa Serbelloni, Bellagio, Italy* (New York: Rockefeller Foundation, 1969), p. 39.

3. For a brief history of the development of the Consultative Group, see Sterling Wortman, "Extending the Green Revolution," *World Development*, 1:12 (December 1973), 45-51.

4. Dana G. Dalrymple, *Development and Spread of High-Yielding Varieties of Wheat and Rice in the Less Developed Nations*, Foreign Agricultural Economic Report no. 95 (Washington, D.C.: Economic Research Service, USDA, 1974).

5. See Carroll P. Streeter, *Reaching the Developing World's Small Farmers* (New York: Rockefeller Foundation, 1973).

6. For a discussion of effects on crop research, see Sterling Wortman, "A New Era in Crop Improvement," H. K. Hayes Memorial Lecture (Minneapolis: University of Minnesota, 1974).

7. The collection of papers was published in *Rice, Science and Man*.

15

Impact of the International System on National Research Capacity: The CIMMYT and IRRI Training Programs[1]

Burton E. Swanson

This chapter examines the impact of the different training strategies being employed by the International Maize and Wheat Improvement Center (CIMMYT) and the International Rice Research Institute (IRRI) to help build research capacities in less developed countries.

It is necessary first to clarify how research capacity is being viewed here. Ruttan and Hayami considered the problem of international technology transfer and differentiated the process into three phases: (1) material transfer, (2) design transfer, and (3) capacity transfer.[2]

The international transfer of wheat and rice technology during the late sixties was largely in the "material transfer" stage. Many developing countries imported substantial amounts of seed during that period.[3] At the same time "design transfer" was also occurring, as national research institutions began to receive new, high-yielding experimental lines and varieties of wheat and rice that were then tested and, in some cases, multiplied and released to agricultural producers.[4] In addition, the international research centers, through their training and outreach programs, were beginning to devote some attention to the third phase of the technology transfer process. However, "capacity transfer," or building a national research infrastructure that can produce scientific knowledge and improved agricultural technology – and adapt it to local ecological, resource, and institutional conditions – has been found to be a slow, difficult, and complex task.[5]

To understand better the institution-building task associated with capacity transfer, it is useful to differentiate research capacity into its two primary functional components: science and technology. Price developed some generalized definitions, during a parallel discussion on technology transfer in space science, that are useful to this discussion: science is taken to be those research activities that result in scientific papers being written and published; technology is "that research where the main product in not a paper, but instead a machine, a drug, a product, or a process of some sort."[6]

The point in making these rather sharp distinctions between science and technology is not artificially to separate one part of this process from the other or to suggest that research workers should work in only one area or the other. Many agricultural research workers function in both knowledge-generating (science) and technology-developing roles and move easily and effectively between them. But in organizing national research institutions there has been a tendency to overlook the technology-development function. Wortman has made this point rather succinctly: "That agricultural research must be undertaken at central experiment stations or in the laboratories of national research organizations or colleges of agriculture is well understood; frequently, however, the further steps of identifying and testing packages of technology in each distinct farming region of a nation and finally at the ultimate experimental site – the individual farm – are erroneously excluded."[7] Because of the importance of this distinction we need to know which hat a research worker wears at any one time in order to understand the thrust of his overall research program and how it contributes to national research capacity.

Before proceeding in this discussion, I wish to present the concept of the *biological architect*, as utilized in this chapter. The biological architect is the high-level research worker who develops improved biological technology. He is the "master builder,"or the "research inventor," who manipulates and integrates new and/or different factors, materials, knowledge, etc., to create improved agricultural technology.[8]

Scientists and biological architects use many common research tools, but differences in research objectives and output demand substantially different types of cognitive behavior. The scientist, in his pursuit of new knowledge, engages primarily in analytical research. *Analysis*, as a cognitive skill, "emphasizes the breakdown of the material into its constituent parts and detection of the relationships of the parts and of the way they are organized."[9] The scientist uses research tools as a means of testing specific hypotheses that result from his analytical inquiry.

The biological architect engages largely in creative thinking and, by trying new materials and methods in different combinations and amounts, attempts to develop new technology or technological components that will better achieve a production objective (such as increased output, reduced risk, or reduced costs). In terms of cognitive behavior such creative thinking may be characterized as *synthesis*.[10] In synthesis, the research worker "must draw upon elements from many sources and put these together into a stucture or pattern not clearly there before. His efforts should yield a product. . . ."[11] The biological architect uses research tools to evaluate his newly created technological combinations or components in order to determine their production potential.[12] Perhaps it is because the biological architect must spend so much of his time evaluating his new technological combinations and components – through the use of relatively simple and routine research trials and tests – that the importance of his intellectual or cognitive contribution has been overlooked or minimized.

Applying this framework to research capacity as it relates to crop technology, we see that in the past it has been the plant breeder who has generally developed improved genetic technology by manipulating different genetic factors to produce improved varieties. The production agronomist has developed and refined improved production recommendations for use by farmers by manipulating new and/or existing factors of production (such as improved varieties, fertilizers, pesticides, cultural practices, etc.) in different ways. In more recent years, as agricultural technology has become increasingly sophisticated and complex, the trend has been to form interdisciplinary research teams which work together in manipulating and integrating different chemical, biological, or mechanical factors of production to build superior agricultural technology.[13]

Before moving into a discussion of the training strategies being used by CIMMYT and IRRI and the impact of these strategies on research worker behavior and research capacity, we should mention one other institutional factor that appears important to the question of science and technology and the process of building national research capacity. This factor is the predominant reward system within agricultural research that appears to be exerting significant and continuing influence on research worker behavior and national research capacity.

Reward Systems in Agricultural Research

One explanation why research institutions in many less developed countries have been ineffective in producing improved agricultural technology is that research workers have tended to concentrate on more theoretical research problems rather than working to solve farmer production problems.[14] An hypothesis that is logically consistent with and will explain this behavior is that agricultural research workers in LDC's have adopted or internalized the normative structure of the public research establishment (the university and the corresponding academic/scientific professions) of industrially developed countries.[15] The following factors serve to support and elaborate on this hypothesis.

First, most research workers in the LDC's with advanced degrees received their academic training in foreign universities, primarily in the United States and Europe. Many of these research workers may have been influenced by the "publish or perish" reward system that is common to large, research-oriented colleges of agriculture in the United States. Furthermore, because advanced research degrees require an "original" research inquiry that contributes to the body of knowledge in the respective discipline, these research projects are frequently highly specialized and theoretical in nature. Both of these factors could tend to orient young research workers toward more theoretical, scientific inquiries.

Second, in the United States the knowledge-generating research function (science) is carried on primarily within the public sector (universities), whereas the greater part of technology-developing research is carried out in private industry.[16] Although research workers who conduct technology-development research have much less opportunity to publish in scientific journals and to gain professional recognition from their colleagues than do their counterparts in the public sector, salary schedules in private industry have traditionally been higher than those in public research institutions. Thus, in the overall agricultural research system of the United States, there is to some extent a trade-off between professional and economic rewards. Research workers, depending on their interests, abilities, and what they consider to be important, have alternative career patterns they can pursue.

On the other hand, in LDC's most if not all of the national agricultural research capability is located within the public sector, generally within a ministry of agriculture or university. Therefore, the opportunity for research workers to select between economic and professional rewards is quite limited. Agricultural research workers receive salaries according to the bureaucratic procedures and criteria being followed by the research institution, not according to the type of research carried on. Therefore, there is no potential within the research institution itself for inducing research workers (through economic rewards) to pursue career patterns oriented toward technology-development research objectives.

Thus, in the absence of an alternative reward system, it appears inevitable that professional rewards will take on increasing importance in influencing the types of research activities being undertaken by research workers in LDC's, particularly given the considerable professional recognition associated with publishing in a prestigious scientific journal with an international clientele.

Third, in most cultures agricultural work is considered a low-status occupation; therefore, there are no positive social rewards to encourage a highly trained research worker to work on practical problems – the results of which would be of direct importance only to a peasant or cultivator and which would not result in any significant professional recognition. Even if individual scientists were motivated to do this type of research, the research organization has no effective way formally to reward these successes.

Finally, the spirit of cooperation is frequently missing from national research institutions, and scientists may tend to think about achieving personal rather than institutional credit. To develop improved agricultural technology requires considerable interdisciplinary cooperation, but the credit accrues to the team, not to the individual. If this team credit is usurped by the research director or the team leader, instead of being shared by the team members, individual research workers will not be encouraged to work together on future endeavors.

All these factors may prompt agricultural research workers in LDC's to concentrate on individual research projects aimed at generating new knowledge where they can receive professional recognition and rewards directly. In most national research systems in LDC's there are too few positive rewards and incentives to encourage research workers to carry out technology development. It is no wonder, as Singer suggests, that the meager research resources of the LDC's (estimated at 2 percent of the world's research and development expenditure) are frequently misdirected to research problems that are more relevant and useful to the rich nations.[17]

The CIMMYT Wheat Training Program[18]

The wheat training program in Mexico is an integral part of CIMMYT's program for making improved wheat technology available to farmers in all major wheat-growing regions of the Third World. Strong national programs are an essential part of this international wheat improvement strategy, both in the process of developing and disseminating improved wheat technology and in dealing with spin-off problems, primarily disease epidemics, that are a potential threat to the precarious food balance in populous nations.

When the wheat revolution began to spread to South Asia and the Middle East in the early sixties, national wheat improvement programs in these areas were generally weak and poorly organized. Training programs were used by CIMMYT to upgrade the technical skills of research personnel in an attempt to build strong, independently functioning national programs that are nevertheless interdependent with other national programs and with CIMMYT for new genetic resources and technical information.

At all levels, CIMMYT stresses a "team" or integrated approach to wheat improvement. At the program level, the emphasis is on interdisciplinary teams. At the international level, the national programs and CIMMYT work together as part of an overall international wheat team systematically to share not only superior germ plasm and new varieties as soon as they are developed but also technical information on these genetic materials.

The wheat training program is directed primarily toward middle-level research workers. The program revolves around three main educational objectives: to impart to trainees the research skills and knowledge needed to run a wheat improvement program; to encourage and develop the trainee's ability to create (synthesize) new forms of wheat technology; and to foster specific types of attitudinal change among trainees.

Technical Research Skills

The first half of the regular eight- to nine-month training program in wheat improvement deals largely with the mechanics of running an efficient, well-organized research program. Trainees learn all the essential research skills and techniques needed to manipulate and evaluate new forms of wheat technology through "on-the-job training."[19] Trainees follow the CIMMYT wheat program through each stage of the growing season (and the varietal development process) with each task or operation first being discussed in the classroom and then demonstrated in the field. After the trainee has had the opportunity to practice the skill and is "checked out" to ensure that he is reasonably proficient, he proceeds to help carry out each research task or operation within the ongoing CIMMYT research program.

Technical Research Ability: Synthesis

Once the trainee learns the methodology and procedures of operating a research program he can give increasing emphasis to the genetic materials passing through the research program. For example, the job of the breeding team is the creation, or synthesis, of new genetic lines and varieties by combining and recombining diverse types of germ plasm. To be effective and efficient in developing improved high-yielding varieties, the trainee must learn and become increasingly familiar with the various genetic characteristics and materials he is attempting to manipulate. For example, an experienced biological architect in the CIMMYT wheat program can walk up to an advanced generation plot — and there are hundreds of such plots — and from visual inspection alone give the approximate pedigree of the line (from several hundred potential parent lines and varieties), give several reasons why the cross was made, and evaluate the line for those visual characteristics. By working side by side with experienced biological architects in the CIMMYT wheat program, and by asking and being asked the question, "why?" trainees soon begin to develop an ability and an insight into the creative process of genetic engineering.

Attitudinal Objectives[20]

There is a common expression used in the CIMMYT wheat training programs: "The plants are talking to you, but you have to use your *eyes* to hear what they are saying." In other words, wheat plants being grown under a variety of different conditions respond differently to those conditions. A good observer is able to detect how plants react to each of these different environmental conditions and, based on all these data, select those genetic lines with the greatest potential.

CIMMYT seems to use a similar selection technique for its trainees. For example, CIMMYT has been criticized by some visitors to the training programs in Mexico for "using" trainees for such routine tasks as inoculating the segregating, or F_2, populations with rust spores. This is a job that CIMMYT needs to have done and requires about ten days to two weeks of hard, back-breaking work, wading through muddy plots (many times in the rain) and injecting two tillers of each F_2 plant with a syringe full of disease inoculum.

After the first morning of this activity, there is no additional technical training value to be accomplished; however, what the CIMMYT staff learns about the "trainee population" during these two weeks is very important. Some trainees can disguise their displeasure at this type of work for a morning or two, but after a week or ten days, trainees are clearly differentiated by their "reaction to hard field work." Some trainees may do the work while CIMMYT staff members are nearby, but then relax under a tree when the latter leave. A few may call in sick for a few days to avoid work. Others, however, are out in the plots getting the work done. It is this last group that CIMMYT particularly wants to identify. CIMMYT believes it is this group that will begin to make up the hard core of working scientists within the national wheat improvement program.

The training program in Mexico is viewed by the CIMMYT wheat team as only the first step in a long-term process of building effective wheat research workers and national wheat improvement teams. Because of this long-run perspective, the training program becomes both a manpower development tool for training skilled research technicians and an "early generation" selection tool for identifying potential biological architects. Trainees are observed in Mexico and again back home on the job. Those who excel in attitude, outlook, intellectual ability, and technical know-how in both working environments are identified as prime candidates for academic fellowships. It is hoped that these individuals, given additional educational opportunities, will become

Table 15-1. Average Number of Research and Production Activities Completed by Those CIMMYT Wheat Trainees Who Were Active in Wheat Improvement Programs during 1972

	Trainees Conducting Each Activity (N = 105)		Average Number Completed per Trainee
Type of Activity	N	%	(N = 105)
Laboratory or greenhouse experiments	13	12.4	1.37
On-station field experiments	56	53.3	5.62
Genetic crosses	56	53.3	227.1
On-station replicated applied research trials	65	61.9	7.24
On-farm replicated applied research trials	44	41.9	4.74
On-farm high-yielding production plots	37	35.2	3.71

key biological architects in their own national wheat improvement programs in years to come.

Results[21]

Approximately 82 percent of CIMMYT's former trainees are still actively engaged in wheat research and production activities, with another 8 percent being indirectly or partly involved in wheat improvement work. Only about 10 percent of former CIMMYT wheat trainees are no longer working in wheat improvement programs.

As shown in Table 15-1, the performance of those former participants who are still engaged in wheat research and production programs (105 trainees of 130 total respondents) is high, as measured by the numbers of research experiments, replicated field trials, genetic crosses, and production plots completed. Furthermore these trainees are emphasizing technology development, as opposed to experimental research, particularly in the area of genetic technology.[22] It appears that the work of CIMMYT trainees, once back home, is quite consistent with CIMMYT's training objectives.

CIMMYT appears to be very effective in producing research workers who can develop improved genetic technology, but it has not had as much success in producing the other half of the wheat improvement team, the workers who can develop the complementary package of practices. This weakness probably stems from the fact that the CIMMYT research program concentrates on widely adapted genetic technology, with production technology being viewed as a location-specific problem. The fact remains that wheat production agronomists are sorely needed in most national programs to develop appropriate production recommendations that will enable farmers to exploit fully and economically the production potential of the superior varieties being developed.

The IRRI Training Programs

IRRI's programs for research training and rice production training make up the major part of its training activities. The impact of these programs on national research capacity is described and analyzed here.

IRRI Research Training Program

IRRI is committed to strengthening national rice research programs, and its leaders see the research training program as a resource that national programs can use in upgrading the technical proficiency of their staff members.

IRRI's research training approach centers on the research project. Trainees are expected to focus on a serious production problem in their home country and to carry out one or more research projects that will generate new knowledge and possible solutions to these problems.

Research trainees work closely with a senior IRRI scientist and frequently carry out projects that are an integral part of their adviser's overall research. In some cases these research efforts result in jointly authored papers that report important research findings and contribute to the growing body of knowledge about rice production in the tropics.

Although some advanced research trainees come to IRRI to learn specific research skills and techniques, or to conduct projects that are beyond the scope of their national programs (because of the lack of adequate scientific equipment), the major objective of the project approach is to provide participants with a solid research experience – teaching them each step involved in planning, designing, executing, and reporting on a research project. At each step of the way, trainees learn by doing. In terms of educational objectives, the major focus of the project approach is on *analysis* and on using research as a tool to generate new knowledge about production problems and their possible solutions and/or to test suspected relationships between production factors. The knowledge that is generated from these projects may be used in the development of improved rice technology that will increase rice production in the tropics.

There are two exceptions to the project training approach: (1) Nondegree trainees in varietal improvement do not conduct research projects but work largely as research assistants in IRRI's ongoing rice-breeding program. Again, these trainees learn by actually carrying out each task involved in a varietal improvement program, and at the same time they become familiar with the germ plasm currently being used in the IRRI program. (2) Agricultural engineering technicians are given short-term training so that they can evaluate and demonstrate IRRI-developed machinery in their own countries under local conditions.

Each research trainee works primarily in a single department at IRRI, but the Saturday Seminar provides the institutional mechanism by which both trainees and staff learn of recent research findings in other departments. By sharing research results with staff members in other disciplines, scientists can frequently gain new insights and perspectives on important rice production problems which may lead to cooperative efforts between departments. Such multidisciplinary efforts demonstrate to trainees how an effective research institute can organize its scientific resources in alternative ways to accomplish different objectives.

The institute is conveniently located next to the University of the Philippines' College of Agriculture, and about one-third of IRRI's research trainees combine an M.S. degree program at the college with research training at the institute. IRRI scientists believe that combining a degree program with a solid research experience enables trainees to perform more effectively after returning home and also gives them the necessary educational credentials to move into leadership positions in their respective national programs. In general, IRRI research trainees are a highly educated group. At the time the participant follow-up survey was taken in 1973, 41 percent of the research trainees had M.S. degrees, 43 percent had either received or were working toward Ph.D. degrees, while only 16 percent of the group were still at the B.S. degree level.

Results

Former IRRI research trainees reacted in very positive terms when assessing their training experience at the institute. IRRI selected both research and teaching personnel from national institutions, and more than 90 percent of all research trainees indicated that the training had been of some or full use to them since returning home. In addition, approximately 71 percent of former trainees (who were working when the survey was taken) were still actively involved in rice research or production programs.

In considering the main educational objective associated with the project training approach (i.e., analysis), it seemed consistent to expect trainees to continue working in experimental (analytical) research after returning home. To test this hypothesis, the work behavior of former trainees was examined to determine which types of research were being emphasized (see Table 15-2). As expected, more knowledge-generating types of research (field, laboratory, and greenhouse experiments) were being carried out than research activities

Table 15-2. Average Number of Research and Production Activities Completed by Those IRRI Research Trainees Who Were Active in Rice Research and Production Programs during 1972

	Trainees Conducting Each Activity (N = 154)		Average Number Completed per Trainee (N = 154)
Type of Activity	N	%	
Laboratory or greenhouse experiments	65	42.2	1.61
On-station field experiments	85	55.2	3.48
Genetic crosses	46	29.9	19.9
On-station replicated applied research trials. . . .	57	37.0	1.55
On-farm replicated applied research trials. . . .	38	24.7	1.17
On-farm high-yielding production plots	39	25.3	1.46

associated with the direct development of rice technology (agronomic field trials and genetic crosses).[23]

Another factor considered in determining the type of research being emphasized by IRRI trainees was the form of research output, particularly since research papers are the primary output from knowledge-generating types of investigation. First, it was found that 106 research trainees had been able to publish (or present) a total of 187 research papers based on the work they completed while at IRRI. Second, trainees were asked if they had been able to publish or present any papers based on research conducted since returning home. In the two-year period (1971-72) 130 former research trainees produced 370 technical papers. Since only 154 trainees indicated they were actively engaged in rice research and production programs when the survey was taken, and only another 5 percent of former trainees were working in other crop research programs, it was concluded that knowledge-generating research is predominant in the work of former IRRI research trainees. By contrast, it was found that of 105 active wheat research workers who were former CIMMYT trainees, 39 had published or presented a total of 101 technical papers during the same period.

IRRI Rice Production Training Program[24]

The IRRI rice production training program was established in 1964 in response to a growing need for competent rice extension specialists who could (1) diagnose serious rice problems, (2) grow a rice crop using the high-yielding rice technology that was being developed by IRRI, and (3) communicate these skills, methods, and techniques to rice producers through efficient extension methods. Later, as it became apparent that IRRI could not begin to train all the rice production specialists needed by national programs, its focus was shifted to "training the trainers" of rice production specialists. The logic of this approach was to create a multiplier effect within each national program, whereby large numbers of field extension workers could be trained to use and demonstrate the new improved rice technology.

The rice production training program is a six-month course conducted during the wet-rice growing season beginning in early June each year. The behavioral objectives of the present program, in addition to the three original objectives as mentioned above, are (1) for trainees to be able to conduct applied research trials to modify the modern rice technology (package of practices) to fit local growing conditions in their home countries and/or regions and (2) to be able to organize and teach in-service rice production training programs for extension personnel in their home country programs.

To achieve these behavioral objectives, trainees spend about one-half of their time in the classroom learning up-to-date knowledge about modern rice production in the tropics and the other half in the field practicing this knowledge and acquiring new skills in rice production and in communicating this technical information. One of the most outstanding features of the training methodology is that the classroom instruction, which imparts extensive technical information about all aspects of modern rice production, is carefully organized around and integrated with the practical field training. What is learned in the classroom is directly relevant to the problems faced in the field. Second, the course is highly efficient in that it is completely organized before the trainees arrive (lectures, field practice, field trips, etc., are all scheduled), so that each hour is accounted for in terms of the instructor responsible and the behaviorial objectives to be achieved. In addition, most technical lectures, field practice exercises, etc., are reproduced and made available to trainees so that each has a rice production training manual and a complete set of technical lecture notes to use both during the training program and after returning home.

Results

Questionnaire responses show that nearly every former rice production trainee was satisfied with the overall training, and approximately 90 percent of the trainees indicated they were making some or full use of the training. The number of respondents who indicated they were in job assignments where they could make direct use of their training was less satisfactory. Although 81 percent of the respondents indicated that they were working in rice production programs, only about 42 percent were in positions directly associated with the stated behavioral objectives of the program.

In regard to the primary behavioral objective of the program, it was found that only nineteen respondents (12.8 percent) were assigned and working as full-time rice production trainers. Although two-thirds of the respondents indicated that they had worked, at one time or another, in organized rice production training programs, in most cases these duties were in addition to their regular job assignments. In addition to the nineteen respondents who are working as rice production trainers, it was found that another thirty-six respondents (24.3 percent) were working as rice extension specialists and eight trainees (5.4 percent) were involved with rice production programs, but as general agricultural extension workers. In all these cases trainees were considered to be directly involved in assignments associated with the behavioral objectives of the program.

Approximately one-fourth of the rice production trainees are research workers in their home countries. In fact, nearly 10 percent of former rice production trainees are currently working in rice breeding. How or why these trainees were selected for this program, rather than for the research training program in varietal improvement, was not established. Although rice production training is probably quite useful and appropriate for many rice research workers, their selection appears inconsistent with the stated objectives of the rice production training program.

The remaining trainees have moved into other types of jobs since returning home (many are in administrative positions), but some are still associated with rice production work. Eleven participants were attending a university when the survey was taken, and probably many of these will eventually return to rice production activities.

Effect of Different Training Strategies on Work Behavior

It was established in this study that the work behavior of the trainees in each of the three groups differed widely following their return to jobs at home. At the same time, however, there were great similarities in the general background of the trainees in the three groups. Therefore, the purpose of this section is to develop an adequate explanation for these differences in work behavior by examining the major independent and intervening variables.

No attempt is made to characterize one training program as better or worse than another; rather, the objective is to learn from the strengths and weaknesses of each training strategy. All training programs included in this study are generally well organized and have been evaluated favorably by former trainees. To establish this point, two major trainee-assessment variables are contrasted in Table 15-3.

There is no significant difference between the ways in which individual

Table 15-3. General Trainee Assessment of Each Training Program

Type of Training Assessment	CIMMYT Wheat Trainees	IRRI Research Trainees	IRRI Rice Production Trainees
Trainee's overall satisfaction with the training program			
Very satisfied	65.4%	55.6%	64.0%
Somewhat satisfied	32.3	35.5	34.7
Neutral or dissatisfied	2.3	8.9	1.3
Total	100.0	100.0	100.0
Trainee's use of training in his present job			
Full use of training	46.4	39.4	48.6
Some use of training	47.2	55.5	42.1
Little or no use of training	6.4	5.1	9.3
Total	100.0	100.0	100.0

trainees either perceive the use of their training or assess their training experience at each center. Therefore, from the viewpoint of the individual trainee, these programs have been equally effective in achieving specified training objectives.

As mentioned above, however, there is considerable difference among trainees in present job performance. Data on trainee work behavior are presented in Figures 15-1 and 15-2 for the three training groups, based on the subgroups of trainees who actually indicated direct involvement (when the survey was taken) in a wheat improvement or rice research and production program. The percentage of trainees indicating such direct involvement for each training group is as follows: CIMMYT wheat trainees, 83.3 percent (N = 105); IRRI research trainees, 71.3 percent (N = 154); and IRRI rice production trainees, 78.8 percent (N = 108).

It is clear from the data presented in Figures 15-1 and 15-2 that CIMMYT wheat trainees are completing more research activities associated with the development of biological technology than are IRRI research trainees. Although this difference in research emphasis was expected, the extent of these observed differences was not anticipated. An analysis of variance test was carried out between the CIMMYT and IRRI research training groups for each type of research activity. The differences in work behavior for the three technology development variables were statistically significant at the .01 level. Since these differences have an important influence on the type of national research capacity that develops, a detailed examination of the factors that contributed to these differences appeared warranted.

To account for these differences, an analysis was made using the concep-

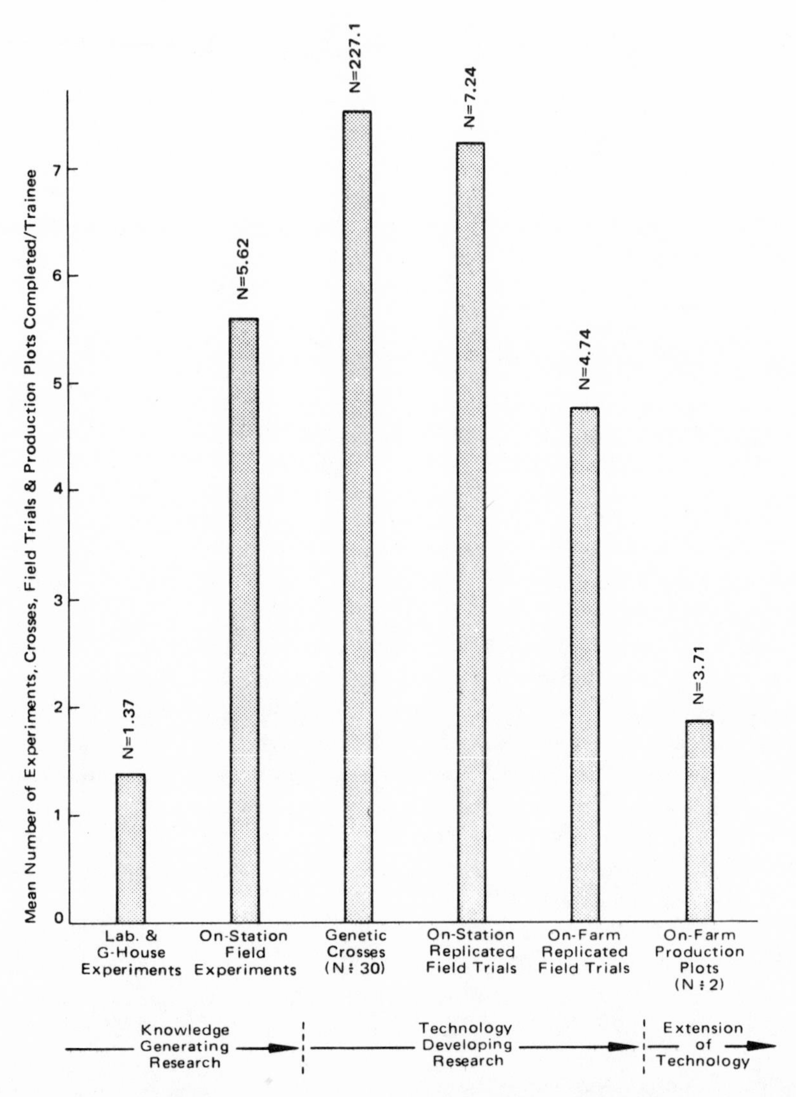

Figure 15-1. Work behavioral patterns of former CIMMYT wheat trainees.

tual model of the training process depicted in Figure 15-3. First, an examination was made of the two major independent variables: personal characteristics of individual trainees; and characteristics of trainees within their home organizations. Then, differences among trainees in respect to the main intervening variable – the actual training strategy and approach employed by each institute – were considered.

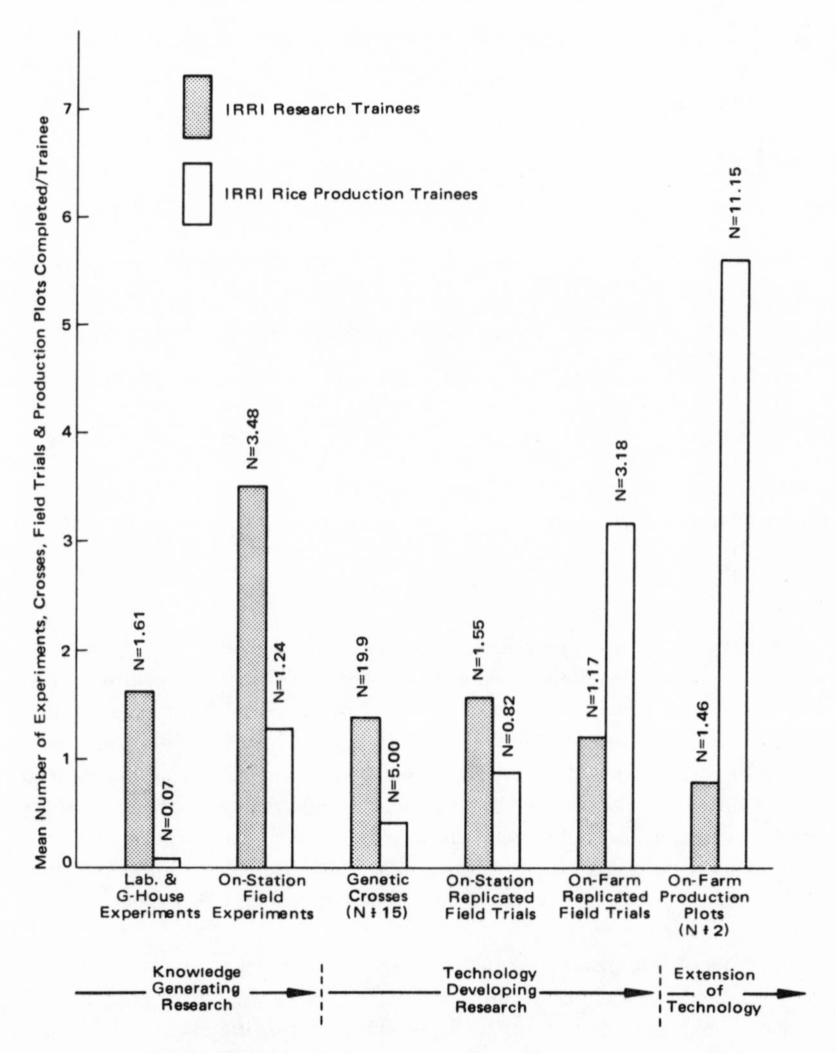

Figure 15-2. Work behavioral patterns of former IRRI research and production trainees.

It should be noted that the comparative analysis which follows concerns only the two research training groups. The work behavior of rice production trainees was included in Figure 15-2 because it provides a more complete profile of the research and extension activities being completed by former IRRI rice trainees, particularly since about one-fourth of former rice production trainees are engaged in research work. However, since this third group was trained to carry out essentially a technology dissemination role, it would be inappropriate to analyze this approach in terms of technology development.

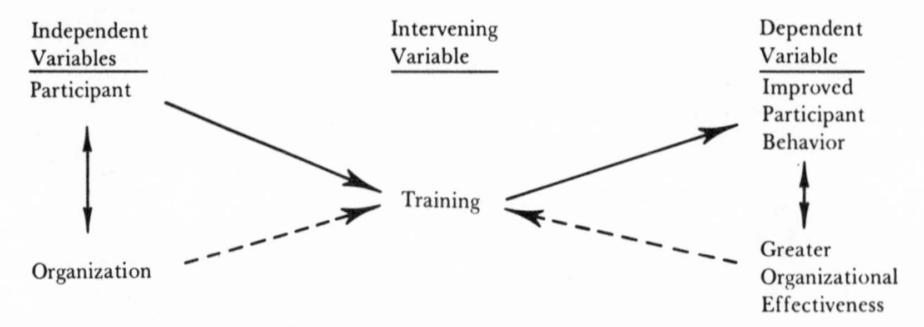

Figure 15-3. A model depicting the role of training in behavioral and organizational change.
Source: R. P. Lynton and Udai Pareek, *Training for Development* (Homewood, Ill.: Richard D. Irwin, and Dorsey Press, 1967), p. 18.

However, data on the two independent variables for the rice production group are included in the following section as a matter of information for the reader.

Personal Characteristics of Trainees

Several key variables describing the personal background and characteristics of trainees were tabulated to determine if there were any significant differences in the types of trainees that were being selected for each training program. In analyzing the data, several different multivariate linear regression models, using key independent variables, were developed in an attempt to predict different types and/or levels of work performance. None of the independent variables used in these analyses were found to have much influence (i.e., explain much variance) on the dependent variable. Table 15-4 summarizes data on these personal characteristics that might be expected to influence work behavior.

In terms of background characteristics, there are two apparent differences between the two research training groups which could affect work performance. The first, as documented in Table 15-4, is the difference in educational level between the two research groups. Although the IRRI research group is more highly educated – and this may influence somewhat the type of research conducted, which does appear to be the case – this factor would not be expected to have a negative influence on the overall amount of work completed.

The second major difference between the two research training groups is nationality and the possible influence of cultural factors on work behavior. The IRRI research group is primarily from countries in South, Southeast, and East Asia, whereas CIMMYT trainees are primarily from North Africa, the

Table 15-4. Personal Characteristics of Trainees in Each Training Group

Personal Characteristics of Trainees	CIMMYT Wheat Trainees (N = 130)	IRRI Research Trainees (N = 234)	IRRI Rice Production Trainees (N = 148)
		In Years	
Mean age when entering the training program	30.6	30.4	32.4
		In Percentage	
Trainees who grew up in rural areas	61.4	59.5	77.0
Trainees' fathers who were engaged in agricultural jobs	47.5	36.7	45.8
Trainees' families whose main source of income was from agricultural sources	47.7	43.2	55.5
Educational level of trainees when entering the training program			
Less than B.S. degree	21.1	1.8	22.4
B.S. degree or equivalent	57.8	62.3	65.0
M.S. or Ph.D. degree	21.1	35.9	12.6
Total	100.0	100.0	100.0
Present educational level of trainees			
Less than B.S. degree	20.5		17.5
B.S. degree or equivalent	41.7	16.2	56.6
M.S. degree or equivalent	27.6	41.2	19.6
Ph.D. degree	10.2	42.6	6.3
Total	100.0	100.0	100.0

Middle East, South Asia, and South America. Because there was insufficient overlap between the two groups to measure the impact of nationality or culture on work behavior, the importance of this influence remains unknown. However, nothing in my experience in agricultural training programs, where I have worked with a wide variety of cultural groups, suggests that cultural background would have a determining influence on the type and amount of work performed.

Characteristics of Trainees in Their Work Organizations

The next set of variables to be examined in attempting to explain differences in work behavior between the two research groups concerns the role of the trainee in his work organization when the survey was conducted. Table 15-5 presents data on certain selected variables that could be expected to influence the dependent variable.

There is no difference between research groups regarding the level of their

Table 15-5. Selected Variables Describing the Roles of Trainees in Their Work Organizations for All Three Training Groups

Characteristics of Trainees in Their Work Organizations[a]	CIMMYT Wheat Trainees (N = 130)	IRRI Research Trainees (N = 234)	IRRI Rice Production Trainees (N = 148)
Types of organizations where trainees are working			
National research organizations	88.0%	56.9%	20.0%
National extension organizations	6.4	1.9	47.9
Agricultural colleges or universities	2.4	27.3	12.1
Other	3.2	13.9	20.0
Total	100.0	100.0	100.0
Level of trainee's position in his work organization			
Policy level	4.0	4.7	2.9
Senior level	41.9	46.5	32.1
Middle level	50.1	45.0	54.0
Lower level	4.0	3.8	11.0
Total	100.0	100.0	100.0
Type of work trainee is engaged in			
Mainly administrative work	11.6	12.1	22.7
Mainly field research work	74.4	44.7	22.0
Mainly laboratory greenhouse research	6.6	24.1	3.8
Mainly field extension work	5.8	3.0	28.8
Mainly teaching or training	1.6	16.1	22.7
Total	100.0	100.0	100.0
Crops worked with by the trainee in his job assignment			
Works only with wheat/rice	71.4	40.7	32.8
Works with wheat/rice and other crops	19.9	31.5	48.2
Does not work with wheat/rice in his present job assignment	8.7	27.8	19.0
Total	100.0	100.0	100.0

[a] At the time the survey was taken.

present positions in their work organizations, but there are other important differences which could contribute directly to the observed differences in work behavior. First, there is a substantial difference in the relative proportions of trainees doing field research work. Since the work behavior of IRRI research trainees tends to emphasize experimental field research (as reported in Table 15-2 and Figure 15-2), the fact that only about 45 percent of the IRRI research group (contrasted with nearly 75 percent for the CIMMYT group) are engaged in field research would be expected to have an important negative influence on the mean level of field research activities completed per trainee. Second, the important difference between the two groups in regard to crops worked with (71 percent of CIMMYT trainees worked full time on wheat research, whereas only 41 percent of IRRI research trainees worked full time on rice research) would also be expected to have some influence on the total amount of wheat/rice research completed.

These findings raise two obvious questions: Why are there such large differences between the two research training groups with respect to these two variables? And, what actual impact are these differences having on the amount of work completed by trainees?

Although there are no clear-cut answers to the first question, certain facts are known. First, IRRI selects trainees from educational as well as research institutions for its research training program. With 16 percent of former IRRI research trainees now working in full-time teaching positions within colleges of agriculture or universities, this group accounts for more than half the difference between the CIMMYT and IRRI research groups. Second, of those IRRI research trainees who are doing research work, more than one-third are engaged in laboratory/greenhouse research, and this group accounts for the remainder of the difference of those trainees doing field research.

Less easy to explain is the fact that only 41 percent of the research workers included in the IRRI research group are working full time on rice research, yet most come from countries totally dependent on rice as the major food staple. The only apparent reason is that the IRRI research group is trained in single-discipline departments and is more highly educated, which again would be an impetus to further disciplinary specialization. Once back home in their national programs, which are concerned with several different crops (rather than a single crop as is the case with IRRI), it is probable that these scientists will be called on to divide their research time among other crops in addition to rice. For example, a pathologist may be assigned to study fungus diseases of several crops rather than work on all different diseases of rice, such as bacterial blight, blast (a fungus disease), and tungro (a virus disease).

Trainees who have gone through the CIMMYT training program, which operates within an interdisciplinary research structure, have tended to continue their focus on wheat improvement rather than specialize on research problems within a particular scientific discipline.

The second question raised was, How much impact are these differences in trainee job assignments between the two research training groups having on the overall work of trainees? To address this question, the procedure was to

Table 15-6. Analysis of Variance of the Work Behavior of Individual Trainees Who Are Working Mainly on Field Research – CIMMYT and IRRI Research Training Groups[a]

Type of Activity Completed	CIMMYT Wheat Trainees (N = 79)	IRRI Research Trainees (N = 72)	t Values	Level of Significance
Laboratory and greenhouse experiments	1.76	0.61	1.1068	n.s.
On-station field research experiments	6.82	4.07	1.7757	n.s.
Genetic crosses	279.5	26.3	5.8646	**
On-station replicated applied research trials	8.95	2.00	4.3683	**
On-farm replicated applied research trials	4.95	1.68	2.4003	*
On-farm production plots	3.17	2.46	0.4226	n.s.

[a] Data reported are the mean level of each activity completed per trainee during the main wheat or rice growing season of 1972.
n.s. = not significant.
* = significant at .05 level.
** = significant at .01 level.

contrast the work behavior of two subgroups of trainees – those mainly doing field research work and those working only on wheat/rice research – with each of the two research training groups to determine if the observed differences in work behavior could be explained by either of these two independent variables or if the differences still persist. Results of these analyses are found in Tables 15-6 and 15-7.

As the data indicate, the observed differences in behavior for those research activities most closely associated with the development of improved genetic and production technology (rows 3, 4, and 5) continue to be present even when we consider just those subgroups of trainees that are doing mainly field research and those doing only wheat/rice research respectively. Therefore, it is concluded that, although a similar proportion of IRRI research trainees are working in field research work and a smaller proportion are only working full time on rice research, these differences in job assignments still do not explain the major differences in observed work behavior between the two research training groups.

To reiterate, it was not possible to account for the differences in work behavior between the two research groups by considering the two main independent variables: the differences in the personal background characteristics of the individual trainees in each group and the characteristics of trainee job assignments in their work organizations. These findings, therefore, direct the

Table 15-7. Analysis of Variance of the Work Behavior of Individual Trainees Who Are Working Only on Wheat/Rice Research – CIMMYT and IRRI Research Training Groups[a]

Type of Activity Completed	CIMMYT Wheat Trainees (N = 81)	IRRI Research Trainees (N = 82)	t Values	Level of Significance
Laboratory and greenhouse experiments	1.72	1.12	0.5778	n.s.
On-station field research experiments	5.89	3.84	1.3358	n.s.
Genetic crosses	261.2	28.0	5.5943	*
On-station replicated applied research trials	8.36	1.52	4.4190	*
On-farm replicated applied research trials	4.43	1.26	2.6425	*
On-farm production plots	3.77	1.99	1.1194	n.s.

[a] Data reported are the mean level of each activity completed per trainee during the main wheat or rice growing season of 1972.

n.s. = not significant.

* = significant at .01 level.

inquiry to the intervening variable – the training itself – to see if the differences in training could account for the observed differences in work behavior.

Differences in Training as an Intervening Variable

The training objectives, methodology, and strategy of the two research training programs being considered here are markedly different. Both training approaches were described earlier. The key points that appear to link these different training approaches directly to the differences in work behavior are as follows.

IRRI's main educational objective in its research project approach to training is analytical skill. In following this type of program the trainee learns firsthand how to design, carry out, and report on a research experiment and in doing so learns how to think analytically – a prerequisite for any successful research worker. Having this ability and skill, however, does not prepare a research worker for all types of research work. In particular, he does not learn the research skills and methods associated with organizing and operating a research program aimed at developing improved agricultural technology, where the primary educational objective is synthesis (an objective that characterizes the CIMMYT wheat training program).

The CIMMYT wheat research program is essentially a highly organized "genetic assembly line" which has standardized procedures and routinized tasks to increase the efficiency and output of a wheat improvement research program, while minimizing error. By systematically and effectively mixing (synthesis) the gene pool through large numbers of genetic crosses and by systematically screening the progeny of these crosses, particularly through the extensive use of early generation testing procedures (evaluation), the biological architects at CIMMYT believe they can maximize the probability of producing superior germ plasm. CIMMYT wheat trainees appear to have learned and to have adopted this research approach to wheat improvement, given their observed behavior.

The conclusion of this study is that the different approaches to research training pursued by CIMMYT and IRRI are having an important and measurable influence on the work behavior of research workers in less developed countries. CIMMYT's training group tends to emphasize and to be very efficient and productive in conducting research aimed at developing improved wheat technology. IRRI research workers tend to emphasize experimental research aimed at generating new knowledge about rice production in the tropics.

Transforming National Research Capacity: A Postscript

As Ruttan and Hayami point out, one of the most serious constraints on the international transfer of agricultural technology is limited experiment station capacity for the production of biological technology.[25] The central theme of this chapter has been an analysis of the ways in which two international research centers have addressed this problem through their training programs. In making this analysis, however, it became increasingly clear that there was an issue involved more basic than just a difference in training objectives and methodologies. More important were the working assumptions made by each center concerning the ability of national research systems to organize their scientific resources. Although this is a complex issue, each center's response to this question provides some valuable insights into the problem.

CIMMYT's response to this question has been based essentially on the premise that the national wheat improvement programs with which it works have not been functioning effectively because they have been unable to organize their scientific resources to solve practical production problems. The problem is twofold.

First is the problem of focus, or research objectives. CIMMYT's response to this perceived institutional problem was to concentrate on one specific production problem – the need for improved genetic technology – and through its training strategy carefully and systematically to build this relatively simple research function into each national program. Thus organizational change was a specific, but implicit training objective.

Second is the problem of functional integration, or the ability of a national program effectively to organize and mobilize its scientific resources around the objective of producing biological technology. If we identify the lack of such ability as a serious institutional problem, it follows that technical personnel, in addition to lacking relevant research skills and methods, would also lack the skills and spirit of working together in organized, cooperative team efforts. More precisely, they would lack the organizational skills and perspective necessary to integrate their research functions around a specific research objective. Thus it was found that CIMMYT trainees in cereal technology spent one-fourth of their time in the wheat-breeding program, so that they would clearly understand and appreciate the functional relationship between their work in cereal technology and the work in the breeding program. Similarly, each training group spent a substantial amount of time working in each of the other research programs that are functionally involved in developing improved genetic technology.

Following the logic of this strategy, it was expected that, once a "critical mass" of trained research workers were present in a wheat improvement program, that program would function effectively. And it was expected that as this happened, each national program would (1) work out its own research structure (i.e., division of labor) to fit local needs and requirements and (2) take over the necessary informal on-the-job training of new personnel. Once this transformation had been achieved, CIMMYT expected national wheat improvement programs to be functionally competent to identify local production problems and meet long-term technological needs. My observations, made while pretesting the survey questionnaire in the field, suggest that this strategy is working.

IRRI's response to the problem of how national research programs organize their scientific resources has been based on a different policy decision. On one hand, IRRI officials and scientists have expressed serious concern that some national programs lack the ability to organize and direct their research resources toward solving local production problems. On the other hand, however, IRRI leaders have expressed a sensitivity and concern that their international activities not dominate the research capability of national programs.[26] By making an explicit policy decision against direct institutional intervention, IRRI placed itself in a position of taking, as a given, the ability of national programs to organize their scientific resources.[27] Furthermore, IRRI's research training strategy, while not resulting from this policy decision, is logically consistent with it.

IRRI is operating on the basis of the same type of institutional relationship that a college of agriculture (as an educational institution) would have with a functionally effective experiment station (as a work organization). The college provides graduate students with a good theoretical background in a field of study, along with a solid research experience in designing, conducting, and reporting on a research project. Once hired by an experiment station for a particular research position, however, a student is given specific research responsibilities that he is expected to carry out, which are functionally integrated with the station's overall research program. Since it is unlikely that the student knows how to carry out these research responsibilities, he will generally be assigned to work closely with a senior scientist in the same program. After a season or two of this informal, on-the-job training, the inexperienced research worker will develop sufficient competence to function independently in carrying out his assigned tasks.

IRRI, by using the research project approach and by training research workers in separate single-discipline departments, does not address institutional problems, such as poorly defined or inappropriate research objectives and the lack of functional integration, through its training strategy. It should be noted, however, that several countries from which IRRI receives substantial numbers of trainees (for example, Japan, Taiwan, and Korea) are assumed to have rather effective research systems at the present time. Therefore, this type of research training may be quite appropriate in meeting their needs. Nevertheless, if these programs are in fact relatively strong and functioning effectively at the present time, there would appear to be little justification for IRRI to expend its scarce training resources on research personnel from these countries, when the important rice-growing countries of the tropics have weak research systems.

The observed work behavior of former IRRI research trainees, as they return home to concentrate on knowledge-generating types of research, raises obvious questions about IRRI's training strategy. Although trainees study production problems that are relevant to home country conditions, it appears that in most cases their research effort—as it contributes to the overall national research capacity—lacks sufficient emphasis on and attention to the problem of producing biological technology.

Conclusion

This chapter is built on the premise that as poor nations of the tropics and subtropics shift to modern agricultural production systems they will need effective research institutions to meet long-term technological needs. Experience suggests that, in some of these national research institutions, technology-development research activities are frequently neglected or are relegated to less qualified and/or less motivated research personnel, while more highly educated scientists concentrate on more theoretical research inquiries.[28] However, if national agricultural research systems are to be relevant and useful to agricultural producers, then knowledge-generating research activities must be, to a large degree, directed by and integrated with the technology-developing research function. Furthermore, this technology-developing research system must be capable of effectively and efficiently transforming new knowledge into improved agricultural technology. Where the national research capacity for producing biological technology is weak or lacking, these national research systems will remain functionally impotent in their ability to solve serious production problems and nations will remain largely dependent on external agencies (particularly the international research centers themselves) for new sources of improved technology. An analysis of the observed work behavior of former IRRI and CIMMYT research trainees supports the proposition that different training approaches and institution-building strategies can have an important impact on the type of national research capacity that develops. Therefore, international centers must be fully cognizant of these potential influences on the technological capability of national research systems when designing their training and outreach programs.

NOTES

1. This chapter was written as part of a research project on international technology transfer systems, supported by the Program of Advanced Studies in Institution-Building and Technical Assistance Methodology (PASITAM) of the Midwest Universities Consortium for International Activities (MUCIA) through a 211(d) grant from the United States Agency for International Development. The original research on which this chapter is based was supported by grants from the International Rice Research Institute, the International Maize and Wheat Improvement Center, and the Land Tenure Center at the University of Wisconsin.

2. Vernon W. Ruttan and Yujiro Hayami, "Technology Transfer and Agricultural Development," *Technology and Culture*, 14:2, pt. 1 (April 1973), 124-125.

3. These international movements of wheat and rice varieties are well documented in Dana G. Dalrymple, *Imports and Plantings of High-Yielding Varieties of Wheat and Rice in the Less Developed Nations*, Foreign Economic Development Report no. 14, Foreign Economic Development Service (Washington, D.C.: USDA in cooperation with USAID, 1972).

4. CIMMYT and IRRI both made available exotic and elite germ plasm to cooperating national programs through this period, and to a differing degree these efforts continue.

5. See Committee on Institutional Cooperation, *Building Institutions to Serve Agriculture* (Lafayette, Indiana: Purdue University, 1968), p. 9; Willard W. Cochrane, *The World Food Problem* (New York: Thomas Y. Crowell, 1969), pp. 218-219; A. H. Moseman, *Building Agricultural Research Systems in the Developing Nations* (New York: Agricultural Development Council, Inc., 1970), pp. 29-30.

6. Derek J. de Solla Price, "The Structures of Publication in Science and Technology," *Factors in the Transfer of Technology*, eds. William S. Gruber and Donald G. Marquis (Cambridge: M.I.T. Press, 1969), pp. 94, 97.

7. See Sterling Wortman, "The Technological Basis for Intensified Agriculture," *Agricultural Development: Proceedings of a Conference Sponsored by the Rockefeller Foundation*, April 23-25, 1969, Bellagio, Italy (New York: Rockefeller Foundation, 1969), p. 31.

8. In terms of a research hierarchy, the *biological architect* is considered to be on at least the same intellectual level as any senior research scientist. In fact, although the biological architect may use less sophisticated research tools, the cognitive demands placed on these individuals are frequently more complex than those problems faced by scientists conducting analytical research. Unfortunately, however, the technology-developing function within the overall research process is generally considered (by the scientific community) to be in an inferior position to the knowledge-generating function. Therefore, less talented research technicians are frequently assigned to these technology-development roles, which may be one reason why many research programs, particularly in LDC's, have been so ineffective in producing improved agricultural technology.

9. Benjamin S. Bloom, ed., *Taxonomy of Educational Objectives, Handbook I: Cognitive Domain* (New York: David McKay, 1956), p. 144.

10. *Ibid.*, pp. 162-172.

11. *Ibid.*, p. 162.

12. See *ibid.*, pp. 185-192.

13. See Ruttan and Hayami, "Technology Transfer," pp. 141-143.

14. For example see, E. O. Heady, "Priorities in the Adoption of Improved Farm Technology," *Economic Development of Agriculture*, ed. E. O. Heady (Ames: Iowa State University Press, 1965), pp. 164-165; Hans W. Singer, "A New Approach to the Problems of the Dual Society in Developing Countries," *United Nations International Development Review*, 3:3 (1971), 24; Lester Brown, *Seeds of Change* (New York: Praeger, 1970), p. 52; C. P. McMeekan, "What Kind of Agricultural Research?" *Finance and Development*, 2:2 (1965), 73.

15. See Burton E. Swanson, "Training Agricultural Research and Extension Workers from Less Developed Countries," Ph.D. dissertation (Madison: University of Wisconsin, 1974), pp. 18-42, for a more detailed analysis of this problem.

16. See H. L. Wilcke and H. B. Sprague, "Agricultural Research and Development by the Private Sector of the United States," *Agricultural Science Review*, 5:3 (1967), 3.

17. Singer, "Dual Society," p. 24. Singer terms this condition "an internal brain drain."

18. The following section is a very brief summary of the training objectives and methodology being employed by the CIMMYT wheat program. For more detailed information on this training program and an evaluation of this strategy, see Swanson, "Training Agricultural Research and Extension Workers," pp. 101-187.

19. For a detailed description of this training methodology see Dugan H. Laird, *Training Methods for Skills Acquisition* (Madison: American Society for Training and Development, 1972), pp. 22-24.

20. See Swanson, "Training Agricultural Research and Extension Workers," pp. 117-123, for detailed information about and examples of how CIMMYT attempts to achieve its attitudinal objectives.

21. The data used in this chapter to analyze the performance of CIMMYT and IRRI trainees were based on an international survey of former participants conducted in 1973. Responses to mail questionnaires were obtained from more than 78 percent of the nearly 700 former trainees who had participated in one of the three training programs included

in this study. A detailed analysis of these follow-up data for each training group can be found in Swanson, "Training Agricultural Research and Extension Workers."

22. Laboratory, greenhouse, and field experiments are defined in this chapter as "knowledge-generating" types of research and genetic crosses and on-station and on-farm applied research trials are classified as "technology-developing" types of research. See Swanson, "Training Agricultural Research and Extension Workers," pp. 14-18.

23. This conclusion is based partly on the assumption that individual research experiments take more time to plan and execute than do applied research trials. The methodology of applied research trials can frequently be standardized and routinized to minimize time spent on planning, experimental design, and analysis.

24. The rice production training program is specifically designed for rice extension workers. The program is summarized briefly here, however, since one-fourth of its trainees were research workers.

25. Ruttan and Hayami, "Technology Transfer," p. 124.

26. See A. Colin McClung, "IRRI's Role in Institutional Cooperation in Asia," *Rice, Science and Man* (Los Baños: International Rice Research Institute, 1972), p. 38.

27. "The basic philosophy of these agreements was quite clear on an important point: they were in no sense intended to lead to the establishment of branch stations of the International Rice Research Institute. Rather, they were designed to foster improvements in the host countries' research and development programs. The degree and pace of IRRI's involvement *would be set by the decisions of local authorities* [emphasis added] and the work would be directed solely towards problems of concern to local rice scientists and rice farmers." McClung, "IRRI's Role," p. 28.

28. Moseman, "Building Agricultural Research Systems," p. 58; and Wortman, "Technological Basis for Intensified Agriculture," p. 21.

Organization and Management
of Agricultural Research Systems

16

Coordinated National Research Projects for Improving Food Crop Production

Albert H. Moseman

The coordinated, multidisciplinary, problem-oriented research project to improve food crop production is a most effective instrument for accelerating the evolution of useful technology. It provides the basic patterns and components necessary for developing more comprehensive national research capabilities to serve the agricultural development objectives of a country.

Agricultural scientists have had many decades of experience with coordinated national research projects in the advanced nations. And, particularly since the green revolution of the mid-1960s, the scientific community has accumulated a number of years' experience with such projects in a number of developing nations. Recent endeavors in research on food crop production demonstrate that the objective of establishing a sustained national research capability in developing countries is difficult to achieve, that insufficient attention has been directed to it, and that special long-term efforts are needed.

Coordinated National Research Projects in Agriculturally Advanced Nations

Agricultural Research in the United States

In the middle 1800s it was recognized that the United States should not depend upon European agricultural research for its development efforts but should build a national capability. The land grants to the respective states and the legislation establishing the USDA in 1862 are critical points in the evolution of our national research system.

For the balance of the nineteenth century and through the early 1900s agricultural research in the United States was rather unstructured. It consisted of independent and isolated studies by research workers in the USDA, the state colleges or experiment stations, and other institutions. The history of research on heterosis, or hybrid vigor, in corn into the early 1920s illustrates this fragmented and individualized approach to agricultural science.

The coordinated national corn improvement research program established in 1925 under the Purnell Act was the first attempt to concentrate resources of the states and the federal government in a fully cooperative effort.[1] This was followed by similar coordinated projects for wheat, oats, barley, etc., during the 1930s.[2]

A number of coordinated national crop improvement research projects in the United States were regionalized, with the wheat research regions developed around the classes of wheat grown in the different parts of the country and with rice research directed specifically to the pattern of rice production suited to California rather than the Gulf states of Texas, Louisiana, and Arkansas.[3] The national corn research program was strengthened, with a new regional component, through the establishment of the special state-federal coordinated project for the South and the Southeast. This regional effort was initiated in 1945, twenty years after the initial coordinated program for corn improvement research was established for the central "Corn Belt" states in 1925.

The coordinated national research projects not only recognized the importance of regional needs but also provided for more precise location-specific research, conducted by the state agricultural experiment stations through studies that were in addition to – yet closely associated with – the total national or regional research efforts.

The present organizational structure of the coordinated national crop improvement research projects is less distinct than it was twenty-five years ago as a result of the increased involvement of the private sector and the increased autonomy of the research of the individual state agricultural experiment stations. The reorganization of the USDA's Research Service in 1972, with its regional administrative pattern, also has tended to obscure and perhaps complicate the relationships between the federal Agricultural Research Service and the state experiment stations as well as between the federal administrative regions themselves. However, the coordinated national research projects – with their regional, state, and localized research focus – functioned effectively during the time when the United States was building its total national capacity for agricultural research and supplied the base for the more autonomous, independent research efforts of the present. Of special significance is their impact in creating an awareness of the need to give continuing, concurrent attention to the many varied hazards or potential restraints faced by farmers. In signaling this need, the projects furnished the base for multidisciplinary team research.

Agricultural Research in Japan

The experience in Japan is similar to that of the United States.[4] The emphasis, early in the Meiji restoration period of the 1860s, was on the introduction of advanced technology from Europe. This was followed by a concerted effort to develop indigenous capabilities during the 1880s. The research in Japan continued into the 1920s on a fragmented basis and depended on farmer innovations for improved varieties.

Japan established national coordinated crop-breeding programs under the Assigned Experiment System for wheat in 1926, for rice in 1927, and for other crops and livestock in subsequent years. This was almost the identical time schedule followed in the United States, and in both Japan and the United States the coordinated national projects provided for research on regional, state, or prefectural problems within the coordinated national structure.

Some Attributes of Coordinated National Research Projects

The coordinated national research projects make the most effective and efficient use of research resources. They facilitate the prompt and continuous interchange of new knowledge and materials among research workers of the central government, the states or prefectures, and the private sector. The projects provide for research on problems of broad national or regional concern as well as on those that are location-specific. The national projects also facilitate the introduction and testing of knowledge and materials from abroad.

The foregoing benefits were common to the research organization pattern that evolved in both the United States and Japan. The development of the coordinated national project structure in the two countries was similar in timing and form but reflected independent judgments, since international contacts and communications on research organization in agriculture were limited before the 1930s.

Coordinated National Research Projects in Developing Nations

Research Resources in Developing Nations

Research in the developing nations before World War II or in the colonial period was strongly commodity-oriented and was carried out primarily in centralized research institutes such as those for rubber in Malaya and sugar in India and Indonesia.[5]

After independence, research in these countries was seriously disrupted, as expatriate scientists departed and new national government leaders gave priority to nonagricultural development. Agricultural research received no significant support for a decade or more in many new nations, up until the late 1960s.

Research in Technical and Economic Assistance Projects

The primary emphasis in technical assistance for agricultural development, as supported by the private foundations, bilateral national programs, and international agencies following World War II, was on the introduction and testing of materials and practices from agriculturally advanced nations. This process was followed for about fifteen years – until the era of the green revolution – even though the earlier experience with agricultural development in the United States and Japan had fully demonstrated that the direct introduction of materials and farming methods from abroad had limited potential.

Even the few cooperative technical assistance efforts that were research-based – such as the Rockefeller Foundation programs in Mexico, Colombia, and Chile – emphasized developing improved technology as rapidly as possible; limited attention was given to indigenous organizational structure or the links between national, state, and university institutions. Headquarters research centers were developed in Mexico at Chapingo, in Colombia at Tibaitata, and in Chile near Santiago, with regional stations in each country. These organizations had a highly centralized leadership structure, the regional or outlying stations serving primarily for testing or evaluation. The lack of trained personnel and a national institutional structure for research and agricultural education at the college or university level in Latin America precluded the development of effective federal-state coordinated research projects during the period between 1940 and 1950.

The cooperative Rockefeller Foundation program in agriculture in India, from its beginning in 1957, focused on developing coordinated national and regional research on maize, sorghum, and millets. Participants included the Indian Council of Agricultural Research, the Indian Agricultural Research Institute, the state governments, and the agricultural universities. The situation in India was different from that in Latin American countries, since India had (1) well-established research institutes, (2) the Indian Council of Agricultural Research which, at the time of independence in 1947, had been functioning for about twenty years, and (3) an emerging agricultural university structure, supported by USAID, that was developing concurrently with the cooperative Rockefeller Foundation-supported projects for crop improvement research.

The coordinated corn improvement research project established in the United States in 1925 and the assigned experiment systems established in Japan for wheat in 1926 and rice in 1927 furnished patterns in these countries for developing research on a coordinated national basis for other crops, livestock, and noncommodity problems. Similarly, the coordinated maize improvement scheme initiated in India in 1957 furnished a pattern for national coordinated schemes for sorghums and millets, wheat, rice, and other crops as well as for noncommodity research projects that have been developed in India in recent years. Many of these are still in the process of being established and are undergoing the usual stresses involved in developing cooperating relationships between participating scientists and participating institutions, but they are well conceived and soundly formulated.

The accelerated wheat production scheme in Pakistan functioned effectively during the period from 1965 through 1969 when it had substantial external funding and external coordinating leadership furnished by the Ford Foundation. However, the coordinated effort deteriorated after 1969 as the result of several factors including (1) diminished interest and support from the central and state governments, (2) the division of the west wing of Pakistan into the four provinces of the Punjab, Sind, Baluchistan, and the Northwest Frontier Province in 1970, and (3) the continuing political stresses and diversions resulting from relationships with India and the formation of Bangladesh in 1971. The strengthening and reorganization of the Pakistan Council of Agricultural Research, now under way with support from USAID, should help in the restructuring of the wheat project. Moreover it should serve as a pilot effort, or pattern, for other coordinated national research schemes in Pakistan.

The wheat research and training project initiated in Turkey in 1969 with cooperative support from the Rockefeller Foundation has made good progress in the development of staff, in the improvement of research facilities, and in providing an interdisciplinary approach to crop improvement and to agronomic and epidemiological research on a national basis. It also has developed effective cooperation with other countries in the region, especially in crop breeding and disease research. It was recognized when this project was initiated that it should be continued for a period of ten years. It appears that this is the minimum time required for the formation of a self-sustaining national coordinated project. A principal uncertainty at this time is whether a suitable institutional base for the project can be provided within the government to furnish the personnel policies, administrative procedures, and stable financial support necessary to ensure continued viability.

Rice breeding and improvement in Thailand has progressed effectively since the Rice Department was established as a separate organizational unit in the Ministry of Agriculture in the early 1950s. Although the Rice Department and the Department of Agriculture were merged in 1972, the base of experiment stations throughout the country, the complement of well-trained personnel, and the working experience that has been gained over the past twenty years should make possible a sustained, productive national rice research project. This project also furnishes a pattern for the strengthening of other agricultural research in Thailand.

The national rice research program in Indonesia, established in 1970, has progressed rather slowly in establishing a coordinated national capability. This program was reviewed in 1974. A new unit that is to be set up to coordinate all research within the Ministry of Agriculture should be helpful in accelerating the development of a coordinated national rice research project in Indonesia.

The Bangladesh Rice Research Institute, established as an autonomous body by act of Parliament in May 1973 (but in the process of formation since 1968), has been adversely affected by the political conflicts attendant upon and following the country's independence, gained in 1971. However, substantial progress has been made in developing the facilities at Joydevpur, in training staff, and in getting a multidisciplinary research program under way. The BRRI is perhaps the strongest of the national research projects in Bangladesh and should be given priority attention so that it may serve as a pattern for other research which will be strengthened under the newly authorized (1973) Bangladesh Council for Agricultural Research.

It is interesting to note that the Department of Agriculture and Natural Resources of the Philippines, which is responsible for national crop improvement and production research, had no research scientist trained to the Ph.D. level on its staff working on either rice research or corn improvement research at the time of the Presidential Decree of November 1972 which set up the Philippine Council for Agricultural Research. This was more than ten years after the International Rice Research Institute was established at Los Baños with a complement of international scientists, and more than fifteen years after Dr. H. K. Hayes and others had been in the Philippines to help strengthen the country's corn improvement research capability. It is expected that coordinated research projects for rice and corn will be set up on a more effective national basis under the Philippine Council of Agricultural Research. The long delay in developing this type of national institutional capability, in a country that has received substantial technical assistance for agricultural development for more than twenty years, points up the need for direct and specific attention to this objective.

Indonesia is another country where external assistance has been furnished to develop corn production over a period of many years but where inadequate attention has been given to the building of an indigenous organization for a sustained national corn improvement research program.

The foregoing examples have been selected to illustrate the importance of devoting specific attention to the building of an organizational or institutional capability in the developing nation – as an integral part of a cooperative technical assistance effort in research to increase crop production. The potentials and restraints in each country will differ, but these are being assessed in many countries and the experience of the developing national food crop research projects in India, Pakistan, Bangladesh, Thailand, Indonesia, and the Philippines, as well as in other developing nations of the world, should be useful in contending with restraints in the building of such capabilities.

International Dimensions of Coordinated National Research Projects

Coordinated national crop improvement projects of the United States have furnished materials, consultation, and other assistance in strengthening coordinated national research projects in other countries. Germ plasm and consulting services were provided by the national rice improvement project of the United States to the cooperative program of the Rockefeller Foundation and the government of Colombia when they initiated the rice improvement research program in that country in 1957. The varietal collection from the United States also composed the base for the rice germ plasm reservoir that has been further developed by IRRI.

The introduction of hybrid corn into Europe following World War II was accomplished through the cooperative support of the United States national coordinated corn improvement program. This involved the furnishing of inbred lines, assistance in setting up procedures for evaluating the various hybrid combinations, and help in developing effective seed production organizations in various European countries.

During its early years, the wheat improvement research initiated in Mexico with the assistance of the Rockefeller Foundation in 1943 utilized the disease-resistant spring wheat varieties developed in the United States hard red spring wheat research project – a regional component of the coordinated national wheat research program. Subsequently, the *Norin* selections which were introduced into the United States from Japan after World War II and which were used to develop outstanding, high-yielding varieties for the Pacific Northwest region were made available to the national wheat research program in Mexico. The short-strawed, high-yielding *Norin* germ plasm performed equally well in the hybrids and the selections produced in Mexico.[6]

The cooperation which extended from the national wheat improvement project in Mexico into the countries of the Near and Middle East, through support from the Rockefeller Foundation and the FAO, carried this germ plasm into these countries in the early 1960s. The inclusion of wheat research as a component of the Rockefeller Foundation cooperative agricultural sciences program in India in 1963 furnished the base for the wheat component of the green revolution in Asia. Since they evolved in the coordinated national wheat improvement research projects in Japan, the United States, and Mexico, it might be more appropriate to refer to the high-yielding varieties of the green revolution period as the "Japanese-United States-Mexican Wheats."

Strengthening Coordinated National Research Projects in Developing Nations

Coordinated national research projects do not emerge automatically from technical assistance support that is geared primarily to the objective of creating new, improved varieties. Experience with intensive or accelerated crop production projects as well as with more recent specific efforts to develop coordinated national research projects has helped to identify some of the restraints to the formation of a self-sustaining capability when external assistance has been terminated.

The National Commitment

National government leaders, including research officers, generally do not understand the kind of organization required to carry on an integrated research effort on a national basis. In countries where agricultural research has been limited or has been carried out as a series of isolated projects assigned to single-discipline specialists, it is usually necessary to operate a coordinated national project for several years in order to demonstrate the professional and administrative relationships involved. In most cases the requisite manpower in the constituent disciplines is not available. Where both manpower and facilities must be developed, a period of eight to ten years is usually required for the establishment of an effectively operating national research project which can produce the kind of results that will attract the commitment and support of government leaders.

The Organizational Base

It is not reasonable to expect a coordinated national research project to develop fully or to remain viable if (1) it is given essentially full autonomy and has no linkage to the central government or (2) it is connected with a college of agriculture, university, or other agency which does not have recognized national responsibility for the subject research. In most countries it is the Ministry of Agriculture which has both the responsibility and the funding authority for research of national scope. There are numerous examples of technical assistance efforts for crop improvement where the institutional linkage has been ill advised; "national research centers" for selected commodities have been set up at isolated field stations or in affiliation with colleges of agriculture under circumstances that hinder the prospects for fulfilling "national" objectives.

Institutional Collaboration

A coordinated national research project requires not only a well-equipped and well-staffed headquarters station but also research facilities in the principal regions of the country where the crop is important. In the United States the presence of state agricultural experiment stations and their substations facilitated the development of well-integrated projects that could give attention to specific localized problems as well as to those of regional and national concern. In the United States a substantial number of federal field experiment stations also have been established to ensure that concerted attention is devoted to the more critical aspects of a given crop improvement effort. The national federal-state research station networks for coordinated national research have been evolving for more than a century.

In developing nations it is essential not only to establish the national research project on the right organizational base but also to utilize the complementary resources in the states or prefectures, the colleges of agriculture, and private research organizations. Where such complementary resources are inadequate they can be strengthened or developed through financial and staff support from the coordinated national program. This was done in the coordinated research schemes for maize, wheat, and rice in India, where the work in the major producing regions was tied into the emerging agricultural university experiment stations. Similarly, the coordinated national research projects that are being established under the Philippine Council of Agricultural Research are planned to include increasing participation by the colleges of agriculture and private research institutes. It is desirable to develop a formal agreement for the cooperative effort which spells out the contributions of the participating organizations, not only to minimize uncertainties but also to furnish a degree of continuity and stability to the project.

Interdisciplinary Collaboration

In many developing countries the crop improvement research has been the primary responsibility of an economic botanist, with little collaboration or participation from scientists in the allied disciplines concerned with diseases, insects, plant nutrition, weed control, or cultural practices. And the role of agricultural economists is still not well appreciated in coordinated interdisciplinary projects, even in the more agriculturally advanced nations.

An interdisciplinary research team will be established only if its leaders (1) identify the nature and relative importance of the particular problems a given country faces in respect to crop improvement and (2) include scientists with appropriate specializations on the team. And the mix of scientists will not remain in proper balance unless continued attention is given to the make-up of the research team.

The multidisciplinary functioning of an integrated research project can be achieved relatively easily in a single institute where all participants are located on one campus and can be brought together frequently. It is more difficult, but equally important, to provide for the interdisciplinary mix of scientists in a research project whose units are geographically dispersed across a nation. At the same time, such staffing may be particularly effective in addressing specific problems. Pathologists can be located in regions where diseases are most prevalent, entomologists can be assigned to stations in areas afflicted with major insect problems, and senior personnel in the various disciplines can be posted at different locations to furnish coordinating leadership to the integrated research.

Coordinating Leadership

A coordinated national project requires a full-time director. Although the project director should not be assigned other unrelated duties, it is desirable that he be a working scientist carrying out significant research in a consituent disciplinary or problem area.

The project director should not be selected on the basis of seniority, but should have the experience, professional capability, and personal attributes that allow him to be accepted by his colleagues. The project director must understand the importance of the component disciplines, ensure that attention is given to all relevant problem areas, and furnish continuing leadership to the national project through the planning, implementation, summarization, and evaluation stages. It is essential that he be able to visit the field and laboratory experiments as frequently as necessary to be informed of the progress of the research underway. He must work effectively not only with his research colleagues but also with the scientific and administrative leadership within the government, with the staffs of nongovernmental institutions that may be collaborating, and with external assistance agencies.

The identification and selection of capable leadership is particularly critical in establishing coordinated national research projects, since the experience with such projects is usually lacking among the scientific personnel in most developing nations. Advanced academic training to a Ph.D. degree does not necessarily impart the traits and competence required.

Personnel and Manpower Development

It is relatively simple to determine the professional and technical manpower needs for a coordinated national research project. The manpower development or training activities should be pursued in a systematic manner in order to provide the numbers of persons in the different disciplines, at the various levels of training required, who will furnish the competence necessary at headquarters and at the principal field locations. Although there is bound to be some loss of staff to other institutions or to other projects, definite targets should be set for the numbers of scientists to be trained to the undergraduate degree, the M.S., and the Ph.D. levels in each of the disciplines and for the technical or supporting staff required to carry on a coordinated national research project at the optimum level. The staff development program should be projected over a ten-year period so that an effective staff complement can be developed.

Facilities

Field stations and laboratories for research in most developing nations usually are not suited for reliable experimentation. The tendency in the past decade for technical assistance or funding agencies to establish independent research projects or to require a research and training component in separate major development schemes has resulted in the setting up, in many developing countries, of research units that are geographically scattered and institutionally isolated. It is still common practice for most technical assistance or lending projects to be developed around one or a few commodities or problem areas, with the research facilities planned only for such specific activity.

More attention should be given to the planning of coordinated national research projects on the various commodities within a national research system or organizational framework so that there may be combined support for mutually necessary facilities at a national headquarters, at selected regional locations, and in the various localities or microecological areas. This would help to avoid the popular tendency to overbuild and duplicate stations, laboratories, and costly items of equipment.

Administrative Management

Developing nations tend to retain the administrative procedures followed during their colonial years. As a result, the recruitment and management of personnel, procurement practices, and other administrative activities frequently are not well suited for national development projects, including research. Governments are reluctant to give any special consideration to salaries, to the promotion of scientific personnel on the basis of performance rather than seniority, or to flexibility in the procurement of equipment and supplies.

It is difficult to establish – and maintain – the concept that in a research organization the administrative functions must be carried out in such a manner that research is facilitated rather than controlled. Any deviations from usual government procedures are difficult to furnish for a single selected project, even one of a coordinated multiinstitutional type which is national in scope. For this reason increasing attention has been given to the establishment of national research organizations on a semiautonomous or autonomous basis, with their own boards of directors and scientific councils offering guidance in administrative matters and on technical programs.

It would be difficult to establish a viable department of chemistry for college-level education outside of the framework of a university. Similarly, it is difficult to establish a single national crop improvement project outside of the institutional framework of a broader research organization. Clearly it is advantageous for the coordinated national research projects on individual commodities to be combined under a unified national agricultural research organization which would supply the type of administrative management that would ensure a reasonable degree of institutional stability.

Funding

Biological research is long term in nature, most projects requiring three to five years before they can produce useful and reliable information and materials. The financial support for research in most developing nations is furnished almost entirely by government, primarily from the central government resources.

It has been difficult to arrange for long-term funding for agricultural research in the agriculturally advanced nations. It is equally difficult in developing countries, although some progress has been made in establishing autonomous or semiautonomous national research organizations with funding handled outside of the regular governmental channels. The Philippine Council of Agricultural Research is closely aligned with the National Science and Development Board, which offers some flexibility in financial management. The Malaysian Agricultural Research and Development Institute has a separate "MARDI Fund," established under the authorizing legislation, which is designed to furnish a high degree of flexibility in the management of funds under the jurisdiction of the governing board. The MARDI Fund has not yet been permitted to function as intended, but it does offer a potentially workable pattern for such national research organizations.

The provision of research funds through special cesses or taxes on the individual commodity is attractive to some, particularly in the case of such cash crops as rubber or oil palm, for which the collection of an export tax is relatively simple. This would be more difficult in the case of food grains or as a source of funding for noncommodity research problems, which are too often neglected. A preferred procedure for funding would be to develop a commitment and responsibility on the part of the national government, with a concomitant recognition that the investment in agricultural research is an essential component in any national development process.

External Support and Collaboration

Although many technical and economic assistance organizations have become increasingly interested in strengthening research in developing countries, the nature and magnitude of such support are still uncertain.

More careful attention should be given, particularly by lending institutions, to the size of investment in a national research project or organization that can be justified within the economic base of the host country. Time is as critical as money in developing a self-sustaining national agricultural research capability, when one considers the number of years required for training scientific personnel, for building and equipping experiment stations and laboratories, and for developing institutional and multidisciplinary cooperation into a compatible operation. The size of the investment in buildings, equipment, expatriate technical or scientific personnel, staff development, and other components should be held to a reasonable minimum to avoid premature disappointment of government officials with the costs and returns aspects of research.

There is usually a need for external, experienced scientific leadership in planning and developing a coordinated national research project. Many developing countries will reject this notion, arguing that such leadership can be supplied by already available senior and experienced persons or by individuals recently trained to the Ph.D. level in overseas universities. Although some of these people may have the capability to furnish effective planning and coordinating leadership, it is commonly found that senior personnel in developing countries are inflexible in adjusting to a teamwork approach while recently returned Ph.D.'s tend to lack research organization and management experience. Scientists who have worked primarily in a strongly discipline-oriented organization face a substantial adjustment if they assume responsibility for a national multidisciplinary research project. In some recent cases where university professors were recruited to furnish coordinating leadership for such a research project they had difficulty in conceptualizing an effort that involved the full working partnership of scientists from several disciplines and tended to revert to an emphasis on research along the single-discipline lines with which they were familiar in the academic setting.

External financial support and some level of technical guidance should be continued for a period of at least eight to ten years in order to ensure the formation of a stable, self-sustaining project. The external input should be substantial for the initial five or six years, after which time there can be a gradual phasing down and out.

There should be continuous dialogue with the appropriate officials in government to ensure not only that the full government commitments of funds and other resources are met but also that necessary action is taken to develop the institutional and organizational structure required for a self-sustaining project when external support is withdrawn.

In addition to the international agricultural research institutes, other international technical assistance organizations—including the IBRD, UNDP, USAID, and other national or bilateral organizations—are giving more attention to the development of national agricultural research organizations and systems as well as to specific coordinated national research projects for selected commodities. It can be expected that, as cooperating technical assistance is increasingly concentrated on this objective, national research capabilities, both on a selected project or commodity basis and on a "national system" basis, will be strengthened in many of the developing countries over the next decade.

NOTES

1. The organization and functioning of the cooperative corn improvement research program is discussed in Herbert K. Hayes, *A Professor's Story of Hybrid Corn* (Minneapolis: Burgess, 1963).

2. The experiences in wheat research that lead to the workers' conferences in 1929 and 1930 which planned the cooperative hard red winter wheat improvement program are reviewed by L. P. Reitz and S. C. Salmon, *Hard Red Winter Wheat Improvement in the Plains*, USDA Technical Bulletin no. 1192 (Washington, D.C.: USDA, 1969).

3. The development of coordinated commodity research systems in the United States is discussed in greater detail in Albert H. Moseman, *Building Agricultural Research Systems in the Developing Nations* (New York: Agricultural Development Council, Inc., 1970).

4. See chapter 2 in this volume. Also Yujiro Hayami in association with Masakatsu Akino, Masahiko Shintani, and Saburo Yamada, *A Century of Agricultural Growth in Modern Japan: Its Relevance to Asian Development* (Minneapolis and Tokyo: University of Minnesota Press and Tokyo University Press, 1975).

5. For information on national agricultural research systems see Albert H. Moseman, ed., *National Agricultural Research Systems in Asia* (New York: Agricultural Development Council, Inc., 1971); John J. McKelvey, Jr., ed., *African Agricultural Research Capabilities* (Washington, D.C.: National Academy of Sciences, 1974).

6. The source and progressive use of the semidwarf wheats are discussed in L. P. Reitz and S. C. Salmon, "Origin, History, and Use of Norin Wheat," *Crop Science*, 8 (November-December 1968), 686-689.

17

Contract Agricultural Research and Its Effect on Management

Tilo L. V. Ulbricht

The Rothschild Report published by the U.K. government in 1971 recommended the application of the customer-contractor principle to some of the work carried out by the research councils.[1] It was a personal report by Lord Rothschild, head of the Central Policy Review Staff in the Cabinet Office, which advises the Cabinet on long-term policy (popularly known as the "Think-Tank"). The effect of the proposals on the Agricultural Research Council (ARC) was that, in the future, most of its funds would come in the form of contracts or commissions for specific applied research from the Ministry of Agriculture, Fisheries and Food (MAFF). To appreciate this report and the stormy debate which followed, some background is necessary.

In common with other research councils, and with many research organizations in developed countries, the Agricultural Research Council expanded greatly after World War II. The number of graduate scientists employed increased from 440 in 1948 to 1,280 in 1970. At the time of the Rothschild Report, the ARC was financed by the Department of Education and Science, which was advised on its budget by the Council for Scientific Policy. This council, in dealing with the research councils, was mainly concerned with strategic and fundamental research, although the ARC in particular had always engaged in applied research also.[2] MAFF, although represented on the ARC's council (its top executive body) and on many of its committees, wanted a more direct say in its affairs, and it was this, among other factors, which led to the Rothschild Report.

At the same time, it would be a mistake to look upon this report and the debate which followed as an isolated phenomenon. It could not have occurred ten, possibly even five years earlier. But since then the rapid increase in the funds devoted to research has been drawing to a close in most developed countries; governments have increasingly had doubts about the benefits that their countries were deriving from all this research; and the public has been increasingly concerned not only about pollution but also about the whole trend of our technological societies. Where are we going? What kind of world do we want? It is because of the absence of generally agreed aims for our societies that there is doubt and conflict about the role of research.[3]

With this perspective, the Rothschild debate, largely conducted in the correspondence columns of the London *Times*, becomes more comprehensible. In this debate three things were confused:

1. the principle that scientists financed by the government should be accountable for what they do and have a responsibility to meet the needs of their country;

2. the application of the customer-contractor principle as a particular means of achieving that accountability; and

3. the wider issue of science policy: how can scientists through their research help to meet their country's needs when the government has no clearly defined long-term policies?

The majority of scientists accepted (1), rejected (2), and were unclear about (3).

Elderly scientists, yearning for the good old days of unfettered expansion and believing in Polanyi's Republic of Science, protested that the research council system had stood the country in good stead and was the envy of the civilized world, including the United States, and that it should not be abandoned in favor of some ghastly government bureaucracy.[4] Outsiders, not knowing the political background, thought the scientists were making a lot of fuss about nothing and that it was time they realized that they had to justify their existence like everyone else. The parliamentary Select Committee on Science and Technology thought that the real trouble was the lack of a coherent science policy and of any mechanism by which one could be formulated.[5] Squeezed out in all this was any serious consideration of the administrative consequences of applying the customer-contractor principle or of possible alternative changes which could improve the management of research in the research councils and, in particular, the procedures by which they decided their resource allocations. The government White Paper essentially accepted Rothschild's recommendations.[6]

It would be misleading to suppose that the changes in the management of agricultural research in the United Kingdom in the last few years can simply be ascribed to the application of the customer-contractor principle. Other contributing factors were the changeover from a period of rapid expansion to one of static or (currently) declining budgets, the change in the public attitude toward science and technology, and the realization that the organizational system for dealing with agricultural research had not kept pace with the increase in personnel and resources during the growth phase.

To deal with that last point, let us make an admittedly exaggerated distinction between adminstration and management. In administration one is concerned essentially with maintaining what exists already. It is often said that the main aim is not to make a mistake, and this implies not taking risks, reluctance to make decisions (any decision involves some element of risk), and delegating work but not responsibility. In such a system the post is more important than the person who occupies it, and it is invariably associated with incremental budgeting – everyone gets a bit more than the year before, and one does not question the rationale of one's present or future resource allocation.

In management one strives to have a clear aim and allied operational objectives: the aim may be to make a profit or to increase it by launching a new product, it may be to control a new crop disease or to save imports by developing a new animal feedstuff, etc. Risks have to be taken, and it is essential that decisions be made; responsibility has to be delegated, and people are more important than the posts they occupy. Such a system is associated with some kind of planned budgeting.

However, the virtues of good administration should not be overlooked. It is a system which makes for stability and it can be remarkably successful when the particular organization's environment is not subject to rapid change. Unfortunately, stability tends to lead to rigidity, and often such a system cannot cope when dramatic changes begin to occur. The history of some of the long-lived cultures and empires, such as Ancient Egypt, Assyria, Rome, Byzantium, and the Ottoman Empire, are examples of this.

It would be true to say, I think, that until relatively recently government-financed and government-controlled research institutions have been administered rather than managed. Although undoubtedly good management ideas and systems have been developed in industry, no one is very clear as yet how best to apply them in the very different environment of government organizations.

It was in response to these circumstances that the Agricultural Research Council decided to set up a planning section in 1971 to advise itself and its chief executive on strategy. What we found was a well-administered and conscientiously run system, but one in which there was little central planning and in which budgeting was incremental. The system could certainly give precise information regarding the budget of a particular institute, but it would be in classical terms of staff costs, equipment, chemicals, animals, library, building overheads, and capital expenditure. It happens that work in many major fields of research is carried out at several of our institutes (we have twenty-nine in all and twelve units attached to universities). For example, eight institutes do some work or other on potatoes, and twelve are concerned with various aspects of grassland research. It was not possible to say what proportion of our total expenditure was devoted to various commodities (pigs, strawberries, wheat, etc.) or to major research areas like nutrition, breeding, disease, or to topics like harvesting, storage, processing, food quality, etc.

We therefore instituted a project system in the Agricultural Research Council. All the research in progress is described in terms of project units (about 3,000 in all, two to three per graduate scientist on average). We devised a system for classifying agricultural research; each project unit is coded in about a dozen fields.[7] Some simply give administrative information (name of institute, department, number of project, etc.), another group defines the agricultural problem to which the work relates, and a further group defines the action being taken to investigate the problem. In contrast to certain unstructured key-word systems, this is a matrix system devised for management purposes, each field having a hierarchy of structured key words.

Concurrently, a project-costing system was introduced. The same project units which are classified, and the information which is stored on computer, are also costed. This made it possible for the first time to look at the existing pattern of our resource allocation – by commodity, research field, or whatever – and to ask ourselves whether this seemed to be a good allocation, taking into account national needs, the current output values of various sectors of the industry and their economic prospects, and so on. Subsequently, MAFF, the Department of Agriculture and Fisheries for Scotland (DAFS), and the Ministry of Agriculture in Northern Ireland all decided to adopt the same project system. Consequently, almost all the agricultural research in the United Kingdom (except that at universities) is now classified in the same manner, and the information is available from one computer.

Retrieval of information from the project system is flexible and can be on either a broad or a narrow basis: questions like, "How much are we spending on cereals research?" as well as "How much are we spending on the mechanical harvesting of cereals?" and "What projects have we on breeding wheat resistant to foliar fungal diseases?" are all answerable.

At the same time that work on the project system was beginning, we suggested that the existing system for reviewing research (which I will not describe) should be revised so that each major commodity or commodity group (e.g., cattle, cereals) and important noncommodity research areas (such as soils research) should be the concern of a committee which would review the research systematically and recommend priorities. A working party was set up by the ARC to devise a scheme and was already at work when the Rothschild Report burst upon the scene. Following this report it was decided that the determination of research priorities should be carried out jointly with the MAFF and the DAFS. The system which was proposed by a new, joint working party of ARC, MAFF, and DAFS and which was accepted is called the Joint Consultative Organization. It is composed of five boards, namely: Animals; Arable Crops and Forage; Horticulture; Engineering; Food.

Each board has twenty or more members, including scientists, members of the industry, officials (MAFF and DAFS), and an economist. These boards examine current research in relation to what research is required and the needs of various interested parties and make recommendations regarding priorities. They take into account not only the research conducted by the ARC but also the research and development conducted by MAFF and DAFS at their own institutes, experimental farms, etc.

Each board has set up a number of committees, as for example:

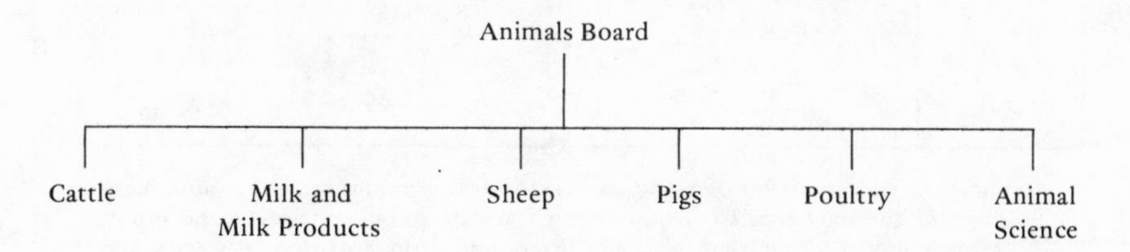

The spectrum of membership of committees is similar to that of the boards, but scientists are more strongly represented, constituting approximately half the membership.

This new system was set up only in 1973, so it is a little early to comment on its functioning. The intention is that each committee will review its own special area – coordinating with others as may be necessary – and report to its board. Each board then produces a composite report which goes to all three of the sponsoring organizations (ARC, MAFF, and DAFS) once a year. The Joint Consultative Organization has an advisory, not an executive function; the boards have not, for example, been allocated budgets which they can disburse. On the other hand, obviously the system can work only if serious note is taken of the advice offered.

The project system has been used to provide the committees with the basic information regarding ongoing research. A framework has been drawn up

	Effort (£000)	Description	Main Objectives
30 The biological characteristics of crop plants	300 Plants in general	Plant physiology and biochemistry	To provide an understanding of the biological, physical, and chemical mechanisms which control plant reproduction, growth, function, and behavior
	301 Grassland and forage crops	Plant physiology and biochemistry in relation to specific crops	To provide means of controlling the reproduction, growth, and function of specific crops
	302 Cereals		
	303 Arable crops		
	304 Vegetables		
	305 Fruit		
	306 Protected crops		
	307 Decorative crops		

Figure 17-1. A portion of a framework for crop production. (The subsequent items in the extreme left-hand column are: The soil as an agricultural input; Natural inputs other than soil; Fertilizers and crop nutrition; Genetics and breeding; Crop protection; Handling and storing crop products; Processing primary crop products.)

which is a simple two-dimensional matrix in which all project units appear (each appearing once only). Part of the framework for crop production is shown in Figure 17-1. Each numbered box in the grid is called a "project area," e.g., "Biological Characteristics – Vegetables." Certain project areas were allocated to each committee to define its area, and so on. In other words, the buildup of the information system is as shown on p. 387.

By this means, it was possible to provide a computer printout to each committee of the project unit with which it is concerned, structured by project area. Specific printouts structured in other ways, as may be required, can also be provided.

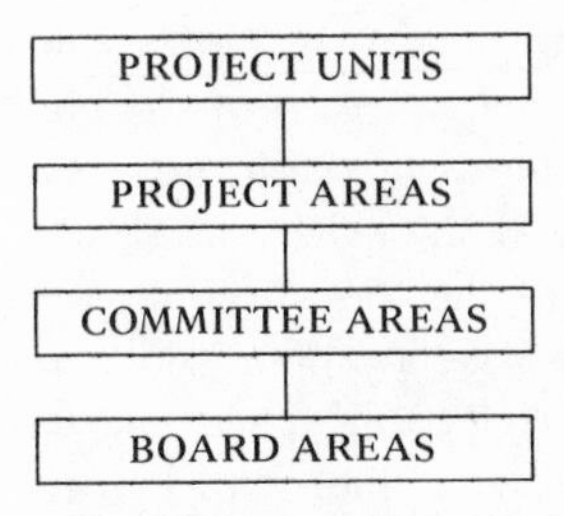

At this point, the following question must be touched on, at least briefly: What criteria are to form the basis of recommendations by boards and committees on research priorities? My Section has been studying available quantitative techniques, such as cost-benefit analysis, for some time. A review we have made of the published material on the cost-benefit analysis of agricultural research and development (R & D) projects has convinced us that, in its present state, cost-benefit analysis does not provide a valid basis for the quantitative planning of agricultural R & D programs.[8] The objections to it are given in a recent paper.[9] They fall under three headings. First, the technical data which are subjected to economic analysis are often so imprecise that corrections stemming from the niceties of economic theory are trivial in relation to the uncertainties in the data themselves. Second, there is a lack of uniformity in procedures, for example, in the way benefit is defined and estimated. This makes comparison of analyses hazardous. Third, the presentation of the results of cost-benefit analyses, particularly in the way initiated by Griliches,[10] we believe to be objectionable. In particular, relating benefit just to R & D expenditure leads to misleadingly high benefit-cost ratios.

The alternative, therefore, is to rely on the informed judgment of the members of the boards and committees. We suggested, however, that this process could be rationalized to some degree by considering priorities at three levels, as follows:

Level (1) Decisions on priorities between commodities and major research areas;

Level (2) Decisions on priorities within one commodity or research area;

Level (3) Priorities between individual project units.

For each level, a checklist of criteria was drawn up. For Level (1), for example, criteria were grouped into economic factors (such as national benefit, output value, value added by commodity sector, import-export considerations, etc.) and social factors (such as regional welfare, consumer welfare, environmental considerations, etc.). The possibility of weighting these different factors and so arriving at a scoring system was considered but rejected on the grounds that it gives a spurious air of precision to what should be recognized as being fundamentally a subjective process.

Clearly, Level (3) is the concern of institute management, Level (2) of the commodity committees, and Level (1) of the boards. The final decisions, including inter-board decisions, rest with the senior management of the three sponsoring organizations, ARC, MAFF, and DAFS.

One of the functions of the Joint Consultative Organization is to give advice on which MAFF can base its commissions to the ARC. In describing the customer-contractor principle, Rothschild had written: "The customer says what he wants; the contractor does it (if he can); and the customer pays." The application of this principle to the financing of research councils in the United Kingdom means that the customer (that is, the government department) must know what it wants. Rothschild recognized this problem and therefore recommended that the appropriate government departments should each have a chief scientist, with an appropriate scientific staff. MAFF now has a Chief Scientist's Organization.

As far as agricultural research is concerned, the customer is MAFF, and the principal contractor is the ARC (which now gets 55 percent of its budget in the form of commissions from MAFF, the remainder continuing to come from the Department of Education and Science). Initially, MAFF has commissioned 55 percent of the ARC's existing ongoing research, but in future its choice of work to be commissioned will be influenced by the advice of the Joint Consultative Organization, just as the ARC takes note of that advice in its own research planning.

The task of drawing up and agreeing on commissions covering something like 30 million dollars' worth of research is one of no mean administrative complexity (even if one puts to one side awkward problems like what to do about capital cost of buildings which will be used for both commissioned and noncommissioned research). How has this task been accomplished?

In the first place, MAFF and ARC were able, after numerous discussions, to agree that the number of commissions must be kept small – in fact, to twenty. These are mostly on a commodity basis (e.g., pigs, cereals, vegetables, etc.). Using our classification system, we were able to produce a printout which allocated each project unit to one or other of these twenty master program areas (which in some cases are identical with or very similar to committee areas, e.g., cereals, vegetables). However, these master program areas were structured not by project areas but in terms of objectives and subobjectives, to which project units were allocated by scientific experts (and the information then added to the classification data on the computer).

A part of a commission is described in Table 17-1. It will be noted that it is a *draft* commission that is illustrated. This is in fact the kind of working document that MAFF and ARC have used, but the formal commissions do not list project units and give costs only down to the subcommission (objec-

Table 17-1. Illustration of Draft Cereals Commission

Aim: To improve the quantity and quality of home-grown cereals				
Objective/Sub-objective		Project Unit No.	Cost (£000)	Project Unit Title
	A. Sub-Commission – Wheat			
A.1	To increase yield/acre and quality of grain			
A.1.A	To provide new high-yielding, disease-resistant, quality winter and spring varieties			
A.1.A.01	Breeding varieties with good agronomic characteristics, including short straw			
A.1.A.02	Establishing durable resistance especially to yellow rust, septomia and mildew			
	etc.	etc.	etc.	etc.
	B. Sub-Commission – Barley			
		etc.		

tive) level, it being recognized that the *management* of research is the function of the contractor, who must be free to change the details of the program in order to meet the stated objectives in the best possible way.

Figure 17-2 provides an extension of our project system diagram. The computer can print out the commissions; it can also produce printouts for each institute showing which project units are commissioned and which are not. Also mentioned in this diagram are project groups; these are departmental or interdepartmental groupings of project units which are often administered as a group (e.g., the "Enteric Diseases Project"). This information can also be computerized, but project groups are essentially for local management, not for central planning.

A further aspect of commissioning is that there must be some kind of review procedure. Since the first commissions have only just been drawn up, this procedure will not begin before 1976. It is envisaged that, in addition to regular annual reporting of the progress of commissions, there will be formal reviews at appropriate intervals (which could vary considerably, depending on the nature of the work) conducted jointly by MAFF and ARC, involving the consideration of a special progress report and discussions with senior scientific staff.

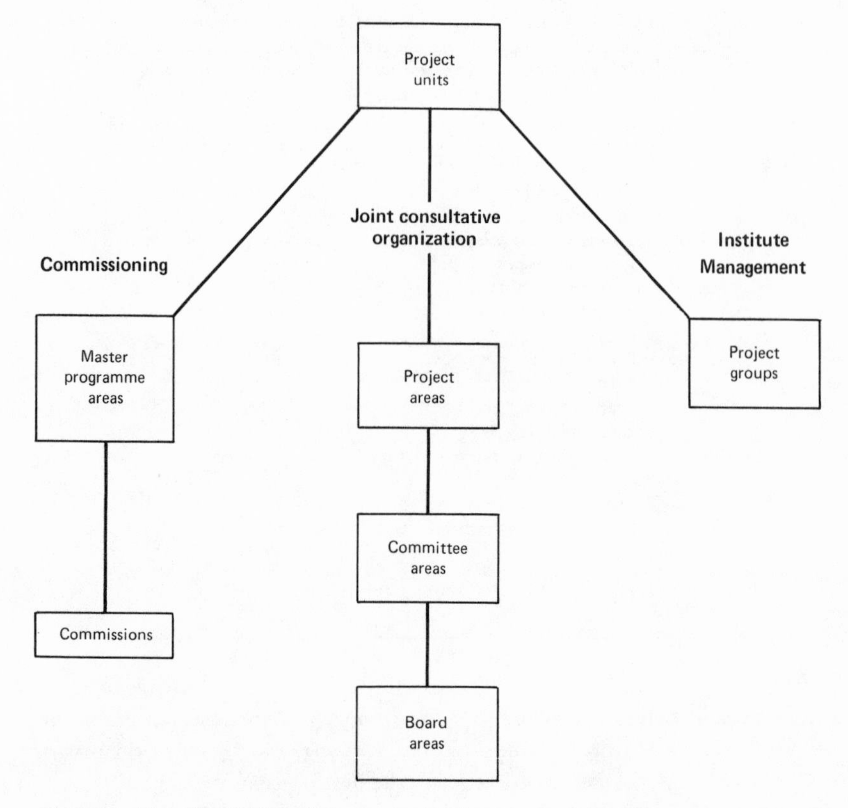

Figure 17-2. The ARC project system.

Conclusions

As indicated earlier, it is not possible to attribute changes in the management of agricultural research specifically to the introduction of the customer-contractor principle, since other important factors have been at work. Indeed, the decision to apply the principle to the research councils may be regarded as deriving from these other factors. In addition, the changes have been very recent and therefore any long-term consequences can only be guessed at.

The principal changes have been a shift away from research administration toward more positive management. The methods used to bring about these changes, as outlined in this chapter, have been: definition of research in terms of specific projects with defined objectives; introduction of project costing; introduction of a system for the systematic review of research programs and determination of priorities and for replacement of incremental by planned budgeting. These changes also provide the basis for improved coordination of research programs in common fields between our own institutes and other organizations (such as the Agricultural Development and Advisory Service of MAFF).

The majority would regard these changes as necessary and desirable. At the same time one must frankly admit a significant drawback: a considerable increase in the administrative load, both at our headquarters and for directors and senior staff at our institutes. This is a question not only of paper work but also of attendance at many more meetings, committees, and boards, meetings to discuss commissions, preparation of papers and reports. This has been accomplished with very little increase in administrative staff. As regards the longer term, some scientists fear that the introduction of contract research will in due course result in an unhealthy over-emphasis on short-term, applied research, with consequent neglect of the strategic and fundamental research which the ARC, at least, recognizes must be the basis for future agricultural development.

Our experience of these changes has also focused our attention on certain other problems. First, there is no clearly defined long-term policy for agriculture in the United Kingdom, an essential input for the effective planning of agricultural research within a framework of financial restrictions.[11] Second, the usefulness of existing quantitative methods of research planning developed in industry, when applied in the context of government research, is limited. My Section is continuing to study this question. Third, it is also apparent that insufficient resources have been devoted in the United Kingdom to the *development* of agricultural research findings.

I think it is important at least to mention these problems, especially as I suspect that the first two may be widely shared.

In concluding this chapter, I would like to address myself to some questions posed by Dr. Richard Nelson in discussions at the Airlie House conference.

1. Is the problem in research management one of criteria (i.e., good selection) or of the generation of good ideas (i.e., entrepreneurship)? The answer is both. The provision of an environment which encourages entrepreneurship and the generation of good ideas must be a major concern of any research organization. Governmental laboratories are handicapped by having to follow rules and regulations on staff, pay, etc., which may be suitable for administrative departments but do not meet the needs of an organization devoted to research. (The international institutes are better off in this respect.)

2. How is selection to be made – by *ex-ante* quantitative evaluation or by process of judgment? As already indicated, we believe that there is no valid basis for the notion that objective quantitative methods yet exist for ranking

research priorities and drawing up an optimum portfolio of projects. The process has to be one of informed judgment using systematic aids, including economic analyses as one input to the decision process.

3. Do we favor planned complementarity in research or pluralism (duplication) and competition? No simple answer can be given to this question. It is a matter of finding the right balance. In most large research organizations there has been wasteful, unconscious duplication, owing to a lack of information and a lack of coordination. Good planning implies good coordination but allows, and may positively encourage, conscious duplication and pursuit of different routes to the same goal.

4. Should publicly funded research be tightly controlled, e.g., by contracts, or is it better to have looser control, with accountability vested in the research director? The fact that the latter system has been found wanting does not necessarily imply that the former constitutes the correct solution. It is reasonable to expect that applied research in agriculture (mostly short term) should be more tightly controlled than strategic and fundamental research (mostly longer term). In effect that is now the situation in the United Kingdom, since 55 percent of the ARC's budget is in the form of contracts for applied and what one might call applied-strategic work; 45 percent of the budget continues to come from the Department of Education and Science, and this is mainly for long-term strategic and fundamental work. To have all or an overwhelming proportion of research controlled by contracts would almost certainly stifle long-term basic work and, in particular, the more imaginative and unpredictable research. Whether contracts are the best mode of controlling applied work no one can yet say. It is a cumbersome system, and one wonders what would have been the return of a comparable investment in staff and time in trying to improve the coordination and planning of research in other ways. But perhaps the inertia in any large organization is such that only strong financial pressures can bring about significant change.

NOTES

1. *A Framework for Government Research and Development*, Command 4814 (London: Her Majesty's Stationery Office, 1971).

2. For definitions of these terms, see T. L. V. Ulbricht, F. H. Dodd, and W. S. Wise, "Six Types of Research," *Nature*, 240 (1972), 427.

3. G. Wansink and T. L. V. Ulbricht, "Mechanisms for Adapting Agricultural Research Programmes to New Goals," paper presented at Second Working Conference of Directors of Agricultural Research, Organization for Economic Cooperation and Development, Paris, November 1972; see also final report of this conference, *Summary Report of the Second Working Conference of Directors of Agricultural Research* (Paris: OECD, 1973).

4. M. Polanyi, "The Republic of Science, Its Political and Economic Theory," *Criteria for Scientific Development*, ed. E. Shils (Cambridge: M.I.T. Press, 1968).

5. *First Report from the Select Committee on Science and Technology, Session 1971-72, Research and Development* (London: Her Majesty's Stationery Office, 1972); *Fourth Report from the Select Committee on Science and Technology, Session 1971-72, Research and Development Policy* (London: Her Majesty's Stationery Office, 1972).

6. *Framework for Government Research and Development*, Command 5064 (London: Her Majesty's Stationery Office, 1972).

7. *Revised ARC Classification System*, Agricultural Research Council, Planning Section Report no. 9 (London: Agricultural Research Council, 1974).

8. *Cost-Benefit Analysis in Agricultural Research*, Agricultural Research Council, Planning Section Report no. 6 (London: Agricultural Research Council, 1973).

9. W. S. Wise, "The Role of Cost-Benefit Analysis in Planning Agricultural R & D Programmes," *Research Policy*, 4 (1975), 246.

10. Z. Griliches, "Research Costs and Social Returns: Hybrid Corn and Related Innovations," *Journal of Political Economy*, 66 (1958), 419-431.

11. See T. L. V. Ulbricht, "Agricultural Research and Science Policy," paper presented at an Experts' Meeting on the Relationship between Agricultural Research and Socioeconomic Policy, and Agricultural Research and Science Policy, Organization for Economic Cooperation and Development, Paris, January 1975.

18

Reforming the Brazilian Agricultural Research System

José Pastore and Eliseu R. A. Alves

Brazil is currently experimenting with a new model for organizing agricultural research: the public corporation. The object of this experiment is to increase the quantity and quality of scientific knowledge relevant to agricultural development. Its main objective is to make the whole research system more sensitive to the demand for technology.

The main organizational agency of the new system is EMBRAPA—the Brazilian Public Corporation for Agricultural Research. This agency operates like any public enterprise, being open to all types of financial and human resources and at the same time ready to "sell" its services to all kinds of clients. The corporation's principal product, of course, is agricultural technology and its primary client is the government. Both federal and state governments establish their priorities in terms of products for export and for domestic consumption. An increase in agricultural productivity is the basic need to be met by the research, extension, and credit complex, and research is the responsibility of EMBRAPA. The initial task of EMBRAPA, then, is to transform the general production goals of the government into research programs geared to increase the productivity of land and labor. Its second task is to organize and improve the skills of the scientific and technical staff who carry out the research programs.

EMBRAPA is not subject to civil service hiring restrictions. It is free to hire whomever is considered qualified for its programs at national and international labor market prices. In order to economize, EMBRAPA is directing its main research programs through national centers. This effort to concentrate financial and human resources on a few, but relevant, products is just beginning. Three national centers have been established to date: wheat, rice, and dairy.

This chapter first offers a brief overview of the trends in Brazilian agricultural development. Second, it illustrates the role of research in agricultural development in Brazil by providing historical background. Third, the basic principles behind EMBRAPA are described and, finally, the main accomplishments of the new research system to date are presented.

Trends in Brazilian Agricultural Development

Land has been abundant in Brazil. For many years it has been Brazilian policy to increase agricultural production through the expansion of cultivated areas. However, this is changing. Although pressures to expand the agricultural frontier continue, there has in recent years been an increasing demand to raise the productivity of land already under cultivation.

During the 1950s the expansion of cultivated area and the increase in farm employment continued to represent the dominant sources of growth of agricultural output in Brazil.[1] In the decade between 1960 and 1970, an increase in land productivity was observed throughout the country, with the exception of the Northeast. At the same time, the rate of labor absorption declined significantly.[2]

The change in trends of agricultural development in this decade was a consequence of several factors. Favorable conditions in the international market and growth of domestic demand stimulated pressures for large increases in agricultural production which exceeded the possibility of growth by expansion of the cultivated area. The availability of good and cheap land for agriculture diminished considerably. These new forces (international and domestic demand for food and fibers) produced a dialogue between official authorities on the one hand and the farmers, industrialists, and, especially, technicians on the other. The result was a revision of basic agricultural policy. Growth through expansion was maintained. However, increase in land and labor productivity was explicitly introduced as a new, additional goal during the late 1960s and early 1970s.

Initially, the emphasis was on disseminating existing technological knowledge from the research institutions to the farmers. The heavy emphasis on agricultural extension services during the sixties can be understood within this framework. This circumstance also explains the high priority allocated to the development of special lines of credit for the purchase of modern inputs as well as the emphasis on minimum price policies to stimulate production and productivity.

An internal crisis for food in the domestic market became an additional and powerful factor for revising agricultural policies, especially in the mid 1960s. The feeding of the large urban centers suddenly became a crucial economic and political goal. Government became aware that inflation plus food shortages were the ingredients for social upheaval and radical political changes.

The initial steps toward the modernization of the agricultural research system were taken at the federal level, within the Ministry of Agriculture, which was under increasing pressure to raise the productivity of the agricultural sector.

The economic forces that entered the picture in the sixties created, in the beginning of the seventies, a favorable atmosphere for profound change in the Brazilian research system. This system has undergone several changes, but none of them has succeeded in providing Brazil with a research system capable of handling agricultural problems. It is our contention that lack of incentives in the economic system has been responsible to a great extent for the failure of the reforms that have been tried.

Historical Background

There were some manifestations in Brazil of the great changes of the eighteenth and nineteenth centuries in the agrarian sciences in Europe. The first Brazilian agricultural research units were created within the atmosphere of European liberalism, which generated a *diffuse model*.

The Diffuse Model of Research

The main feature of this model is that each research unit tries to diversify its activities, researching many different products and attempting to generate a wide array of technologies.[3] It represents an adequate system for organizing research in an environment with special characteristics. Among these characteristics are the following:

1. Availability of abundant resources for research. The abundance of resources destined for research indicates that the society has already recognized the importance of research in the modernization of agriculture. Furthermore, mechanisms have been developed to provide agricultural research with sufficiently generous and flexible budgets to meet its needs.

2. Predominance of a liberal philosophy, which accepts the behavior of the scientists as individuals and provides an atmosphere of freedom in the choice of research projects.

3. Existence of a critical mass of farmers sufficiently organized to interact with researchers and administrators and to make the problems they face ex-

plicit. From this interaction, pressure develops to allocate adequate resources to research. This pressure also prevents the scientist from becoming alienated from the real world and concerned only with his particular problems.

The pressure from farmers, together with the individual orientations of scientists, results in a research system which seeks to generate diversified information covering a vast range of subjects and large numbers of crop and animal enterprises. There will be many lines of research, some seeking to economize on land, some on labor.

The tendency is to develop what is possible in such a broad range of areas, given the limitations of time and money. The individual interests of the scientists are satisfied because they have a wide range of choice with respect to areas of research. At the same time, this system guarantees that the desires of the majority of farmers, particularly those in a position to influence the research institutions, will be satisfied. When an individual farmer seeking information on how to improve the efficiency of his farm comes in contact with the universe of knowledge generated, it is likely that he will find the information he needs.

The diffuse model generates a large amount of information, only part of which crystallizes into new technology. This makes the model expensive and thus practical only in wealthy societies which can invest large quantities of resources in research. For example, the model has been in use in Europe and the United States for some time and more recently in Japan.

In the developing countries two of the ingredients essential to the functioning of the diffuse model are in short supply. First, resources for research are scarce. Second, the low cultural level of farmers, together with difficulties of transportation and communication, make the establishment of a dialectic difficult. Nevertheless, many researchers have adopted the individualistic approach from the developed countries through training abroad and through the scientific literature.

Conditions in the developing countries, therefore, alienate research from the current agricultural situation and lead to a dispersion of research among many crop and animal enterprises. Since human and financial scientific resources are limited, this dispersion of effort reduces the efficiency of research. The farmer finds only limited and incomplete information available, which does not permit the elaboration of a production system. Hence it is necessary to modify the diffuse model in such a way that the knowledge generated meets certain defined guidelines.

Historical Pattern in Brazil

In Brazil (with the exception of São Paulo and Rio Grande do Sul) human and financial resources are extremely limited. In addition, an organized mass of farmers does not exist to sensitize Brazilian authorities to the sector's needs. As a consequence, the imported diffuse model cannot be expected to succeed in Brazil.

Until recently, the Brazilian agricultural research system had gradually developed an individualistic orientation; research topics and methodology were viewed as being the exclusive property of the investigators themselves even though research was completely financed by public money. Research priorities were aimed at sacred themes, and the directing of science and technology toward the solution of the entrepreneurs' problems was considered heretic thinking.

Scarce resources tended to be allocated to a wide variety of research topics defined by the researchers who, not rarely, were more eager to duplicate an investigation recently published abroad than to solve the farmer's problem. The style of working in research was a "one-man venture"; research tended to be designed in such a way that research teams were not used. The government's research investments were mainly an "act of good faith" rather than a goal-directed effort. This type of social background pervaded both the agricultural colleges and the more applied research units, namely, the agricultural experimental stations and institutes.

In short, the Brazilian agricultural research structure seems to have been negatively affected by two forces. On one hand, owing to the relative abundance of land and labor, there was little pressure for research to develop technology which economized on these factors. On the other, extremely individualistic research patterns were imported from developed countries.

The picture began to change at the beginning of the 1970s. Pressure developed to increase agricultural production in order to meet the increased domestic and international demand for food and fibers as well as the political need for feeding the urban population. These forces have created a new atmosphere for shifting from a diffuse research organization model to one in which concentrated research efforts predominate.

Changes in Brazilian Research

The role of science and technology in increasing agricultural productivity became one of the central concerns of Minister of Agriculture Dr. Luis Fernando Cirne Lima in late 1971. In early 1972, he called a meeting of all state secretaries of agriculture and agricultural experiment station directors and made it clear that the central government desired to modernize the research system to accomplish the newly defined national goals. At the same time, he nominated a special committee to recommend reforms in the agricultural research system.

The report of this committee pointed out the strengths and weaknesses of the federal research units. The positive aspects can be summarized as follows: (1) A geographically dispersed network of research units was available to the federal government and covered practically the whole nation. (2) Equipment and basic infrastructure were considered reasonably adequate for most of the units, with a total investment in land, building, laboratories, and other facilities totaling about $300 million. (3) Sixteen technical journals were available for publishing the results of agricultural research. (4) There was a small but well-qualified group of researchers whose talents could be better used by the units if their administrative load were assumed by other professionals. (5) A relatively well-defined consciousness of the need for an integrated research policy for the agricultural sector was present in most of the researchers.

The negative aspects, unfortunately, were overwhelming: (1) The basic national needs in respect to agriculture were unknown to most of the research personnel. (2) There was little interaction between research personnel and farmers. (3) The existing adminstrative structure inhibited the recruitment, training, and promotion of well-qualified personnel. (4) A complete lack of internal communication among units and individual researchers was evidenced by large numbers of parallel projects on unimportant products. (5) The lack of suitable programming and evaluation mechanisms permitted researchers to undertake individual activities of doubtful value. (6) Of 1,902 individuals considered to be formal researchers, only 10 percent could be considered professionals, with some kind of graduate training in research. (7) The salary policy did not permit the government to compete in the professional labor market; there were no means to hire and promote qualified personnel quickly or to demote unqualified persons. (8) Higher salaries given to administrators reduced researchers' incentives to argue for their projects. (9) There were inadequate mechanisms for obtaining and managing financial resources which came solely from the federal government. (10) All the existing facilities were underutilized.

The committee recommended a public corporation as the best institutional means to remedy these defects. The Congress, on December 7, 1972, created EMBRAPA as a public corporation to coordinate and administer research in agriculture and animal husbandry. EMBRAPA started operating on April 26, 1973.

The Basic Principles of the Present Brazilian Model

The basic tenet of EMBRAPA is that applied agricultural research should be guided by the concrete needs of the national society as expressed in government policies and in the concerns of farmers, extension agents, and industry. Execution of applied research directed toward immediate needs is seen as the province of the technological research institutes. More fundamental research is seen as the province of the universities. There is not a rigid division of labor between the two types of research institutions. To a great extent, however, the comparative advantages of each are utilized in the two types of research.

Research units under the Ministry of Agriculture are generally in the first category. Their main effort should be directed toward generating technology which can be readily incorporated into the production system. This implies that emphasis should be given to creating technological packages that achieve technical and economic efficiency.

In addition to these general principles, six other ideas have been used as guidelines in reforming the existing research apparatus. First, the transfer of foreign technology to the agricultural sector is considered a valid means of improvement but of limited importance in many instances. The transfer of specific materials and of certain packages (i.e., poultry technology), however, is looked upon as an opportunity to capitalize on some other country's investments. Also, training abroad and imports of personnel are very helpful in the Brazilian situation.

Second, given the scarcity of financial and human resources for research activities, efforts should be concentrated on regional projects. This should help to overcome the difficulties of transferring technology among different ecological and economic regions throughout the country.

Third, the private sector should participate in the development of most of the research projects.

Fourth, the agricultural research system should have more administrative flexibility including the freedom (a) to obtain additional resources through contracts and agreements; (b) to pay researchers wages at market rates; and (c) to carry out an aggressive training program, including basic training and graduate work.

Fifth, a closer relationship should be developed with the extension services and the agricultural input industries to speed the dissemination of knowledge throughout the country.

Sixth, knowledge from the international institutes and from other foreign research centers should be adapted and spread throughout the country. The research system should seek technical packages which decrease the farmer's risk. This means that an economic investigation should be systematically included in the agronomic investigations.

The development of EMBRAPA implies the concentration of relatively large financial and human research resources on a limited number of products. The challenge that this model presents is that of defining priorities and responding to changing circumstances.

This type of orientation implies a number of problems:

1. Since resources are scarce, it is necessary to limit the number of production system prototypes developed and the number of commodities researched. Clearly, priorities must be established, but this means that some groups of farmers may not receive the benefits of research.

2. There are problems of allocation of resources between research with immediate applicability and that with applicability in the long run.

3. It may be difficult to use the concentrated research model to develop systems of production adequate to the needs of the small farmer who combines various enterprises in his operation.

4. The concentration of effort requires an appropriate institutional system. It is unlikely that research institutes which work on a large number of commodities and are organized on the basis of disciplinary departments such as soils and plant improvement will have a high degree of success in developing production systems. In this type of environment, given the individualistic tradition to which researchers are accustomed, pressures will develop that cause departure from the established priorities and areas of concentration. These pressures arise from the departments which seek to develop an area of specialization, as is common in the developed countries, and from researchers that have dedicated their lives to commodities not considered to be of national priority. It should be noted that the organization of research in institutes of this type is a consequence of the requirements of the diffuse model. In rejecting this model, it is also necessary to modify the institutional arrangements which made it possible.

Agricultural Research under EMBRAPA

EMBRAPA concentrates on applied research, to generate improved technology for agricultural development. However, it is not EMBRAPA's responsibility to perform all agricultural research in the twenty-five Brazilian states. As a consequence, two important roles have been defined for EMBRAPA. On one hand, it has the responsibility of creating and/or supporting the state research systems. On the other hand, it is responsible for creating and implementing commodity-oriented national research centers.

Supporting the State Systems

Agricultural research at the state level is very heterogeneous in Brazil. The southern states possess relatively mature research systems. EMBRAPA plans to continue supporting their activities. At the same time, it expects them to adopt more flexible administrative units (corporation-type agencies) to facilitate coordination between the state and EMBRAPA.

There are many other states, however, which have no research tradition whatsoever, although many of them have been receiving research funds from the central government. In these states, EMBRAPA is helping the state governments to create their own capabilities. The main support up to now has been in training massive groups of research personnel as well as aiding the state secretaries of agriculture in organizing their own state corporations.

The National Centers

These centers are defined in terms of basic national needs for the agricultural sector. The main strategy is to concentrate funds and talents on a few, relevant products in specific regions. Wheat, sugarcane, corn, beans, soybeans, rice, coffee, rubber, livestock, and dairy have been defined as the crucial agricultural products for the country. Among the key resource areas to be developed through national centers, EMBRAPA has included "cerrado," semiarid agriculture, and humid-tropical agriculture.

State agencies can link themselves directly with the national research centers, particularly when they are located in the state where a given center is located.

The most important results obtained in the 1973-74 period are the following:

1. EMBRAPA replaced the National Department for Agricultural Research of the Ministry of Agriculture. The year 1973 was transitional and the corporation actually assumed the operation of research activities in 1974.

2. The realized budget of the old system in 1973 amounted to $14 million (United States currency, exchange rate of December 1973). In 1974 EMBRAPA expended about $25 million in research activities (exchange rate of December 1974). The planned budget for 1975 was estimated at $65 million (exchange rate of December 1974).

3. The old system was overcrowded with bureaucratic personnel. The corporation was, by law, allowed to select the personnel best suited for its work. It chose 3,422 (data of January 1975) out of 6,705 employees of the old system.

4. The training of personnel forms one of EMBRAPA's most important programs. The program's current goal is to enable 1,000 researchers to acquire the master's and/or doctoral degree in Brazilian and foreign universities. The program is financed by both Brazilian and foreign funds. Included in the latter is a USAID loan to the Brazilian government in support of the training of researchers at universities in the United States. At present, 500 researchers are studying for the M.S. or Ph.D. degree in various universities. Under the old system, 10 percent of researchers held graduate degrees. The aim of the present program is that at least 80 percent of EMBRAPA's researchers will hold the master's or doctoral degree.

5. National centers for the most significant products of Brazilian agriculture will be in operation by the end of 1976. Actually, the national centers for dairy cattle, rice, and wheat are already in operation. Three national centers for the development of natural resources will be in operation in 1976, one in the area of cerrado, another in the semiarid region of the Northeast, and a third – the Center for Tropical Agriculture – in the Amazonas region.

6. EMBRAPA is strengthening institutional linkages with Brazilian and foreign universities, with the international research centers, and with development banks to obtain technical and financial support for its program.

7. Three states have already reformulated their research systems according to the federal model. Their research projects are supported to some extent by EMBRAPA funds. In other states an institutional arrangement has been established with the purpose of strengthening their research capability and creating conditions favorable to the future conversion to the corporation system.

NOTES

1. A. C. Pastore, E. R. A. Alves, and J. B. Rizzieri, "Inovação induzida e os limites a modernização na agricultura brasileira," *Alternativas de desenvolvimento para grupos de baixa renda na agricultura brasileira*, 2 vols. (São Paulo: Instituto de Pesquisas Economicas, 1974).

2. São Paulo has traditionally been an exception to the general Brazilian pattern. In the state of São Paulo the increase in agricultural output has for several decades been almost entirely the result of increases in land productivity. See H. W. Ayer and G. E. Schuh, "Social Rates of Return and Aspects of Agricultural Research: The Case of Cotton Research in São Paulo," *American Journal of Agricultural Economics*, 54 (1972), 557-569; R. Evenson, "The Contribution of Agricultural Research to Production," *Journal of Farm Economics*, 49 (1967), 1415-25; Z. Griliches, "Research Costs and Social Returns: Hybrid Corn and Related Innovations," *Journal of Political Economy*, 66 (1958), 419-431; Z. Griliches, "Sources of Measured Productivity Growth: United States Agriculture 1940-1960," *Journal of Political Economy*, 71 (1963), 331-346; W. L. Peterson, "Return to Poultry Research in the United States," *Journal of Farm Economics*, 49 (1967), 656-669.

3. The term *diffuse model* is used to describe the institutional pattern that Hayami and Ruttan have referred to in their discussion of induced innovation. See Y. Hayami and V. W. Ruttan, *Agricultural Development: An International Perspective* (Baltimore: Johns Hopkins Press, 1971), pp. 53-63 and 82-85. The induced innovation model implies that government and private research agencies tend to concentrate their effort to generate the type of technology that saves the scarce and hence expensive factors of production. In this sense, the main lines of scientific and research policies really reflect the relative prices of land and labor in the case of agriculture. Institutional reform, on the other hand, is made possible and stimulated by the new opportunities opened up by changes in the relative prices of land and labor and by the increase in the demand for food.

19

Private Sector International Agricultural Research: The Genetic Supply Industry

S. M. Sehgal

There have been many instances, throughout the last three decades, of the transfer of elite cereal varieties and associated crop production technology from one agricultural zone to another of similar latitude. The International Maize and Wheat Improvement Center in Mexico (CIMMYT) has been responsible for transferring high-yielding varieties (HYV's) of wheat from the subtemperate/subtropical zone of Mexico to the subtemperate/subtropical zone of South Asia and the Near East. The International Rice Research Institute (IRRI) in the Philippines has been responsible for transferring HYV's of rice from the Philippine tropics to other areas within the tropical belt. The genetic supply industry has been instrumental in transferring high-yielding hybrid maize varieties from temperate North America to temperate Europe and hybrid sorghum varieties from subtemperate North America to subtemperate Mexico, Argentina, Australia, and South Africa. In all these cases the impact on total grain production has been dramatic.

High-Yielding Varieties

Since the HYV's of wheat and rice formed the basis of the green revolution, it is appropriate to outline some of the salient features of these two discoveries.

Wheat

The high-yielding varieties of wheat were developed largely by Norman Borlaug, using standard breeding techniques of hybridization and selection, at a small experiment station outside Ciudad Obregón in the state of Sonora in northwest Mexico. The station is located in Mexico's subtropical or semitemperate zone. The facilities at the station at the time Borlaug took over were, by modern institute standards, rather primitive.

The work was not intended for application in South Asia or any other region of the world; its purpose was to increase yields and to combat the rust which was plaguing wheat cultivation in Mexico at the time. Between 1945 and 1949, four rust-resistant varieties were developed which were widely planted. In later years, semidwarf wheats were developed by crossing the new rust-resistant varieties with Japanese dwarf wheat, *Norin* 10, and the new semidwarf varieties were released for cultivation in the early 1960s.

The seed was sent from Mexico to Pakistan and India in 1962. The purpose of sending seed to India was to screen these wheats against rust. In India the importance of the semidwarf wheats was first realized by Dr. M. S. Swaminathan in a small plot located at one corner of his nursery at the Indian Agricultural Research Institute, New Delhi. In Pakistan, too, the potential importance of the semidwarf wheats was first recognized in a small patch of land at the experiment station at Lyallpur.

Subsequently, an all-out effort was made to transfer these varieties and the related technology from the semitemperate zone of Mexico to the semitemperate zone of South Asia.

Rice

The first modern high-yielding rice variety was developed not by IRRI but in Taiwan. Called Taichung Native 1, it was a short-season, semidwarf variety. The IRRI scientists, recognizing the importance of Taichung Native 1, distributed it to several tropical areas of the world during the mid 1960s. The first high-yielding IRRI rice variety was IR-8; since its appearance several other IRRI varieties have been released. Although none of these newer varieties represent any improvement in yielding ability over IR-8, they do offer improved grain quality, disease resistance, and, to some extent, insect resistance.

It is interesting to note that the dramatic increases in wheat and rice yields which accompanied the green revolution have been matched or exceeded in several developed, temperate zone countries. For example, in 1972 the average *wheat* yield increases over the 1961-65 average in two leading developing nations were as follows: India, 64 percent; Pakistan, 43 percent. By comparison, yields rose in several developed nations as follows: France, 56 percent;

Table 19-1. Area, Production, and Average Yield of Wheat in Some Important Wheat-Growing Countries

Country	Years	Area (in 1,000 ha)	Production (in 1,000 MT)	Yield per Hectare (in 100 kgs)	Yield Increase (%)
Bulgaria . . .	1948-52	1,432	1,776	12.4	
	1961-65	1,222	2,213	18.1	46[a]
	1972	961	3,582	37.1	105[b]
France	1948-52	4,264	7,791	18.3	
	1961-65	4,265	12,495	29.3	60[a]
	1972	3,958	18,123	45.8	56[b]
India	1948-52	9,290	6,087	6.6	
	1961-65	13,402	11,191	8.4	27[a]
	1972	19,139	26,410	13.8	64[b]
Pakistan . . .	1948-52	4,218	3,685	8.7	
	1961-65	4,984	4,152	8.3	– 5[a]
	1972	5,797	6,890	11.9	43[a]
United States	1948-52	27,756	31,065	11.2	
	1961-65	19,432	33,040	17.0	52[a]
	1972	19,135	42,045	22.0	29[b]
U.S.S.R. . . .	1948-52	42,633	35,759	8.4	
	1961-65	66,622	64,207	9.6	14[a]
	1972	58,492	85,950	14.7	53[b]

Source: FAO Production Yearbook, 1972.
[a] Increase in yield over 1948-52.
[b] Increase in yield over 1961-65.

Table 19-2. Area, Production, and Average Yield of Rice in Some Important Rice-Growing Countries

Country	Years	Area (in 1,000 ha)	Production (in 1,000 MT)	Yield per Hectare (in 100 kgs)	Yield Increase (%)
India	1948-52	30,092	33,383	11.1	
	1961-65	35,587	52,752	14.8	33[a]
	1972	36,019	57,950	16.1	9[b]
Indonesia. . .	1948-52	5,876	9,441	16.1	
	1961-65	7,036	12,393	17.6	9[a]
	1972	7,983	18,031	22.6	28[b]
Japan	1948-52	2,996	12,736	42.5	
	1961-65	3,281	16,444	50.1	18[a]
	1972	2,581	15,281	59.2	18[b]
Philippines	1948-52	2,350	2,767	11.8	
	1961-65	3,147	3,957	12.6	7[a]
	1972	3,112	4,415	14.2	13[b]
Thailand . . .	1948-52	5,211	6,846	13.1	
	1961-65	6,394	11,267	17.6	34[a]
	1972	6,571	11,660	17.8	> 1[b]
United States	1948-52	752	1,925	25.6	
	1961-65	705	3,084	43.7	71[a]
	1972	736	3,875	52.7	21[b]

Source: FAO Production Yearbook, 1972.
[a] Increase in yield over 1948-52.
[b] Increase in yield over 1961-65.

United States, 29 percent; the Soviet Union, 53 percent (Table 19-1).[1] Similarly average *rice* yields per hectare, during the same time period, increased as follows in developing nations: India, 9 percent; Indonesia, 28 percent; Philippines, 13 percent; and Thailand, 1 percent. By comparison, yields increased 18 percent in Japan and 21 percent in the United States (Table 19-2).

Maize and Sorghum

During the past forty to fifty years, vast improvements in maize and sorghum yields have taken place in almost all temperate countries and a few tropical countries where hybrid seed of these two crops is used.

The introduction of hybrid maize to Western Europe after World War II and to Eastern Europe in the late 1950s and early 1960s revolutionized maize production in Europe, as it had in the United States before World War II.[2] In Bulgaria, France, Italy, Romania, and Yugoslavia, the national average yields in 1961-65 were substantially above the 1948-52 average levels. The same was true of 1972 yields as compared with the 1961-65 level (Table 19-3).[3]

The transfer of grain sorghum hybrids from the United States to Mexico, Argentina, Australia, and South Africa took place in the late 1950s and early

Table 19-3. Area, Production, and Average Yield for Maize in the United States and in Major European Countries Where Hybrids Are Used

Country	Years	Area (in 1,000 ha)	Production (in 1,000 MT)	Yield per Hectare (in 100 kgs)	Yield Increase (%)
Bulgaria . . .	1948-52	737	720	9.8	
	1961-65	632	1,601	25.3	158[a]
	1972	689	2,974	43.2	70[b]
France	1948-52	332	452	13.6	
	1961-65	914	2,760	30.2	122[a]
	1972	1,880	8,190	43.6	44[b]
Italy	1948-52	1,253	2,306	18.4	
	1961-65	1,108	3,633	32.8	78[a]
	1972	892	4,802	53.8	64[b]
Romania . . .	1948-52	3,089	2,495	8.1	
	1961-65	3,308	5,853	17.7	118[a]
	1972	3,197	9,817	30.7	74[b]
United States	1948-52	29,856	74,308	24.9	
	1961-65	22,933	95,561	41.7	67[a]
	1972	23,237	141,568	60.9	46[b]
Yugoslavia	1948-52	2,297	3,078	13.4	
	1961-65	2,474	5,618	22.7	69[a]
	1972	2,383	7,940	33.3	47[b]

Source: FAO Production Yearbook, 1972.

[a] Increase in yield over 1948-52.

[b] Increase in yield over 1961-65.

Table 19-4. Area, Production, and Average Yield for Sorghum in the United States and Other Important Countries Where Hybrids Are Used

Country	Years	Area (in 1,000 ha)	Production (in 1,000 MT)	Yield per Hectare (in 100 kgs)	Yield Increase (%)
Argentina	1948-52	77	73	9.5	
	1961-65	856	1,359	15.9	67[a]
	1972	1,564	2,502	16.0	> 1[b]
Australia . . .	1948-52	57	75	13.3	
	1961-65	154	228	14.8	11[a]
	1972	639	1,228	19.2	30[b]
Mexico	1948-52				
	1961-65	205	452	22.1	
	1972	965	2,593	26.8	21[b]
South Africa	1948-52	283	180	6.4	
	1961-65	296	295	10.0	56[a]
	1972	380	556	14.6	47[b]
Spain	1948-52	5	4	7.3	
	1961-65	9	19	21.1	189[a]
	1972	44	177	40.2	91[b]
United States	1948-52	3,087	3,897	12.6	
	1961-65	4,909	13,912	28.3	125[a]
	1972	5,410	20,556	38.0	34[b]

Source: FAO Production Yearbook, 1972.
[a] Increase in yield over 1948-52.
[b] Increase in yield over 1961-65.

1960s. This introduction of better yielding United States-bred hybrids revolutionized grain sorghum production in each of these countries.

Before the introduction of hybrids into Mexico, total sorghum grain production in Mexico was negligible, whereas in 1972 over 2.5 million tons of sorghum grain were produced (Table 19-4). Furthermore, Mexico recorded national average yields comparable to those in the United States. Mexico did not make significant improvements in average yields in subsequent years, however, whereas average yields in the United States kept going up owing to considerable improvement in farming practices.

Striking increases in production also took place in the Southern Hemisphere (Table 19-4). In Argentina over 2½ million tons of sorghum grain were produced in 1972 as compared with less than 100,000 tons before the introduction of hybrid sorghum. From 1948-52 to 1961-65 yields increased significantly in Argentina and in South Africa. Because of its strict quarantine requirement, Australia was four to five years behind in utilizing the United States hybrids, and, as a consequence, improvement in average yield in 1961-65 in that country was more modest.

Some of the European countries which can grow sorghum recorded even higher gains than the countries in the Southern Hemisphere.

Genetic Supply Industry

The genetic supply industry has, to a great extent, been responsible for bringing about these increased yields by developing, multiplying, and distributing hybrid seed to the farmers.

The scientific basis for hybrid maize and hybrid sorghum, as well as for a host of other hybrid plants and animals, is the phenomenon known as "heterosis" or hybrid vigor. Heterosis can be defined as the increased vigor occurring in the progeny of crosses among inbred parents, varieties, or races. It has been said that this phenomenon, more than any other, has revolutionized the agriculture of the United States.

The genetic supply industry has attempted to exploit this phenomenon to the greatest extent possible within the existing body of scientific knowledge on the subject. There are two primary reasons for this interest.

First, until the United States Plant Variety Protection Act was promulgated in 1972, there was no law to protect the varieties produced by private breeders of open pollinated varieties. The hybrids, because of their built-in protection, offered security to private breeders. In other words, the hybrids offered what is known in the trade as "proprietary" varieties.

Second, since hybrids had superior performance over the open pollinated varieties, the seed could be sold to the farmers at a price which assured breeders far greater and far more certain profits than if they were sold the seed of commonly available varieties. Since the true hybrids are available only through the original breeder or his distributor, the breeder has repeat customers year after year if the product performance is satisfactory to the farmer. The combination of these two factors – the large benefits that accrue to the users of genetically improved seed stock and the proprietary nature of many seed stocks – made the genetic supply industry flourish in the United States.

Until recently, the overseas research and development work of the United States industry has remained more or less limited to the transfer of temperate varieties and technology to temperate areas of the world. For example, all the major United States seed companies are active in Western Europe as well as in other temperate areas of the Southern Hemisphere.

Although there has been a great deal of interest on the part of the genetic supply industry in contributing its know-how to the developing countries, few efforts to accomplish this have been made to date.

Pioneer Hi-Bred International is unique in this respect. In 1964, it established the Tropical Research Station in Jamaica, West Indies, primarily to develop maize hybrids adapted to the lowland tropics. The station is located in the lowlands, at 18° N latitude.

Improved populations and varieties, some of which were collected by Dr.

W. L. Brown while he was a Fulbright scholar at the College of Tropical Agriculture, St. Augustine, Trinidad, were used as the station's foundation stock and source-breeding materials. Also, several improved breeding populations were received through the courtesy of the Rockefeller Foundation and CIMMYT in Mexico.

The classical inbred-hybrid method of maize improvement was used to develop hybrids, this method having previously been successfully employed by breeders, both commercial and private, in the United States.

The work on maize was carried out during the period from 1964 to 1968 on an average of ten acres of land and with an annual budget of less than $15 thousand. (The station's total annual budget was $30 thousand, a little over half of which was spent on sorghum research.) In 1966-67 the first experimental hybrids were entered in tests throughout the Caribbean and Central America, and, in 1968, after only four years of operation, the station released for commerical cultivation in the Caribbean and Central America two yellow hybrids, X304 and X306. Two years later, it released two white hybrids, X101A and X105A, for commercial use.

Table 19-5 shows the performance of the yellow hybrids in several Caribbean countries in 1968-69. These hybrids gave significantly higher yields than local varieties. In some instances, the yield increases were more than double.

Trials were also conducted in parts of Central America, West Africa, East Africa, and South East Asia. As in the Caribbean, the hybrids from Jamaica were among the highest yielding hybrids in the tests.

It is a well-known fact that, regardless of the success of breeders in breeding improved varieties and hybrids, the impact of such developments on agricultural production is zero unless the same varieties reach the farmer, not only in quantity but in a state which maintains the original genetic potential of the variety.

Among the cereals, the nature and requirements of seed production and distribution vary greatly depending upon the species involved and its mode of reproduction. In wheat and rice, both of which are self-pollinated species, the multiplication of seed is a rather simple process. All that is needed to reproduce a variety in large quantity is simply to grow and harvest the crop while exercising care to maintain varietal purity by avoiding mechanical mixtures of seed. However, in the case of maize, which is almost completely cross pollinated, and sorghum, which is partly cross pollinated, the situation is entirely different. To maintain purity of hybrids, fertilization must be controlled; consequently, the methods of large-scale production of quality seed are much more sophisticated than are those used with self-pollinated species.

There is little doubt that the large-scale, successful introduction of high-yielding varieties of wheat and rice into several developing countries in recent

Table 19-5. Performance of Hybrid Corn versus Local Corn in the Caribbean in 1968-69

Variety	Yield (in quintals per hectare at 15.5% moisture)						
	Jamaica	Haiti	Dominican Republic	Grenada	Barbados	Trinidad	Guyana
Hybrid X306	54.7	59.6	63.4	39.2	30.6		45.2
Hybrid X304	56.2	44.1	55.9	35.2	40.4	54.2	47.7
Local Variety . . .	24.9[a]	32.9[b]	38.9[c]	17.1[d]	19.1[e]	27.9[f]	30.1[g]

[a] Jamaica selected yellow.
[b] Jeremie.
[c] Frances.
[d] Grenada corn.
[e] Barbados corn.
[f] Economic botany selection.
[g] Charity.

years was, to a considerable extent, a result of the reproductive mechanism of the species. On the other hand, there is little doubt that the failure of the international maize program of the institutes is largely attributable to a failure to develop satisfactory systems of seed production and distribution. The quality of breeding that has gone into international maize improvement programs is comparable to that characterizing the self-pollinated cereals, yet the impact of increased productivity has been negligible compared with that of wheat and rice.

As a profit-oriented private company, we realized that the breeding of new hybrids and the multiplication and marketing of seed must be closely coordinated in order to bring the results of research to the farmers in the shortest possible time. And since research must be financed from profits, the seed must be sold at a price that provides adequate profit to sustain research.

With these objectives in mind, we established modest seed production and distribution facilities, first in the Caribbean, in 1968-69, and later in Central America, in 1970-71. Our four years of experience in marketing hybrid seeds in Central America have shown the following:

1. If a hybrid performs, farmers will buy the seed year after year. And, contrary to widespread belief, small as well as large farmers can and will buy hybrid seed: anywhere from 10 percent, in Panama, to 80 percent, in Nicaragua, of our hybrid seed customers are small farmers.

2. There is no doubt that hybrids are bred to take advantage of better than average farming conditions. However, even under average growing conditions they do perform significantly better than the local varieties, thus contributing to increased yields at the farm level.

With our tropical research and distribution of seed well under way in Central America, we extended our activities to other developing areas of the world, including Brazil, India, and, more recently, the Philippines. The results obtained in each country are similar to those obtained earlier in Central America and the Caribbean. In Brazil, we have two yellow maize hybrids under seed production, and seed is being successfully marketed there.

In India, we started with limited seed production in 1974, and we are scheduled for greater production in 1975-76. The hybrids being produced in India are X102 and X104. These were developed at our Hyderabad Research Station from inbred lines supplied by the parent company from the United States and Jamaica. Tests conducted in Andhra Pradesh during the 1973-74 season revealed that these varieties outyielded the best available local hybrids (Ganga-5 and Deccan) by 17.4 to 31.0 percent at each of nine locations.

All our overseas research stations undertake research on hybrid sorghum as well as hybrid maize, and most of our overseas producer-distributor organizations which produce and market hybrid corn seed are also marketing hybrid sorghum seed.

In this chapter, I have attempted to point out what can be done in crop improvement research with limited resources of manpower, capital, and land. The approach of our company has been what is called a "small experiment station" approach. The philosophy behind this method is that described by Wallace and Brown in their book *Corn and Its Early Fathers*.[4] As they have put it:

> [T]here are dozens of plantbreeders who can point to the fact that when they were living very close to their plants, seeing them every day, and spreading attention thickly over a small area, they got many times greater a return per hundred square feet than they did when working with large numbers of plants covering acres of land.
>
> The modern trend in *plant breeding* is in exactly the opposite direction. The present emphasis is directed toward doing things in a big way, toward the use of large numbers and "coordinated research" . . . toward the use of large areas of land, and in many cases, routine types of investigation and thought. The work accomplished is often measured in terms of budget size, of the number of pollinating bags used, or the number of acres devoted to yield testing . . .
>
> The point we are making is that lots of land, equipment and power can never produce scientific advancement in plant breeding or anything else unless ideas are big enough to match. And unfortunately, when the equipment, land and manpower pass a certain point of immensity, the men who are supposed to do the scientific thinking tend to become mere administrators, making the wheels go around, keeping records, compiling tables, but not thinking often enough or hard enough about the next fundamental step forward.

The early HYV's of wheat and rice were developed when research facilities were, by modern standards, rather "primitive."

Conclusions

The goals of research work at the international institutes and in the genetic supply industry are the same: that is, to help farmers to produce more per unit of land area than they are now producing. In our company there is a common saying: "What is good for the farmer is good for us"; therefore we have a strong interest in the farmer's welfare. However, the decisions that set the scope and direction of private research are motivated by a strong vested interest in contrast to those which allocate and direct research at the institutes.

There are several areas in which some sort of relationship between the institutes and the private sector already exists and in which exchange of information has been occurring. However, there is a need to strengthen this relationship considerably. For example, one of the areas in which cooperation between the two could be expanded is in the exchange of breeding materials. It is well known that the effectiveness of a breeding program is, to a large degree, dependent upon the extent to which elite breeding materials are present in a breeder's nursery. It is therefore highly desirable to establish an exchange of breeding materials between the two on a more or less regular basis.

Another area in which the private sector and the institutes can expand their cooperation is in variety evaluation work. At present, there are several regional test programs in which institute scientists are directly or indirectly involved, for example, the Central American Cooperative Project for the Improvement of Basic Food Crops (PCCMCA) in Central America, the Inter-Asian Corn Program (IACP) in Asia, the East African Maize Variety Trials (EAMVT) in East Africa, and the West African Uniform Maize Trials (WAUMT) in West Africa. The private sector could benefit greatly if it could have its varieties/hybrids evaluated and obtain meaningful information through the worldwide contacts of the institute scientists.

Seed production and distribution form another area in which cooperation between the institutes and the genetic supply industry can play a significant role in increasing the productivity of farmers.

The common method of wheat and rice improvement in the temperate countries is, and has been, hybridization and selection. In the late 1950s and early 1960s there was a general trend to breed short, early, fertilizer-responsive varieties of several crops, including important cereals. CIMMYT and IRRI scientists successfully exploited these techniques to evolve high-yielding varieties.

For maize and sorghum, however, the most widely and successfully employed method of improvement is the inbred-hybrid method. In other words, improvement of these two crops is based upon exploitation of controlled heterosis. Much of the maize breeding now being done by the international institutes, however, emphasizes the development of improved synthetics and populations and largely ignores the use of hybrids. The reason most often given for this approach to breeding is that farmers in most developing countries do not employ sufficiently advanced technology to make good use of hybrid seed. As we have pointed out earlier, we have not found this to be true. A more realistic reason is the failure of the institutes to work out satisfactory systems of seed production and distribution of hybrid varieties in the developing countries. Instead of making an all-out effort to resolve the problem of seed production and distribution systems, the institutes have discontinued work on an effective and proven breeding methodology.

The hybrid approach to improving corn and sorghum offers tremendous flexibility because a hybrid is the product of at least two parents in the case of a single cross, three parents in a three-way cross, and four parents in a double cross. By combining lines (parents) with different qualitative traits with regard to disease and insect resistance and the like, one can have great diversity among the hybrids which are put on the market.

Regardless of the methods employed, the objectives of the international institutes and the private sector, as mentioned earlier, are the same – that of improving agricultural productivity. There is every reason, therefore, for increased cooperation between the two groups. Both are now engaged in varietal improvement, and both will no doubt continue this activity. It would seem that the institutes are much better prepared than the private sector to provide the extension service so badly needed in the Third World. The private sector, on the other hand, is much better equipped to provide those services associated with seed production and distribution.

Agricultural research need not always be expensive in manpower, capital, or land. Increased cooperation between public and private sectors can further reduce these costs and increase efficiency.

NOTES

1. In addition it may be noted that yields rose 67 percent in Hungary and 64 percent in Romania.

2. United States-bred hybrid and/or inbred lines. The most widely spread United States hybrids were W240, W255, W275, W355A, W416, W464 among the early maturity group; Iowa 4417, Wisconsin 641AA, Nebraska 301 among the medium maturity group; and U.S. 13, Ohio C92, and Kansas 1859 among the late maturity group.

Several United States inbred lines were used extensively in hybrid combinations with the European lines. These are WF9, 38-11, C103, Hy, OH43, OH7, OH51, W22, K148, K150, N6, and M14 for the Danube Plain and other regions of southeastern Europe. For the northern European production zone, early maturity lines such as W153, W37A, W19A, W41A, W59E, W9, WD, A374, A375, M13, and 1205 were commonly used.

3. Moreover, yields rose substantially in both periods in Czechoslovakia, Hungary, and Spain.

4. H. A. Wallace and W. L. Brown, *Corn and Its Early Fathers* (East Lansing: Michigan State University Press, 1956).

20

A Systems Approach to Agricultural Research Resource Allocation in Developing Countries[1]

Per Pinstrup-Andersen and David Franklin

Effective agricultural development requires the interaction of farmers and rural institutions working within constraints imposed by their socioeconomic and ecological environments. For the effective allocation of their scarce human and financial resources, institutions such as those involved in public agricultural research must take into consideration the needs of farmers as well as overall national, social, and economic goals. Decision-making on agricultural research resource allocation has received less attention than has farmer decision-making. As a consequence, there is a shortage of both useful data and effective techniques for their analysis.

While improved productivity and increased production may be the immediate goals of applied agricultural production research, they are at the same time the means to reach some final goals such as improved human nutrition, a more equitable income distribution, and increased foreign exchange earning. Agricultural research institutions are presumed to seek ways to produce more and/or better food, feed, and fiber at a reduced per unit cost and in such a way as to maximize the contribution of agriculture to the achievement of ultimate social and economic goals. Hence, there is a need for effective means to assist in predicting the relative contributions and costs of alternative research activities in order to establish research priorities and allocate available research resources.

This chapter suggests a systems approach to the collection and analysis of information expected to be useful for establishing such means.[2] The first part offers a brief discussion of means and ends in agricultural research. This is followed by an outline of a scheme for data collection and analysis; the chapter terminates with a discussion of some of the information-generating efforts currently under way in CIAT.

Means and Ends in Agricultural Research

A clear understanding of the distinction between final and immediate goals on the one hand and means to reach these goals on the other is essential to appreciate fully the need for improved tools for research management and to assure that such tools are relevant for establishing research priorities. For example, while increased production may be an immediate research objective, it is not a final goal of agricultural research but rather a means to reach some final goals such as improved income distribution or improved nutrition. In a similar fashion, improved income distribution, although it may be a final development goal, does not serve as a working objective for the agricultural scientist.

To help clarify the distinction between means and goals, Figure 20-1 outlines the process by which applied agricultural production research may contribute to the achievement of social and economic goals. Successful applied agricultural research produces knowledge and/or improved material, e.g., seed. The knowledge and improved material may be fed back into the research process for further work, or it may be released to the farmers as new technology. There are three – and only three – potential direct contributions of such technology: (1) increasing *technical efficiency*, a measure of output per unit of input where both output and input are expressed in physical terms (e.g., production per hectare) of at least one resource; (2) changing the *characteristics and composition of products* and developing new products (e.g., developing plant types more suited to mechanization and improving the amino acid composition in the protein of a given crop); and (3) reducing *production risk*. Any other contribution will be indirect, that is, it must come about as a consequence of one or more of the three direct contributions.

There are three potential results of the direct contributions listed above: (1) changing the composition and quantity of the aggregate supply of food, feed, and fiber; (2) changing the composition and quantity of the aggregate resource demand, e.g., increased or decreased employment; and (3) changing the composition and quantity of aggregate domestic farm consumption. Any of these results may contribute to the achievement of national development goals through changes in elements such as farm income and its distribution among groups of farmers, relative resource earnings, consumer real income

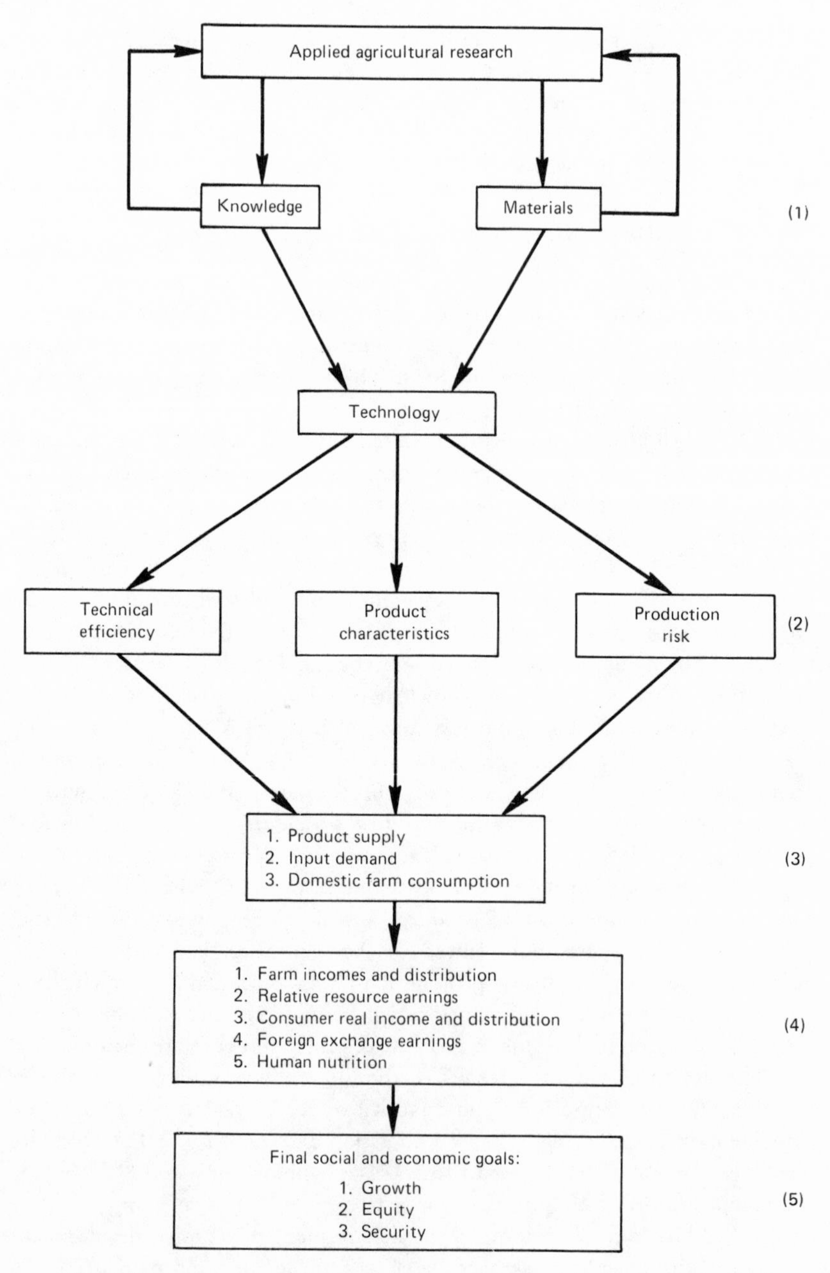

Figure 20-1. Illustration of the potential outcomes and implications of agricultural research.

and its distribution among consumer groups, foreign exchange earnings, and human nutrition.

Viewing agricultural research and its potential outcomes and implications as a process reduces the confusion over means and ends. The first level of outcomes (marked by (1) in Figure 20-1) is clearly a set of means, except when research is carried out for its own sake. The second level represents the working objectives for the agricultural production scientist. For research management and society as a whole, however, this level expresses alternative approaches to the goals shown in the fourth level. The third level in Figure 20-1 represents the vehicle by which activities meeting the scientist's working objectives influence the achievement of the final goals. In other words, changes in product supply, input demand, and domestic farm consumption are not themselves goals but are means to final goals.

Two conclusions may be drawn from the discussion above. First, the working objectives for the agricultural production scientist must be expressed in terms of technical efficiency, desired product characteristics, and/or production risk. The specific working objectives and the most effective technology to reach these objectives should be determined on the basis of national development goals. Concurrence between the technology specification received by the scientist and the technology which results in maximum contribution to the achievement of social goals is the responsibility of research management.

Second, research management needs information for research resource allocation that is capable of both translating national development goals into working objectives for the agricultural production scientist and helping the production scientist select the most effective technology to reach the working objectives.

A Suggested Information System

An effective information system for the allocation of resources in applied agricultural production research must be capable of providing research management with reliable data that will make possible the establishment and the periodic review of research priorities in such a way as to maximize the expected contribution from research to the achievement of national development goals. The system should also provide a frame of reference within which project priorities can be established and individual projects can be accepted or rejected without great time delays. Extreme care must be taken to avoid a system that imposes heavy bureaucratic procedures on the production scientists.

The system should be sufficiently comprehensive to improve currently available methods. However, the decision on how much should be spent on

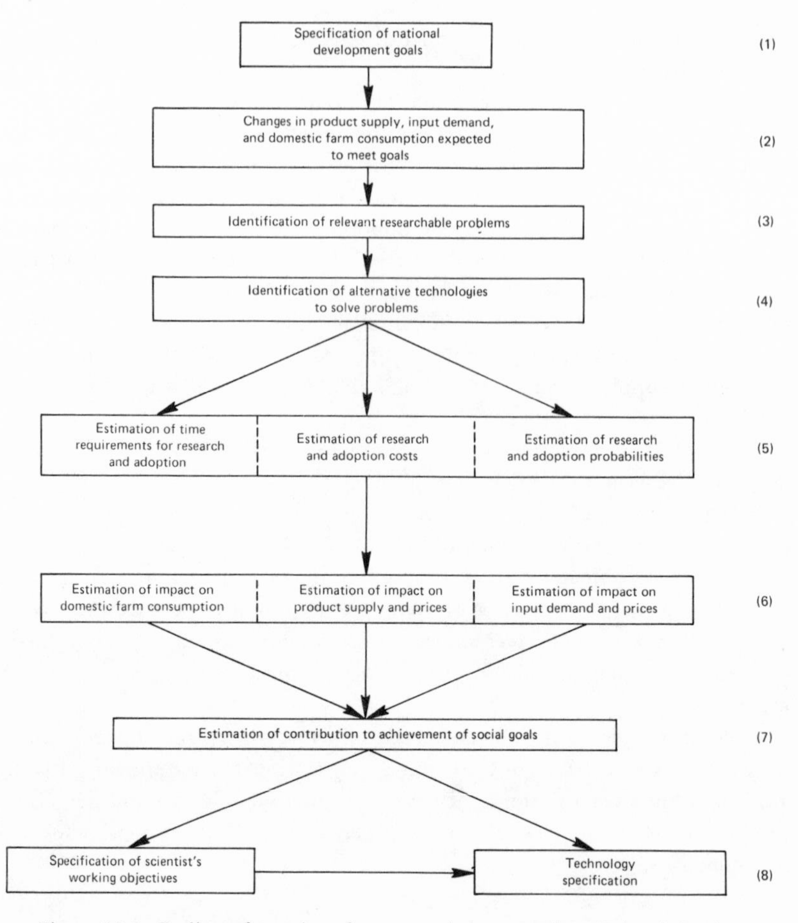

Figure 20-2. Outline of a series of steps needed to translate national development goals into working objectives and technology specification.

achieving such a system must be based on the same principles as those used to allocate resources among alternative agricultural research activities.

Before the data requirements and the conceptual model are discussed, it may be useful to illustrate a series of steps necessary to translate national development goals into working objectives for use by the scientist and in technology specifications. The illustration is shown in Figure 20-2.

It is essential that the development goals be clearly specified. The changes in product supply, input demand, and domestic farm consumption expected to meet some or all of these goals should be identified. Then the researchable problems, the solution of which is expected to accomplish such changes, must be identified. At this point no attempt should be made to quantify the expected contributions to development goals.

Let us assume, as an example, that one of society's goals is to increase protein intake among protein-deficient groups of the population. It may be expected that – among other activities – increased production of grain legumes, animal products, and high protein cassava may make a contribution. The researchable problems limiting production of these commodities, e.g., a particular disease in field beans, the nonavailability of a high protein cassava variety, etc., should then be identified.

It is important that the problems limiting the achievement of established objectives be identified independently of possible solutions, i.e., a "technology-free specification of the problem" should be outlined. For example, if the problem is one of low yields, it should be expressed in terms of the factors causing low yields, such as lack of insect resistance, rather than specified as a problem of developing an insect-resistant variety. This is because alternative solutions to the problem of the lack of insect resistance do exist. As such, the technology-free specification of the problem provides an implicit measure of the potential value of assembling technology to solve a particular problem. The technology-free specification of the problem has to identify the farmer's needs and convert these needs into a specification of the parameters and constraints that must be satisfied by the technological innovations.

When the relevant researchable problems are identified, the alternative technologies expected to solve the problems should be specified. Then the cost, probability, and time requirements of both research and farm adoption should be estimated for each proposed technology. Based on these estimates, as well as on the nature of the problem, the structure and performance of the production sector, and the input and product-market relationships, it is now possible to estimate the impact of solving each of the problems on product supply, input demand, and domestic consumption. The last step before specifying the scientist's working objectives and the technology to be developed refers to a quantitative estimation of the contribution of alternative research efforts to the achievement of national development goals.

Data Requirements and Sources

From the broad framework presented above, it is now possible to specify the data requirements and the possible sources of these data. An exact specification of data requirements is not attempted.[3] Four sources of data are discussed: the farm sector, the market sector, the research sector, and the government.

Farm sector data. Allocation of resources in applied agricultural research is frequently made without sufficient knowledge about the existing problems and their relative economic importance in the production process. Communication between the farm sector and the research institute is often deficient, and the needs at the farm level for problem-solving research may not be well known by the researchers.

Farmers in most developing countries—except perhaps those who maintain large commercial operations and/or are members of efficient producer associations—tend to have severe difficulties in communicating their research needs to the research institutes because of institutional and social barriers. Because of this situation, some research may be irrelevant to actual farm problems, and research results may not be adopted.

There is urgent need for a system that will provide a continuous flow of information to the production scientists and other persons who make decisions on the increase in production, productivity, and risk likely to result from such research activities as developing resistance to specific diseases and insects, improving cultural practices, improving plant types, and changing plant responses to nutrients. Furthermore, information is needed on the farmers' preferences with respect to new technology and on how these preferences may be changed so that attention can be given to the development of technology with a high probability of adoption.

Such a system can easily be developed where there is a continuous feedback of information from the farmer through the extension service to the research agency. Unfortunately, such feedback is rare in developing countries, and it is not likely to take place on a national scale in the very near future. In the meantime, the essential information can probably best be obtained through organized surveys, including field observations. In addition to surveys, it may be necessary to carry out controlled experiments to determine the degree to which each of the various researchable problems leads to a reduction in yield. While field surveys will provide information on the area affected by each of the researchable problems and some indication of their yield-depressing impact, controlled experiments on yield losses will provide more exact information on yield-reducing effect. Together the two data sources offer a sound basis for estimating production and productivity impact of research on each of the problems specified. The impact of research on risk can be estimated from survey data on the past occurrence and severity of problems (e.g., pests, climate, etc.) and the resulting yield variances.

Market sector data. Information on the structure and performance of product and input markets is essential to predict the contribution of alternative research efforts to the achievement of development goals.

Existing and expected future product-demand relationships may be very unfavorable to the expansion of the supply of certain commodities while favorable to the expansion of others. In this regard, demand elasticities are needed to estimate the expected impact on prices and on the distribution of benefits between producers and consumers. In the case of new products or drastic changes in traditional products, it is important to predict consumer preferences either before research is initiated or at as early a stage in the research as possible. Although a certain change in a traditional product may make it "better," using some objective measure such as nutritional value, the consumer may find it less acceptable than the original product. A number of cases could be cited where "good" products have been developed through research, only to find that they were unacceptable to the consumer. Had the consumer preferences been checked out at an earlier stage, a considerable amount of research resources might have been saved.

Instead of allocating research resources to fit existing product market relationships, it is frequently possible to change the market relationships to fit the research results. For example, consumer preferences may be changed or new markets may be found. It is important to predict how these relationships would behave in the case of supply expansion if adequate public policy measures aimed at facilitating the necessary changes are to be recommended.

The impact of new technology on input demand will depend on the particular technology developed. Hence, before the decision is made on the type of technology to develop, information should be obtained on existing and expected future input-supply relationships.

Research sector data. Data are needed to estimate the costs and the time requirements of research as well as the likelihood of achieving desired results. Because of the very nature of research, its outcome can rarely be predicted with great precision. It is argued here, however, that efforts to achieve at least some crude predictions of outcomes, on the basis of existing scientific knowledge, are likely to make resource allocation in applied agricultural research considerably more efficient.

Government sector data. Development goals may be classified under three general headings: growth, equity, and security. Although specific development goals may differ considerably among countries, all three of these general goals are usually found in some form.

The development goals must be clearly defined and, if possible, the socially acceptable trade-offs among them should be specified.

At present, research management tends to have very limited information on these issues, and research priorities tend to be based exclusively on the objectives of increasing production and productivity.

The Conceptual Model

Figure 20-3 is a conceptual model for an information system for resource allocation in applied agricultural research. The figure outlines the relation-

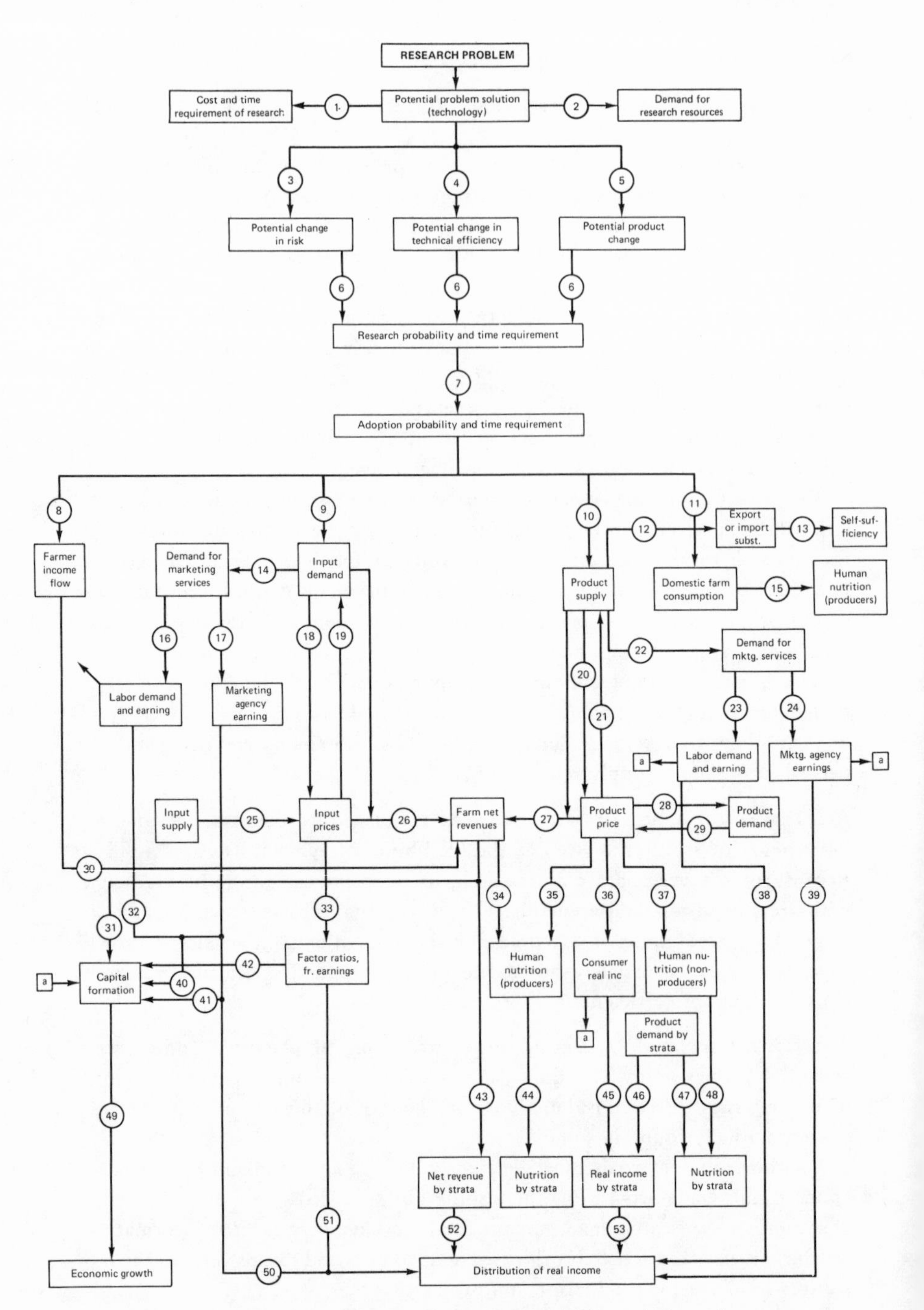

Figure 20-3. Flow diagram for an analytical model for an information system for resource allocation in applied agricultural research.

ships determining the expected contribution of alternative research efforts to the achievement of selected development goals. It also outlines some of the implicit relationships we believe should be considered when decisions are made regarding resource allocation in applied agricultural research. It is not suggested here that a quantitative model incorporating all these relationships be constructed. Rather, what is intended by presenting the model is to make explicit the entire range of relationships, so that when a particular subset of relationships is analyzed the assumptions about the excluded relationships are made explicit.

The following social goals are considered in the model: (1) economic growth; (2) more equitable income distribution; (3) increased productive employment; (4) increased net incomes to small farmers; (5) a more even cash flow to farmers; (6) improved human nutrition; (7) higher degree of self-sufficiency in basic foods; and (8) increased foreign exchange earning. The model may be changed to accommodate a different set of goals. Implicit in each numbered line is a causal relationship between change in one variable and change in another.

The contribution of new technology to the achievement of development goals depends heavily on existing public policy. Hence, existing policy should be clearly specified, and it may be useful to apply the model to allow for alternative policy measures.

Selected CIAT Activities

The remainder of this chapter discusses some of the recent CIAT efforts aimed at developing and field testing simple methodologies for generating the information discussed above. Although the information obtained from these efforts is expected to be useful for CIAT and the national research agencies in the countries where the empirical testing is carried out, the primary purpose of the work is to develop simple methodologies for use by national research agencies in Latin America.

The CIAT work is discussed under three headings: single commodity analyses; multi-commodity analyses; and a systems engineering methodology for small farms. The discussion is limited to selected illustrative projects. A description of all the CIAT activities in this area is beyond the scope of this chapter.

Single Commodity Analyses

This type of work is relevant when a decision has been made to research a specific commodity either indefinitely or for a certain minimum time period. Although the amount of research resources allocated to a certain commodity may be gradually increased or decreased over time, the low mobility of research resources may not permit rapid and large changes in the relative emphasis given to research on particular commodities. Hence, the single commodity analysis may be appropriate, at least for the short run.

In the case of a single commodity, information is needed on the commodity itself as well as on its interaction with other commodities in both production and consumption. The current CIAT single commodity data collection and analysis focus on the farm sector.

The single commodity approach attempts to identify the factors associated with low productivity in a specific crop. It then proceeds to (1) identify researchable problems expected to improve productivity and production, (2) estimate the impact of solving each of the problems on productivity and production, (3) estimate the costs, time requirements, and adoption probabilities of research for each problem and each technology, and (4) estimate the impact of alternative research efforts on product supply, input demand, domestic farm consumption, and farm sector income and its distribution on farm size. Such projects are currently under way for maize, cassava, and beans. Basic data are collected from agroeconomic surveys and agrobiological experiments.

Agroeconomic surveys. The agroeconomic survey attempts to transmit to production scientists and research management the farm-level demand for applied agricultural research, through establishing a direct link between the farm and the research agency. The survey describes the production process and focuses on identifying factors which limit production and productivity and on estimating their relative importance. Although highly interrelated, these factors may be classified as primarily agrobiological, socioeconomic, or institutional. Given the purpose of the survey, emphasis is placed on agrobiological and related economic factors.

Most of the data related to the *agrobiological factors* are obtained from direct observation in the farmers' fields. The occurrence and severity of disease, insect damage, and weeds are noted. Furthermore, existing cropping systems, cultural practices, soil quality, availability of water, plant type, and general plant development are described, and yields and yield variance are estimated. The farmer's perception of the agrobiological problems is compared to field observations, and an effort is made to discover his attitudes toward solutions to the problems (new technology). In this endeavor, emphasis is placed on obtaining some indication of the farmer's *objectives* – including the relative importance to the farmer of income, risk, and home consumption – to help identify technology with a high expected rate of adoption.

With respect to *economic factors*, data are sought on (1) the use of purchased inputs such as chemical fertilizers and insecticides, (2) labor use and production costs by production activity, and (3) gross and net revenues obtained from the crop.

The information sought in respect to *institutional factors* focuses on certain aspects of input and product-market relationships as well as on the availability and use of credit and technical assistance.

A small team of agronomists and economists provides the *data collection mechanism*. After having received an intensive training course in diagnosing farm-level production problems, the team visits each of a selected sample of farmers three to four times over the period of a complete crop cycle. Field collection of data on agrobiological issues takes about half the time spent on each farm, while the other half is used to interview the farmer.

Training of the field team is one of the most critical factors in assuring high quality data from the agroeconomic survey. Making a correct diagnosis in the field, for example, in distinguishing among the symptoms of certain diseases or types of insect damage, in most cases requires considerable expertise. Before initiating an agroeconomic survey, the agronomists on the CIAT field teams spend a certain amount of time with a group from each discipline represented on the relevant CIAT commodity team. This instruction is supplemented in some cases with training from professionals from national research and extension agencies. Most of this initial training takes place in the field.

Agrobiological experiments. The agroeconomic survey provides an estimate of the area affected by each of the problems identified. Furthermore, it gives an indication of the yield-depressing effect of a problem. However, it is frequently difficult to estimate, with a great deal of accuracy, the yield impact from survey data. Hence, controlled experiments are carried out to help quantify the impact of the problems on yields.

Results. The work described above is in its preliminary stages, and before the real value of these efforts for research resource allocation can be established more time is needed to terminate the first round of data collection and analysis. At this stage, however, it appears that in planning their future research CIAT agricultural production scientists have found valuable both their participation in project planning and the training and supervision of field agronomists and the information they have gained from the distribution of preliminary project findings.

Multi-Commodity Analysis

As opposed to the analysis described above, the multi-commodity approach assumes that the choice of commodities for research and the relative priority among those commodities are not determined a priori. Hence, in addition to the data collected for a single commodity, information is needed on the relative contribution to development goals of research on alternative commodities.

In this area, CIAT is currently undertaking a project whose initial objective is to develop and test a methodology to estimate the impact on human nutrition of increasing the production of each of a number of foods. The empirical testing is currently being done for the city of Cali in Colombia. In addition to the issue of the impact on human nutrition, the project provides information on the impact of alternative production expansions on consumer real income by income strata and may at a later stage be extended to include data on the impact of such expansions on farm sector incomes and distribution.[4]

The methodology is based on a simulation model using as basic data a set of price elasticity matrices (one for each of five income strata) as well as current food prices, quantities consumed, and protein and caloric intakes. The model facilitates the estimation of the impact of alternative agricultural research efforts on human nutrition. The model forms a part of the conceptual model shown in Figure 20-3, estimating the coefficients indicated in the figure by the numbers 20, 21, 28, 29, 36, 37, 45, 46, 47, 48, and 53.

A Systems Engineering Methodology for Small Farms

This approach centers on the farmer and his goals and is considered complementary to the commodity-oriented approaches discussed above. It involves the development of models for the small farm; the small farm system is one in which the farm family and others living on the farm assemble individual enterprises into production, consumption, and marketing systems in which biological and physical factors interact with social, political, and economic systems. Such systems engineering models of the small farm help to explain the dynamic behavior of the farm system as a function of its input and output relationship with external systems (the biological, ecological, and institutional environment) and make it possible to identify the agricultural technologies which will be most effective in stimulating changes in the performance of the individual farm systems. In particular, by being centered on the farm as a system, these models will, it is hoped, identify the principal limitations to the generation of well-being, income, and marketable agricultural surpluses in what we earlier called a technology-free specification of the problem, i.e., a specification of the problem independent of possible technologies for its solution. The relationships explored by these efforts correspond to the numbers 8, 25, 26, 27, and 34 in Figure 20-3.

The systems engineering methodology for small farms is currently being applied by the Small Farm Systems Program of CIAT in its collaborative

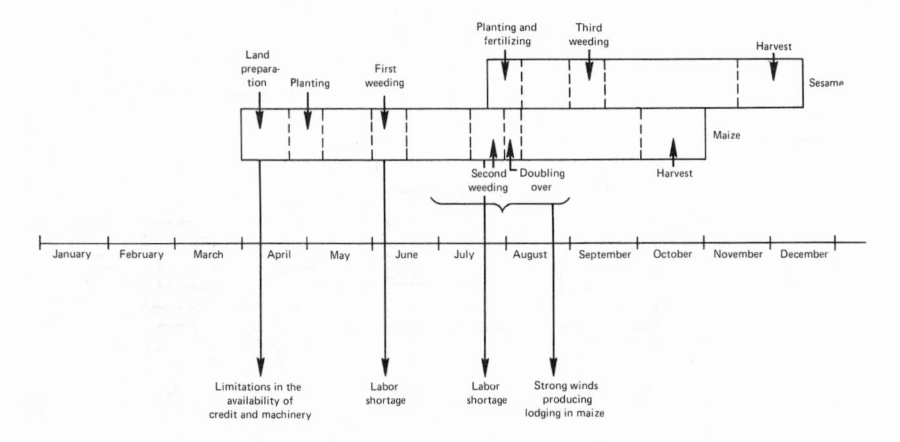

Figure 20-4. Cropping cycle on some farms in southern Guatemala.

work with the Institute of Agricultural Sciences and Technology of Guatemala (ICTA). Before discussing the expected utility of this methodology for research resource allocation, we will describe briefly the overall structure of the models.

The collaborative project is being carried out in an agrarian zone in a southern coastal region of Guatemala. Figure 20-4 shows the principal activities of the agricultural cycle for that zone. A schematic representation of a general model for the small farm system is presented in Figure 20-5, while Figure 20-6 is a reduced version currently being utilized for the study of the farm system in the zone. The behavior of the small farm system is being studied as a function of the principal inputs for the system: credit, prices, availability of machinery and labor, and climate. This is a limited set of input factors, and the principal concern at this time is to understand the behavior of the small farm system when confronted with climatological risk and the interaction of this risk with other inputs.

The farmers in this zone currently utilize almost no modern factors of production, and it is speculated that this situation is due primarily to risk aversion. Delays in the credit system and lack of confidence in the support prices create a situation in which institutional factors do not help to absorb the risk. There are serious delays in the availability of machinery, and a seasonal labor shortage exists owing to competition with the large plantations. The primary purpose of the model is to analyze whether in fact the dynamic interactions of institutional and climatological factors are the principal limitations to production and farm incomes.

The principal function of subsystem Z_1 in Figure 20-6, denominated

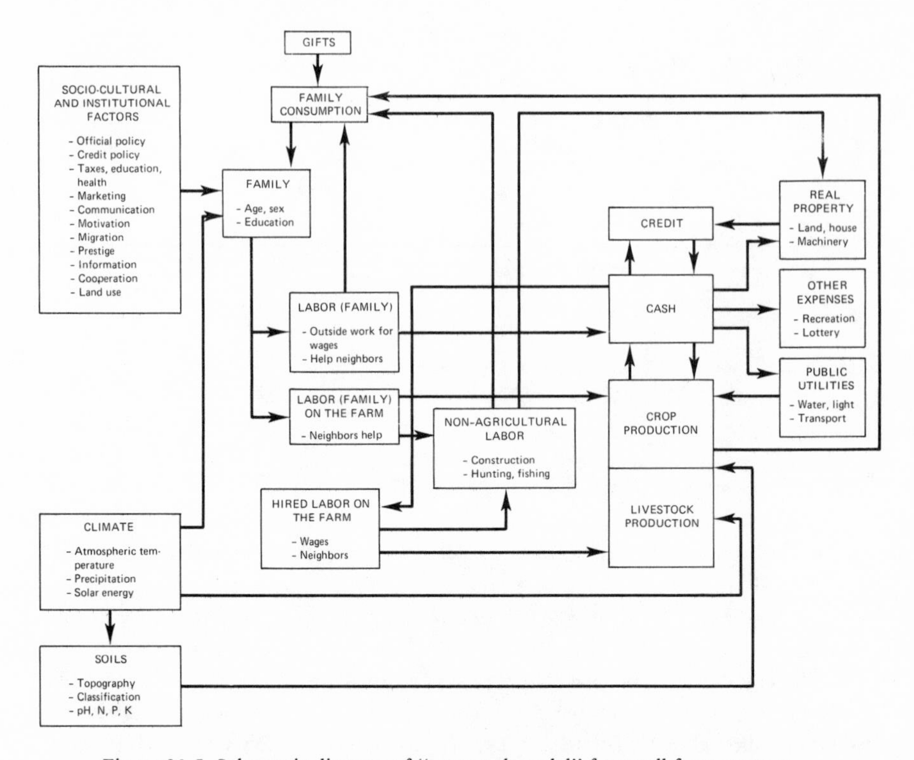

Figure 20-5. Schematic diagram of "a general model" for small farm systems.

"Cash," is to keep account of and allocate the cash flow to the different activities of the family, including the purchase of family consumption goods, factors of production, and payments to credit. It is in this subsystem that the criteria for farmers' decisions are studied.

Subsystem Z_4, "Crop production," is linked to the external inputs of machinery and climate and to the "Cash" and "Soil" subsystems. The evaluation of technological alternatives for production is carried out within this subsystem. The "Family consumption" subsystem represents the need for on-farm consumption of the various products produced on the farm as well as for the purchase of foodstuffs and nonfoods. This subsystem helps to estimate the family nutritional situation.

The technical coefficients used in the model are the best estimates on the behavior of each of the subsystems that have been provided by technical experts. The structure of the model was derived from information gathered through frequent visits to the zone by the members of the CIAT Small Farm

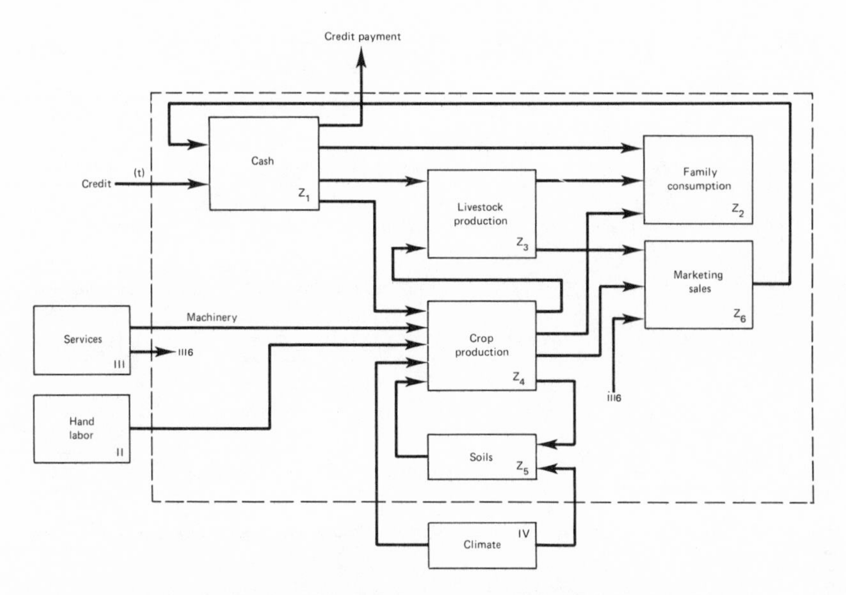

Figure 20-6. Reduced model of a farm system in southern Guatemala.

Systems Program and represents the synthesis of insights available on the behavior of small farms in that zone.

A number of agronomic experiments and a socioeconomic survey are being carried out by the CIAT Small Farm Systems Program to test the technical coefficients available at present and the behavior and predictive ability of the model.

It is not suggested that this model as it now stands represents the total reality of agriculture in the zone. The purpose of developing and utilizing the model is to illustrate some of the principal structural relationships in the physical, biological, and economic environment and to demonstrate the possible utility of such a model.

The model as a research guide. It is expected that this model will be useful in estimating the likely outcomes of alternative research, public policies, and institutional changes. With specific reference to the likely outcomes of alternative extension and research policies, the model evaluates a number of proposed technological packages. These packages are evaluated with regard to their expected impact on family nutrition, family income, risk (as measured by income and production variance), and labor utilization. Preliminary results from this work are shown in Figures 20-7, 20-8, 20-9, and 20-10.

Figures 20-7 and 20-8 present production trajectories generated by the

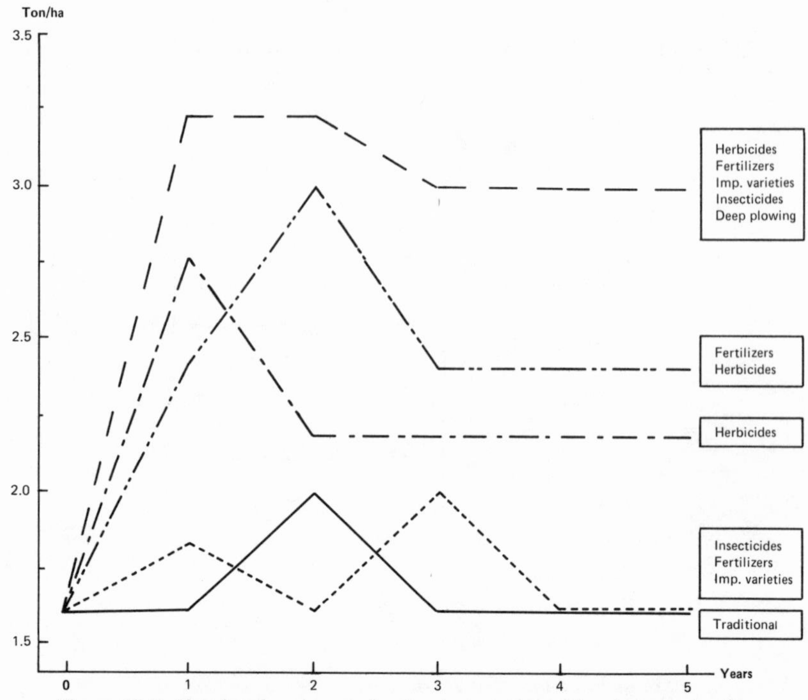

Figure 20-7. Simulated maize production trajectories with prices varying from U.S. $70 to U.S. $120 per metric ton.

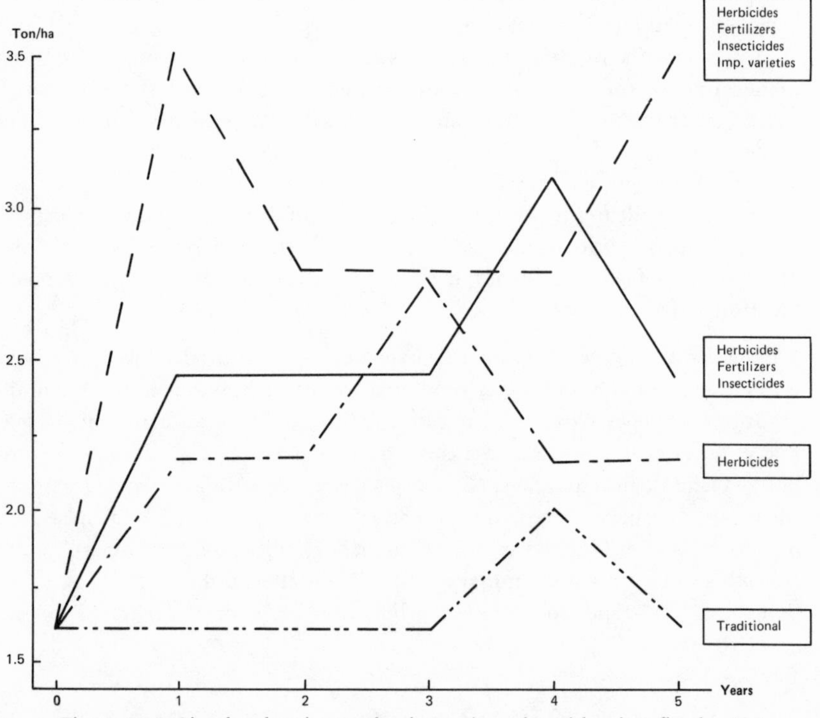

Figure 20-8. Simulated maize production trajectories with prices fixed at U.S. $120 per metric ton.

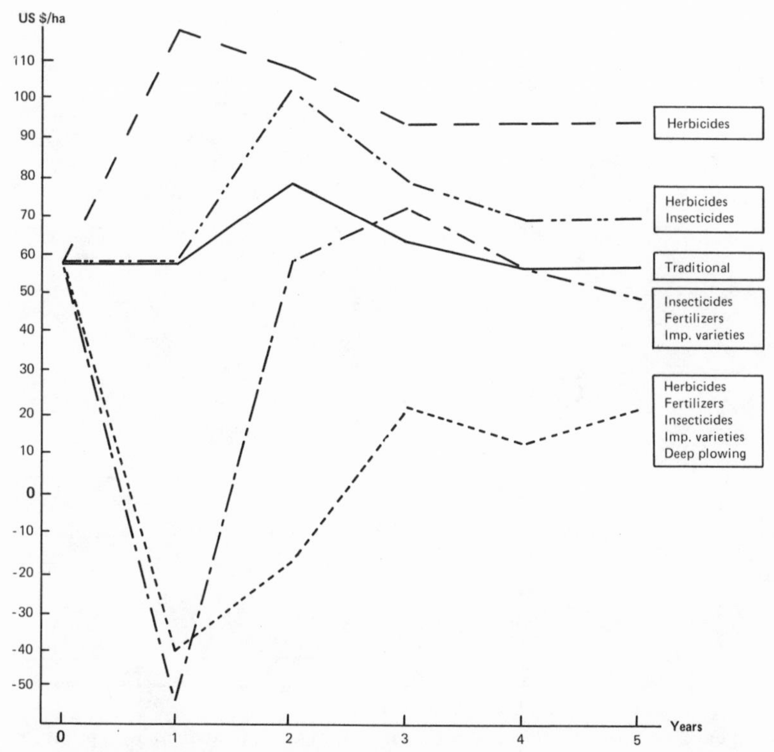

Figure 20-9. Simulated annual net income from maize production with prices varying from U.S. $70 to U.S. $120 per metric ton.

model over a simulated five-year period. Each production trajectory is identified with the production package which was simulated. Figure 20-7 presents production under the assumption that prices fluctuate between $70 and $120 per metric ton throughout the year, as is now the case in the zone. Figure 20-8 presents the production trajectories under support price. Comparison of the graphs indicates that price stability can be a means by which the adoption of technological packages is stimulated.

Figures 20-9 and 20-10 present the net family income trajectories for some of these technological packages under the two sets of price assumptions. Figure 20-9 is illustrative of the risk that is involved under a situation of unstable prices and unstable weather conditions. In particular, two of the so-called "production packages" are so costly that when risks are taken into consideration they would generate negative net income for at least one year. Traditional, or subsistence, farmers cannot tolerate this kind of risk. Another salient

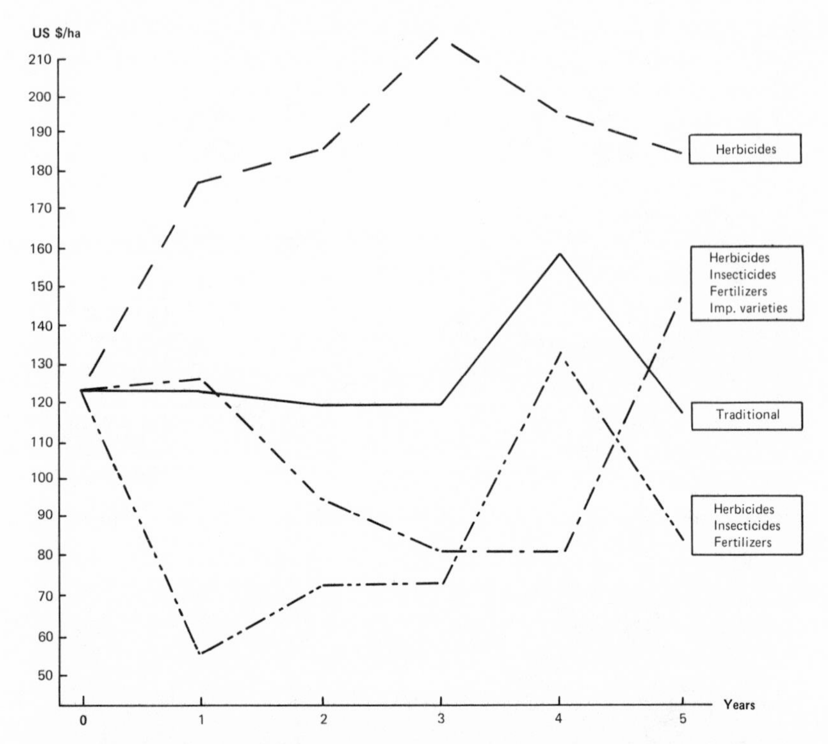

Figure 20-10. Simulated annual net income from maize production with prices fixed at U.S. $120 per metric ton.

feature of the four graphs is that the traditional production package, which utilizes few modern factors of production, produces the lowest yields but tends to be better for net income than some of the more complicated production packages. The traditional system has the lowest income variance. A comparison of Figures 20-9 and 20-10 would illustrate the potential value of an effectively functioning market and price support system.

It appears that the package expected to make the largest contribution to the income goal is that referring to the use of herbicides. Before this finding, ICTA did not have any work planned on weed control for that zone. However, as a result of the finding, a professional has now been sent for training in weed control, and the collaborative CIAT-ICTA work for the coming agricultural season will involve extensive research on weed control methods and the economic evaluation of different weed control techniques.

In addition to this immediate though preliminary outcome of the model, it is expected that sensitivity analysis will reveal its further utility in decision-making on research resource allocation. This analysis will involve the estimation of the response of the system to variation in the parameters and coefficient. The experimental work will be focused on measuring with precision those technical coefficients which appear to be sensitive to variation. If, for example, the model were to indicate sensitivity for insect damage, intensive research on insect control would be recommended. On the other hand, if the model is not sensitive to variation in these technical coefficients, such research would have a lower priority. Thus the model can be utilized to establish research priorities both in the farmer's field and at the experiment station.

The above-mentioned systems simulation efforts are expected to be utilized for the agroeconomic survey results to achieve some of the analyses suggested by the conceptual model presented in this chapter.

NOTES

1. Acknowledgment is due to John Fitzsimons for his valuable suggestions on this chapter.

2. A discussion of the systems approach in general may be found in a number of books including Stanford L. Opener, ed., *Systems Analysis* (Middlesex, England: Penguin Modern Management Readings, 1973); C. W. Churchman, R. L. Ackoff, and E. L. Arnoff, *Introduction to Operations Research* (New York: Wiley, 1957); A. W. Wymore, *Systems Engineering Methodology for Interdisciplinary Teams* (New York: Wiley, in press); Preston C. Hammer, ed., *Advances in Mathematical Systems Theory* (University Park: Pennsylvania State University Press, 1969). The usefulness of this approach for research resource allocation is discussed in John L. Dillon, "The Economics of Systems Research," paper presented at the Agricultural Systems Research Conference, Massey University, Australia, November 20-22, 1973.

3. A more complete discussion of data requirements may be found in Per Pinstrup-Andersen, "Toward a Workable Management Tool for Resource Allocation in Applied Agricultural Research in Developing Countries," revised version of paper presented at the Ford Foundation meeting for Program Advisors in Agriculture, Ibadan, Nigeria, April 29-May 4, 1974.

4. The results of this research are reported by Per Pinstrup-Andersen et al. in "The Impact of Increasing Food Supply on Human Nutrition: Implications for Commodity Priorities in Agricultural Research and Policy," *American Journal of Agricultural Economics*, 58 (May 1976), 131-142.

21

Models and Methods Used to Allocate Resources in Agricultural Research: A Critical Review[1]

C. Richard Shumway

Introduction

The objectives of this chapter are to describe systems of resource allocation now being used by United States public agricultural research organizations, either routinely or experimentally, and to suggest other procedures potentially useful to the administrator in ranking research proposals and allocating resources to them.

Current Decision Process

The greater part of agricultural research conducted in the United States is performed by public organizations, primarily the USDA and the various state agricultural experiment stations. While the decision process for allocating resources among research efforts varies considerably between organizations, the following approach is reasonably typical.[2] The management system consists of the administrator, department heads, and professional scientists. In the project funding process the administrator judges project outlines submitted to him by department heads, who have previously analyzed proposals prepared by a larger number of scientists, most of whom are tenured or career staff. Decisions about individual projects are made largely on a year-to-year basis by each scientist acting as an entrepreneur on behalf of his own professional life; if he chooses projects that are not rewarding to the funders, to other scientists, and to the administration, he suffers. A great weight is placed on him for selecting the appropriate projects. The decision function involves not the selection and funding of specific projects so much as the employment of specific scientists who will be attached to the research organization for a considerable period. Emphasis is on maximum freedom for the scientist to select projects and undertake desired research with minimum control or direction from administrators. Furthermore, if experiment station administrators are not willing to support specific research, the scientist there is normally free to seek grant funds from outside sources.

Why Modified Decision Processes Are of Interest

Soon after World War II, as public investment in research began a two-decade period of rapid growth, investigations of alternative approaches to the management of research were reported for the first time. The stimulus for such work appears to have come primarily from corporate managers concerned with the rate of return from investments in research as opposed to other activities and from government funders anxious to prevent waste from the rapidly growing budget. Since the availability of scientists was lagging behind, it was a seller's market. Researchers had little difficulty securing funds for interesting problems.

Things are now different, however. The cycle appears to have peaked and the supply-demand relationship has changed. Federal research and development expenditures are leveling off; substantial decreases have occurred in several areas, including atomic energy and space.[3] Nonfederal expenditures have continued to rise, but not fast enough to offset even the effects of inflation.[4] With greater competition among scientists for available funds, internal pressures for more intensive investigation of management processes have surfaced.

Attention is focusing more directly on the research administrator, an important link between funding sources and scientists. It is he who must account for current expenditures of funds while assuring significant achievements over time from a highly uncertain production process; he must allocate support funds even though he cannot program breakthroughs; he must keep scientists' morale high and give them sufficient latitude to perform at their best at the same time they are competing more strenuously for limited resources. He can solve the increased competition problem simply by allocating support funds evenly among all researchers or in proportion to salary. He can also reduce the size of his research staff through attrition and/or termination of untenured or noncareer scientists. Or, he or his designates can evaluate the merits of individual research proposals and specify which will be funded and for how much.

Of pertinence to this last alternative, an impressive array of management techniques have been generated in the last few years, the objectives of which are to help the administrator to be internally consistent and/or to increase his likelihood of receiving essential information to form a valid judgment. Some of these structured methods are technically referred to as "decision models." However, virtually all that are applied to problems of ranking and resource allocation in research are more appropriately classified as "information systems." They are mechanisms or logics for generating information for the decision-maker(s) rather than for automatically effecting decisions.

Examples of Structured Information Systems in Agricultural Research

Several new concepts in the management of agricultural research resources have recently been implemented or experimented with in the USDA and at state experiment stations. Some have already been well publicized, e.g., USDA's Planning, Programming, and Budgeting System (PPBS), Iowa's Review Panel System, California's Academic-Responsive Budgeting System, and Minnesota's Resource Allocation Information System.[5] Florida's system for establishing accountability in resource allocation in agricultural research was discussed in a paper presented at the American Agricultural Economics Association meeting.[6] Therefore, attention will be given here to two recent applications of structured methods. One organization is a component of the USDA Agricultural Research Service; the other is a state agricultural experiment station. A similar numerical approach (i.e., scoring models) was applied by both organizations to develop a rank order of research activities.

Agricultural Research Service Resource Allocation Experiment in Livestock Research[7]

Before fiscal year 1972, the deputy administrator of the Agricultural Research Service (ARS) in charge of livestock research initiated a comprehensive evaluation of all research activities under his jurisdiction. Approximately $1 million was withheld from initial allocations to permit expansion of research efforts in those areas judged to be of highest priority.

The research program was divided into two echelons of effort – projects and program activities (aggregates of research projects). Research projects were further subdivided into two groups – ongoing projects and proposed new starts.

Three panels evaluated research efforts of each type – program activities, ongoing projects, and proposed new starts. Panel A, consisting of eight members (assistant division directors, branch chiefs, and one laboratory director), evaluated animal science research efforts. Panel B, consisting of seven members (assistant division directors and the directors of the three largest laboratories), evaluated veterinary science research efforts. Panel C, consisting of five members (the deputy administrator, two division directors, the planning, programming, and review officer and his staff assistant) evaluated research efforts in both areas. Members of the first two panels considered only those research efforts that related to their respective organization's programs plus other selected efforts. Members of the third panel evaluated everything.

A scoring model was used by each panel to evaluate research efforts. The scoring model is a ranking approach that formally incorporates the decision-maker's subjective trade-offs and decision criteria into the model framework. A primary assumption is that a few criteria can be established which, when properly related, will specify the desirability of a decision alternative. The set can consist of both quantitative and qualitative criteria so long as each is independent of the others. A discrete scale is developed for each criterion with sufficient range to include all efforts being evaluated and with only enough categories to discriminate between those that differ significantly relative to the criterion. An overall score is calculated for the effort by summing the product of criteria weights and scores over all criteria.

Different criteria were considered in evaluating different types of research effort. For example, five criteria were used to evaluate research activities, six for ongoing projects, and eight for proposed new starts (see Table 21-1). Criteria weights were specified by members of Panel C, including the deputy administrator and the planning, programming, and review officer. Panelists then evaluated research efforts with respect to each criterion by distributing efforts evenly across a five-point discrete scale. An average score for each effort was obtained by giving equal weight to each of the three panels and equal weight to each member of an individual panel.

Considerable information on each research effort was available to all panel members. For the research activities, information included identification of the current state of the art, technological objective, promising research approaches, consequences of attaining the technological objective, magnitude of potential benefits, probability of success, and research resources required over a ten-year period. For ongoing research projects, information included the location of work, research activity contributed to, major achievements, publications, and research resources used in the previous year, redirection, objectives, plan of work, and resources required for the next year, initiation date of work related to the project, and planned duration. For proposed new starts, information included location of proposed work, research activity contributed to, ongoing project related to, justification for the proposal, objectives,

Table 21-1. Criteria for Evaluating Livestock Research in the Agricultural Research Service (ARS)

Evaluation Level	Criteria	Criteria Weights
A. Research activities	1. Benefit/cost ratio	40
	2. Extent to which research meets national, ARS, livestock research, and division goals	15
	3. Contribution to knowledge	15
	4. Urgency	15
	5. Inadequate research results are expected elsewhere	15
	Total	100
B. Ongoing projects	1. Priority of corresponding research activity	20
	2. Importance of corresponding specific research activity goal(s) identified as being contributed to by this project	20
	3. Past achievement (last fiscal year) relative to cost	15
	4. Goals set for the next fiscal year are specific, realistic, and worth achieving	15
	5. Urgency	15
	6. Cost relevance for this project	15
	Total	100
C. Proposed new starts	1. Priority of corresponding research activity	15
	2. Importance of corresponding specific research activity goal identified as being contributed to if this proposal is funded	20
	3. Priority of ongoing project to which this proposal relates	15
	4. Goals set for the next fiscal year are specific, realistic, and worth achieving	10
	5. Urgency	10
	6. Cost relevance for this proposal	10
	7. Adequacy of plan of work	10
	8. Suitability of location	10
	Total	100

specific goals for the next fiscal year, plan of work, expected duration, resources required, and impact of this proposal on other projects.

From this evaluation, all 98 research activities, 210 ongoing projects, and 134 proposed new starts were rank ordered. The results were provided to the deputy administrator for his guidance in allocating funds initially held in reserve. Results were also disseminated to all livestock research managers and to scientists at their discretion. Because the nature of the scoring model necessitated the evaluation of specific aspects of each research effort, information was available to division directors, branch chiefs, and scientists concerning

Table 21-2. Livestock Research Activities Ranking in Top 15 Percent in ARS Experiment

Rank	Research Activity
1	Beef cattle reproduction
2	Diagnosis of foreign animal diseases
3	Foot and mouth disease
4	Swine reproduction
5	Dairy cattle selection and breeding
6	Pork quality
7	Dairy cattle reproduction
7	Beef quality
9	Beef cattle feed efficiency
10	Calf scours and enteric diseases
11	African swine fever
11	Animal waste research
13	Pesticide residues
14	Dairy cattle feed efficiency

areas in which specific efforts were judged by their superiors to be deficient. The results of this exercise were not used to terminate all low-ranking research efforts. However, they did provide an informational base from which some resources were reallocated from low-ranking to high-ranking efforts. Research activities ranking in the upper 15 percent include those listed in Table 21-2.

North Carolina Experiment on Research Priorities[8]

Public concern about increasing costs and complexity of agricultural and related research, coupled with leveling off of research support and stronger demands for accountability, led to a general evaluation of agricultural research priorities and of research resource allocation at the North Carolina Agricultural Experiment Station beginning in 1972. The immediate goal of this examination was to determine which research problem areas (RPAs) should be given greater emphasis at the station in the next five years.

A joint administration-faculty effort was mounted to conduct an exhaustive review of all research programs and projects at the station and to explore possible redirections for the future. All members of the research faculty and much of the extension faculty participated in the study. Outside scientists evaluated faculty task force recommendations. Recommendations included funding and scientist reallocations and additions at the RPA and sub-RPA level. Narrative support for such changes was further evaluated by numerical assessments of the recommendations based on a small number of important criteria. A set of scoring models was applied in this part of the evaluation.

Twenty task forces, each composed of five to ten research and extension faculty members and occasional representatives of state agencies, reviewed the entire research program of the station. One task force was appointed for biological sciences and technology, five for animal research, six for plant research, three for environmental and natural resources research, and five for food-fiber-people-economics research. Each one was responsible for evaluating a portion of the total program, generally crossing departmental lines. They recommended quantitative changes in resource use (both money and scientists) as well as timing for such changes, and they rated each recommendation according to prespecified criteria.

Following submission of the task force reports, eighteen extramural panels, each consisting of three scientists not associated with North Carolina State University and chaired by a representative of the Cooperative State Research Service, evaluated the task force recommendations. They rated those recommendations, using the same criteria considered by the task force, and also made independent recommendations for resource allocation.

To this point the evaluations were interdisciplinary in nature. The twenty-three academic departments then reacted to the task force recommendations and extramural panel reviews in a disciplinary context, developed a third set of five-year recommendations, and rated the task force recommendations.

The scoring models were developed beginning with the list of evaluation criteria used in the National Program of Research for Agriculture.[9] The station administration consolidated and restructured them to improve apparent independence and relevance. These proposed criteria varied between the four major research areas. They were submitted through the mail to members of the Research Planning Advisory Committee, composed of department heads, first for revision of the criteria sets and then for specification of importance weights. Revisions, weightings, and explanations of reasons for them were developed by each member of the committee, summarized by the administration, and resubmitted to the committee members using an interactive Delphi procedure.[10] This approach was repeated twice to permit modification of initial opinions based on the convincing anonymous arguments of others. The four criteria sets, with the approximate wording developed by the committee, are listed in Table 21-3 together with their average weights from the final Delphi round. To permit comparison across major research areas, the criteria weights for each area were standardized by the committee to sum to 100.

Without knowledge of the weight attached to each criterion, each member of the three groups – task forces, extramural review panels, and department heads – independently scored task force recommendations for increased RPA resources on a five-point scale. Some participants rated most RPAs in the upper two scoring intervals while others dispersed them more evenly among all intervals. Since each person did not score all RPAs, it was possible that those RPAs scored by the former rated higher than those scored by the latter only

Table 21-3. Criteria for Evaluating Research Problem Areas at the North Carolina Agricultural Experiment Station

Research Area	Criteria	Criteria Weights
A. Biological sciences and tech- nology	1. Urgency – basic information needed to aid in solution to threat or problem.	20
	2. Cost relevance – expected long-term benefits in relation to costs.	15
	3. Degree to which similar research is not now being conduct- ed or not likely to be conducted elsewhere (higher scores if inadequate research results expected elsewhere).	15
	4. General importance and potential for contribution to knowl- edge. Higher scores to be assigned for greater scientific merit and potential for contribution to faculty development and improved academic performance.	50
	Total	100
B. Animals and plants	1. Extent to which proposed research is consistent with sta- tion, regional, and national goals in agriculture and forestry. Consider economic value of the crop or animal enterprise and its products to people of North Carolina.	35
	2. Cost relevance – expected benefits in relation to costs.	20
	3. Extent to which similar research of adequate quality is not being conducted on this commodity elsewhere (higher score for RPAs and sub-RPAs for which adequate results are not likely to be available elsewhere), and degree of urgency of need for research results.	20
	4. Potential for contribution to knowledge.	25
	Total	100
C. Environ- ment and natural resources	1. Extent to which proposed research is consistent with sta- tion, regional, and national goals in natural resource develop- ment and conservation.	35
	2. Cost relevance – expected benefits in relation to costs.	15
	3. Extent to which similar research of adequate quality on this resource is not being conducted elsewhere (higher scores for inadequate research elsewhere) and whether or not there is (1) a threat to natural resource, (2) public pressure, or (3) a critical need for environmental protection.	15
	4. Potential for contribution to knowledge.	20
	5. Extent to which the research will aid in meeting broader pub- lic service commitment of the school and university, beyond traditional statutory charge of the experiment station.	15
	Total	100
D. Food- fiber- people- economics	1. Extent to which recommended research is consistent with station, regional, and national goals of promoting and pro- tecting public health and improving family living; potential for improving quality of life and developing rural communi- ties in North Carolina.	35
	2. Cost relevance – expected benefits in relation to increased costs of research in these areas, resulting from these recom- mendations.	20
	3. Extent to which similar research of adequate quality is not being conducted elsewhere (higher scores for inadequate re- search elsewhere) and whether there is (1) public support for research to evaluate the impact of improved agricultural tech- nology, (2) a threat to public health, or (3) a need for infor- mation to support new processing industries.	20
	4. Potential for contribution to knowledge.	25
	Total	100

because of personal differences in using the subjective model. While there are also legitimate reasons for different average scores between participants, no objective procedures were available to make such interpersonal judgments. Therefore, to permit comparison across participants, each person's scores were given equal weight by adjusting proportionally so that the *average* overall score for all RPA resource increases evaluated by any individual was the same.

Recommendations for resource increases in ninety RPAs were evaluated with the scoring models. Project rankings were obtained from the adjusted scores according to two standards: average score by all raters and average score minus one standard deviation of all raters. Although this latter standard forces the rank to respond to some extent to the degree of variability among participant scores, it is of course arbitrary. The standard deviation is computed based on the assumption of a normal distribution. For the normal distribution, approximately 83 percent of observations lie above the *mean minus one standard deviation*. If the observations are skewed to the right, more than 83 percent of them would lie above this point; if skewed to the left, less than 83 percent would. For other distributions the percentage may also be different.

The second standard caused RPAs for which there was much sampling variability, difference of opinion, and /or variability in the basic predictability of the area to be ranked lower than when ordered according to the first standard. RPAs with little difference ranked higher in the second list. Ten RPAs that ranked in the top 15 percent on both lists are as follows (not in order of rank):

Appraisal of soil resources
Control of diseases, parasites, and nematodes affecting forests
Control of diseases of livestock, poultry, and other animals
Genetics and breeding of forest trees
Improvement of biological efficiency of field crops
Improvement of biological efficiency in production of livestock, poultry, and other animals
New and improved forest products
Improvement of economic potential of rural people
Improved income opportunities in rural communities
Improvement of rural community institutions and services

As expected, there was considerable difference of opinion between raters and between groups of raters. Between groups, the rank order correlation between the department heads' scores and the extramural panels' scores was highest at only .45. The correlation between the task forces and extramural panels was comparable at .42 while the correlation between the task forces and department heads was .24. Furthermore, an analysis of variance suggested that the variation of scores between groups was not significantly different from the variation between raters of the same group. There was no evidence that the three groups came from different populations of opinion holders, only that the population was extremely diverse. The coefficient of variation of scores for individual RPAs was frequently in excess of 20 percent.

Alternative Methods

The first part of this chapter has described a currently typical resource allocation decision process and has presented a pair of case studies in the experimental application of a more structured mechanism. Attention will now turn to a review of various other methods reported in the research management, utility theory, and behavioral science literatures that could help the research administrator to be internally consistent when ranking research efforts and allocating resources to them. They do not assist directly in formulating the "correct" judgment about a particular effort, although some stimulate opinion changes through additional information. These techniques vary widely in the amount of time required to implement them, processing costs incurred, and information generated. Important questions selectively addressed in this section include: (1) How quickly can judgments be obtained using the technique? (2) What processing costs are incurred? (3) How many opinions are listened to? (4) How does the technique affect the risk of a bad decision? (5) What types of information can be obtained from the process (i.e., ordinal or cardinal ranking of projects, benefit/cost ranking, recommended allocation, summaries of specified characteristics)? (6) How does the process handle funding options, multiple constraints, and uncertainties? (7) With how many techniques is it compatible? (8) How have users judged the value of its products?

Project Ranking

Methods of ranking projects using both single and multidimensional measures of project benefit will be introduced first.

One-dimensional ranking methods. Several techniques, including Q-sort, paired comparisons, successive ratings, and successive comparisons, are included in the category of *individual participant comparative methods*.[11] In each of these techniques a single judge compares the overall subjective worth of one item with another or with a group of items. When only a few items must be evaluated, methods in this group are among the simplest procedures for systematic comparison. An implicit assumption is that each item is independent and mutually exclusive of all others.

With Q-sort, projects are divided into hierarchical categories on the basis of their expected benefits, or on any other standard basis. Typically, essential information about a project is written on a card and the cards are then sorted into piles. No quantitative values are assigned to any category, but each may be divided into additional categories until no significant differences in anticipated benefits are discernible among its projects.

With paired comparisons, a complete ordering of projects is obtained and verified. All possible pairs are examined, and the project within each that has the higher expected benefit is identified. Again, no quantitative value is assigned to any project – only an ordinal ranking. After projects are ordinally ranked, successive ratings and/or successive comparisons can be applied to establish and verify a relative cardinal ranking. Benefit-cost ratios can be computed from these rankings and used as a rational method of allocation if the funding decision is strictly of the "go/no go" type.

With successive ratings, an arbitrary base number is assigned to the highest ranked project. Numbers are given to each subsequent project in accordance with its anticipated benefits relative to the top one. These values are verified by covering them and repeating the comparison relative to the lowest ranked project. If significant differences appear in the two sets of numbers, the procedure is repeated until consistent ratings are obtained when the scale is anchored to both the highest and the lowest benefit projects.

With successive comparisons, initially assigned values are refined by comparing the value of one high benefit project with portfolios of lower benefit ones. The number of projects in the portfolio is successively reduced until the single high benefit project is preferred to the portfolio. The logic of this method is structured to establish bounds on the cardinal ranks of all projects and to identify inconsistencies in earlier assigned values.

A final comparative approach, dollar metric, permits the calculation of absolute benefit-cost ratios.[12] Paired comparison are used first to identify the preferred project from each possible pair. From an estimate of the expected cost of each project, the participant specifies how much the cost of the preferred project could increase before the other would be chosen. He repeats this procedure for all pairs. Next, he determines how much the cost of the least preferred project could increase (or decrease) before he would be indifferent toward the choice between funding the project or not having a project in this research area. This base figure permits the specification of anticipated benefit in dollar terms. By identifying actual expected cost, he can calculate a benefit-cost ratio with the numerator and denominator in the same units.

Features of the comparative ranking methods are summarized in Table 21-4. Also included are two rankings of the methods by ease of use and user satisfaction with the product of the method. Rankings of the methods are based upon the reactions of middle managers who used four of them during

Table 21-4. Comparative Ranking Methods

Category of Comparison	Q-sort	Paired Comparisons	Successive Ratings	Successive Comparisons	Dollar Metric
Type of rank obtained	Categories	Ordinal	Cardinal – relative scale	Cardinal – relative scale	Cardinal – absolute scale
Potential uses in research evaluation	Grouping of projects, program areas, criteria, or objectives	Ranking of projects, program areas, criteria, or objectives	Ranking of projects, program areas, criteria, or objectives	Ranking of projects, program areas, criteria, or objectives	Ranking of projects, program areas, criteria, or objectives
Ranking methods it can be used with	All	Q-sort, dollar metric, group and multidimensional methods	Q-sort, paired comparisons, successive comparisons, group and multidimensional methods	Successive ratings, group and multidimensional methods	Q-sort, paired comparisons, group and multidimensional methods
User satisfaction with ranking	4	3	2	1	n.a.[a]
Ease of use	1	2	3	4	n.a.

Note: Ordering of methods is based upon my experience in working with twenty middle managers who applied the methods in a large governmental research and development organization. 1 = highest ranked method (most satisfaction or easiest).

[a] Not applied.

a workshop on benefit measurement. Those methods which were easier, faster, and more natural to use also resulted in project rankings which were less useful and/or less satisfactory to the person applying them. Several expressed most satisfaction with the ranking generated by using successive comparisons. However, each concluded that the thought process of comparing one project with portfolios was unnatural and preferred successive ratings even with a slightly less satisfactory ranking.

Even after the individual's subjective rankings are elicited, the problem remains concerning which person(s) to listen to and how to process multiple opinions. Using *group-determined measures of benefit* opens up a number of options. The decision-maker can identify one "best" judge and listen exclusively to him. Alternatively, to increase the likelihood of obtaining a more nearly "correct" opinion, he can identify several experts and accept a simple or weighted average of their opinions as the best estimate. Following this course also permits the degree of variance among their opinions to be determined; this is an important by-product. A third course permits interaction between participants. However, when that occurs, independence of opinion is partially muted, and the ability to use statistical measures of variance is restricted. There are several forms of group interaction.

In the first approach, called "committee or round table," the group meets together, airs differences of opinion, and concludes with a "group opinion." Minority opinions may also be expressed.

In the second, or "chain-of-command" approach, interaction takes place between adjacent links. If a superior disagrees with a subordinate's opinion, the subordinate's view may be totally suppressed in the next step when the superior interacts with his superior. Most typically, only a single opinion exits from the highest link in the chain, and that is often formed through interaction with one other person. That individual in turn presented only one opinion although a wide variety may have surfaced in the interaction that preceded it. The Iowa Agricultural Experiment Station is using a variant of this approach, in which panels are included as three links in the chain to evaluate five-year research alternatives.[13]

A third approach, "Delphi," is a formalized method designed to promote consensus without obscuring variants.[14] (See also earlier reference to the use of Delphi in the North Carolina experiment.) It consists of a series of individual interrogations to a group of experts, interspersed with information and opinion feedback. Some questions inquire into the reasons for previously expressed opinions. A collection of such reasons is then presented to each respondent, who is invited to reconsider his earlier estimate. Delphi attempts to improve the committee approach by subjecting views of individual experts to each other's criticism in ways that avoid face-to-face confrontation. It provides anonymity of opinions and of arguments advanced in their defense.

Each of the group methods can be used to obtain ordinal or cardinal rankings of projects, program areas, criteria, or research objectives. They can be used singly or in conjunction with comparative and/or multidimensional methods. However, while several of the comparative methods make a logical system when used in succession, few of the group methods do. I have ranked the methods relative to three criteria (see Table 21-5). Like the comparative methods, none is consistently high or low with respect to all criteria. The one which requires the least time from all judges combined also permits the fastest rendering of a final decision because only one person need be involved, but it results in the highest risk of a "bad" judgment being made. In addition, measures of variant opinions are normally restricted to the group average and Delphi methods. And, although opinion variation can be calculated at each round of Delphi, a statistical measure of variance satisfies the independency of observation requirement only in the first round.

Table 21-5. Group Ranking Methods[a]

Evaluation Category	"Best" Judge	Group Average	Committee	Chain of Command	Delphi
Other ranking methods with which each group method can be used	Comparative and multidimensional	Comparative and multidimensional	Comparative and multidimensional	Committee, comparative, and multidimensional	Comparative and multidimensional
Theoretical risk of "bad judgment"	5	3	2	4	1
Total time required of all judges	1	2	3	2	3
Speed of obtaining judgment	1	3	4	2	5
Estimate of variance among judges possible?	no	yes	not usually	not usually	yes
Special problem areas	How to identify "best" judge		Most respected or persuasive member's opinion likely to carry excessive weight	Subordinate opinions are subdued in subsequent interactions	

Note: Evaluation of methods is by the author. The highest ranking is 1, indicating least risk, least time requirement, and fastest judgment.

[a] All methods are flexible as to type of rank obtained – category, ordinal, and/or cardinal type of rank. The potential uses in research evaluation of all these methods include ranking of projects, program areas, criteria, and/or objectives.

Multidimensional ranking methods. The basic objective of *information systems* is to provide important information to the administrator in a form that will assist him in choosing among decision options. No attempt is made to place projects in overall rank order, but measurements for each dimension of benefit are listed. Several information systems have been designed for research project evaluation, including the one developed by Fishel for the Minnesota Agricultural Experiment Station.[15]

The *benefit contribution models* approach is similar to the scoring models illustrated earlier in that several factors are specified which can be used to judge the expected merits of a decision alternative. However, instead of specifying general evaluation criteria, overall goals of the organization are divided into independent, mutually exclusive objectives. The relative importance of

Table 21-6. Comparison of Multidimensional Ranking Methods

Category of Comparison	Minnesota Information System	Scoring Model	Benefit Contribution Model
Type of rank obtained	Partial cardinal orderings	Overall cardinal orderings	Overall cardinal orderings
Potential uses in research evaluation	Quantitative attribute display of projects	Ranking of projects, program areas and/or objectives	Ranking and funding of projects and/or program areas
Other ranking methods each multidimensional method can be used with	Comparative and/or group	Comparative and/or group	Comparative and/or group
Special properties		Diagnostic capacity for identification of project's weakest area(s)	Ranking may be conditional upon assumed funding pattern

such objectives and the contribution of a project to each of them are estimated. Total benefit of a project may be computed by product summation.

Considerable flexibility for evaluating project interrelationships is permitted. For example, the structural relationship between projects and objectives may be defined such that the contribution of one project is considered to be a function of which other projects are funded. In this case it is not possible to obtain a simple ordering of benefits or benefit-cost ratios without specifying the funding level of interrelated projects.[16]

A summary comparison of three multidimensional ranking methods, including scoring models, is contained in Table 21-6. Scoring models have been used more than the others for ranking research projects. However, partial-ordering information systems analogous to the Minnesota Experiment Station system have also been used in an informal way.

These multidimensional methods are compatible for use in combination with comparative and group methods; however, the combinations vary. For example, partial orderings provided by an information system may be used as input data for obtaining an overall ordering utilizing the comparative and/or group methods. However, with scoring and benefit contribution models, other methods are used to weight the scoring criteria or research objectives, and the overall ranking is determined by the multidimensional model.

Approaches for Optimizing Resource Allocation

Benefit-cost analysis. Benefit-cost ratios are frequently used as a basis for maximizing expected benefit from a given investment. To use these ratios exclusively, there must be only one constraint on resources (e.g., total research budget for one period) and only one positive funding option for each project. The decision consists merely of funding the highest ranked projects until funds are exhausted. Problems may arise because of project indivisibilities (i.e., when there is money left but not enough to fund the project next in order), but a satisfactory solution can generally be obtained by a manual check of nearby alternatives. However, when there are multiple constraints, time periods, and/or project funding options, simply allocating funds based upon magnitude of the benefit-cost ratio is not so appealing.

Optimization models specifically for research. A number of specific optimization models have been developed for resource allocation in research and development when one or more of the complicating factors exist. Some have been designed for research evaluation, but many are appropriate only for development projects. Each purports to analyze quantified subjective data in a prespecified manner and to suggest the most appropriate allocation of available resources, given the model assumptions. They vary greatly in scope and procedure. Virtually all focus on the allocation of funds, and some address the issues of manpower and facility allocation as well. Some are deterministic models; others incorporate stochastic elements. Most are static; a few are dynamic. Some focus only on economic evaluations; others are not restricted in the type of variable they can consider. A few attempt to derive the optimal research budget, but most treat the research budget as an upper restraint on the decision problem.[17] No allocation model developed specifically for agricultural research allocation has been reported, although two partially developed models have been presented. The algebra for an agricultural experiment station allocation model has been discussed by Paulson and Kaldor.[18] However, the specific structure of the model and the method of solution were not defined. The mathematical structure for a fund and manpower allocation problem in a university department has been identified by Cartwright.[19] The complexity of the problem resulted in its being left in conceptual form.

Since a large number of models have been described in the references above, this discussion will focus on two recent models representing different approaches to the allocation problem. Both include operational computerized models for allocating money to research projects. The first was developed for and has reportedly been implemented by an industrial research organization. The second model was developed for a defense research and development organization and is now being used there experimentally. Both are being used primarily for *applied* research evaluation.

Atkinson and Bobis, authors of the *simulation model*, seek to attain maximum expected profit through optimal allocation of a fixed five-year research budget among projects and over time.[20] The approach is dynamic in that it solves the problem of distributing the funds among the five years. It incorporates stochastic elements, measures costs and benefits in economic units only, and utilizes a simulation procedure to obtain solutions. Expected economic value of a research product is computed as a function of the year the project is completed and the probability of technical, legal, engineering, and commercial success. Year of completion is a stochastic function of project funding in that and each previous year. The selection of projects, optimal rate of funding, and annual allocation of the total research budget are determined through an iterative procedure using only point estimates of potential payoff. The point estimates are subsequently replaced with distributions and randomly sampled by simulation to estimate the range in payoff possible from the selected strategy.

The *network model*, by Baker et al., is both static and deterministic, but it incorporates noneconomic as well as economic variables and permits direct interaction between decision-maker and computer.[21] An optimal allocation of a single-period research budget is sought through the use of a network program. Separable programming can also be used. A piecewise-linear benefit function approximates the nonlinear function. It is maximized subject to a set of budgetary constraints. The benefit function is completely flexible in the type and form of variable it can consider. The constraint set includes an upper limit on total budget and upper and lower constraints on the funding of individual projects, program areas, technologies, and performing organizations. Interactive opportunities are built into the program to permit the participant to conduct sensitivity analyses by varying data inputs which are least certain and by asking "what if" types of questions.

The simulation and network models represent different philosophical and mathematical approaches to allocation. Being designed for an industrial laboratory where economic profit is the motivating force, the simulation model is comprehensive in the types of market decision variables incorporated. Both stochastic variables and dynamic allocation decisions are considered. The network model was prepared for a government research and development organization where nonmarket goals are at least as important as economic ones. The budgeting process earmarks some funds for specific types of research; therefore, multiple constraints must be considered. The model is static and the variables deterministic although stochastic elements can be evaluated through interaction.

In another respect both models represent a similar view of resource allocation among research efforts. A project can be expanded or reduced in scope, and progress can be speeded up or slowed down depending upon the resources allocated to it. Significant uncertainties exist about cost and potential payoff from any given effort. Both models view the allocation problem as a complex

Table 21-7. Comparison of Resource Allocation Optimization Methods

Category of Comparison	Benefit-Cost Analysis	Atkinson-Bobis Simulation Model	Baker et al. Network Model
Variables in benefit function	Flexible, generally economic	Economic returns	Flexible
Form of benefit function	Point	Functional relationship between date of research success and benefit	Linear or nonlinear approximated by piecewise linear segments
Number of project funding options permitted	1	∞	∞
Number of constraints permitted	1 (budget)	1 (5-year budget)	Finite number (budget)
Method of handling uncertainty	Manual sensitivity	Simulation	Sensitivity analysis through interaction
Number of periods over which funding is optimized	1 period	Up to 5 periods	1 period
Speed of use	Fastest	Slow relative to benefit cost	Slow relative to benefit cost
Equipment required for solving allocation problem	None	Computer	Computer
Cost of using method (data sources, computer, analyst)	Least	High relative to benefit cost	High relative to benefit cost
Special properties	Computer can be used to reduce manual time requirements		Tabulates requirements for nonbudgetary resources also

one and permit direct consideration of multiple funding options and the effects of uncertain parameters.

A summary comparison of optimization methods is included in Table 21-7. Depending upon the combination of methods used in the evaluation system, the benefit-cost approach is generally the fastest and easiest to use. One of the simplest combinations consists of using successive ratings to order the projects and then dividing by expected costs. When alternatives need to be considered beyond what is feasible with a benefit-cost approach, a host of other models is available. The two described require more data processing equipment and analytical time than the benefit-cost approach does, but they also provide more information.

Although some aspects of the latter models are identified rather specifically, considerable flexibility is inherent in both. The benefit function of the simulation model consists of economic returns only. However, the function could be expanded to include noneconomic variables also. Each of the models is flexible as to the combination of ranking methods that can be used to measure benefit. For example, a scoring model has been used to weight research objectives as part of a benefit contribution model that estimates benefits of projects to which an allocation of funds is subsequently proposed by the network model.

Framework for Investment in Management Tools

Conceptually, the decision framework for optimal investment in management is the same as for any other investment option: purchase units of management until the marginal returns added per dollar spent is equal to the marginal returns from the dollar invested in the next best alternative. Since administrative costs typically come out of the research budget, this framework is also valid for research management investment decisions. Management approaches are in direct competition with research projects for resources. There, the management approaches which add the most to the expected value of research (because of a better decision) per dollar spent should be engaged until the value gained from another dollar's investment is no greater than if invested in the marginal research project. This would equalize the marginal rate of return from alternative lines of investment. Hence, the amount of management used in different organizations could vary also. If all the research options in one organization are expected to yield a very high rate of return, its investment in management tools should be restricted to those that will yield at least as high a return through improved decisions. Another organization, with lower and more varied rates of return expected on research options, could afford to invest in more management.

The major problems in implementing such a framework are not conceptual. They are measurement problems: how to determine which management approach is dollar-for-dollar better than another and how to measure value from management tools in the same units as research projects. While much work has been undertaken to develop new ways for subjectively ranking projects and allocating resources to them, far less has been done to determine which is the best management approach to use. In fact, only one paper appears to have addressed this problem directly: Souder has reported the application of a scoring model to an evaluation of twenty-six project selection models.[22] Opinions were solicited from administrators and management scientists concerning the status of the models with respect to five criteria – realism, flexibility, capability, ease of use, and cost. The criteria were subjectively weighted and an overall score derived for each model. The only model included in that evaluation that was also discussed in this chapter, the Atkinson-Bobis model, tied for the highest score.[23]

Laboratory costs to obtain reasonably *objective* measures of costs and benefits of research management tools would probably be prohibitive. Controlled experiments would be necessary over lengthy periods of time sufficient to observe the benefits resulting from research projects selected with different tools. One is probably limited for practical purposes to *subjective* measures. Souder's subjective scoring approach provides a useful comparison of a number of management methods. However, it is not sufficient for determining the optimum level of investment in management. Factors of both benefit (e.g., realism and capability) and cost (e.g., dollar costs and ease of use) were included in the score. To determine the optimal investment in such management methods, benefits and costs must be separated and measured in the same units as for research projects. The dollar metric approach previously discussed has potential as an initial mechanism for subjectively placing management tools and research projects in comparable units of measurement. Then the marginal principle could be applied to both sets of options to select the optimum portfolio.

Summary

Considerable research has been conducted to develop management tools for possible substitution for or complementary use in subjective decision processes. Case examples in the application of such methods in public agricultural research organizations have been cited. Other literature has been selectively surveyed to emphasize the breadth of methodological developments for research project selection and resource allocation. A partial comparative evaluation of the selected methods is included with the survey. Applicability of the marginal principle for determining the optimum investment in research management has been emphasized and a rudimentary framework for implementing that principle suggested. Many research challenges remain in the area of methodological development. Particularly great is the need for innovative thought and practical experimentation to compare costs and benefits of alternative subjective decision processes.

NOTES

1. This chapter is Technical Article 11986 of the Texas Agricultural Experiment Station. Portions have been taken from C. R. Shumway, "Allocation of Scarce Resources to Agricultural Research: Review of Methodology," *American Journal of Agricultural Economics*, 55 (November 1973), 557-566, and from C. R. Shumway and R. J. McCracken, "Use of Scoring Models in Evaluating Research Programs," *American Journal of Agricultural Economics*, 57 (November 1975) (copyright © 1975 by the American Agricultural Economics Association). My sincere appreciation is extended to Ralph J. McCracken and

Ernest L. Corley for very substantial contributions to this paper. The two experiments in numerical evaluation of research reported here were separately conducted by them. Their permission to report on those experiments is gratefully acknowledged.

2. Thanks are due Dale Hoover for his clear articulation of this process. Although the terminology is for a state experiment station, the essential features are not much different from many USDA research units.

3. United States Bureau of the Census, *Statistical Abstract of the United States* (Washington, D.C.: United States Government Printing Office, 1968), p. 526; 1969, p. 524; 1971, p. 510; 1972, p. 522.

4. United States Bureau of the Census, *Statistical Abstract of the United States* (Washington, D.C.: United States Government Printing Office, 1971), p. 508.

5. W. L. Fishel, ed., *Resource Allocation in Agricultural Research* (Minneapolis: University of Minnesota Press, 1971).

6. J. R. Conner and V. C. McKee, "Florida's System for Resource Allocation in Agricultural Research," paper presented at the American Agricultural Economics Association Meetings, College Station, Texas, August 1974.

7. Dr. Ernest L. Corley designed and implemented the evaluation procedures used in the Agricultural Research Service experiment. This section has been developed from information he supplied.

8. This study was initiated in 1972 by the North Carolina Agricultural Experiment Station, J. C. Williamson, Jr., director, with the assistance of the Cooperative State Research Service. The help of Ralph J. McCracken in developing this section is gratefully acknowledged.

9. USDA and Association of State Universities and Land Grant Colleges, *A National Program of Research for Agriculture* (Washington, D.C.: United States Government Printing Office, 1966).

10. N. Dalkey and O. Helmer, "An Experimental Application of the Delphi Method to the Use of Experts," *Management Science*, 9 (April 1963), 458-467.

11. Basic references for these techniques include J. P. Guilford, *Psychometric Methods* (New York: McGraw-Hill, 1954); R. D. Luce and Howard Raiffa, *Games and Decisions* (New York: Wiley, 1957); C. W. Churchman, R. L. Ackoff, and E. L. Arnoff, *Introduction to Operations Research* (New York: Wiley, 1957).

12. See E. A. Pessemier and N. R. Baker, "Project and Program Decisions in Research Management," *R & D Management*, 2 (October 1971), 3-14.

13. Fishel, *Resource Allocation*, pp. 326-343.

14. Dalkey and Helmer, "Delphi Method."

15. Fishel, *Resource Allocation*, pp. 344-381.

16. See A. B. Nutt, "An Approach to Research and Development Effectiveness," *IEEE Transactions on Engineering Management*, EM-12 (September 1965), 103-112, and B. V. Dean and L. E. Hauser, "Advanced Material Systems Planning," *IEEE Transactions on Engineering Management*, EM-14 (March 1967), 21-43, for applications of benefit contribution models to resource allocation among defense research and development efforts.

17. For detailed references see N. R. Baker and W. H. Pound, "R & D Project Selection: Where We Stand," *IEEE Transactions on Engineering Management*, EM-11 (December 1964), 124-134; M. J. Cetron, Joseph Martino, and L. A. Roepcke, "The Selection of R & D Program Content – Survey of Quantitative Methods," *IEEE Transactions on Engineering Management*, EM-14 (March 1967), 1-13; R. W. Cartwright, "Research Management in a Department of Agricultural Economics," Ph.D. thesis (Lafayette, Indiana: Purdue University, 1971); N. R. Baker and J. R. Freeland, "Recent Advances in R & D

Value Measurement and Project Selection Methods," paper presented at the Forty-first National Meeting of the Operations Research Society of America (ORSA) in New Orleans, April 1972; and T. E. Clarke, "Decision-making in Technologically Based Organizations: A Literature Survey of Present Practice," *IEEE Transactions on Engineering Management*, EM-21 (February 1974), 9-23.

18. Arnold Paulson and D. R. Kaldor, "Evaluation and Planning of Research in the Experiment Station," *American Journal of Agricultural Economics*, 50 (December 1968), 1149-1162.

19. Cartwright, "Research Management."

20. A. C. Atkinson and A. H. Bobis, "A Mathematical Basis for the Selection of Research Projects," *IEEE Transactions on Engineering Management*, EM-16 (February 1969), 2-8.

21. N. R. Baker, W. E. Souder, C. R. Shumway, P. M. Maher, and A. H. Rubenstein, "A Budget Allocation Model for Large Hierarchical R & D Organizations," *Management Science*, 23, no. 1 (September 1976).

22. W. E. Souder, "A Scoring Methodology for Assessing the Suitability of Management Science Models," *Management Science*, 18 (June 1972), B526-543.

23. Atkinson and Bobis, *IEEE Transactions*.

Economic and Social Factors
in Research Resource Allocation

22

Environmental Constraints, Commodity Mix, and Research Resource Allocation[1]

Martin E. Abel and Delane E. Welsch

The purpose of this chapter is to show how the allocation of research resources among commodities and the effects of such allocations on the output mix depend upon (a) the initial production conditions, (b) the nature of the research production functions, (c) the nature of the demand relations for the commodity outputs, (d) relative factor endowments, and (e) the existence of different types of environmental constraints. The basic model used is a two-factor, two-product model in which certain kinds of technical change are introduced. This model is presented and discussed in the next section. The third section deals with the effects of technical change and of demands for the outputs on the product mix. The role of factor endowments is the topic of the fourth section, which is followed by a discussion of the effect of certain types of environmental constraints on the allocation of research resources and the output mix. The policy implications of the analysis are discussed in the fifth part of the chapter.

The Basic Model

To analyze certain questions concerning the benefits to be derived from the diversification of agricultural production, we need a theoretical model which will enable us to trace through changes in production functions, factor endowments, and relative product prices on output, income, and factor rewards.

A simple, but useful model for looking at the influence of technical change on the output mix is the standard two-factor, two-product model of production.

Let us start by assuming that a region (either an area within a country or a country trading in a larger, world market) produces two goods, q_1 and q_2, with two homogeneous factors of production, L and K, where L is the labor input and K is the land (capital) input. Total factor supplies are assumed to be fixed.

Production of our two goods is given by the Cobb-Douglas production functions

$$q_1 = \tau_1 L_1^{\alpha} K_1^{1-\alpha} = \tau_1 L_1 \left[\frac{K_1}{L_1}\right]^{1-\alpha} \quad (1a)$$

$$q_2 = \tau_2 L_2^{\beta} K_2^{1-\beta} = \tau_2 L_2 \left[\frac{K_2}{L_2}\right]^{1-\beta} \quad (1b)$$

which reflect constant returns to scale. τ_1 and τ_2 are indices of technology. In addition, the fixed supplies of labor and land (capital) are represented by

$$L_1 + L_2 = \bar{L} \quad (2a)$$

$$L_1 \left[\frac{K_1}{L_1}\right] + L_2 \left[\frac{K_2}{L_2}\right] = \bar{K}. \quad (2b)$$

Furthermore, we assume that the factors of production are fully employed.

We can derive the expression for the slope of the production possibility curve, which is

$$-\frac{d\left[\frac{q_1}{\bar{L}}\right]}{d\left[\frac{q_2}{\bar{L}}\right]} = \frac{\tau_1}{\tau_2}(bR)^{1-\alpha}(aR)^{\beta-1}[a+(b-a)\ell]^{\alpha-\beta}\left[\frac{a+\alpha\ell(b-a)}{a+(b-a)(1-\beta+\beta\ell)}\right] \quad (3)$$

where,

$$R = \left[\ell\left[\frac{K_1}{L_1}\right] + (1-\ell)\left[\frac{K_2}{L_2}\right]\right] = \frac{\bar{K}}{\bar{L}}$$

$$\ell = \frac{L_1}{\bar{L}}$$

$$a = \frac{\alpha}{1-\alpha}$$

$$b = \frac{\beta}{1-\beta}.$$

The reader is referred to Johnson and to Abel, Welsch, and Jolly for detailed derivations of the production possibility curve and methods for solving for the outputs q_1 and q_2, given the product prices.[2]

We can consider two possibilities with respect to the influence on product prices of changes in the output levels of our producing region. One is a competitive environment in which both product prices, p_1 and p_2, are given to the region and do not vary with changes in q_1 and q_2. The other is where changes in either q_1 or q_2 influence the levels of market prices. In the first case, the region will face straight line iso-revenue curves. In the second case the iso-revenue curves will be convex to the origin over the relevant range of output. A fuller discussion of the price (revenue) side of the model is contained in Abel, Welsch, and Jolly.[3]

Our model assumes Cobb-Douglas production functions to be relevant throughout the full range of production – from complete specialization in q_1 to complete specialization in q_2. We would like to make two points about this assumption. First, there is no need to suppose that the agricultural production world is fully represented by Cobb-Douglas functions. Other forms of production functions, such as quadratic or CES production functions, may be more appropriate in some circumstances. Second, there is no reason to expect a particular form of production function to hold over the full range of possible factor substitution. At best, any given form may be a good approximation over a given (and sometimes small) range of resource substitution between the two production functions. At the extreme ranges of substitution between q_1 and q_2 the production possibility curve might exhibit either a complementary or a supplementary relationship in the production of q_1 and q_2.

The model presented above has some interesting properties. Most important is that the production possibility curve will have little curvature for a wide range in values of the production elasticities α and β. (This result will hold over the range in output variation for which the Cobb-Douglas production functions are good approximations of the real world.) This has been clearly demonstrated by Johnson and can be easily verified by evaluating equation (3) for alternative values of α, β, and ℓ.[4] From this result, it follows that the sensitivity of the output mix of q_1 and q_2 depends very much on whether the producing region operates as a price-taker or whether changes in the outputs of the region influence product prices. This is illustrated in Figure 22-1. One can easily see how slight variations in the product price ratio, P, would cause large changes in the output mix along the production possibility curve $f(q_1^0, q_2^0) = 0$.

On the other hand, when our region faces downward-sloping demand curves for one or both products, a high degree of stability in output mix is as-

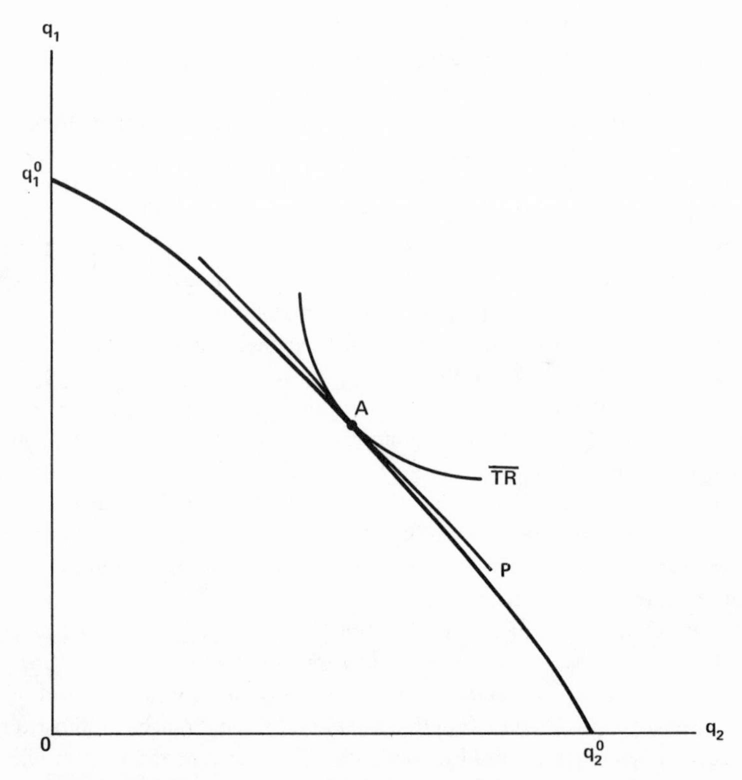

Figure 22-1. Equilibrium in production with either fixed relative prices or downward sloping demand curves.

sured. Exogenous shifts in the demand curves for the two products of our region will result in a rotation of the conic section represented by the iso-revenue line $\overline{TR}$ in Figure 22-1. The less the curvature of the iso-revenue lines, the greater will be the effect of exogenous shifts in the demand curves on changes in the output mix. In other words, as the price elasticities of demand approach infinity, the situation we assume to prevail under a competitive framework, the curvature of our iso-revenue line approaches a straight line, and the effect of a given rotation of the iso-revenue line on changes in the output mix increases.

Technological Change

We now wish to examine the consequences of certain types of technological change in the context of our two-commodity, two-factor world. National research leaders are faced with the question of the allocation of research resources among commodities. Even if research administrators follow the Hayami-Ruttan prescription of generating technological change of a type which is consistent with relative factor endowments and (undistorted) relative factor prices, they are still faced with the question of how best to allocate research resources among commodities.[5] As we shall see, the decision on how research resources are allocated depends not only on characteristics of the research production functions, but also on the nature of the demands for the final products. Three alternative situations are analyzed.

Situation I

This situation is presented graphically in Figure 22-2. The following assumptions are employed.

1. The initial production possibility curve, $f(q_1^0, q_2^0) = 0$, is a straight line which implies $\alpha = \beta$.

2. If q_1 and q_2 are measured in terms of the same physical units, complete specialization in q_1 results in greater output than does complete specialization in q_2.

3. Our producing region can face either fixed prices or downward-sloping demand curves for its outputs.

4. There is a fixed research budget which can be allocated between generating changes in τ_1 or τ_2. Thus, we are concerned with determining the optimum allocation of research resources subject to a research budget constraint.

5. The research production functions for τ_1 and τ_2 exhibit constant returns to scale. For simplicity, we assume the research production functions are of such a nature as to make $q_1^0q_1^1 = q_2^0q_2^1$. The latter assumption implies that the two research production functions yield identical absolute increases in production for equal research expenditures on τ_1 and τ_2. The analysis can be modified in appropriate ways for alternative assumptions about $q_1^0q_1^1$ and $q_2^0q_2^1$; e.g., a given budget increases efficiency in equal proportions for q_1 and q_2.

The implications of our assumptions are:

1. Allocation of all research resources to increasing τ_1 results in a new production possibility curve $f(q_1^1, q_2^0) = 0$. Similarly, allocation of all research resources to increasing τ_2 results in a new production possibility curve $f(q_1^0, q_2^1) = 0$. Under the assumption of constant returns to scale in the research production function, linear combinations of research expenditures trace out an innovation possibility frontier which is convex to the origin. The innovation possibility frontier represents the highest output combinations attainable from alternative allocations of a fixed research budget. We can illustrate this result in the following way. Assume that research resources are equally divid-

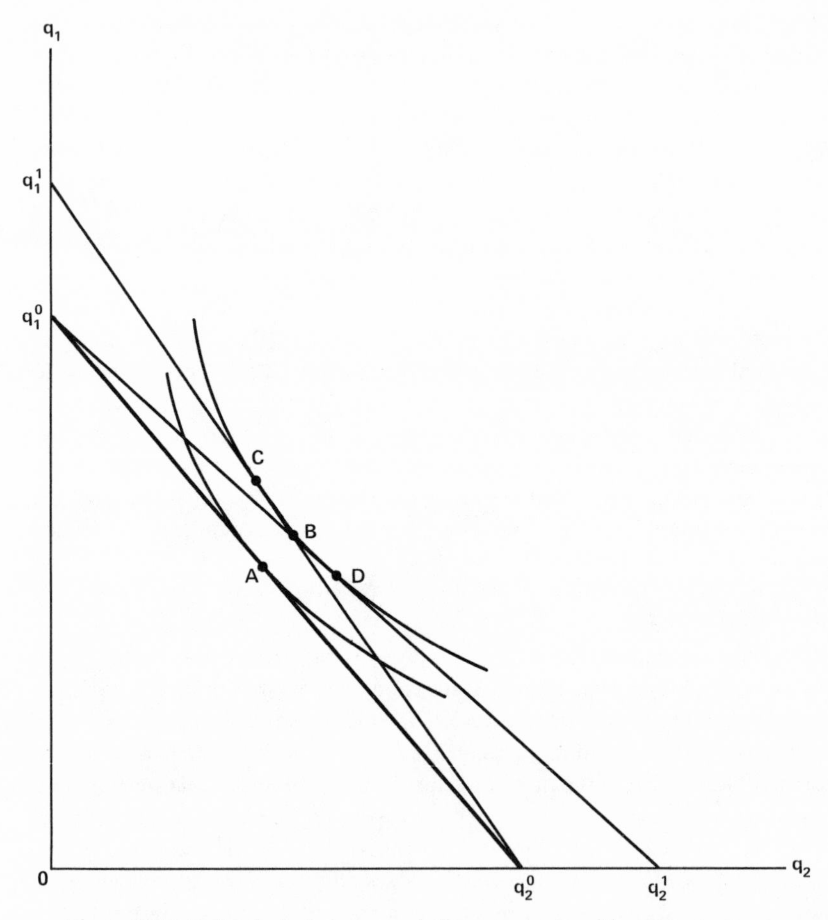

Figure 22-2. Technological change and the output mix with constant returns in research – linear production possibility curves.

ed between increasing τ_1 and τ_2. We get a new production possibility curve such as $f(q_1{}^2, q_2{}^2) = 0$. The line segment CD represents higher levels of output than are attainable from either $f(q_1{}^1, q_2{}^0) = 0$ or $f(q_1{}^0, q_2{}^1) = 0$. If one rotates line $f(q_1{}^2, q_2{}^2) = 0$ to reflect alternative combinations of research resources one can see that this traces out an innovation possibility frontier which is slightly convex to the origin.

2. If the producing region faces fixed prices, it pays to specialize completely in research, and there will be complete specialization in production of either q_1 or q_2. If product prices are such that their initial result is complete special-

ization in q_1 at level $0q_1^0$, our producing region would benefit most from investing all research resources in increasing output of q_1; i.e., generating the new production possibility curve $f(q_1^1, q_2^0) = 0$. The reader can verify that even with a range in relative prices which would result in production of either $0q_1^1$ or $0q_2^1$, total output would be greater at $0q_1^1$ and, therefore, increasing τ_1 is superior to increasing τ_2. If prices are given but result initially in specialized production of $0q_2^0$, then the converse of the situation above holds with respect to technical change. (This would not necessarily hold if $f(q_1^0, q_1^1) = 0$ were sufficiently different from $f(q_2^0, q_2^1) = 0$.)

3. If the region faces downward-sloping demand curves, not only will the region produce a combination of q_1 and q_2, but also the highest level of production is obtainable from allocating research resources to increasing both τ_1 and τ_2. In Figure 22-2 we show that, given the iso-revenue line, the highest level of output is achieved at B, which is on the new production possibility curve $f(q_1^2, q_2^2) = 0$. Furthermore, the more price inelastic the demand curves, the more convex to the origin will be the iso-revenue curves and the smaller will be the effect of technical change on the changes in the output mix.

Situation II

In this case we modify situation I by assuming that decreasing returns to scale prevail in the research production functions. This is probably the most realistic assumption about returns to scale in research. Decreasing returns could arise in two possible ways. First, the static research production functions could exhibit decreasing returns to scale because the stock of "basic" knowledge from which the research activities draw is fixed at any point in time. We assume that our research activities are not directed toward expanding the supply of "basic" knowledge. Second, if one views research as a probabilistic search process, decreasing returns in the research production functions are likely to prevail.[6]

All the remaining assumptions in situation I hold in situation II. The results are illustrated in Figure 22-3.

The implications of our assumptions are:

1. Allocating all research resources to increasing τ_1 results in the new production possibility curve $f(q_1^1, q_2^0) = 0$. Similarly, allocating all research resources to increasing τ_2 gives us $f(q_1^0, q_2^1) = 0$. Linear combinations of research resources on τ_1 and τ_2 will trace out an innovation possibility frontier which is convex to the origin, but less convex than in the case of situation I. We can illustrate this in the following way. Because of decreasing returns in both our research production functions, $q_1^0q_1^2 > 1/2\, q_1^0q_1^1$ and $q_2^0q_2^2 > 1/2\, q_2^0q_2^1$. The line segment BC in Figure 22-3 is relatively longer than CD in

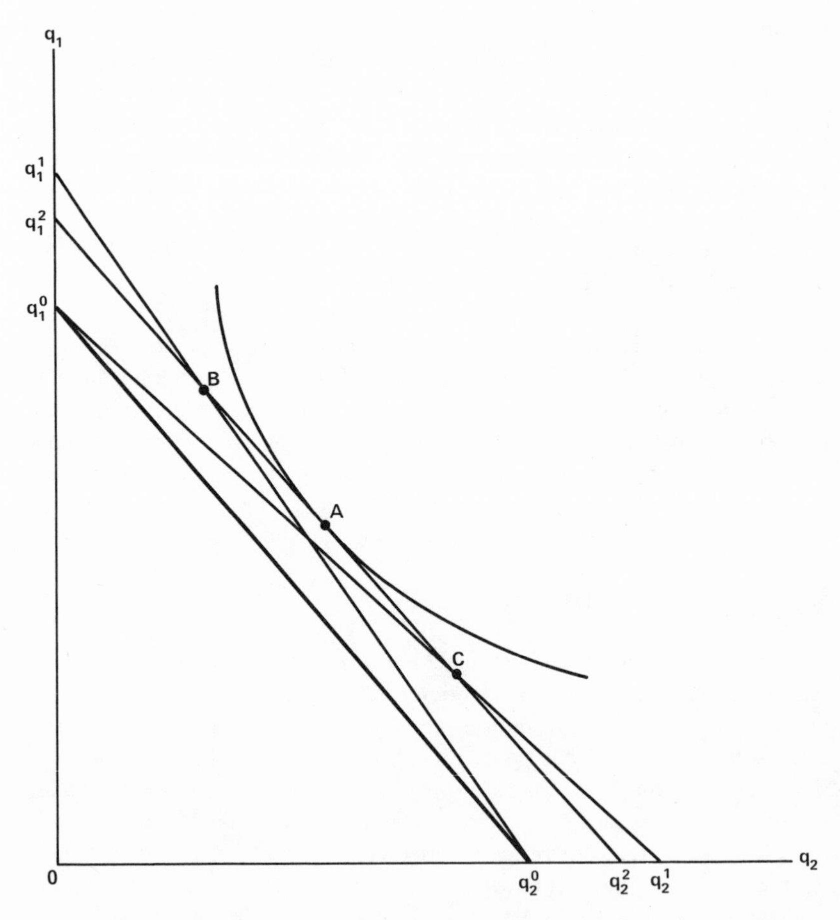

Figure 22-3. Technological change and the output mix with decreasing returns in research – linear production possibility curves.

Figure 22-2. If one rotates line $f(q_1{}^2, q_2{}^2) = 0$ to reflect alternative combinations of research resources, keeping in mind that decreasing returns to scale in the research production functions result in successively smaller increments in τ_1 or τ_2 for successive absolute increases in research resources of a given size, one can see that this traces out an innovation possibility frontier which is convex, but less so than in Figure 22-2.

2. If the producing region faces fixed prices, it pays to specialize completely in research, and there will be complete specialization in production of either q_1 or q_2. This result is the same as that obtained in situation I.

3. If the region faces downward-sloping demand curves for its products,

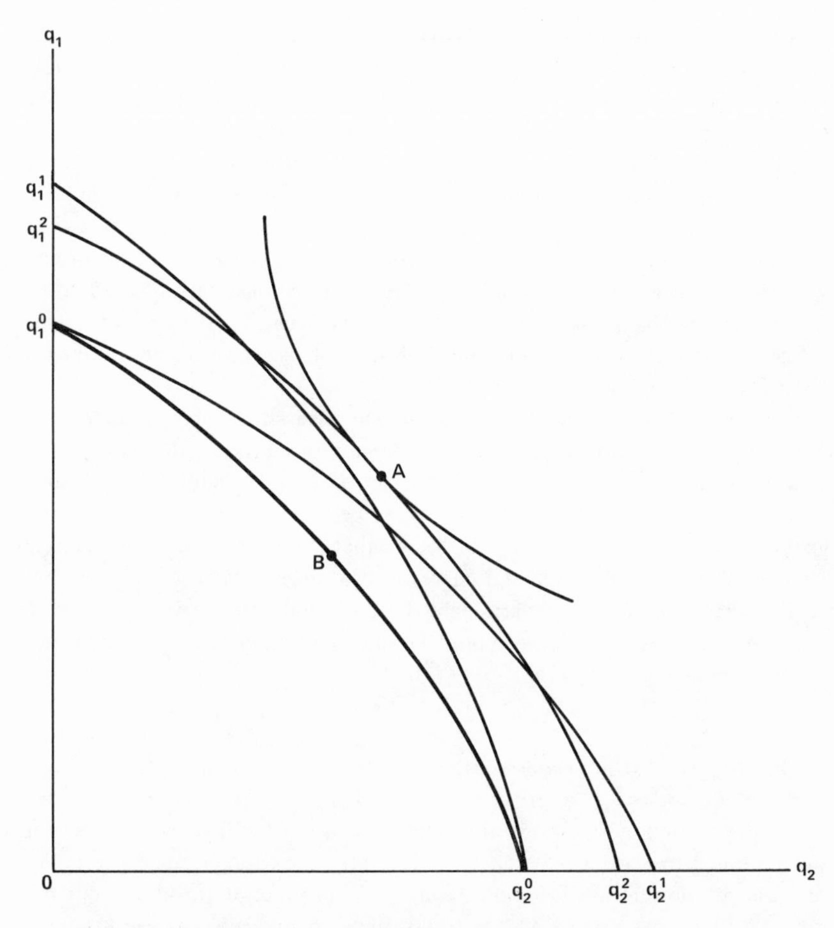

Figure 22-4. Technological change and the output mix with decreasing returns in research – concave production possibility curves.

not only will the region produce a combination of q_1 and q_2, but also the highest level of production is obtainable from allocating research resources to increasing both τ_1 and τ_2. In Figure 22-3 we show that, given the iso-revenue line, the highest level of output is achieved at A, which is on the new production possibility curve $f(q_1{}^2, q_2{}^2) = 0$.

Situation III

In this case we make the same assumptions as in situation II except that we now assume the initial production possibility curve, $f(q_1{}^0, q_2{}^0) = 0$, is concave to the origin. The results of these assumptions are shown in Figure 22-4.

The implications of our assumptions in this situation are as follows:

1. With given prices, the region would completely specialize in the production of q_1 or q_2 only if the terms of trade were sufficiently in favor of one output or the other. Otherwise the region would produce some combination of q_1 and q_2. The more concave the production possibility curve, the more likely it is that there would not be complete specialization in production.

2. Alternative combinations of research resources for increasing τ_1 and τ_2 will trace out an innovation possibility frontier which is concave to the origin. This can be shown by the same procedure suggested in situation II. As in the previous case, the production possibility curve $f(q_1{}^2, q_2{}^2) = 0$ is the one which results from allocating one-half of available research resources to each commodity.

3. In this situation, it might pay to allocate research resources to increasing both τ_1 and τ_2, regardless of whether the region faced fixed product prices or downward-sloping demand curves. This can be seen in Figure 22-4. Assume that relative prices are such that the price line for fixed prices would be tangent to $f(q_1{}^2, q_2{}^2) = 0$ at A. Also assume that the iso-revenue line resulting from downward-sloping demand curves is also tangent to $f(q_1{}^2, q_2{}^2) = 0$ at A. In either case, the highest attainable level of production results from an allocation of research resources to both τ_1 and τ_2 which generates the new production possibility curve $f(q_1{}^2, q_2{}^2) = 0$.

Situation IV

One might also wish to consider the case where the research production functions exhibit increasing returns to scale.[7] Increasing returns may prevail if the research production functions are S-shaped and the fixed research budget is small enough to restrict research activities to the increasing returns portion of the research production function. If the initial production possibility curve is a straight line, as in Figures 22-2 and 22-3, the new innovation possibility frontier representing alternative combinations of research expenditures on q_1 and q_2 will be convex to the origin. If, on the other hand, the initial production possibility curve is concave, the new innovation possibility frontier could be less concave, a straight line, or convex, depending on the degree of increasing returns in the research production function. Increasing returns to research will result in complete specialization in research activity so long as the new innovation possibility frontier is convex. This will be so whether or not the region faces given prices or downward-sloping demand curves for its products.

Resource Endowments and Environmental Constraints

We can also use our model to illustrate how different resource endowments affect both the output mix and the allocation of research resources. Let us

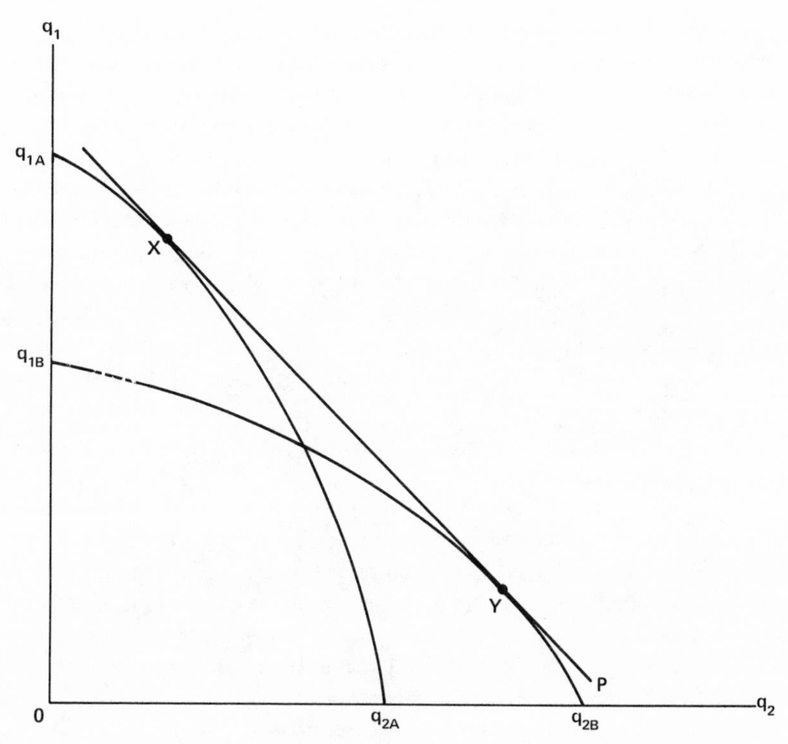

Figure 22-5. Relative factor endowments and the allocation of research among products.

assume that (a) there are two regions, A and B, producing the same two outputs q_1 and q_2; (b) the production function for each output is the same in both regions; (c) the production of q_1 is more intensive in the use of land (capital) relative to labor than the production of q_2; and (d) one region, A, has relatively more land than labor compared with the other region, B.

The initial situation is illustrated in Figure 22-5. The production possibility curve for region A is $f(q_{1A}, q_{2A}) = 0$ and that for region B is $f(q_{1B}, q_{2B}) = 0$. Since the production of q_1 is relatively more land (capital) intensive than the production of q_2 we would expect region A to favor the production of q_1. With both regions facing the same fixed relative prices, P, the output mix of region A would be at point X and the output mix of region B at point Y in Figure 22-5. The results are as one would expect. Region A, which has an abundance of land (capital) relative to labor, produces more of q_1 than q_2, and region B, which has an abundance of labor relative to land (capital), produces more of q_2 than q_1.

Employing the same type of analysis concerning technological change as was used in the previous section and assuming the same fixed relative prices, P, in both regions as shown in Figure 22-5, one can verify that (a) in region A it would pay to invest a higher proportion of the research budget in increasing τ_1 than in increasing τ_2, and (b) in region B it would pay to invest a higher proportion of the research budget in increasing τ_2 than in increasing τ_1. However, the results may change as relative product prices change. If the price of q_2 is significantly higher relative to the price of q_1 then in the situation illustrated in Figure 22-5 region A would allocate more resources to increasing τ_2 than τ_1. With sufficiently strong product price incentives in favor of q_2 both regions A and B would allocate proportionately more of their fixed research budgets to τ_2 than to τ_1. The reverse would be true with sufficiently strong price incentives in favor of q_1.

In addition to the role of demand conditions for the final products and the nature of the research production functions, variations in relative factor endowments and in relative factor intensities with respect to the outputs also play important roles in determining the allocation of research resources. For example, under the product price assumptions illustrated in Figure 22-5 the labor "rich" region will allocate relatively more research resources to the labor intensive commodity, and the labor "poor" region will allocate relatively more research resources to the land (capital) intensive commodity. (See chapter 24 for empirical support for this proposition.)

In the paper on which this chapter is based we used the model to examine how several environmental constraints affect the allocation of research resources and the output mix. We examined the effects of four types of physical or institutional (economic) situations: (1) heterogeneity in the quality of at least one factor of production; (2) restrictions on the use of certain technologies; (3) restrictions on the output of one commodity; and (4) improvement in the quality (productivity) of one or more inputs. The results can be summarized as follows:

1. Heterogeneity in the quality of factors increases the likelihood that it is profitable to allocate research resources to increasing factor productivity for both commodities.

2. Where restrictions are placed on the use of certain technologies, the optimum allocation of research resources depends heavily on final demand conditions. If relative prices more *strongly* favor the production of one product, say, q_2, then research resources should be allocated more to increasing τ_2 than τ_1. As relative prices move more in favor of the other product, say, q_1, the relative mix of research resources will move in favor of increasing τ_1.

3. When an output restraint for one commodity is binding, it may still pay to devote some research resources to increasing factor productivity for that

commodity. However, the general effect of the restraint is to cause a reallocation of research resources to increasing factor productivity for the unrestrained commodity.

4. Investments may be made to improve the productivity of one of the inputs, say, land. As an example, consider a situation in which irrigation is feasible but there is little control over the application of water in individual fields. The effect of improvements in the irrigation system which result in full water control in individual fields will, under given product prices, result in a switch from complete specialization in the initial crop to complete specialization in an alternative crop.

Some Implications

Our analysis shows that the optimum allocation of research resources among commodities and its effect on the output mix of a region depend upon the initial production conditions (concavity of the production possibility curve and the relative size of q_1 and q_2 with complete specialization in the production of each), the extent to which there are either increasing or decreasing returns to scale in research, whether the producing region faces given prices or downward-sloping demand curves for its outputs, and changes in relative factor endowments. To decide on the optimum allocation of research resources among commodities, research administrators need information on all four aspects of the problem.

If the production possibility curve is relatively flat and the region is a price-taker, we would expect significant shifts in the output mix as a result of changes in relative output prices. Furthermore, the allocation of research resources depends heavily on relative product prices and return to scale in research. Research resources would be devoted entirely to increasing the production of q_1 if (a) prices initially favor complete specialization in the production of q_1, (b) there are constant or increasing returns to scale in research, and (c) there are identical production functions for τ_1 and τ_2. Research would strengthen the tendency toward complete specialization in production. On the other hand, if the production possibility curve is concave, both q_1 and q_2 would tend to be produced, except in the case where the region faced fixed prices which were of such an extreme nature as to dictate complete specialization in production. Except for the extreme case, research resources would be allocated to increasing both τ_1 and τ_2.

Even if the production possibility curve is relatively flat over a wide range of variation in q_1 and q_2, we may still observe a high degree of stability in the output mix even with technological change, because the region faces downward-sloping demand curves for its outputs. The more price inelastic the demand curves, the more convex the iso-revenue lines and the less sensitive is the output mix to technological change. Furthermore, even with downward-sloping demand curves, it would still pay to devote all research resources to one commodity if the combination of (a) the slope of the initial production possibility curve and (b) returns to scale in research resulted in an innovation possibility frontier which was either a straight line or convex.

A region might face downward-sloping demand curves for its products either because of short-run rigidities in parts of the marketing system or because changes in output levels of a region were sufficient to change prices throughout the marketing system. There is evidence that significant changes in the production of one crop can cause temporary distortions in the relative price structure of a region compared with prices in a larger marketing area. Lele, in her study of sorghum grain marketing in western India, found that distortions in intermarket price differentials arose when the volume of grain production and marketings pressed against the supply of transport services.[8] Jolly, in a study of corn and soybean price behavior in southwestern Minnesota, found that the margin between central market prices and local prices was a function of the level of output and the output mix in the local region.[9]

Yamaguchi, and Yamaguchi and Binswanger, in a study of the effect of technical change and population growth on the economic development of Japan, observed patterns of production and price behavior consistent with our model.[10] In looking at the agricultural and nonagricultural sectors (equivalent to our two commodities), they found (a) a very flat production possibility curve and (b) a high degree of stability in the output and consumption mixes, because the demand curves for the outputs of both sectors were downward-sloping and especially price inelastic in the case of demand for agricultural products.

In a situation with downward-sloping market demand curves, intervention in the markets for q_1 and q_2 by government (or other groups) in the form of price support measures or trade restrictions can yield results similar to the competitive model, i.e., intervention can result in a higher degree of specialization than would result from a market solution. (This does not automatically follow, because governments can also set the relative support prices in ways which will shift the terms of trade against the commodity experiencing the technological change.) Furthermore, price support programs or trade restrictions can also affect the allocation of research resources to the extent that product price behavior is important in determining such allocations.

The question of which commodity should receive research resources depends very much on society's developmental objectives and policies. For example, suppose it is the primary concern of policy makers to increase the incomes of producers, and relative prices are unimportant. Then one rule which could be followed is to increase the production of the commodity with the highest price and income elasticities. In this way one would tend to minimize the extent to which a shift in the terms of trade tends to counteract the effect of technological change. On the other hand, suppose one of the commodities is a wage good, it has lower price and income elasticities than the non-wage good, and it is the policy makers' desire to keep the price of the wage good as low as possible. In this case, it would make sense to invest research resources in bringing about technological change in the wage good, i.e., we want to maximize the shift in terms of trade against the wage good. These are but two of many possible situations.

We should be cognizant of the fact that the price elasticity of demand which a region or country faces depends on both domestic and export demand parameters. It is possible for the domestic demand curve to be quite price inelastic while the export demand curve facing the country or region is quite price elastic, e.g., the case of corn in Thailand. In such a situation it would be important for the country or region to follow price policies which did not exclude domestic production from entering export markets, if the policy objective were to minimize the adverse effect on terms of trade for corn of a change in output. On the other hand, if the name of the game were to keep domestic prices as low as possible, then export barriers might be erected, e.g., the case of the rice premium in Thailand.

Finally, we explored in our earlier paper the implications of four environmental situations for the allocation of research resources and for the resulting output mix. In each situation our model gives us useful insights. Demand conditions for the products play an important role in allocating research resources in each environmental situation considered.

Heterogeneity in the quality of factors of production imparts convexity to the production possibility curve. Regardless of demand conditions, heterogeneity in factors will tend to cause research resources to be allocated to both commodities. In the case of restrictions on the use of certain technologies in the production of one of the commodities, the optimum allocation of research resources depends heavily on final demand conditions. Restrictions on the level of output of one commodity should cause a reallocation of research resources to increasing factor productivity in the other commodity. However, it may still be profitable to allocate research resources to both commodities even when the output restraint is binding. Improving the quality of one factor can also have a significant effect on the output mix, with the nature of final demand conditions again playing an important role.

Conclusions

We have constructed a relatively simple theoretical model which shows that the allocation of a fixed research budget between research on two commodities and the effects of such allocations on the output mix of a region depend on the initial production conditions, the presence of economies or diseconomies of scale in research, the nature of the demands for the outputs of the region, changes in relative factor endowments, and the existence of certain types of environmental constraints. Research administrators require information on all these aspects of the problem in order to determine the optimum allocation of research resources.

Our analysis indicates that there is nothing inherently good or bad about diversification of production. Changes in output mix must be evaluated in terms of a country's resource endowments and developmental objectives. Environmental considerations clearly play an important role in effecting an optimal allocation of research resources. Price policies also have a significant effect, not only on the allocation of traditional resources among commodities in a region but on the allocation of research resources.[11] Falcon has argued cogently that agricultural price policies should be consistent with national development objectives.[12] Unfortunately, price and market policies designed in response to short-run food procurement needs or political pressures from producers often are not consistent with the price policies that would facilitate achievement of longer term development goals.

NOTES

1. This chapter draws heavily on Martin E. Abel and Delane E. Welsch, "Microeconomics of Technology and the Agricultural Output Mix," Staff Paper P74-16, Department of Agricultural and Applied Economics, Minneapolis, University of Minnesota, August 1974.

2. Harry G. Johnson, "Factor Market Distortions and the Shape of the Transformation Curve," *Econometrica*, 34 (July 1966), 686-698; Martin E. Abel, Delane E. Welsch, and Robert W. Jolly, "Technological Change and Agricultural Diversification," Staff Paper P73-10, Department of Agricultural and Applied Economics (St. Paul: University of Minnesota, January 1973).

3. Abel, Welsch, and Jolly, "Technological Change."

4. Johnson, "Factor Market Distortions."

5. Yujiro Hayami and Vernon W. Ruttan, *Agricultural Development: An International Perspective* (Baltimore and London: Johns Hopkins Press, 1971).

6. Robert E. Evenson and Yoav Kislev, "A Model of Technological Research," processed, August 1971.

7. Robert E. Evenson, "Economic Aspects of the Organization of Agricultural Research," *Resource Allocation in Agricultural Research*, ed. Walter L. Fishel (Minneapolis: University of Minnesota Press, 1971).

8. Uma J. Lele, "Market Integration: A Study of Sorghum Prices in Western India," *Journal of Farm Economics*, 49 (February 1967), 147-159.

9. Robert W. Jolly, "The Derived Demand for Specialized Inputs by a Multi-Product Firm: An Examination of Corn and Soybean Buying by Minnesota Country Elevators," processed, 1973.

10. Mitoshi Yamaguchi, "Technical Change and Population Growth in the Economic Development of Japan," Ph.D. dissertation (Minneapolis: University of Minnesota, 1973); Mitoshi Yamaguchi and Hans P. Binswanger, "The Role of Sectoral Technical Change in Development: Japan 1880-1965," Department of Agricultural and Applied Economics Staff Paper P74-7 (St. Paul: University of Minnesota, April 1974).

11. The role of price in the allocation of resources among crops in developing countries was highlighted by Raj Krishna, "Farm Supply Response in India-Pakistan: A Case Study of the Punjab Region," *Economic Journal*, 73 (September 1963), 477-487, and subsequently by many other analysts. See also Raj Krishna, "Agricultural Price Policy and Economic Development," *Agricultural Development and Economic Growth*, ed. Herman M. Southworth and Bruce F. Johnston (Ithaca: Cornell University Press, 1967).

12. Walter P. Falcon, "The Green Revolution: Generations of Problems," *American Journal of Agricultural Economics*, 52 (December 1970), 698-710.

23

Relating Research Resource Allocation to Multiple Goals

John W. Mellor

In most countries, technical change is necessary for agriculture to play a positive role in economic development. With limited land area and classic diminishing returns to the use of other inputs, it becomes more and more costly to meet the increasing demand for food unless research provides new, higher yielding crop and livestock technologies. The need for effective agricultural research is strongly reinforced in those many low-income countries in which productivity of agricultural labor and even capital have already been driven to very low levels by past population growth. Such research is further necessitated as social objectives accelerate the demand for food by fostering increased employment of low-income people who wish to spent the bulk of their added income on food.

This latter "distributional" pressure on food production capacity will become even greater if international and national demands for more jobs and broader participation in growth are to be met. The dramatic effect on the demand for food of change in income distribution is illustrated by Indian data which show that while the top 5 percent in the expenditure distribution spend only 2 percent of increments to income on food grain, the bottom 20 percent spend 59 percent.[1]

The increasing societal need and demand for agricultural research results lends urgency to achieving efficient allocation of scarce research personnel and institutional resources. This scarcity also requires that attention be given to organizing and conducting research to expand research capacity. This need for expansion results in part from past failures to recognize the importance of agricultural research and past and present deficiencies in agricultural education. Thus a current objective of research programs must be to make their own growth self-sustaining, through developing their own personnel and institutional resources.

The crucial task of determining public policy for optimal allocation of agricultural research resources cannot rely on the market price mechanism alone. Three particular problems arise in addition to the usual one of the extent to which prices reflect the underlying supply and demand for the commodities from which research derives its returns. First, the relationship between quantity of research resources and productive output is poorly understood. That is, once a problem is defined – even a straightforward one such as increasing yield per acre of a particular crop in a particular area – there is little knowledge of how much research investment of what type will give how much return. Second, agricultural research has a unique capacity to affect variables, such as health and nutritional status, which because of factors such as consumer ignorance may not be given their true societal value in market prices. Third, in a world of unequal distribution of productive assets and income, new technology affects the distribution of income by changing both the relative returns to owners of various productive resources and the prices of goods consumed in unequal proportion by various income classes. Even though society may accept the initial inequality in income and assets, it may not find increased inequality agreeable. We may view determination of public research policy as a process of explicitly creating the demand for research of certain types rather than simply responding to existing demand through estimates of market forces and their effects. In this alternative view, policy may call for an allocation of research resources quite contrary to that suggested by the market relationships.

Research resource allocation is a problem in multivariate analysis: it is a problem of multiple goals and multiple instruments complicated by a lack of information about the relationships between dependent and independent variables. Perhaps it is the bewildering complexity of the task which accounts for the tendency of decision-makers in research resource allocation to follow a simple sequential procedure, emphasizing only one or a few objectives at a time but gradually pyramiding to a meeting of multiple goals. In this approach to decision-making in a context of multiple goals, it is important to set priorities for an appropriate sequence of actions to meet those goals. Even this approach is complicated because societal objectives may be reached through diverse instruments, of which allocation of various types of research resources is only one. The trade-offs in efficiency among the various policy instruments must be carefully considered in allocating research resouces, and they will certainly affect the sequence of efforts to achieve specific objectives.

The problem of multiple goals presents two major tasks: first, to categorize the goals with sufficient detail and specificity to give them operational meaning; and, second, to set priorities among objectives, reflecting the probabilities of success and the comparative costs of alternative research lines for reaching the various objectives and, perhaps, outlining a schedule for time phasing and ordering.

This chapter attempts to shed light on research decision-making first by a discussion of some specific aspects of societal objectives, then by a brief statement about the nature of the research resources to be allocated, and finally by an analysis of specific instruments of research policy in terms of implications for their choice and manipulation. The analysis concludes with a hypothetical research policy sequence and a statement of research needs for improving research policy. (For a theoretical treatment of some of the issues discussed in this chapter, see chapter 22.)

Societal Objectives of Agricultural Research

Society's objective in research is presumably to increase its own welfare. That objective is advanced by research allocations which maximize the per capita stream of welfare over some time span from the initial stock of resources. This concept allows for augmenting the per capita stock of resources and using research resources specifically to generate more research resources, perhaps in the short run conflicting with research output. Since welfare is likely to be defined in per capita terms, we can say that the further objective of research is to raise the productivity of the stock of labor, subject to the constraint of that labor's working conditions and of fixed resources such as land. Given existing concern about questions of mechanization, employment, and income distribution – to which this chapter will return – it is worth emphasizing this basic concern with labor productivity. At the same time, it must be noted that in densely populated countries research policy may still focus on the intermediate objective of increasing the productivity of land.

Under pressure to simplify choice by concentrating on a simple objective function, the agricultural science research worker often focuses first on yield per acre or on a crop output index. In doing so he is apt to have in mind a set of market weights. Once yield is markedly increased, various subobjectives are pursued, usually sequentially, including reducing the use of inputs such as chemicals and hence costs and reducing variance of yields. Apparently substantial gains in research efficiency are achieved through such simple statement and pursuit of objectives in sequential order. And the resulting loss in economic efficiency may be slight, if price relationships correctly reflect society's costs and values. The demand for simple objectives undoubtedly traces from the very large number of variables with which the agricultural scientist deals – from a wide range of soil and climate variables on the input side to physical output, variance in output, and consumption quality on the output side. Without even considering social factors the agricultural scientist works in a multiple-goal environment.

Beyond the well-known, general market imperfections, two specific problems arise in the use of price-weighted output as an objective function. First is the possibility that consumers may be ignorant of the relationship between a product and aspects of its welfare function such as health and nutrition. Second is the fact that relative prices result from a distribution of income and assets that itself may be inconsistent with societal values. This in turn reflects a broader concern with what society perceives to be a maldistribution of income, often intensified by technical change. Needless to say, the use of "international" rather than domestic prices substitutes a different set of societal structures which may be equally arbitrary and equally inconsistent with current national objectives.

With respect to the effect of consumer ignorance of price relationships, an important caveat is in order. The perception of utility may differ between rich and poor, and it may well be the former, not the latter, who are in error. For example, the poor in low-income countries are unlikely to pay more for increased protein content in food because, contrary to widespread belief, their diet is far more deficient in calories than in protein.

Distribution of income is of increasing concern to society. The fact that technical change has the potential profoundly to influence income distribution enhances the importance of enlightened research policy. However, research policy is only one of many instruments for altering income distribution. The emphasis on the distribution question arises from the divergence between societal objectives with respect to income distribution, the actuality of productive-asset distribution, and the apparent inability of society to redistribute either assets or income directly. On the one hand, inequality of distribution of assets and income may be reinforced by technological change which biases returns toward one production factor. On the other hand, those very biases offer an opportunity to change income distribution in a desired direction. Society may wish to allocate research resources so as to influence factor shares in the desired direction or at least to know how innovation will affect income distribution so that appropriate complementary policies may be instituted. Choice of crops, regions, and disciplines of emphasis all will affect the distribution of gains among regions, between landowners and laborers, between men and women, and among other class groupings.

It must be recognized that, although there is necessarily a loss in efficiency in attempting to meet distributional objectives through research allocation rather than through other means, all societies appear to face this problem. To attempt to equalize income distribution by breeding labor-intensive crop varieties rather than redistributing the land is analogous to moving the piano to the piano stool. Indeed one wonders why, when the social scientist complains about the skewing of income distribution by the green revolution, the biological scientist hasn't pointed out that better income distribution could be rather simply effected through land redistribution.

In more sophisticated terms, new technology changes the distribution of income because it increases the relative productivity and demand for various factors of production. If all factors were equally owned or if all income were equally divided, no skewing of income would result from change in technology. It is the imposition of change in factor shares on an existing inequality which creates the problem. Clearly, to divert research attention from increasing output per unit of input to additional concern with the bias in factor shares interferes with the accomplishment of primary objectives.

The distributional problem is particularly acute with respect to labor shares. Since their income is directly related to the amount of work needed to be done, laborers are particularly vulnerable to new technologies which increase labor productivity and thus reduce needed man-hours, as do almost all yield-increasing innovations. Such innovation may decrease the demand for labor and result in both a lower wage and less employment. The return to land rises of course since it is the residual claimant after payment to labor. If land were distributed in proportion to labor, then what an individual lost on return to labor would be gained in return to land. A system of taxes on land or on technology-associated inputs could be devised to have the same effect. It should also be noted, as above, that society's welfare objective almost certainly includes raising the productivity and reducing the onerousness of labor – creating a conflict if the very act removes work opportunity for a segment of the population, because of differences in asset distribution and power.

The points made above apply also to the distribution of income among regions. Even if income and assets are equally divided within a region they may occur unequally among regions. In such a case, new technology may intensify interregional disparities. In this context, mainland China's rising rate of regional disparity is especially instructive. Within the commune, mechanization apparently has little effect on income distribution, but since each commune reaps benefits of its own land and labor, those communes where technological input is concentrated will experience greater increase in incomes. In this system, as in others, it is difficult to redistribute assets, labor force, or income. In China, too, then, the second best solution – effecting income redistribution through allocation of research resources – may need to be considered.

To summarize, the widely accepted societal objective of increasing total human welfare is subject to the constraint of the distribution of benefits among and within regions. Moreover, welfare is inadequately measured by the price mechanism and nutritional concerns may often be understated. Finally, because of our imperfect knowledge of the effect of research inputs, only very general policies on research resource allocation can be considered.

Before proceeding to a discussion of the resources to be allocated and the instruments of research policy, we may note that actual decisions on research policy are made in a bureaucratic framework in which the primary objective may be one of maintaining power, with consequent potential for departure from general welfare objectives with respect to effects of decisions on the growth of income and its interpersonal and interregional distribution. Indeed, depending on the sources of the political and bureaucratic power, there may be little or no actual concern for the substance of objectives as stated in the preceding paragraphs. An analysis of decision-making in that context would give much more weight to the specific nature of administrative structure and power relationships than does this discussion.

The Nature of Agricultural Research Resources

The agricultural research resources to be allocated are the physical plant and trained personnel and, most important, the institutional structure into which plant and personnel are organized for a productive purpose. Because of the potentials for transfer of research results between indigenous and foreign research units, the nature and quantity of research resources available for indigenous purposes are a complex composite of what now exists plus what will be coming into existence. So little is known of how to train competent personnel and how to build institutions that the most effective way to promote growth in the research resource base is almost certainly through developing the research structure itself.

What this means, in practice, is that the research resources to be allocated should not be viewed in a static sense. In a world of less than infinite discount rates, the very mode of allocation of research resources will strongly influence the quantity of such resources to be allocated in the future. In this context, we need to consider several specific questions.

To what extent can time be saved in the short run by use of foreign technicians, and will such input reduce or increase institutional capabilities in the long run?[2] Will regional diversification of research slow the growth of the system by fractionization, or will it speed growth by forcing decentralization and encouraging diffusion of responsibility and consequent growth in capacity? How far will a close relationship with an international system expand results and personnel resources? Will such a relationship effect institutional development, or will it have deleterious consequences through setting up conflicts between national and international perceptions and objectives, by providing less encouragement for important indigenous capabilities? The problem of finding answers to these questions – the dynamics of which may be much more important than the status of allocating current resources – underscores the difficulty of decision-making in this area. We can see how these questions may be decided on the basis of political rather than technical considerations, how the sequential method may be chosen in preference to a multidirectional approach.

The Instruments of Research Policy

The three prime instruments of research policy are (1) allocation among crop and livestock systems; (2) choice of emphasis, within a farming system and among geographic regions, on the productivity of the various factors of production; and (3) selection of research disciplines. The latter, of course, interacts with the others.

From the preceding sections of this chapter it can be seen that, beyond market-price based criteria, analysis for each of these areas of allocation choice should be made with respect to effect on (1) the degree to which the supply of those food commodities most heavily weighted in the consumption patterns of the poor will support increased employment and rising real incomes of low-income people; (2) the demand for labor and the distribution of actual employment among various social classes; (3) the level of total output and producers' net income; (4) variation in producers' net income; and (5) the nutritional composition of the food supply. Through these instruments, research and resultant technological change may have a profound effect on the structure of the society. Understanding of the relationships is important in pointing the way to optimal allocation of research resources. Such knowledge can be even more important in assisting societies to adjust to the stresses which typically accompany technological changes no matter how beneficial such change may ultimately be.

Choice of Crop and Livestock Systems

Research results affect the choice of farming systems by increasing the efficiency and reducing the cost of production of one system compared with another. Technological change is likely to increase production both directly, with a given set of resources, and indirectly, by increasing relative profitability and thereby transferring resources from other enterprises. Farming systems vary in the extent to which they contribute to various objectives. Four specific characteristics are discussed below to illustrate the nature of such variation and its relationship to research resource allocation.

Food grains as basic wage goods. Allocation of research resources to food grains facilitates employment growth through expansion of the supply of basic wage goods. In India, as noted earlier, the two lowest deciles in expenditure class spend 59 percent of increments to income on food grains (Table 23-1). Without a large increase in imports and an effective system of rationing, an employment program which increases incomes of low-income laborers must be matched by rapid increase in domestic food grain production. If it is not, sharp inflationary pressures will bring about a demand, on the part of the politically powerful urban middle class for cessation of the program.[3] Such an effort to improve employment justifies an initial major emphasis on basic food grains in research resource allocation. Now if rapid technological change in food grain production is not matched by an increase in the incomes of low-income consumers, deficiency of demand will result in reduced relative food grain prices which on the one hand directly raise real incomes of the poor but on the other hand may remove the price incentive needed to stimulate continued application of high input technological methods. Clearly, there is a high degree of interdependence between technological progress in food grain production and high employment policies. Policy in respect to one must be meshed with policy on the other.

Employment content and factor shares. While research emphasis on food grains helps to release the most basic wage goods constraint to employment, it tends in itself to provide only a small portion of employment at the new equilibrium level. Although geographically concentrated breakthroughs in food grain production may provide substantial aggregate increase in employment in those areas, they augment the capacity to support increased employment by a vastly greater amount.

For example, in a typical case of the new wheat technology in India, only 10 percent of the added gross income produced by the high-yielding variety is expended for labor, while 67 percent is allocated to family-owned capital and land, and 23 percent is paid out for other inputs, of which fertilizer represents a major component (Table 23-2).

Table 23-3 illustrates two important aspects of employment in food grain production. First, there is considerable variability in the labor share of increased output. Second, typically only a small share – roughly 5 percent to 15 percent – is paid as wages. It follows then that food grain production, al-

Table 23-1. Division of Incremental Expenditure among Expenditure Categories, by Rural Expenditure Class, India, 1964-65

	Rural Expenditure Class by Decile[a]						
Expenditure Category	Bottom Two (landless)	Third (under 1 acre)	Fourth and Fifth (1-5 acres)	Sixth, Seventh, and Eighth (5-10 acres)	Ninth (10-15 acres)	Lower One-Half of Tenth (15-30 acres)	Upper One-Half of Tenth (30 + acres)
Mean per capita monthly expenditure (Rs.) . . .	8.93	13.14	17.80	24.13	30.71	41.89	85.84
Allocation of additional rupee of expenditure							
Agricultural commodities	0.79	0.69	0.59	0.52	0.46	0.40	0.33
Food grains	0.59	0.38	0.25	0.16	0.11	0.06	0.02
Nonfood grains	0.20	0.31	0.34	0.36	0.35	0.34	0.31
Milk and milk products	0.07	0.11	0.12	0.13	0.13	0.12	0.09
Meat, eggs, and fish.	0.02	0.03	0.03	0.03	0.03	0.03	0.02
Other foods	0.01	0.05	0.07	0.09	0.10	0.12	0.16
Tobacco	0.01	0.01	0.01	0.01	0.01	0.01	0.01
Vanaspati . . .		0.01	0.02	0.02	0.02	0.02	0.01
Other oils . . .	0.05	0.05	0.04	0.04	0.03	0.02	0.01
Sweeteners. . .	0.04	0.05	0.05	0.04	0.03	0.02	0.01
Nonagricultural commodities	0.21	0.31	0.41	0.48	0.54	0.60	0.67
Textiles.	0.09	0.08	0.07	0.08	0.07	0.06	0.07
Cotton textiles	0.09	0.08	0.07	0.06	0.06	0.05	0.03
Woolen textiles				0.01	0.01	0.01	0.02
Other textiles				0.01			0.02
Nontextiles	0.12	0.23	0.34	0.40	0.47	0.54	0.60
Footwear. . . .		0.01	0.01	0.01	0.01	0.01	0.01
Durables and semidurables	0.01	0.01	0.01	0.02	0.02	0.03	0.05
Conveyance	0.01	0.01	0.02	0.02	0.03	0.05	0.10
Consumer services	0.02	0.02	0.02	0.03	0.03	0.04	0.06
Education . . .	0.01	0.01	0.02	0.03	0.03	0.05	0.11
Fuel and light	0.07	0.07	0.06	0.05	0.05	0.04	0.03
House rent . . .		0.01	0.01	0.02	0.03	0.04	0.08
Miscellaneous		0.09	0.16	0.22	0.27	0.28	0.16
Total	1.00	1.00	1.00	1.00	1.00	1.00	1.00

Source: John W. Mellor and Uma J. Lele, "Growth Linkages of the New Foodgrain Technologies," *Indian Journal of Agricultural Economics*, 28:1 (January-March 1973), 35-55. The data are reported in B. M. Desai, "Analysis of Consumption Expenditure Patterns in India," Cornell University-USAID Employment and Income Distribution Project, Department of Agricultural Economics, Occasional Paper no. 54, Ithaca, Cornell University, August 1972. The source for the data is National Council on Applied Economic Research, *All-India Consumer Expenditure Survey, 1964-65*, vol. 2 (New Delhi, 1967).

[a] Roughly corresponding agricultural holdings are in parentheses.

Table 23-2. Allocation among Inputs of the Increased "Payments" from a High-Yielding Wheat Variety, Aligarh District, Uttar Pradesh, India, 1967-68[a]

Inputs	Allocation of Payments[b]		Allocation of Incremental Payments[c]		Percent of Increment in Gross Value of Production of Incremental Payments
	Traditional Variety	High-Yielding Variety			
Gross value of production	653	1,115	462	(71)	100
Payments to:					
All inputs except labor	573	989	416	(73)	90
Family land and capital	380	690	310	(82)	67
Inputs other than family land and capital	193	299	106	(55)	23
Fertilizer	37	76	39	(110)	8
Labor	80	126	46	(59)	10
Family labor	54	91	37	(68)	8
Hired labor	26	35	9	(35)	2

Source: Adapted from R. S. Dixit and P. P. Singh, "Impact of High Yielding Varieties on Human Labor Inputs," *Agricultural Situation in India*, 24:12 (March 1970).

[a] The traditional variety of wheat in Aligarh District yielded 7.5 quintals per acre, while the high-yielding variety yielded 14.8 quintals per acre, an increase of 96 percent. "Payment" is defined here as either the amount of income paid out for inputs such as fertilizer or the value of income assigned to inputs such as family labor.

[b] In rupees per acre.

[c] In rupees per acre, with percent of increment in parentheses.

though necessary to relax a wage goods constraint, has not normally provided a full solution to the employment problem. And consequently, not only is there insufficient effect on the social welfare objective, but the production objective is brought into question as well, since the increase in demand for food grains will not be commensurate with the increased supply. For that balance, one requires the indirect or linkage effects of the food grain production increase.[4] It is here that one must recognize the options in dealing with problems of policy. Emphasis on increasing the secondary employment effects of new production technology may be more effective than attempts to direct research so as to increase the primary employment effects.

A major potential for increasing employment within agriculture lies with increasing emphasis on and shifts among various nonfood grain crops. Schluter shows, for Surat District, India, that sugar can return 50 percent more income to labor per acre than rice and nearly four times as much as improved varieties of wheat (Table 23-4). Similarly, among unirrigated crops, both groundnut and cotton return more to labor than sorghum (but less than rice), while

Table 23-3. Division of Increased "Payments" between Labor and Other Inputs, Various High-Yielding Varieties and Areas, India

Area	Increase in Gross Value of Output		Increase in Labor "Payments"		Percentage of Increased Output to Labor[a]	Percentage of Increased Output to Other Inputs[b]	Percentage Increase in Labor "Payments" for a 1 Percent Increase in Gross Value of Output
	Rupees per Acre	Percentage Increase	Rupees per Acre	Percentage Increase			
					Wheat		
Aligarh, U.P.	462	71	46	58	10	90	0.8
Varanasi, U.P.	620	65	11	15	2	98	0.2
Udaipur, Rajasthan	343	43	18	13	5	95	0.3
Punjab	450	100	56	42	12	88	0.4
					Kharif Paddy		
West Godavari, Andhra Pradesh	269	38	32	17	12	88	0.4
East Godavari, Andhra Pradesh	216	33	20	13	10	90	0.4
Uttar Pradesh	1,100	200	67	92	6	94	0.5
Tamil Nadu	550	100	33	20	6	94	0.2
Laguna, Philippines	374	72	3	3	1	99	0.0
Sambalpur, Orissa	404	95	36	28	11	89	0.3
					Rabi Paddy		
West Godavari, Andhra Pradesh	562	86	39	16	7	93	0.2
East Godavari, Andhra Pradesh	761	153	39	30	5	95	0.2
Tamil Nadu	625	100	46	21	7	93	0.2
Gumai Bil, Bangladesh	948	208	302	125	32	68	0.6
					Bajra		
Kaira, Gujarat	300	85	39	27	13	87	0.3
Average	532	97	52	35	9	91	0.3

Source: John W. Mellor and Uma J. Lele, "Growth Linkages of the New Foodgrain Technologies," *Indian Journal of Agricultural Economics*, 28:1 (January-March 1973), 35-55, Table II.

[a] Labor "payment" is defined as physical labor input (family and hired) in man-days at a constant wage.

[b] Other inputs "payments" defined as gross value of output minus share to labor.

Table 23-4. Returns to Labor and Land per Acre for Unirrigated and Irrigated Crops over a Six-Year Period in Surat District, India

Crop	Returns to Labor[a]	Returns to Land[b]
Unirrigated, 1966/67-1971/72		
Rice (with fertilizer)[c]	129	202
Groundnuts	108	207
Cotton (without fertilizer)[c]	65	230
Sorghum	50	163
Irrigated, 1971-72[d]		
Improved rice	242	507
HYV rice	265	608
Improved wheat	82	237
HYV wheat	68	473
Sugarcane	383[e]	2219

Source: Michael G. G. Schluter, "Interaction of Credit and Uncertainty in Determining Resource Allocation and Incomes on Small Farms, Surat District, India," Cornell University-USAID Employment and Income Distribution Project, Department of Agricultural Economics, Occasional Paper no. 68, Ithaca, Cornell University, February 1974, chapter III, Table 7, p. 14.

[a] Estimated at the market rate. The number of man-days used for each operation is multiplied by the appropriate wage rate, with no distinction between the hired and family labor inputs. For unirrigated crops, we assume a general wage of 2.00 rupees per man-day; for weeding, we assume 1.50 rupees. For rice, we use 3.00 rupees per day; for rice weeding, irrigation, and fertilizer application, 2.50 rupees. For wheat and sugarcane, we use 2.00 rupees per day but 3.00 rupees for planting, as this occurs during a peak period.

[b] Estimated as a residual. From gross returns, subtract average total variable costs in 1971-72, including the imputed value of family labor. The only cost not subtracted is interest charges. Gross returns include the value of the by-product.

[c] Most farmers in the unirrigated zone use fertilizer on rice, but no farmer uses it on cotton.

[d] For irrigated crops we use data for 1971-72 rather than for the six-year period, owing to the difficulties of obtaining time-series yield data for the new varieties.

[e] Over 50 percent of this goes to migrant workers from Maharashtra who are employed by the factory for harvesting.

groundnut returns to labor two-thirds more than cotton. Data on vegetables or on livestock such as dairy would be even more striking.

Donovan shows, through a linear programming analysis, the substantial effect of cropping pattern changes on labor requirements in Mysore State, India, and he stresses the role of seasonal labor bottlenecks and the possibilities of clearing those bottlenecks offered by migration and mechanization.[5] Similarly, Desai and Schluter show dramatically how changes in cropping patterns, specifically between groundnut and cotton, alter the total demand for labor in Surat District, India.[6] The same point could be made in respect to change between jute and rice in Bangladesh.

It is of course crucial to recognize that production patterns must be in balance with demand and hence that efforts to increase employment by changing the structure of production must be matched by shifts in demand. Such demand shifts may occur as indirect effects from expenditure of rising incomes, or they may be the product of specific public policy. Some of the most labor-intensive agricultural enterprises such as sugar, vegetables, milk, and certain other livestock commodities have highly elastic demand. On the one hand, demand will grow rapidly for such commodities as incomes rise, creating a favorable environment for increasing production and employment. On the other hand, if supply is inelastic, rising production costs and hence prices will readily shift consumption to other commodities which may or may not provide as many jobs. Consequently it is important to provide the institutional changes needed to facilitate ready increases in production and marketing, while research which increases production efficiency and the supply elasticity can greatly facilitate growth in employment in this sector.

In a context of rapid growth in incomes, a favorable environment is provided for increasing production of demand elastic, potentially labor-intensive commodities such as sugar, vegetables, and livestock products. Since relaxing the wage goods constraint through increased food grain production may be effective in creating such an environment, a simple research allocation strategy follows: first, stress food grain production; then set in motion a high employment program, which creates a favorable demand environment; finally, accentuate labor-intensive crops and livestock in research programs.

After taking advantage of the increased income from new food grain technology as a means of increasing demand for labor-intensive production, one may pursue additional steps. First, export demand may be stimulated for labor-intensive agricultural commodities. This could be further facilitated by the introduction of technological change which decreases cost of production. Second, internal demand structure may be influenced by subsidy and educational programs. The key point is that, while agricultural research may appropriately emphasize labor-intensive commodities, it must be backed up by policies which increase or shift demand accordingly. Research policy must be seen in complementary play with other policies.

Risk and uncertainty. Allocating research resources in such a way as to decrease the variation in yields in a context of varying weather and management practices will reduce risk and uncertainty. Lessening the probability of particularly poor results not only raises average returns but reduces variance as well, and this may be seen as part of the general effort to increase production efficiency.

However, variation in results has implications beyond the effect on average production. In general, society prefers present income to future income and so risk involves a loss through unexpected deferral of income. Perhaps more important, low-income farmers tend to avoid risk, and hence the amount of variation in results affects the distribution of benefits and the adoption of innovation.

Schluter's analysis of Surat District, India, shows clearly that lower income farmers choose the lower profit and less labor-intensive but more certain enterprises. Thus the farmers with lower income tend to choose cropping patterns with a lesser deviation in income. More specifically, on the basis of a MOTAD model Schluter shows that for each 100 rupees of increased income from change in cropping pattern there is a sacrifice of 100 rupees in increased variance.[7] It follows that it would be useful to small farmers to reduce variance of income in the more labor-intensive and profitable crops. Indeed, because the more profitable crops are more labor-intensive, it follows that, except for risk aversion, small farmers should have a comparative advantage in those crops.

Although variation in income arises from fluctuations in both price and yield, fluctuation in yield is the more important factor. For example, Schluter shows for Surat District that the coefficient for variation for yield is greater than that for price by a factor of 4 for rice, 3 for sorghum, 1½ to 2 for cotton, and 1½ for groundnuts (Table 23-5). Further, price variation tends to be inverse to yield variation and therefore tends to reduce overall variation, given the fluctuation in yield induced by weather.[8] Thus, as Schluter's data confirm, in the complex real world of many conflicting forces, price stabilization schemes may increase income instability, at least for lower income producers.

It follows then that the burden of reducing variation in income must fall more on production than on price policy factors. Within production policy there are many variables: expenditure on extension effort may improve management and reduce crop failure; expenditure on irrigation may reduce dependence on favorable weather; expenditure on credit programs or insurance schemes may provide greater staying power. However, a research program designed to reduce variation in results may be an efficient means of dealing with the problems. The greater the concern with reducing income disparities and increasing the participation of small farmers, the more emphasis will be given to research on reducing variation in results. Such emphasis may contribute initially to the most widely grown high-yielding varieties of crops and

Table 23-5. Coefficients of Variation for Yield, Price, and Revenue under Alternative Price Stabilization Policies for Major Unirrigated Crops, Surat District, India, 1966/67-1971/72

			Revenue		
Crop	Yield	Price	No Stabilization	Partial Stabilization[a]	Complete Stabilization[b]
Rice					
Low fertilizer	0.40	0.09	0.36	0.39	0.40
High fertilizer	0.42	0.09	0.36	0.39	0.42
Cotton					
Low fertilizer	0.42	0.24	0.30	0.36	0.42
High fertilizer	0.51	0.24	0.43	0.47	0.51
Sorghum	0.39	0.13	0.43	0.40	0.39
Groundnuts	0.37	0.26	0.42	0.34	0.37

Source: Michael G. G. Schluter, "Interaction of Credit and Uncertainty in Determining Resource Allocation and Incomes on Small Farms, Surat District, India," Cornell University-USAID Employment and Income Distribution Project, Department of Agricultural Economics, Occasional Paper no. 68, Ithaca, Cornell University, February 1974, chapter VII, Table 30, p. 44.

[a] Gross returns with price deviations from the mean reduced by 50 percent in each year.

[b] Gross returns with prices fixed at the mean levels.

then spread to the high profit, demand elastic, labor-intensive enterprises which have such a substantial potential to increase employment.

Nutrition aspects of the production pattern. Finally, a word on nutrition. Nutrition may deserve special emphasis in research allocation because consumer ignorance may result in too little expenditure for nutritious foods, while research may be able to increase nutritive content of foods at low cost and high convenience.

Unfortunately it is all too often forgotten that the key trade-off in research allocation vis-à-vis nutrition is likely to be between yield and nutrients other than calories. Thus the critical question may be what is most detrimental to the health of a particular population – insufficient calories or lack of other nutrients, such as protein or vitamins.

For most of the world's low-income people, food grains are usually the basic source of calories. It is only after expenditure is sufficient to meet basic calorie needs that consumption may properly stress protein and other nutrients in the cropping pattern. Just as the objective of increasing employment calls for first increasing food grain production and then more labor-intensive crops, so improving nutrition calls for food grains first and then other crops. As noted earlier, it is a common error to minimize calorie deficiency relative to protein deficiency.[9] As long as calories are deficient, research should not emphasize other nutrients at the expense of yield.

In this context, however, one should note the interaction of improved nutrient content with labor intensity. Although nutritious foods like livestock products may be inefficient means of providing nutrients such as protein and may in any case be sold to higher income persons, they may also provide the demand for labor which, in turn, provides the purchasing power for the poor to meet their more basic food needs. Thus, research that accentuates income generation for the poor may do much for their state of nutrition through market processes.

Again it should be noted that if low-income persons are to be encouraged to devote scarce land and other productive resources to production of vegetables to improve home nutrition, then it is important to reduce the risks involved in their production. Moreover, there is a clear trade-off between allocating resources to research to reduce yield variance and allocating resources to such activities as extension for teaching cultivation methods that reduce variation in results.

Choice of Emphasis within an Enterprise Category

Once decisions are made on the relative research emphasis among various enterprises, there remain allocational questions on the relative weight of objectives for the individual enterprises. In the broader sense the research objective is to increase physical efficiency – output per unit of input. But since various factors of production are distributed unevenly, raising the productivity of such factors unevenly will affect the distribution of income. And the extent to which the reduction of variation in results is highlighted will affect rates of adoption among income classes and hence the distribution of benefits. Similarly, relative accent on approaches that raise or lower the use of purchased inputs will affect income distribution.

Choice of Discipline

Relative emphasis on various academic disciplines in staffing a research institution is an important policy instrument and is likely to have substantial effect on the distribution of benefits. The most interesting set of considerations in this regard is probably that which arises with respect to engineering, since that is particularly related to mechanical innovations which have been most associated with displacement of labor. The allocation among disciplines is, however, very complex, even from an employment point of view. For example, a general production problem is that of seasonal labor bottlenecks. This is particularly a problem of intensive multiple-cropping programs, but even societies with large numbers of very low-income people suffer clear labor constraints to production at seasonal periods.[10] Now, increasing total employment may require mechanization to break seasonal bottlenecks. It is by no means certain, however, that such effort will not overshoot the mark, reducing total earnings of the poor. Ideally, research resource allocation would stress increase in efficiency, including labor efficiency, using redistribution of assets and income to deal with income distribution problems. Unfortunately, that is not necessarily how policy priorities get set. The conflict calls for detailed economic analysis of seasonal labor supply/demand balances to facilitate research on how to break seasonal labor bottlenecks without displacing labor and thereby reducing incomes of the poor. This is an area where cooperation between social scientists and production scientists can be particularly effective.

Presumably reduction of the arduousness of work is a widespread human objective. Technical innovation which accomplishes this is desirable on egalitarian grounds only if the benefits are distributed to those formerly performing the arduous task. In a communal society machine hulling of rice relieves the low-income women of a very difficult task and releases their time for better child care and other welfare-increasing activities. However, if income and assets are unequally distributed, displacing hand pounding of rice may simply eliminate the primary source of income of a low-income group with an increase in net income of a higher income entrepreneurial class and a higher income consuming group. A similar point could be made for many other rural activities. Again, conflict between stated societal objectives and societal realities greatly complicates research resource allocation.

As in the case of allocation among crops, a decision regarding one crop must be made on the extent to which various nutritional elements will be emphasized. Again, the main trade-off appears to lie in the choice between increasing protein content or increasing yield, or calorie production. As we have seen, the Western approach has tended to accent protein. Clearly, more research is needed, both on the effects of diet and health among the poor and on the extent to which foods naturally contain a sufficient balance of nutrients as long as the total quantity is adequate.

The Regional Problem

The choice of enterprise to emphasize and the approach to research on any one enterprise profoundly affect the distribution of income among geographic regions. This is the result of not only regional deficiencies in soils, topography, and climate, but also economic and social differences – including income differences – which affect the acceptability of various types of innovations. In the long run, the regional allocation of research resources may be one of the most powerful influences on regional power and equity.

In countries with generally adequate food supplies, such as Thailand, it may be regional disparities of food which are most constraining to increased equity. Even mainland China has had great difficulty in meeting the problems of regional inequity. Trade-offs in research may be substantial, with emphasis on regions with poor production resources resulting in high costs of research and low increases in productivity. And again, the trade-offs in achieving objectives with unequal investment in education or with direct income transfers should be considered in allocating research resources.

An Example of a Research Policy Sequence

The criteria for research policy are, of course, dependent on the objectives stated for research policy, the research resource availabilities, and the political acceptability of alternative means of reaching objectives. The latter, of course, reflects a broader set of objectives, both explicit and implicit, and the weights on these objectives dictated by the power structure within the society. Pursuing objectives in a sequential manner recognizes that priorities can be established and that effective research requires clearly understood guidelines, which are possible only if objectives are separated and sequenced. Then research policy must set the priorities which order the sequence of effort – that is, not only the order of objectives but the ease of meeting particular objectives given the realities of the tasks and the resources. In particular it should be recognized that adding criteria to research both reduces productivity and reinforces the need for pursuing objectives sequentially rather than simultaneously.

The process may be exemplified by assuming that we are faced with a society attempting to increase participation of the poor in growth, but without major redistribution of income or assets. Such a society might first emphasize increasing yields per acre of the basic food grains in the most responsive areas of the country. This would release the wage goods constraint to employment and increase the supply of basic calories. As success is met on the basic food grain front, geographic area coverage of the effort would be expanded to less productive regions. Soon, coverage would be extended to additional crops, with priority given those which are particularly labor-intensive in production. The latter effort would be coordinated with expansion of demand either through exports, rising domestic incomes, or subsidies. In each case, however, the conditions of success require expansion of basic food grain production. The reason for this is obvious in the case of domestic markets, but even raising income from exports requires a large supporting increase of domestic food production.[11] Throughout each of these thrusts research would stress reduction in year-to-year yield variation because of the special risk aversion of lower income farmers.

Research Needs for Improved Research Direction

Because of our minimal background of knowledge and experience, the allocation of research resources offers in its own right an area ripe for research. Research in the social sciences to support optimal decision-making in allocation of production research resources needs to emphasize definition of societal objectives; sources and means of income by income class; the consumption pattern of various income classes; and the factor share bias of innovations.

Societal objectives have been defined in this chapter in terms primarily of level and distribution of income with added attention to uncertainty of income and nutritional status. However, little is known about the importance of absolute income as compared with relative income in determining well-being, yet that distinction may be crucial to determining optimal policy. Similarly, if research is to be keyed explicitly to societal objectives, is is important to know whether the same forces which prevent land reform, presumably because redistribution of income is not a prime objective of society, will prevent effective impact of research results which would effect redistribution of income. Or could it be a stated objective to effect income redistribution through research only as long as such effort is not effective? It may be inconvenient to make societal objectives explicit, but that may be essential to effective public sector decision-making.

Only when it is known how much income each income class generates through what production means can research policy be designed to improve income distribution. Innovation affects income distribution through its effect on demand for factors of production and supply of consumer goods. Generally the poor provide labor, and relative increase in demand for labor increases their income in relation to that of others. Effective policy demands knowledge of the precise nature of labor supply and demand, including elasticity of labor supply. Such knowledge provides a basis for estimating returns to increased demand for labor and probable wage behavior. Similarly, consumption patterns of the low- and high-income classes differ greatly. Research concentrated on goods the poor consume will benefit them relative to the more well-to-do. Despite this obvious fact little detail is available on the precise nature of consumption patterns of the poor.

Finally, knowledge is needed of the effect of various innovations on the demand for payment to various factors of production. What is needed is knowledge not only of the average amount of labor used and yield increase but of the variation in functional relationships between input and output and the variance in those relationships. Then optimal choice of technology can be made in the context of varying risk and uncertainty and levels of input.

We close on a research-oriented note in this chapter on research allocations. Social scientists could contribute substantially to increased efficiency in this most crucial area of development. To do so, they must verse themselves in the technology of agriculture to grasp somewhat more fully the constraints that delineate the art of the possible. They must recognize the impracticality of a general equilibrium approach in this extraordinarily complex area and settle for numerous partial approaches and consequent doubt of whether or not they have even an approximation of a correct answer. Finally, they must immerse themselves in the operational context of the real world so that they may choose economically among the many possible "partial" problems in allocating their own potentially valuable resources.

NOTES

1. John W. Mellor and Uma J. Lele, "Growth Linkages of the New Foodgrains Technologies," *Indian Journal of Agricultural Economics*, 28:1 (January-March 1973), 35-55.

2. For a full discussion of this point see Uma Lele, *The Design of Rural Development: Analysis of Programs and Projects in Africa* (Baltimore: Johns Hopkins University Press, 1975).

3. Uma J. Lele and John W. Mellor, "Jobs, Poverty and the Green Revolution," *International Affairs*, 48:1 (January 1972), 20-31.

4. For a full treatment, see Mellor and Lele, "Growth Linkages," pp. 35-55.

5. Graeme W. Donovan, "Employment Generation in Agriculture: A Study in Mandya District, S. India," Cornell University-USAID Employment and Income Distribution Project, Department of Agricultural Economics, Occasional Paper no. 71, Ithaca, Cornell University, June 1974.

6. Gunvant M. Desai and Michael G. G. Schluter, "Generating Employment in Rural Areas," *Seminar on Rural Development for the Weaker Sections* (Bombay: Indian Society of Agricultural Economics, 1973).

7. See Michael G. G. Schluter, "Interaction of Credit and Uncertainty in Determining Resource Allocation and Incomes on Small Farms, Surat District, India," Cornell University-USAID Employment and Income Distribution Project, Department of Agricultural Economics, Occasional Paper no. 68, Ithaca, Cornell University, February 1974.

8. *Ibid.*

9. See the discussion of this controversy in Uma J. Lele, "The Green Revolution, Income Distribution and Nutrition," *Proceedings – Western Hemisphere Nutrition Congress III*, ed. Philip L. White (Mount Kisco, N.Y.: Futura, 1972).

10. M. Habibullah shows this clearly for Bangladesh in *The Pattern of Agricultural Unemployment: A Case Study of an East Pakistan Village* (Dacca: Dacca University, Bureau of Economic Research, 1962).

11. For a detailed analysis of this point, see Lele, *The Design of Rural Development*. For a somewhat different set of points, emphasizing the dependence of labor-intensive exports on relaxing the food grains constraint to employment, see John W. Mellor and Uma Lele, "The Interaction of Growth Strategy, Agriculture and Foreign Trade – The Case of India," *Trade, Agriculture and Development*, ed. G. S. Tolley (Cambridge: Ballinger, 1976).

24

An Empirical Test of an Economic Model for Establishing Research Priorities: A Brazil Case Study

J. P. Ramalho de Castro and G. Edward Schuh

Increased attention has been given in recent years to the management and allocation of research resources according to specified criteria or to some sense of priority.[1] This increased concern is in part a result of the growing recognition that research is indeed an economic activity involving the organization of scarce resources. But, in addition, it reflects the growing consensus that technical change is a key element in the development process.

This chapter reports a modest attempt to formalize the economic concepts available for establishing research priorities, to utilize data to test the model, and to bring to bear as much empirical information as possible on the problem of establishing research priorities for a particular economy. The basic model draws on the concepts of consumer and producer surplus, the neoclassical theory of production, recent work on models of induced technical change, and a two-sector general equilibrium model. Important themes in the analysis are (1) that the specification of goals is a critical element in the establishment of research priorities; (2) that an understanding of the distribution of benefits from technical change is a key to understanding the extent to which these goals are met; and (3) that in a country the size of Brazil regional differences are important.

The Model

The starting point of the model is to view new production technology as having an instrumental role in attaining a larger set of goals and objectives. Technical change in agriculture can be understood in the context of the contributions agriculture can make to general economic development. These contributions include (1) supplying food for the total population; (2) supplying capital to the economy, especially for expansion of the nonfarm sector; (3) supplying labor for the expansion of nonfarm activities; (4) supplying exchange earnings in order that imports critical to the development process can be purchased from abroad; and (5) providing a market for the products of the nonfarm sector. The development and distribution of new production technology can be a powerful means of strengthening agriculture in regard to each of these factors. Hence, the demand for new production technology can be derived from this framework.

There are rather obvious conflicts between some of these contributions. For example, if all of agriculture's surplus earnings are siphoned off as capital for the nonfarm sector, agriculture's potential as a market for goods and services produced by the nonfarm sector is greatly reduced. In addition, the relative importance of each contribution will depend on the stage of development of the economy, the particular development model the government uses as a basis for economic policy, and the specific policy measures the government uses to implement its policy.[2]

For analytical purposes we specified four alternative sets of goals for the research program: (1) to increase the total net income of the agricultural sector; (2) to increase employment and income of workers in the agricultural sector; (3) to increase consumer welfare by providing food at lower real prices; and (4) to maximize the contribution of agriculture to the growth of the economy as a whole. These goals are somewhat broader than the contributions that agriculture can make to economic development, although in most cases a direct linkage can be made to the latter. The advantages of these particular specifications are that they are operational in terms of the more formal model that will be presented below and that they are commonly held goals for agricultural research.

It should be obvious that to attain each of these goals may require different "kinds" of technical change. Hence, the choice of technology becomes an important issue. In addition, the kind of technology chosen will have different income distribution consequences, as will the choice of product on which to focus the research program. Therefore, we next turn to a discussion of the theory that will help to answer these questions.

Induced Technical Change

The basic idea of the induced innovation theory is that innovations and their resource-saving directions depend on economic conditions prevailing within a given economy rather than being determined exogenously to it, as is frequently assumed. Hayami and Ruttan have extended the theory of induced innovation from its micro-formulation to an understanding of technical change at the sectoral level.[3]

Two elements of the model developed by Hayami and Ruttan are of interest from our standpoint. First, they recognize that technical change can take alternative routes in its resource-saving effects and that the particular route it takes is conditioned by relative factor scarcities. This idea gives rise to the concept of an efficient path for technical change and suggests the importance of allocating scarce research resources so that technical change is directed along such an economically efficient path. Hayami and Ruttan argue that for agricultural technology to be an efficient source of growth, it must be directed toward easing the barriers to output expansion posed by inelastic factor supplies.

The second element of the Hayami-Ruttan model is the instrumental role it assigns to individual "technologies" in facilitating the substitution of one input for another. Hayami and Ruttan view biological improvements, such as the development of varieties that are more responsive to fertilizer, as a means of facilitating the substitution of fertilizer for land. Mechanization, on the other hand, is a means of facilitating the substitution of land for labor in the sense that it permits a substantial change in the man/land ratio. In both cases there is an infusion of capital into the production process, but the contribution in facilitating substitution among the primary inputs of land and labor may be as important as the contribution of the additional input qua input.

We use the Hayami-Ruttan framework in an *ex-ante* sense as a normative model of what direction technical change *ought* to take if it is to be an efficient source of growth. Our analytical-empirical task is to identify the resource constraint that is inhibiting output expansion so that production technology can be directed toward easing such constraint. This information will enable us to determine whether the research program should concentrate on biological or mechanical innovations.

Distribution of Benefits between Consumers and Producers

Although the notion of an efficient path for technical change (in the resource dimension) can serve as an important basis for allocating research resources, it alone is not sufficient. Technical change has important income distribution consequences, and these in turn are important in determining the extent to which a given research program will attain specified goals. One issue is the extent to which the benefits of technical change redound to the consumer or to the producer. A second issue is the extent to which the producer

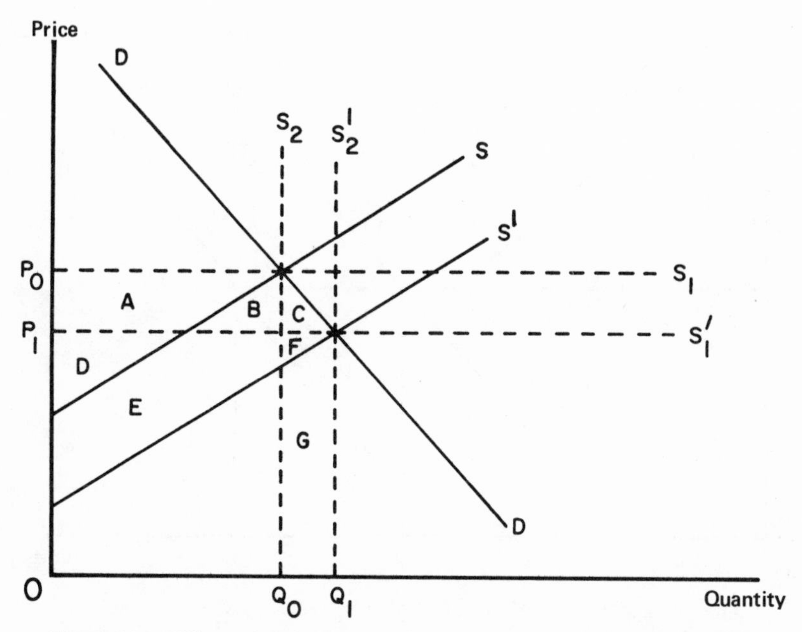

Figure 24-1. Effects of a shift in the supply curve as a result of technological change in a closed economy.

surplus redounds to particular factors of production. In this and the following sections we turn to a consideration of these issues.

The concept of economic surplus is a useful tool for analyzing the distribution of benefits between consumers and producers.[4] The basic analytical model for a closed economy is given in Figure 24-1. Suppose that before some technological innovation the equilibrium price and quantity are P_0 and Q_0. In addition, suppose that the supply curve were perfectly elastic. The gain to society from a technological change which lowered the supply curve to S_1' would be the gain in consumers' surplus, A + B + C. If the initial supply curve were perfectly inelastic, on the other hand, a shift in the supply curve from Q_0S_2 to Q_1S_2' would result in both a change in producers' surplus (F + G – (A + B)) and a change in consumers' surplus (A + B + C). However, if the supply curve is positively sloped, the net gain will be B + C + E + F, since the change in producers' surplus is (E + F) – A, while the gain in consumers' surplus is A + B + C.

For our purposes we want to extend this model to that of an open economy with traded products. The distribution of benefits in this case can be un-

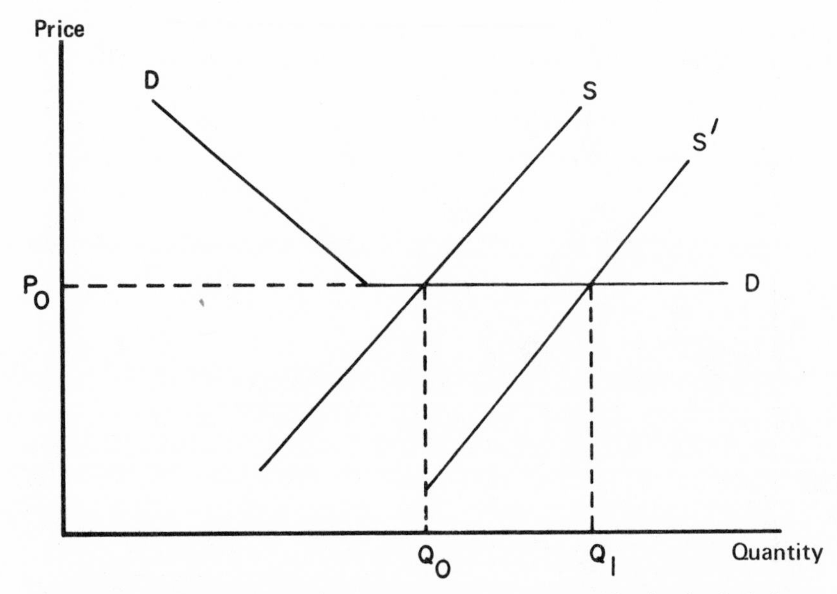

Figure 24-2. Effects of a shift in the supply curve as a result of technological change in an open economy.

derstood by means of Figure 24-2. The figure is drawn on the assumption that the country is relatively unimportant in world markets so that its exports will not affect world prices. This assumption can be relaxed, although it was not necessary to do so in our research.

With a shift in the supply curve, the elastic demand curve implied by an open economy will not allow for gains in consumers' surplus if the product has traditionally been exported. If the product was not exported before the technical change, there may be some gain in consumers' surplus as a result of the initial shift of the supply curve. This would occur, for example, if before the technical innovation the domestic price were above the world price and "protected" either by transportation costs or by trade policy.

It can be seen from Figure 24-2 that all economic surpluses will accrue to the producers if the product has been traditionally exported. If this is the first time the product has been exported, however, and if the internal price was previously above the world market, the producers will share some economic surplus with the consumers for the initial shift in the supply curve.

These concepts are useful tools for analyzing the return to investments in research, as illustrated among others by Griliches, Peterson, Ayer and Schuh, and papers presented at the Airlie House conference.[5] And if rate of return criteria were to be used for establishing research priorities, the basic framework would provide a means of selecting the products to which researchers should give priority in their research efforts.

For our purposes, however, these concepts provide a means of setting priorities in terms of more specific goals established by policy makers. Knowledge of the demand and supply elasticities will provide a basis for determining whether the flow of benefits from a given technical change will be realized as a producer surplus or as a consumer surplus. Then, depending on whether the policy makers prefer to favor the producer or the consumer, research resources can be allocated accordingly.

It should be noted that in the open economy case the export multiplier becomes an additional mechanism through which social gains are realized. These gains are derived from the increased exportable surpluses made available by the new technology. Although recognized where relevant, the gains realized through the export multiplier are not explicitly accounted for in the empirical work reported below.

Technological Change and the Neoclassical Theory of Distribution

Another important aspect of the present study is to determine how the benefits of technical change are distributed among the factors of production. The neoclassical theory of distribution provides the means for analyzing this problem. At the macroeconomic level one distributional question of interest concerns the behavior of aggregate relative shares in response to technical change. In fact, one customary way of defining technological bias is by whether the relative share of labor increases, remains unchanged, or declines as technological change takes place.

The problem in using the pure neoclassical theory is that the analysis is restricted to two factors. One possible way to proceed is to specify the aggregate production function in separable form, where the degree of substitutability among inputs is assumed to be greater than one within the postulated subfunctions, but less than one between subfunctions. More specifically, assume

$$Y = F(f(L^*, K_L^*), g(T^*, K_T^*)$$

where Y is the aggregate agricultural output, T and L are land and labor inputs, respectively, K_L is laboresque capital (mechanical), and K_T is landesque capital (biological, chemical, and agronomic).[6] The asterisk indicates that the factor in question is measured in "effective" units, e.g., $L^* = t_L L$, where t_L is an index of nonneutral technological change which increases the quality, or "effective units," of the nominal input L (labor). The further assumptions of fixed input prices, homogeneity, and weak separability permit the definition of price indices for the subfunctions on the basis of which the optimization process can be performed in two separable stages.

Elasticities can be derived which indicate the effect of nonneutral technological change on the income to individual factors, on the functional distribution of income, and on the employment of individual factors of production. Since the magnitudes of the estimated elasticities of substitution were not consistent with the a priori restrictions, this part of the analysis was prejudiced.[7] However, the parameters of the production function are of interest in their own right and will be presented below.

General Equilibrium Considerations

The analysis above assumes a partial equilibrium framework. However, to provide a complete view of the effects of technological change in one sector of the economy on employment and the returns to factors, general equilibrium considerations must be taken into account. The problem that arises in the general equilibrium framework is that technical change affects the level of output, which in turn can affect the relative prices among products. The change in price of the product can lead to a redeployment of resources which in some instances can more than offset the direct effect of the production technology on factor productivity.

Empirical Results

The empirical results are presented in three sections: (1) estimates of the total potential gains from an assumed technologically induced shift in the supply curve for selected crops, together with estimates of the distribution of these gains between producers and consumers; (2) estimates of the parameters of the underlying production function; and (3) data on recent trends in factor prices as a basis for determining relative factor scarcity.

Total Gains from Assumed Shifts in the Supply Curve and Their Distribution between Consumers and Producers

At some point in the decision process allocation decisions on research money are made on a crop basis. (The livestock sector was excluded from the present study). Policy makers decide that X amount of money will be allocated to crop Y and that some crops will receive attention while others will not. Our assumption is that decision variables involved in this process include some judgment about the expected flow of total benefits from a given research effort and some notion of who will receive the benefits of the research.

Two criteria were considered in selecting crops for this part of the analysis. The first was their relative economic importance as measured by value of total output, total area planted to the crop, and the geographic spread of the crop over the country. The second criterion was the size of the price elasticity of demand, since this parameter is important in determining the relative distribution of benefits between consumers and producers.

An important determinant of the size of the demand elasticity is whether the product is exported or not. Therefore, it was decided to choose some products that have only a domestic market and others that either have been exported in the past or have the potential to be exported. By this means the analysis would consider a rather wide range of values for the structural parameters.

The crops chosen according to these criteria were cotton, sugarcane, corn, rice, edible beans, and manioc.[8] Cotton and sugarcane are traditional exports from Brazil. Rice and corn have been exported occasionally but for the most part on a much smaller scale than cotton and sugarcane. Edible beans and manioc are traditional staple foods which for the most part have not been exported. During the period 1966-70 these six crops accounted for 46 percent of the total value of output from crops and 74 percent of the total area in crops. Moreover, each of them was grown over a rather wide area in Brazil.

Available estimates of the demand and supply elasticities for these crops are summarized in Tables 24-1 and 24-2. As the tables indicate, there has been more work on the supply side than on the demand side, and on the supply side there were more estimates available for the state of São Paulo than for Brazil as a whole or for other states.

Given the range in the estimates of the parameters, three alternative estimates for each crop were selected on the supply side – low, medium, and high – and two were selected on the demand side – low and high. On both the demand and supply side it was necessary to specify arbitrary values in some instances in order to have a desired range in the parameters. For sugarcane and rice a relatively elastic response to price on the demand side was assumed since both have considerable potential in world markets. (The available estimate for sugarcane is based solely on the domestic market.) In the case of manioc, for which no estimates of the demand elasticity were available, it was assumed from a priori knowledge that it is an inferior good and that it therefore would be expected to have a low elasticity of demand.

The elasticities chosen are presented in Table 24-3. It would have been desirable to have more precise estimates of these parameters, but econometric work is not yet very far advanced in Brazil. Consequently, the results which are presented below have to be interpreted as illustrative of the effects of demand and supply elasticities in determining who receives the benefits from technologically induced shifts in the supply curves.

Since the analysis is cast in an *ex-ante* rather than an *ex-post* framework,

Table 24-1. Selected Supply Elasticities for Brazil, State of São Paulo, and State of Goias from Various Authors

	Brazil						São Paulo							Goias	
	Pastore		Paniago		Thompson		Pastore		Toyama & Pescarin		Brandt		Ayer & Schuh	Villas	
Products	SR[a]	LR[b]	SR	LR	SR	LR	SR	LR	SR	LR	SR	LR	LR	SR	LR
Cotton	.19	.63					1.22	2.03	.37		.69	1.57	.944		
Sugarcane	.16	.16					.12	.12	.27	.39					
Rice.	.31	1.17	.31	1.74			.61	1.96	.42	.69	.62	4.10		.30	2.34
Corn	.15	.57			.15	.58			.83	3.32	.45	2.55			
Edible beans	.14	.15					.37	.37	.31	.43	.10	.31			
Manioc	.11	.96					.26	.47							

Source: A. C. Pastore, "A oferta de produtos agricolas no Brasil," *Pesquisa e planejamento*, 1:2 (December 1971), 171-234. E. Paniago, "An Evaluation of Agricultural Prices for Selected Food Products: Brazil," Ph.D. thesis, Purdue University, 1969. R. L. Thompson, "The Impact of Exchange Rate Policy and Other Restricted Policies on Corn Exports in Brazil," M.S. thesis, Purdue University, 1969. N. K. Toyama and R. M. C. Pescarin, "Projeçoes da oferta agricola do Estado de São Paulo," *Agricultura em São Paulo* 17:9-10 (September-October 1970), 1-97. S. Brandt, M. Barros, and D. D. Neto, "Relaçoes area-preço de algodao no Estado de São Paulo," *Agricultura em São Paulo*, 12:1-2 (January-February 1965), 31-38. H. W. Ayer and G. E. Schuh, "Social Rates of Return and Other Aspects of Agricultural Research: The Case of Cotton Research in São Paulo, Brazil," *American Journal of Agricultural Economics*, 54:4 (November 1972), 557-569. A. T. Vilas, "Estimativas de funçoes de oferta de arroz para o Estado de Goias e suas implicaçoes economicas, periodo 1948-1969," M.S. thesis, Federal University of Vicosa, Brazil, 1972.

[a] Short-run elasticity as implied by a Nerlove-type distributed lag model.

[b] Long-run elasticity as implied by a Nerlove-type distributed lag model.

Table 24-2. Selected Elasticities of Demand for Brazil and the State of São Paulo from Various Authors

Product	Elasticity	Author
Cotton	– 5.3[a]	Ayer & Schuh
Sugarcane	– .56[b]	Martini
Rice.	– .10	Paniago
	– .16	Mandell
Corn	– .66	Thompson
Edible beans	– .32	Paniago

Source: H. W. Ayer and G. E. Schuh, "Social Rates of Return and Other Aspects of Agricultural Research: The Case of Cotton Research in São Paulo, Brazil," *American Journal of Agricultural Economics*, 54:4 (November 1972); E. Martini, "Açúcar no Brasil: Produçao, procura e preço," M.S. thesis, Federal University of Vicosa, Brazil, 1964; E. Paniago, "An Evaluation of Agricultural Prices for Selected Food Products: Brazil," Ph.D. thesis, Purdue University, 1969; P. I. Mandell, "A expansão da moderna rizicultura: Crescimento da oferta numa economia dinamica," *Revista Brasileira de economia*, 26:3 (July-September 1972), 169-236; R. L. Thompson, "The Impact of Exchange Rate Policy and Other Restricted Policies on Corn Exports in Brazil," M.S. thesis, Purdue University, 1969.

[a] Elasticity for the State of São Paulo.

[b] Demand elasticity for sugar.

an arbitrary shift of 10 percent was assumed for the supply curve. One of the first results obtained is the flow of total benefits by crop that would result from the specified shift. Based on the average value of output in the 1966-70 period, the annual flow of benefits from a 10 percent shift to the right in the supply curve is estimated to be: rice, $157 million; corn, $145 million; sugarcane, $106 million; manioc, $92 million; edible beans, $88 million; and cotton, $88 million. These data represent gross benefits since neither the cost of obtaining the supply shift, the value of complementary inputs such as fertilizers, nor adjustment costs are considered. The gross benefits give some notion of the relative payoff from investments in research under the assumptions that the cost of obtaining a given supply shift is the same among crops and that complementary inputs and adjustment costs are ignored. It is worth noting, moreover, that the gross flow of benefits is determined largely by the relative economic importance of the crop. The elasticities have less influence.

Data on the relative distribution of these benefits among consumers and producers according to various assumptions about the structural parameters are summarized in Table 24-3. The results support the notion that technological shifts for crops that have a relatively high price elasticity of demand (e.g.,

Table 24-3. Percent of Benefits to Consumers and Producers Resulting from a Specified Shift in the Supply Curve, Based on Two Demand and Three Supply Elasticities for Each of Six Crops, Brazil

Crop	Demand Elasticities	Percent of Benefits[a]					
		Consumer	Producer	Consumer	Producer	Consumer	Producer
		(.19)		(.94)		(1.57)	
Cotton	– 2.00	9	91	32	68	42	58
	– 5.30	4	96	15	85	23	77
		(.10)		(.60)			
Sugarcane	– .56	16	84	52	48		
	– 2.50	4	96	19	81		
		(.31)		(1.17)		(2.34)	
Rice.	– .16	20	80	88	12	94	6
	– 1.50	17	83	44	56	61	39
		(.15)		(.58)		(3.32)	
Corn	– .30	33	67	66	34	92	8
	– .66	18	82	47	56	83	17
		(.15)		(.31)		(.43)	
Edible beans . . .	– .32	32	68	49	51	57	43
	– .50	23	77	38	62	46	54
		(.11)		(.47)		(.96)	
Manioc	– .10	52	48	82	18	91	9
	– .30	27	73	61	39	76	24

[a] Supply elasticities are given within parentheses.

crops with export potential) will tend to favor the producer, while for crops with a relatively low price elasticity of demand (e.g., necessities) the shifts tend to favor the consumer. However, the results also illustrate that it is the relative magnitude of the elasticities that is important. If the supply elasticity were larger than the demand elasticity, regardless of the absolute size of the demand elasticity, the consumer would tend to receive a larger share of the benefits. (The extreme cases of perfectly elastic or perfectly inelastic demand and supply curves are ruled out of this comparison.)

The case of rice illustrates this relationship rather well. Even with a relatively low price elasticity of demand, producers receive a major share of the benefits from a shift in the supply curve if the supply elasticity is low. Similarly, if the supply elasticity is high, consumers can receive a (comparatively) large share of the benefits even if the demand curve is relatively price elastic. To benefit consumers, therefore, one would want to induce shifts in the supply curve for crops that have a relatively low price elasticity of demand and a relatively high supply elasticity (manioc may be a case in point). To benefit producers, on the other hand, one would want to shift the supply curve for crops that have a relatively high price elasticity of demand and a relatively low supply elasticity (cotton, for example).

An additional inference that could be drawn is that, if the crop has a high price elasticity of demand, the research effort might be directed to increasing the supply elasticity if it were desired to increase the share of benefits going to the consumer. This might be done by increasing the geographic area of adaptability of a particular crop.

Estimates of the Production Function

The estimates of the parameters of the separable production function provided important insights about the stage and character of modernization in Brazilian agriculture, even though they did not turn out quite as expected.* To estimate the parameters of this function it was assumed that the aggregate production function is strongly separable into two subfunctions that are homogeneous of degree one and which belong to the CES class of production functions. One subfunction contained labor and the "laboresque" capital variable (mechanization) as independent variables, and the other contained land and the "landesque" capital variable (fertilizers). It was expected that the elasticity of substitution between the two variables in each subfunction would be greater than one and that the elasticity of substitution between the two subfunctions would be less than one.

Two time series of data for Brazil as a whole were available to estimate the parameters of the labor subfunction. One data series referred to tractors on farms and covered the period from 1950 to 1971. This was used as a proxy for laboresque capital, with the price of a "typical" tractor unit used as the price. The flow of tractor services was estimated as the *combination* of an opportunity cost rate of interest and depreciation. (The opportunity cost of capital was assumed to be 10 percent, and the depreciation charge was 5 percent. The latter assumes straight line depreciation and an expected life of twenty years. Sensitivity analysis indicated that the results were quite stable under alternative measures of the flow of services.) Estimates of the labor input were made from data on the total agricultural labor force. The daily wage rate for the cash, daily-paid worker of São Paulo was used as the price of labor.

The second time series provides estimates of the stock of horsepower on farms and is available only for the period from 1962 to 1971. This data series provides an alternative measure of laboresque capital. It was also possible to estimate an average price per unit of horsepower. In this case the "stock" of horsepower also represents the flow, and hence the problem of estimating the flow of services is reduced. Again, the price of the "flow" of horsepower was

* The authors' detailed discussion of the production functions utilized has been omitted to shorten and simplify the presentation. For details, see the work cited in n. 7. – Ed.

Table 24-4. Regression Results for the Labor Subfunction with Time Series Data, National Model, Brazil

	Proxies	
Coefficients	Tractor (1950-71)	Horsepower (1962-71)
Constant term	– 12.588*	– 2.174*
Price variable	3.740	– .605
	(4.231)[a]	(– 10.681)
R^2	.46	.92
D.F.	20	8

* Significant at the 1 percent level.
[a] The numbers in parentheses are t values.

estimated by assuming a twenty-year life for the tractor. The labor input and price of labor were defined in the same way when this data series was used.

The statistical results with the two sets of data are presented in Table 24-4. In both cases they are reasonably good. The coefficients of determination are relatively high, and the coefficients for the price variable are statistically different from zero at the 1 percent level. However, the price coefficient has opposite signs in the two equations. (The R^2 in the horsepower specification is also substantially higher than it is in the tractor specification.) The sign when data for the longer period (and the tractor variable) were used was contrary to a priori expectations, while the expected sign was obtained when data for the shorter period (and the horsepower variable) were used.

One interpretation of these results is that about 1960 there was what Brown would call a technological turning point.[9] To test this hypothesis, the 1950-71 series was disaggregated into two periods, one extending from 1950 through 1961 and the other from 1962 through 1971. The statistical results for the two periods are presented in Table 24-5. They support the hypothesis of the existence of a turning point.

The evidence for a technological turning point is rather strong, since the negative coefficient for the more recent period is obtained with both concepts of the laboresque capital. However, a problem still remains. When the capital variable is measured as the value of tractor services, the coefficient of the price variable is larger than one and hence consistent with a priori expectations. When the flow of horsepower services is used, the coefficient is less than one and hence not entirely consistent with expectations. Despite this problem, the results with the horsepower measure were used in further analyses, in part because horsepower would seem to represent a conceptually "cleaner" measure of the services provided by laboresque capital, and in part because the use of this measure results in a larger coefficient of determination for the estimation equation (.92 in contrast to .48).

Table 24-5. Regression Results for the Labor Subfunction with 1950-71 Time Series Disaggregated into Two Components, National Model, Brazil

Coefficients	1950-61	1962-71
Constant term	− 10.500*	5.570*
Price variable	2.754	− 2.311
	(12.510)[a]	(− 2.902)
R^2	.93	.48
D.F.	10	8

* Signficant at the 1 percent level.
[a] The numbers in parentheses are t values.

Both the notion of a turning point and the relatively small coefficient for the price variable are plausible results. Up until 1960 the level of mechanization was indeed low in Brazil.[10] Moreover, the most important use of tractors was for the power-demanding land preparation operation, which is believed by most authorities on the subject to increase the demand for labor rather than to be labor displacing. The increased demand for labor comes about by increasing the crop area and yields.[11]

During the 1960s there was considerable mechanization in Brazil. Moreover, a start was made toward mechanizing the harvesting operation, especially for crops like wheat, soybeans, and cotton, and, to a lesser extent, sugarcane. The mechanization of the harvesting operation is generally believed to be strongly labor displacing.

This difference in both the extent and the kind of mechanization seems a plausible explanation of the turning point. The finding of a coefficient no greater than one is probably explained by the fact that as late as 1971 mechanization in general was fairly limited and in particular was not yet widely used in the harvesting operation. As mechanization becomes more widespread, however, and as it is extended to the harvesting operation through greater use of combines and harvesters, the substitutability of capital for labor will probably increase. This has important implications for the establishment of research priorities.

To estimate the parameters of the land subfunction, data were available on the total quantity consumed and the respective prices for each nutrient (N, P, K) plus the area in crops in Brazil.[12] Unfortunately, data were available on land values or land rentals only since 1966. Therefore, only the price of fertilizer was used as an independent variable in estimating the elasticity of substitution. This specification assumes that the price of land does not vary systematically with the price of fertilizer.

Given that the results for the labor subfunction suggested a turning point in the production technology at some time in the early 1960s, the equations were estimated for two different time periods: 1950-60 and 1961-70. In addition, separate models were estimated for each nutrient and for the aggregate of the three.

The statistical results (Table 24-6) were consistent with those for the labor subfunction in that they also indicated the existence of a technological turning point in the early 1960s. When the equations were fitted with data from the 1950-60 period, in no case were the coefficients of the price variable significantly different from zero, and the coefficients of determination were close to zero. These results were consistent with the hypothesis that the elasticity of substitution between fertilizer and land is zero, or that they are complements in production.

When data from the 1961-70 period were used, however, the coefficients of the price variable were all significantly different from zero at usually accepted levels, and the coefficients of determination were relatively high. The coefficients all have the expected signs, and for nitrogen and potash the coefficients (elasticities of substitution) are greater than one. This suggests that fertilizer was becoming a good substitute for land.

In conclusion, the time series data provide evidence for a technological turning point in both components of the production function. The explanation for these turning points seems to be that during the decade of the 1950s the modern inputs of fertilizer and machinery were still used at relatively low levels. The complementarity of fertilizer with land was not quite so high as was the complementarity of mechanization with labor. In fact, fertilizer application may have been doing little more than replacing nutrients removed by crops. During the decade of the 1960s, however, the use of both fertilizer and mechanization appears to have reached the point where they were land- and labor-substituting, respectively. This change in structure has very important implications for research policy and the establishment of research priorities.

Because of data limitations, only one estimate of the elasticity of substitution between the two subfunctions was made. The statistical results were reasonably good. The coefficient of determination was relatively high (.74), and the coefficient of the input/output ratio had the expected sign and was significant at the 1 percent level. The size of the elasticity of substitution (.90) as estimated from the input/output ratio was consistent with a priori expectations (less than 1).

Trends in Factor Prices

Trends in factor prices should give some notion of relative factor scarcity and hence some notion of the direction in which research should be focused in the factor-factor dimension. It was originally hoped to be able to estimate

Table 24-6. Regression Results for the Land Subfunction, Time Series Data, 1950-70, Brazil

Item	N		P_2O_5		K_2O		Total	
	1950-60	1961-70	1950-60	1961-70	1950-60	1961-70	1950-60	1961-70
Constant term	2.004	2.994	1.060	2.395	.462	3.587	.869	3.341
Price of fertilizer	– .778	– 1.025**	– .202	– .717*	– .079	– 1.237**	– .010	– .933**
	(– .860)[a]	(– 7.328)	(0.437)	(– 2.729)	(– .105)	(– 3.424)	(– .015)	(– 5.751)
R^2	.07	.85	.02	.45	.00	.56	.00	.79
Degrees of freedom	9	8	9	8	9	8	9	8

* Significant at 5 percent level.
** Significant at 1 percent level.
[a] Numbers in parentheses are t ratios.

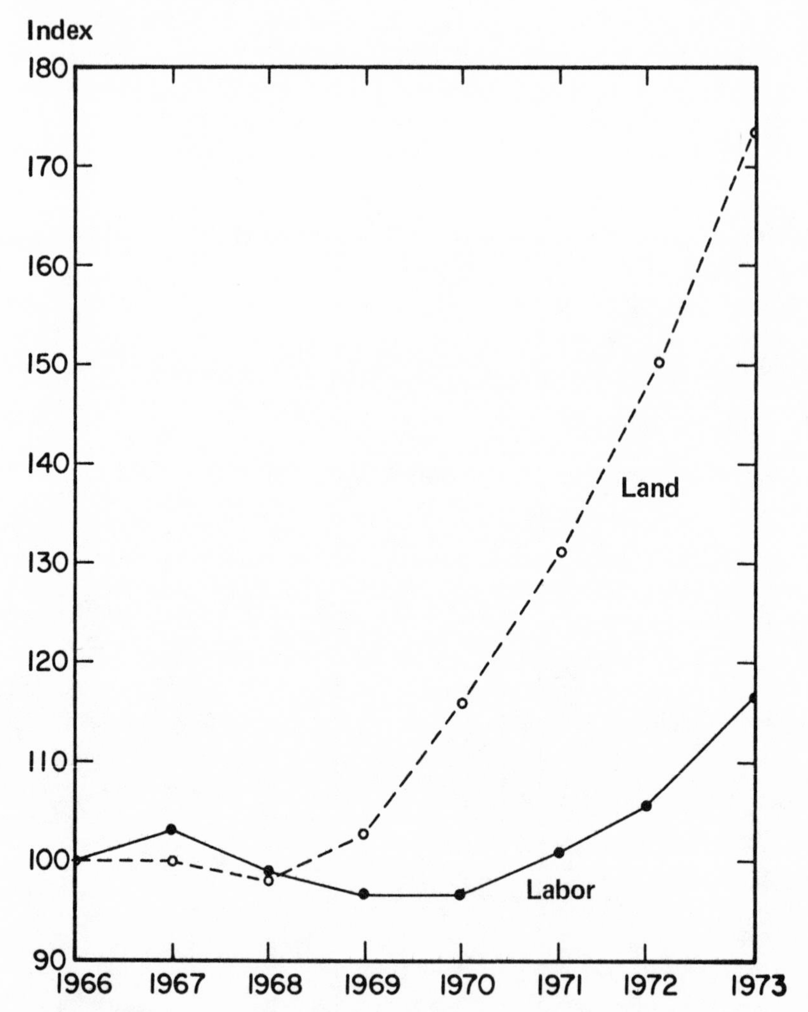

Figure 24-3. Recent trends in factor prices in real terms, Brazilian agriculture, 1966-73 (prices deflated by the cost of living index; data for 1973 refer to the first semester).
Source: Fundaçao Getulio Vargas, *Conjuntura Economica*, various issues.

implict prices through knowledge of the production function, but, because the lack of time series data on land values precluded the identification of a key parameter, this was not possible. Instead, trends in the prices of the individual inputs were considered.

Data on the real price of land and labor, the two primary inputs, are graphed in Figure 24-3. The data indicate that between 1968 and 1973 the price of land had been increasing faster than the price of labor, although the prices of both have been increasing since 1970. These data suggest that in recent years land has become increasingly scarce in relation to labor. Therefore, a tentative conclusion is that the agricultural research program should give special attention to the development and adoption of land substitutes.

This conclusion should be tempered with a certain degree of caution, however, because Brazil is basically not a land-scarce country. It still has large areas of unsettled land, and the government is making sizable investments to open up new areas, especially in the vast Amazon region. Thus it is possible that the relative scarcities of land and labor may change, especially if the recent rapid rates of industrialization continue and if new areas of fertile land become available to the economy as a result of improvements and extensions of the transportation system.

The trends in the prices of the close substitutes of labor and land – tractors and fertilizers, respectively – provide additional insights into what direction research should take. In general the weighted price of fertilizer has been declining relatively more than the price of tractors. The real price of fertilizers in the aggregate declined some 35 percent from 1966 to 1970, while the price of tractors declined on the order of 25 percent. Among the plant nutrients, nitrogen has experienced the greatest decline in price, followed by potassium, with the price of phosphorus (an important nutrient under Brazilian conditions) declining the least.

These data abstract from the recent upsurge in fertilizer prices. However, if this is of a temporary nature, the data on price trends suggest that research on the land subfunction – the development of improved varieties, increased knowledge about pesticides and fertilizers, etc. – should receive high priority. It should be noted that recent efforts of the Brazilian government to strengthen its agricultural research arm are therefore in the right direction. Moreover, the large road-building programs designed to open up new areas are also consistent with the need to ease what appears to be a growing land constraint.

The decrease in fertilizer prices which was occurring until recently makes the development of fertilizer-responsive varieties an attractive means of easing the emerging land constraint. If the current economic boom continues, however, the mechanization of agriculture may also become an increasingly important aspect of the development process.

Finally, the use of modern inputs has increased fairly rapidly in recent years. The fact that this increase appears to be the result of changing factor prices provides support for the model of induced technological change from which the present analytical model was developed.

Economic and Policy Implications

The results presented above do not provide conclusive recommendations on research priorities since they do not take account of the expected costs of making a given technological advance. However, they do suggest certain emphases for consideration in allocating a given research budget among the many alternatives faced by the decision-maker. The particular emphasis chosen will depend on the goals held for the research program.

The following discussion is organized in four parts. First, the focus of the research in the factor-factor dimension is dealt with. Then some general equilibrium matters are discussed, followed by a consideration of research priorities in the product-product dimension. Finally, a crude evaluation of the potential for technological change in the six crops studied is presented as a proxy for the cost of making technological advances with them. An implicit assumption throughout the analysis is that the rate of return to investment in agricultural research will be high.

The Factor-Factor Dimension

In the context of the Hayami-Ruttan induced innovation model the correct technological path for a country to choose is that which eases the particular factor scarcity that is constraining output expansion, with the factor scarcities of major interest being those of the primary inputs, land and labor. The trends in relative factor prices are important, since these would indicate which resource was most inelastic in supply.

Data on the trends in relative factor prices were summarized in the previous section and provide strong clues on the direction that research should take in Brazil. There was evidence of a growing relative factor scarcity of both land and labor in recent years, but the price of land was increasing at a much higher rate than the price of labor. This suggests that greater emphasis should be given in the research program to the land subfunction than to the labor subfunction. Of particular interest would be research which helps to bring more land into production (soil research, for example) or which facilitates the replacement of land by land substitutes. Soil research might be focused on problem soils such as the *cerrados*,[13] while biological research might focus on the development of varieties that are more responsive to fertilizer.

A legitimate question might be raised whether so much emphasis on raising land productivity is justified when labor productivity is so low and higher earnings are not likely to be realized until the productivity of labor is increased. These alternatives are not mutually exclusive, however. If one recognizes the well-known relationship between labor productivity and the productivity of land and the land/man ratio,

$$Y/L = (Y/T)\,(T/L),$$

(where Y = gross output, L = labor, and T = land), it becomes clear that increasing the productivity of land may raise the productivity of labor without having a large labor-displacing effect. To focus more directly on raising labor productivity by increasing the land/labor ratio (by increasing mechanization, for example) may have a strong labor displacement effect.

Considerations such as these may have special relevance in the Brazilian Northeast, where the land frontier is almost closed, some 63 percent of the labor force is still in agriculture, and the absorptive capacity of the nonfarm sector is still quite limited. Increasing the productivity of land may be the only way that agricultural research can improve the welfare of the rural population without having strong labor-displacing effects. Ultimately, of course, other measures are needed to solve this problem, including greater investments in schooling and training, the reduction in factor price distortions, and more generalized industrialization within the region. These may be more effective means of solving the problem of low returns to labor than agricultural research per se.

This discussion of the special problems of the Northeast brings to the fore the larger question of regional disparities in a country as large as Brazil. For example, in contrast to the Northeast, the frontier of the Central West is characterized by a relative abundance of land and a relative scarcity of labor. Sanders has argued that one of the reasons for mechanization on the Mato Grosso and Goias frontiers is the cost of obtaining and controlling seasonal labor.[14]

Trends in the prices of land and labor in selected regions are presented in Figures 24-4 and 24-5. Land prices have been increasing at a faster rate in the old regions (Northeast, East, and South) than in the more recently opened region, the Central West. There is less disparity in the trends in real wages, but in the last two years there was a sharp upward movement in both the West and the East. The latter region is close to the industrial heartland of Belo Horizonte, Guanabara, and São Paulo. The rapid industrialization of recent years has drawn a lot of labor away from agriculture in this region, with the result that the real wage has started to rise.

To summarize, the discussion above suggests two basic conclusions if output growth is the primary goal of technology policy. First, in the aggregate, primary attention should be directed to the land subfunction in order to ease what appears to be a growing land constraint to output expansion. However, given the regional diversity in Brazil, a case can be made for regional differences in research policy. In the Northeast, East, and South, major emphasis should be given to raising land productivity. In the West and East, however, a labor constraint is emerging. Moreover, if the economy continues to expand at the rapid rates of the recent past, more attention may need to be given generally to the problem of labor scarcity.

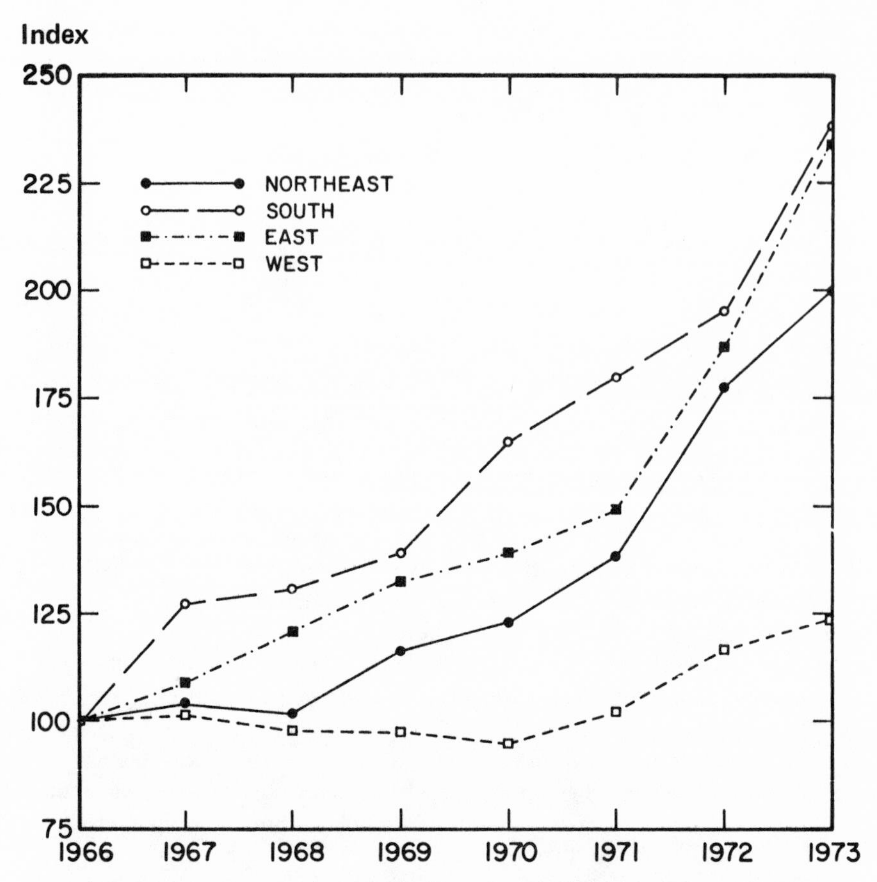

Figure 24-4. Recent trends in land prices in selected regions, real terms, Brazil, 1966-73 (prices deflated by the cost of living index; data for 1973 refer to the first semester).

General Equilibrium Considerations

The analysis above has been cast in a partial equilibrium framework. However, a technological breakthrough is expected to lead to a shift of resources from one crop to another as well as from agriculture to the nonfarm sector. A two-sector general equilibrium framework is useful for analyzing these shifts and for drawing implications regarding the allocation of research resources.[15]

The contribution of the general equilibrium model is to introduce the price elasticity of demand for the product into the analysis so that conclusions can be drawn about changes in the distribution of income between the two fac-

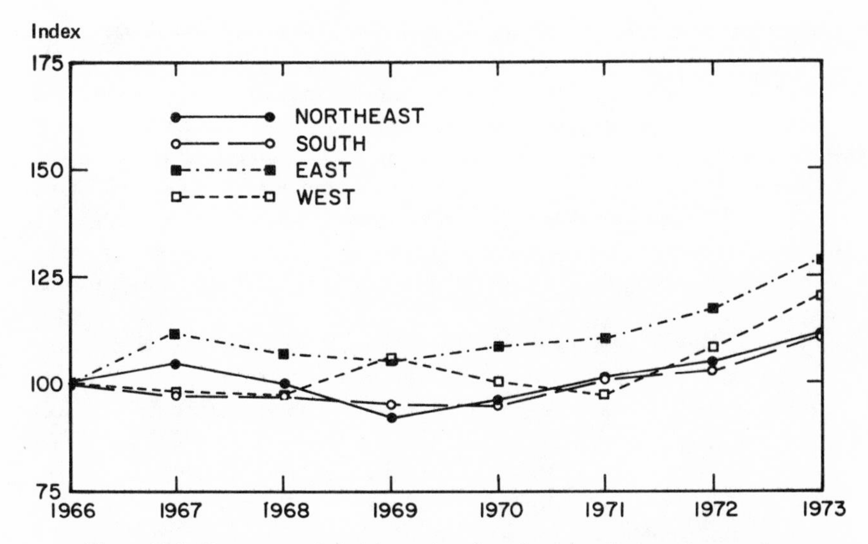

Figure 24-5. Recent trends in labor prices for selected regions, real terms, Brazil, 1966-73 (prices reflected by the cost of living index; data for 1973 refer to the first semester).

tors of production, labor and capital. The discussion will be conducted in two steps. First, the agricultural sector will be analyzed in relation to the nonagricultural sector. Second, groups of crops, classified according to their price elasticity of demand, will be considered in relation to the rest of the economy.

Consider the introduction of a land-saving (or land-augmenting) technical change into agriculture. The estimates of the elasticities of substitution in the production function were .625 between labor and mechanization, .933 between land and fertilizer, and .900 between the land subfunction and the labor subfunction. Since the labor subfunction was made up primarily of labor (the level of mechanization being quite low), we have essentially two estimates of the elasticity of substitution between labor and other inputs: .625 and .900. The latter is probably more appropriate for the present analysis, for it indicates the elasticity of substitution for a broader category of inputs.

In a closed economy the price elasticity of demand for agricultural output in the aggregate tends to be low. Although to the best of our knowledge no estimates are available for Brazil, most analysts would judge this elasticity to be in the range of .4 to .6. Under these conditions a decrease in labor incomes is expected from the technical change, and a redeployment of labor out of the sector will be the result.

If the economy is opened, the demand elasticity will be much larger, with the size depending on the extent to which the small-country assumption applies and on the relative importance of exports. In principle the elasticity of demand in the aggregate could be sufficiently large so that no decline in labor income would result nor would a redeployment of labor out of agriculture be required. However, in the past Brazil has discriminated severely against its agricultural sector by means of restrictive trade policies (export quotas) and a greatly overvalued currency. Consequently, technical change would have been strongly labor displacing, almost independently of what technological path was chosen.

During the late 1960s and early 1970s trade policy shifted to overt export promotion, with an exchange rate that was near equilibrium. But because of the convulsion in commodity markets in recent years, restrictive practices have been imposed again to protect the domestic consumer. This is being done at the very time that the research infrastructure is being strengthened. It should be noted that, although such policies do benefit consumers, they force a major adjustment problem onto the labor force.

The results are similar at the lower level of aggregation. Products such as rice, corn, edible beans, and manioc have relatively low price elasticities of demand. A technological change for any one or all of these crops will release labor to the rest of the agricultural sector and to the nonfarm sector, and adjustments will be required if a new equilibrium is to be reached. If the research program is focused on export products such as cotton and sugarcane, however, the expected outcome will be an increase in the return to labor and an increase in employment, other things remaining equal.

The moral to this story is that even technical changes which are not directly labor displacing can cause the displacement of labor as a result of the output effects and their effect on product price. To the extent that the research program focuses on export products and free trade is promoted, this problem need not arise. But to limit the research program to export crops is to forgo the benefits to consumers from a more generalized technical change. If these benefits are desired, then complementary policies to facilitate the labor adjustment problem are required.

The Product-Product Dimension

As the analysis above suggests, the choice of product for research emphasis is an important determinant of who will receive the benefits from a given expenditure on research. This is because the price elasticity of demand for the product is an important determinant of the distribution of benefits between the producer and the consumer, as well as the functional distribution of income. The present section contains a more systematic analysis of this choice, cast in the framework of four alternative sets of goals. (The role of the supply elasticity is submerged in this discussion.)

If policy makers choose *to increase the income to the agricultural sector* as their primary goal, the products that should be considered are those with a large price elasticity of demand. Among the six products treated in this study, cotton and sugarcane would be the obvious choices, since they are traditional export products. But products such as corn, of which Brazil is a marginal exporter, would also have considerable potential. And more generally, products which are already exported or which would have export potential with technological improvements should receive attention.

The same set of conclusions would apply if the goal were *to increase the income and employment of agricultural labor*. If the technological change is in the land subfunction (i.e., the development of more fertilizer-responsive varieties and the increased use of fertilizer), then it is sufficient that the price elasticity of demand for the product be greater than .9 in order for the specified goal to be attained. Obviously labor-displacing technologies such as mechanization should be avoided, with an important caveat for that mechanization which does increase the demand for labor.

If the goal is *to increase consumers' welfare*, the choice should stress those products which have a low price elasticity of demand. And it may be desirable to emphasize those products that are consumed by low-income groups. Of the six products considered in this study, corn, edible beans, and manioc would receive high priority, as would rice. As development proceeds in Brazil, corn is declining in importance as a human food, but it may eventually become important as a feed input for the production of the livestock products. Rice is consumed by all income groups, but as long as the goal is to benefit the consumer, it should receive attention.

The attainment of the goal of *enlarging agriculture's contribution to general economic development* is a bit more complicated than achieving the other three goals. It depends in part on easing the constraints which prevent the economy from realizing its potential. The choice of product for research emphasis can contribute to this end, as seen by recalling the five contributions that agriculture can make to the general development of the economy: (1) to keep the price of food low so that wage pressures are diminished – a stimulus to industrialization, (2) to increase the supply of exchange earnings, (3) to supply capital for the expansion of the nonfarm sector, (4) to provide a market for the products from the nonfarm sector, and (5) to supply labor for the expansion of the nonfarm sector.

The product choice to facilitate contribution (1) is the same as that for benefiting the consumer. Certain agricultural products are wage-goods, and these should receive emphasis. Among the six products considered, emphasis should be given to corn, rice, edible beans, and manioc.

Contributions (2), (3), and (4) will be attained by the same product choice that will increase the income to agriculture in the aggregate. To the extent that the choice focuses on export products there will be less competition with the nonfarm sector through the product market and a larger net gain to the economy.

Finally, to release labor from agriculture (5), two approaches can be taken. The first is to concentrate on labor-displacing mechanization. The second, and more important in the present context, is to focus the research effort on those crops with a low price elasticity of demand. Since these products tend to be those that are wage-goods, such an approach would make a double contribution to the labor market. In addition to releasing labor directly from the agricultural sector, it would also keep nominal wages down in the nonfarm sector by helping to lower the price of wage-goods.

The Potential for Technical Change

Our analysis up to this point has completely neglected the cost side of the question. The assumption has been implicit that comparable investments in research directed to each of the crops would produce comparable results. This is not likely to be the case.

In the absence of data on the costs required to obtain a given technological advance or data on the expected rate of return to research on particular crops, other rather crude indicators can be considered. For example, the immediate potential for yield increases might provide a crude proxy for the costs required to obtain yield increases. One measure of this potential would be a comparison of yields in Brazil with those in other countries. Another indicator would be the ease with which international technology can be adapted to Brazilian conditions.

A comparison of yields in Brazil with those from selected other countries for each of the six crops considered shows that, in general, Brazilian yields are quite low by international standards and have shown little tendency to increase. On the basis simply of the differentials between Brazilian and these foreign yields, we may hypothesize that yield increases will be easiest to obtain with corn, followed by rice, cotton, sugarcane, edible beans, and manioc, in decreasing order. This consideration of existing yield differentials, although a crude indicator, may provide a first approximation of the costs involved in increasing yields.

With respect to the transfer of technology, the most promising potential would appear to be with those crops that have received attention from the international centers. Not only have these centers concentrated on increasing the ecological adaptability of the crops they work with, but they also have the capacity to deal with these crops under ecological and economic conditions that are at least somewhat similar to those in Brazil. By this criterion, rice, wheat, corn, and manioc may offer the lowest cost advances.

Some Concluding Comments

The "model" presented here does not lend itself to formalization in a system of equations. We believe, however, that it points out the directions in which analysis might proceed to develop such a formal model. One of the most serious deficiencies at the present time is the lack of input-output data on the research process itself – data which would indicate how much it would cost to make a given technological advance. In making more precise judgments about research priorities, however, some estimation of these costs could be achieved by working closely with knowledgeable biological scientists and administrators.

The analysis also points up the importance of certain key parameters as the basis for establishing research priorities. The state of knowledge for many of these for countries like Brazil is rather deficient. As the basic econometric work proceeds, however, improved judgments about research priorities can be made.

The question of research priorities is strongly dependent on policy makers' goals for both the agricultural sector and the research program itself. These in turn will depend on the stage of development of the economy, the particular development model used as a basis for policy, and the particular measures used to implement the policy. Economists would contribute greatly by identifying for policy makers what these goals might be.

Finally, research priorities cannot be established in isolation from other policy measures. Economic policy, for example, can either cancel out the expected goals of a research program or reinforce them. Similarly, economic policy can provide alternative means of easing output constraints, as illustrated by the road-building program in Brazil.

Economic policy is also important as a means of offsetting some of the social costs of the research program. Technical change which prematurely releases labor to the nonfarm sector can impose serious costs on particular groups in the society. These costs should enter the calculus of research priorities. Moreover, to the extent that effective programs are implemented to deal with them, the benefits of the research program will be larger.

NOTES

1. An important landmark on the way was the Minnesota Symposium, the predecessor to the present conference. See Walter L. Fishel, ed., *Resource Allocation in Agricultural Research* (Minneapolis: University of Minnesota Press, 1971).

2. For more detail on these issues see G. Edward Schuh, "Some Economic Considerations for Establishing Priorities in Agricultural Research," paper presented at Ford Foundation Seminar of Program Advisors in Agriculture, Mexico City, November 1972.

3. Yujiro Hayami and Vernon W. Ruttan, *Agricultural Development: An International Perspective* (Baltimore: Johns Hopkins University Press, 1971).

4. For an excellent review of the concepts of economic surplus and their use in applied research, see J. M. Currie, John A. Murphy, and Andrew Schmitz, "The Concept of Economic Surplus and Its Use in Economic Analysis," *Economic Journal*, 81:324 (December 1971), 741-799.

5. Zvi Griliches, "Research Cost and Social Returns: Hybrid Corn and Related Innovations," *Journal of Political Economy*, 66:5 (October 1958), 419-431. W. L. Peterson, "Returns to Poultry Research in the United States," *Journal of Farm Economics*, 49:3 (August 1967), 656-669. H. W. Ayer and G. Edward Schuh, "Social Rates of Return and Other Aspects of Agricultural Research: The Case of Cotton Research in São Paulo, Brazil," *American Journal of Agricultural Economics*, 54:4 (November 1972), 557-569.

6. For the derivation of a two-stage production function, see K. Sato, "A Two-Level Constant-Elasticity-of-Substitution Production Function," *Review of Economic Studies*, 34:2 (April 1967), 201-218, and Alain de Janvry, "A Socioeconomic Model of Induced Innovations for Argentine Agricultural Development," *Quarterly Journal of Economics*, 87:3 (August 1973), 410-435.

7. The interested reader can find the analytical equations in J. P. Ramalho de Castro, "An Economic Model for Establishing Priorities for Agricultural Research and a Test for the Brazilian Economy," Ph.D. thesis (Lafayette: Purdue University, 1974).

8. The important coffee crop was excluded from the analysis because government intervention in this sector has been great and because Brazil still occupies a dominant position in world coffee markets. The small-country hypothesis would not be valid.

9. Murray Brown, *On the Theory and Measurement of Technological Change* (London: Cambridge University Press, 1966), chapter 5.

10. The 1960 Agricultural Census indicated that 76 percent of Brazilian farms were using only human power, while less than 1 percent were using some mechanical power. See G. Edward Schuh, *The Agricultural Development of Brazil* (New York: Praeger, 1970), p. 154.

11. These points were suggested by J. H. Sanders in "Mechanization and Employment in Brazilian Agriculture, 1950-1971," Ph.D. thesis (Minneapolis: University of Minnesota, 1973), p. 14, and Appendix F.

12. In fitting the models it was assumed that all fertilizers were used for crop production. This is a reasonable assumption, given that fertilizer is seldom used on pastures in Brazil. Knight has noted that even in Rio Grande do Sul, which has a fairly advanced agriculture and one in which modern beef production is important, fertilizer is used on pasture only for experimental purposes. See Peter Knight, *Brazilian Agricultural Technology and Trade: A Study of Five Commodities* (New York: Praeger, 1971).

13. The cerrado soils are for the most part highly leached out latosols with a high de-

gree of acidity, low nutrient levels, and apparently some problems of toxicity. They cover large areas of Brazil, and very little is known about them.

14. Sanders, "Mechanization and Employment."

15. For a succinct treatment of the general equilibrium model, see Harry G. Johnson, *The Two-Sector Model of General Equilibrium* (Chicago: Aldine Atherton, 1971).

25

Measuring the Impact of Economic Factors on the Direction of Technical Change

Hans P. Binswanger

This chapter looks at recent theoretical and empirical advances in the theory of induced innovation, with respect particularly to the direction of technical change. The first section discusses the mechanisms of induced innovation and the potential roles of economic factors and fundamental biases in determining the direction of technical change. The second section explores the question of whether the agricultural sectors of different countries have developed along technological paths of different factor intensities or whether the observed factor ratio differences simply reflect ordinary substitution adjustments to differences in factor prices. The third section assesses the relative importance of factor prices and fundamental biases in determining the direction of technical change.

Mechanisms of Induced Innovation

In agriculture, the term *induced innovation* has been used by Hayami and Ruttan essentially to indicate that factor scarcities or factor prices influence the direction of technical change for a particular commodity.[1] It is hypothesized that technical change is directed toward saving the progressively scarce or more expensive factors, i.e., saving proportionately more of the scarce factor than of the abundant factor per unit of output at constant factor prices. (Many technical changes reduce input requirements of all factors per unit of output and therefore "save" all factors. In economics, however, we define a technical change as labor saving only if it reduces labor requirements to a greater degree than it reduces the requirements of other factors.) In this chapter, consistent with recent advances in the theory of induced innovation, the term *induced innovation* is used more broadly than before and includes the response of the rate and direction of technical change to final demand conditions and to factor scarcities.[2]

The response of the rate of technical change to final demand conditions has been empirically well established by Schmookler, Griliches, Lucas, and Ben-Zion.[3] As a result we can concentrate here on the factors which induce biases in the direction of technical change.

The factor intensity of agriculture as a sector of the economy is determined, in the absence of research, by two sets of forces: the choice of the commodity mix of output and the choice of technique for each commodity. Intrinsically different factor ratios are associated with different agricultural commodities. Vegetables, tree crops, and livestock typically have a higher labor/land or labor/capital ratio in all economic environments than do most grain crops. The theory of comparative advantage states that countries with high labor/land or labor/capital ratios will concentrate on the more labor-intensive agricultural commodities and that agricultural output as a whole will be produced in countries with lower land/labor or capital/labor ratios than in countries with small labor endowments. Moreover, for each commodity produced, countries with low wage rates will use more labor-intensive techniques than countries with high wage rates; this tends also to decrease the sectoral land/labor and capital/labor ratios.

If a country with relatively abundant labor decides to subsidize capital (e.g., through tractor subsidies, subsidized credit, tax preferences, or overvalued exchange rates), it will distort both the choice of commodities and the choice of technique for each commodity such that the aggregate capital/labor ratio increases. Whether this increase comes primarily from a change in commodity mix or from a change in the technique of each commodity depends on the elasticity of substitution among factors in the production of each commodity. If the elasticities of substitution are low, the aggregate factor ratio changes primarily as a result of the change in commodity mix. If the elasticities are high, however, the aggregate factor ratio changes primarily as a function of a change in the technique used for each commodity.[4]

Research has an impact on the agricultural factor ratios through its effects both on the commodity mix and on optimal factor ratios within each commodity. Research on a labor-intensive commodity which results in neutral technical change for this commodity will, at existing factor and goods prices, increase the profitability of the commodity compared with all other commodities. Cultivators will tend to produce more of the commodity which experiences rapid technical change by reallocating land and capital resources to the improved commodity. If they can hire more labor at the going rate, the aggregate land/labor intensities of agriculture will decrease despite the neutrality of the technical change in the particular commodity.

Abel and Welsch have shown (in chapter 22) that for a labor-intensive country it makes sense to concentrate research resources on the labor-intensive commodities in which the country has comparative advantage. Even before research, such a country or region will have produced relatively more of the labor-intensive commodity owing to its comparative advantage. The factor productivity increase, therefore, would apply to a larger volume of output in the labor-intensive commodity than in the capital-intensive commodity, and the rate of return to research on the former would be higher. This is the first mechanism of induced innovation by which factor scarcities affect agricultural factor intensity.

In addition, research can alter factor intensities of production of a given commodity. I have recently worked out a microeconomic model of this type of induced innovation mechanism.[5] The model is based on a view of technological research and discovery similar to that of Kislev's (see chapter 10). Kislev's model asks the question, what quantity of resources shall be allocated to a single yield-increasing line of research? In my model it is assumed that, to increase the productivity of a given commodity, researchers have a choice of several research strategies or research lines. Each line of research has different implications for the factor ratios of the commodity, and the researcher chooses the combination of research lines which leads to the maximum reduction of costs of production at the existing factor prices. It can then be shown that the higher labor prices are in in relation to land or capital prices, the greater will be the allocation of research resources to those lines of research which tend to save labor and the smaller the allocation of resources to the lines of research which tend to save capital or land. This means that the higher the cost of labor, the higher will be the labor-saving bias of technical change.

Sometimes, because they do not look at factor prices in making their research decision, researchers and research administrators claim that the mechanism just described is not realistic. However, for such a framework to operate it is not even necessary that the research resource allocator be conscientiously trying to save the relatively more expensive factors. If he makes his research resource allocation decision on the basis of expected benefits from a research project, he will automatically tend to favor those research lines which either favor the high-priced commodity or save the high-priced factor. The reason for this is that the payoffs of any research are the product of increases in output and reductions in input requirements (owing to the innovation) weighted by the prices of the output and the inputs. Since the prices enter as weights, projects which result in increases in output of expensive commodities or which save primarily the expensive factors will have higher payoffs than projects which affect primarily inexpensive outputs and inputs.[6]

What has been said so far establishes the potential responsiveness of the factor ratio to factor prices through a mechanism of induced innovation. However, this mechanism might be empirically irrelevant if research possibilities themselves were severely biased and allowed the researcher little choice in research strategy. Research possibilities might be such that research projects on capital-intensive commodities and in capital-intensive directions would always be easier to carry out than research projects on labor-intensive commodities and in labor-intensive directions. In such a situation, induced innovation at best could only partly offset the fundamental bias of innovation possibilities. Efficient labor-using paths for technical change would not exist. On the other hand, if technical change possibilities were completely neutral toward commodities and toward biases within commodities, and if the potential research lines had a wide variety of factor-saving implications, there would be many different paths of efficient technical change from which countries could choose according to their factor scarcities. For example, considerable attention has been given recently to the effect of fertilizer availability and price on the attractiveness of studying biological nitrogen fixation. Technical change evolving from such exploration would be commercial fertilizer saving. Whether such technical changes will occur in the future will be determined both by the price of fertilizer and by biological research possibilities. If for some biological reason, nitrogen fixation cannot be brought about in crops other than legumes, and if there is a biological barrier to the amount of nitrogen fixation by legumes, technical change possibilities would preclude a substantial fertilizer-saving bias from this source even if fertilizer prices rose considerably. Such problems could also make induced labor-using biases empirically irrelevant.

A Two-Factor Test of Induced Innovation in Six Countries

The empirical tests described in this and the following section have been inspired by and are partly based on the work of Hayami and Ruttan.[7] Because the tests look at the responsiveness of aggregate factor ratios or factor shares to economic forces, changes in commodity mix cannot be distinguished from changes in factor mix for given commodities. Of the two tests reported, the first is based on a comparison of technical change paths among six countries

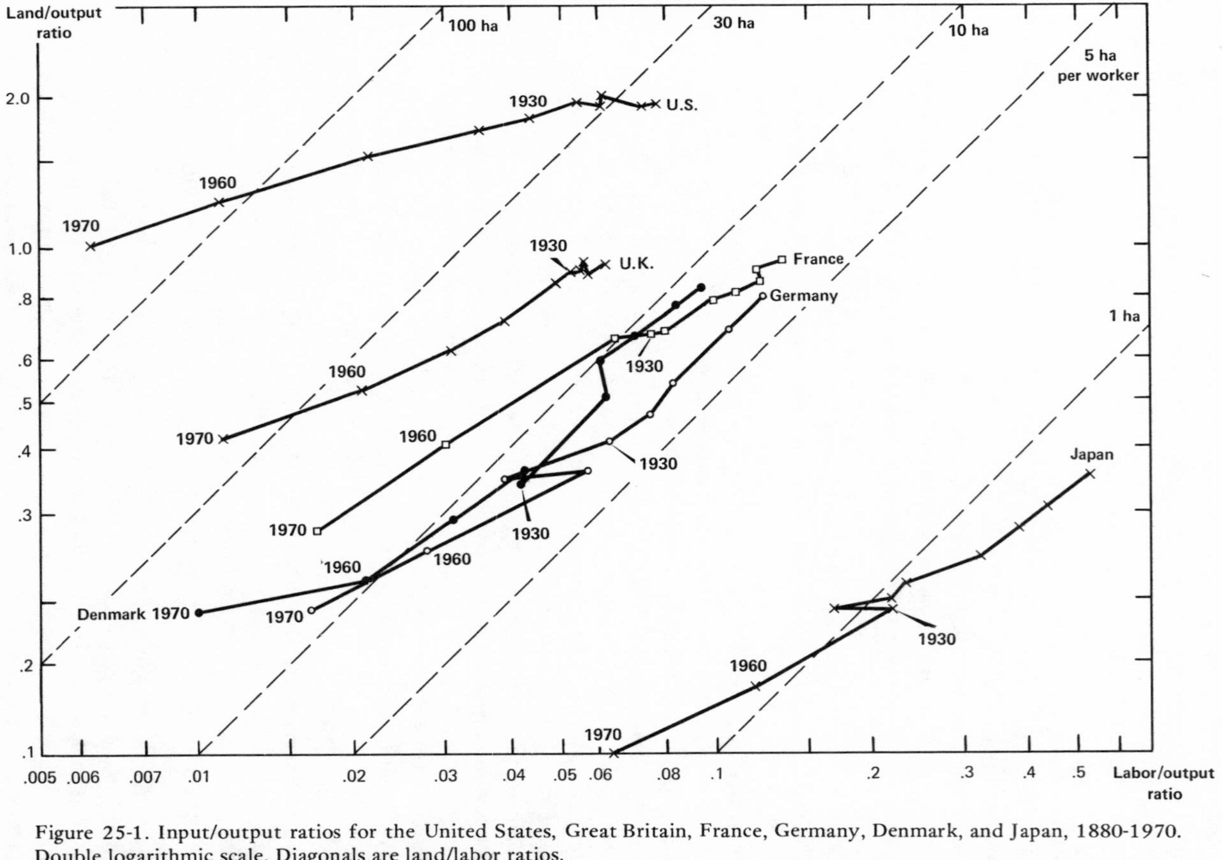

Figure 25-1. Input/output ratios for the United States, Great Britain, France, Germany, Denmark, and Japan, 1880-1970. Double logarithmic scale. Diagonals are land/labor ratios.
Source: Vernon W. Ruttan et al., "Factor Productivity and Growth: A Historical Interpretation," in *Induced Innovation: Technology, Institutions and Development*, eds. Hans P. Binswanger and Vernon W. Ruttan (Baltimore: © The Johns Hopkins University Press, in press).

and addresses the question of whether the paths are really different, i.e., whether there is a wide array of choices with respect to factor ratios. The second and more powerful test shows that fundamental biases of technical change possibilities explain a substantial part of the observed biases. It also reveals, however, that factor prices are capable of offsetting or increasing fundamental biases substantially; it thus demonstrates the empirical relevance of the induced innovation hypothesis.

Figure 25-1 traces the paths of input-output combinations for six countries from 1880 to 1970 in full logarithmic scale. Constant land/labor ratios are represented by the diagonal lines. Movements of the input-output combination toward the lower left-hand corner are technical advances. Equidistant inward movements correspond to equal rates of productivity improvements. From this figure we can see that all countries experienced substantial productivity advances during the period. However, for the United States the rate of productivity increase between 1930 and 1970 far exceeded the rate of productivity increase before 1930. In Great Britain there was practically no technical change before 1930.

The differences in land/labor ratios are enormous, as can be seen in Figure 25-2. The land/labor ratios are plotted in semilogarithmic scale. (Equal *slope* of lines indicates equal *rate* of increase in factor ratios.) The United States ratio exceeded the Japanese ratio by a factor of 30 in 1880 and by a factor of 102 in 1970; i.e., the differences in land/labor ratios increased substantially over time.

Figures 25-1 and 25-2 cannot by themselves answer the question whether the increases in factor ratios over time are caused by biases of technical change or by decreases in the price of land relative to that of labor. Neither can they answer the question whether the different land/labor ratios among countries reflect that the countries were using technologies with inherently different factor ratios or whether the factor ratio differences simply reflect factor substitution owing to price differences within a technology having identical factor intensities.

It is useful at this point to state the problem that must be solved in any test of induced innovation. In Figure 25-3, assume that the labor/land factor ratio in Japan can be represented by a line from the origin through P and that the labor/land factor ratio in the United States can be represented by a line from the origin through Q. Assume also that the slope of the line BB represents relative factor prices in Japan, where land is expensive in relation to labor, while the slope of CC represents the relative factor prices in the United States where labor is expensive in relation to land. If the substitution possibilities of the available agricultural technology can be represented by an isoquant map with little curvature, such as I_0^* and I_1^*, the differences in factor ratios be-

Figure 25-2. Agricultural land in hectares per male worker, 1880-1970, semi-log scale.
Source: Vernon W. Ruttan et al., "Factor Productivity and Growth: A Historical Interpretation," in *Induced Innovation: Technology, Institutions and Development*, eds. Hans P. Binswanger and Vernon W. Ruttan (Baltimore: © The Johns Hopkins University Press, in press).

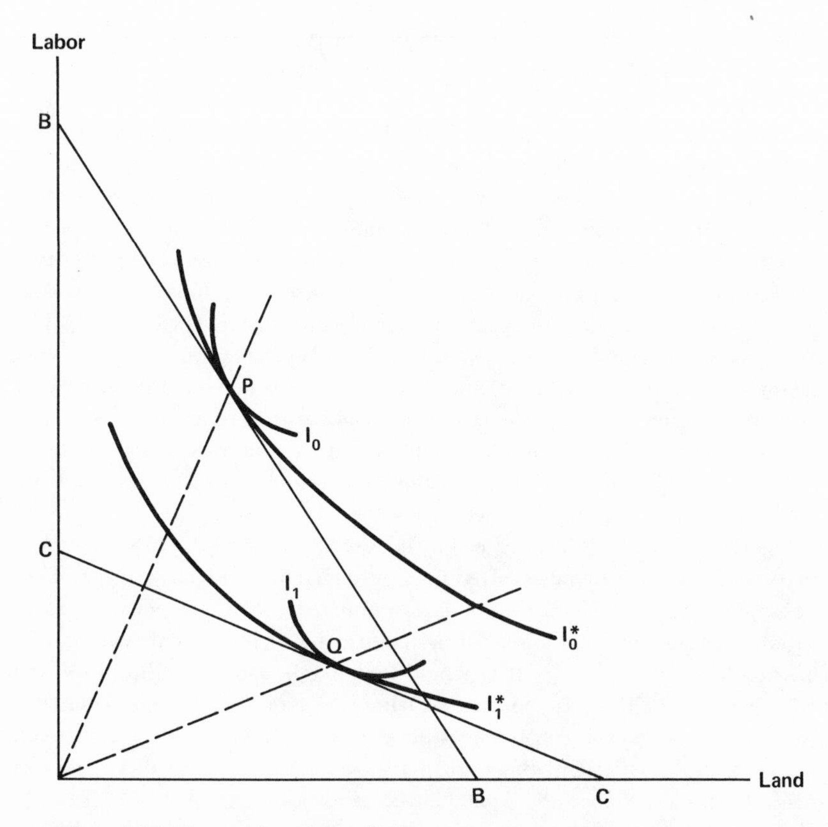

Figure 25-3. The problem of identifying causes of factor ratio differences.

tween Japan and the United States could be explained by simple substitution along production functions with equal factor-intensity characteristics (but not necessarily equal efficiency) caused by the differences in relative factor prices. If, however, the possibilities of substitution between labor and land are represented by I_0 in Japan and I_1 in the United States, the points P and Q would not represent alternative factor combinations along production functions with equal factor intensity. The difference between the isoquant set $\{I_0, I_1\}$ and the isoquant set $\{I_0^*, I_1^*\}$ is that the elasticity of substitution of the latter is much larger than that of the former. Therefore, the two-factor test to differentiate between technological factor intensities and simple factor substitution builds on the concept of pairwise elasticity of substitution between land and labor. The test involves an implicit assumption that the production process is separable between land and labor and all other factors. This assumption implies that the land/labor ratio is independent of changes in the prices of other inputs. To the extent that the land/labor ratio is influenced by other factor prices – the price of farm machinery, for example – the assumption of separability does introduce a bias into the simple induced innovation test used in this chapter.[8]

The test is somewhat complicated by absolute efficiency difference between periods or countries. Referring again to Figure 25-3, let P and Q represent the input-output combinations in two different time periods in the same country (or two different countries at the same time). Further, let BB and CC represent the respective factor/price ratios in the two time periods (or the two countries). Q has a higher land/labor ratio than P and a lower relative land price. Now construct a homothetic isoquant map which is tangent to BB at P and tangent to CC at Q. These would be the two production functions if the factor ratio change between P and Q could be explained by ordinary factor price substitution and neutral technical change over time (or equality of factor intensities of the production functions in two countries at a given time). Pairwise elasticity of substitution of this isoquant map can be measured from the factor ratios and factor/price ratios (see the mathematical footnotes in Tables 25-2 and 25-3), and we call it the *Necessary Elasticity of Substitution* (σ_N) to explain factor ratio differences by differences in relative prices. Suppose now that we have an econometric estimate σ' of the elasticity of substitution between land and labor for agriculture and that isoquants with σ' tangent to BB at P and to CC at Q would look like I_0 and I_1 instead of I_0^* and I_1^*. The shift of I_0 to I_1 represents a labor-saving technical change over time (or true factor intensity differences between two countries at a given time). Hence, if σ_N is sufficiently larger than σ' we can reject the hypotheses of both neutral technical change and equal factor intensities.

Estimates of a pairwise elasticity of substitution between land and labor are not available in the literature. However, I have provided estimates of a full set of factor demand elasticities for the United States, and Mundlak has developed a methodology to compute pairwise elasticities of substitution from factor demand elasticities.[9] Using Mundlak's method, I have computed an elasticity of substitution between land and labor (σ') to be 0.67. Thus a 10 percent change in the labor/land price ratio leads to a 6.7 percent change in the land/labor use ratio. For reasons given in footnote 10, $\sigma' = .67$ is a central value for the elasticity of substitution, but depending on the nature of the price rises it may exceed it or fall short of it.[10] Furthermore, a statistical error is attached to the estimate. Therefore, the critical value for rejecting the hypothesis of neutral shift or identical factor intensities will be taken as twice the estimated value, i.e., the hypothesis will be rejected if σ_N exceeds 1.34 for any comparison of points with different factor intensities.

This test is first applied to each country individually over time to see whether biases occurred along the path of each country. Japan is not included in

Table 25-1. Land/Labor Ratios and Land Price/Labor Price Ratios

Year	U.S.	Great Britain	France	Germany	Denmark	Japan
			Agricultural Land/Worker (A/L)			
1880 . . .	25.4	14.7	7.0	6.34	8.91	0.659
1930 . . .	40.5	19.7	8.8	6.46	8.17	0.908
1960 . . .	109.5	23.3	13.4	8.83	10.21	1.131
1970 . . .	160.5	33.5	16.2	12.20	17.92	1.573
			Price of Agricultural Land/Daily Wage Rate[a] (P_A/P_L)			
1880 . . .	181.0	995.0	780.0	967.00	381.00	1874
1930 . . .	115.0	189.0	264.0	589.00	228.00	2920
1960 . . .	108.0	211.0	166.0	378.00	166.00	2954
1970 . . .	108.0	203.0	212.0	244.00	177.00	1315

Source: Vernon W. Ruttan et al., "Factor Productivity and Growth: A Historical Interpretation," in *Induced Innovation: Technology, Institutions and Development*, eds. Hans P. Binswanger and Vernon W. Ruttan (Baltimore: © The Johns Hopkins University Press, in press).

[a] Price of land is price of agricultural land without buildings, except for Denmark, where buildings are included in the price. Price of labor is daily wage rate without room and board.

this time-series test because the comparability of its factor price data over time is doubtful.

Next the test is applied to all the countries at a given point in time to see whether the factor intensities of their production functions differ. Denmark is excluded from the comparison because its factor price series is based on a land price concept different from those of other countries.

The data on factor intensities and relative prices are given in Table 25-1.

The Paths of the Land/Labor Ratio in Five Countries

Table 25-1 shows that between 1880 and 1930 in the United States the land/labor ratio rose from 24.5 to 40.5, while the corresponding price ratio fell from 181 to 115. In Table 25-2 it is shown that the necessary elasticity of substitution to explain the drop in the land/labor ratio (A/L) by simple price effects is 1.03. This falls short of the critical value of 1.34. Therefore, the hypothesis that technical change was neutral during this period cannot be rejected. However, the necessary elasticity of substitution jumps to 16.5 for the period 1930-60, and in the sixties the relative price ratio does not change, while the land/labor ratio continues to rise. Therefore, change has been strongly biased in a labor-saving direction since 1930.

In Great Britain, France, Germany, and Denmark the necessary elasticity was very low between 1880 and 1930, which implies neutral technical change. In Denmark the changes in factor ratios and relative prices even imply labor-

Table 25-2. Necessary Elasticity of Substitution (σ_N)[a] to Explain the Interperiod Changes in Land/Labor Ratios by Price Effects within Selected Countries[b]

Time Period	U.S.	Great Britain	France	Germany	Denmark
1880-1930	1.03	0.16	0.20	0.04	e
1930-60	16.50*	d*	0.90	0.70	0.70
1960-70	c*	9.43*	d*	0.74	d*

* Significantly labor saving.

[a] $\sigma_N = \frac{\text{Percentage change in land/labor ratio between two periods}}{\text{Percentage change of labor price/land price ratio}}$

with geometric means as a basis for the two percentage changes, i.e.,

$$\sigma_N = \frac{(A/L)_{i+1} - (A/L)_i}{(P_L/P_A)_i - (P_L/P_A)_{i+1}} \times \sqrt{\frac{(P_L/P_A)_{i+1}\,(P_L/P_A)_i}{(A/L)_{i+1}\,(A/L)_i}}$$

where i = 1880, 1930, 1960, 1970.

[b] The critical ratio to reject hypothesis of neutral technical change is 1.34.

[c] No price change, technical change labor saving.

[d] Price ratio and land/labor ratio rise, which implies labor-saving technical change. (No common isoquant map can be constructed through P and Q in Figure 25-3 in this case.)

[e] Land/labor ratio declines very slightly but price declines much more, which implies labor-using technical change.

saving technical change, and that may be the case also in the other European countries where the necessary elasticities of substitution are exceedingly small.

Between 1930 and 1960 Great Britain experienced labor-saving technical change, while technical change remained neutral in the other European countries. In the 1960s technical change in Great Britain, Denmark, and France was labor saving but it remained neutral in Germany. Labor-saving technical change (relative to land) is therefore a recent phenomenon for Europe, and in Germany it never occurred at all.

The Cross Section Comparison of Factor Intensities

The comparison of factor intensities and factor prices across countries is shown in Table 25-3 with the necessary elasticities to explain the factor ratio differences among the countries. All the necessary elasticities of substitution of the comparison between the United States and Japan exceed the critical value of 1.34. Therefore Japan and the United States must have used technologies with different factor intensities throughout the period. This confirms the induced innovation hypothesis because the high wage country did use a more labor-intensive technology than the low wage country.

Table 25-3. Test of Differences in Technological Paths: Necessary Elasticity of Substitution[a] to Explain Differences in Land/Labor Ratio by Price Ratio Differences[b]

Item	1880	1930	1960	1970
U.S. vs. other countries				
Japan	2.08*	1.35*	1.93*	3.12*
Great Britain	0.29	1.47*	2.50*	2.70*
France	0.87	1.96*	5.79*	4.13*
Germany	0.80	1.16	2.42*	4.00*
Japan vs. Europe				
Great Britain	7.01*	1.21	1.24	2.04*
France	3.26*	0.92	0.79	1.39*
Germany	4.13*	1.29	1.00	1.28
Great Britain vs. continental Europe				
France	c*	2.47*	c*	17.12*
Germany	c*	0.98*	1.71*	5.72*
Continental Europe				
France vs. Germany	0.46	0.38	0.50	2.02*

* The paths of the two different countries differ significantly in land/labor intensity.

[a] $$\sigma_N = \frac{(A/L)_i - (A/L)_j}{(P_L/P_A)_j - (P_L/P_A)_i} \times \sqrt{\frac{(P_L/P_A)_i\,(P_L/P_A)_j}{(A/L)_i\,(A/L)_j}}$$

where i and j are the different countries in the same year.

[b] Critical value to reject hypothesis of equal technology: 1.34, i.e., twice the value of σ for equiproportional changes in P_A and P_L.

[c] Denotes cases where the country with the higher land/labor ratio also has the higher land price/labor price ratio. Such behavior is possible only if the country with the higher land/labor ratio employs a more land intensive technology, i.e., the hypothesis of equal technology is rejected. No common isoquant maps can be constructed through points P and Q in Figure 25-3.

In 1880 the United States and the European countries were apparently on essentially the same production function. This is not implausible given the state of mechanical technology in the late nineteenth century. By 1930, however, differences between the European and United States production functions had clearly emerged, as indicated by the increases in the necessary elasticities of substitution. Even in 1930, however, the hypothesis of similar production functions cannot be rejected for the United States-Germany comparison. Because after 1930 technical change became extremely labor-saving in the United States, European technology differed very strongly from United States technology in 1960 and again in 1970.

The differences in factor ratios between United States and European agriculture thus reflect differences not only in relative factor prices but in the production functions available to United States and European farmers.

In the second section of Table 25-3 the test is carried out for Japan in comparison with Europe. In 1880 European technology was much less labor intensive than Japanese technology. But neutral or labor-using technical change in Europe together with neutral or possibly labor-saving technical change in Japan brought the technologies much closer by 1930. By 1970, however, labor-saving technical change in the United Kingdom and France had become strong enough to make their technologies clearly different from Japanese technology again. In Germany the technical change remained neutral, however, and by 1970 it seems likely that differences in factor ratios between Japanese and German agriculture are accounted for primarily by differences in factor prices.

The third section of Table 25-3 shows that British technology was almost always less labor intensive than French and German technology, except for Germany in 1930, when the test fails. The fourth section indicates that Germany and France were using the same technology until 1960. Thereafter labor-saving technical change in France moved the two technologies apart.

Implications for the Induced Innovation Hypothesis on the Basis of the Time Series and Cross Section Comparison

We can distinguish essentially four paths for technology:

1. The United States, starting from a position similar to Europe, developed away from Europe in a strongly labor-saving direction.

2. Great Britain also experienced strong labor-saving technical change after 1930, but its technology remains much more labor intensive than the technology of the United States. British technology was also less labor intensive than the technology of continental Europe and Japan.

3. Continental Europe experienced essentially neutral technical change, or possibly even labor-using technical change, throughout the period, except for France and Denmark in the 1960s.

4. Japan started from an extremely labor-intensive position. Its technical change must have been either neutral or possibly slightly labor saving.

The tests therefore confirm the induced innovation hypothesis. Indeed, the United States, with the highest relative wage rate, developed on the least labor-intensive path, while Japan, with the lowest relative wage rate, had the most labor-intensive technology. The European countries, with labor price levels between those of the United States and Japan, also used technologies with intermediate factor intensities. Furthermore, given the often high values for the necessary elasticities of substitution, the technology differences must have been quite substantial.

Although the tests performed in this section clearly establish that the paths of factor ratios of technology are guided by factor scarcities, there are some observations which are not consistent with the simple version of the induced innovation hypothesis. For example, the United States followed a more labor-saving path with respect to its initial land/labor ratio than did the four European countries despite the fact that in the United States the decline in the price of land relative to that of labor was less rapid than it was in Europe (see Figure 25-4). A simple induced innovation framework which assumes that all biases are caused exclusively by factor prices would predict that Europe should have more labor-saving technical change than the United States.

There are several factors that may account for the less than complete consistency between the induced innovation hypothesis and the observed difference in factor price and resource use ratios. One possibility is that there are fundamental biases in innovation possibilities in the labor-saving direction which were only partly offset, in the case of Japan, by technical change induced by the rising relative price of land. A second possibility is that the effect of changing relative prices on the direction of technical change may sometimes be offset by the low cost of borrowing from countries with differing factor price ratios. Finally, differential rates of growth in demand may induce technical change through changes in the factor/product price ratios.

The impact of technology transfer on borrowing from countries with different factor/price ratios may be particularly important for the countries with extreme differences in factor/price and use ratios such as Japan and the United States. If a country starts the process of modernization from an extremely labor-intensive position, as Japan did in the 1880s or as some of today's developing countries are doing, the only technologies which it can transfer or borrow from other countries will be more labor saving than those which would be induced by its own factor endowments and price ratios. Similarly, if a country starts the process of modernization from an extremely labor-extensive position, as in the case of the United States in the post-Civil War period, it is likely that the technologies which it borrows will be more land saving than those which would be induced by its own factor endowments and price ratios. In such situations it is unlikely that the inducement process will be able to offset more than partially the combination of fundamental bias and transfer bias. At this stage we are not able to provide quantitative estimates of the effect of the fundamental and transfer biases.

A Many-Factor Test of Induced Bias in United States Agriculture

The test discussed in the last section established the existence of different paths of technology. This section provides a more powerful test of induced

Figure 25-4. Price of land relative to daily wage rate, 1880-1970.
Source: Vernon W. Ruttan et al., "Factor Productivity and Growth: A Historical Interpretation," in *Induced Innovation: Technology, Institutions and Development*, eds. Hans P. Binswanger and Vernon W. Ruttan (Baltimore: © The Johns Hopkins University Press, in press).

innovation in United States agriculture. The methodology used overcomes one flaw of the earlier test in which a many-factor production process was treated as if it were a two-factor process. As a consequence, the test neglected the influence of prices other than of land and labor on the land/labor ratio. In this section a true many-factor framework is employed.

The test is based on directly measured biases of technical change instead of factor ratios. Biases are measured for the United States agricultural sector from 1912 to 1968, and for five factors (land, labor, machinery, fertilizer, other inputs). Once these biases are measured they can be compared with the trends in factor prices to see whether factor-saving biases correspond to rising factor prices and factor-using biases correspond to declining factor prices.

One problem with testing the inducement mechanism is that factor prices may be endogenous to agriculture. Suppose technical change biases were very responsive to factor price changes, and suppose, in the absence of technical change, that the land price has a tendency to increase. If such an increase were followed by technical change in a land-saving direction, the land price would rise less than it would in the absence of technical change, and in an extreme case it might even fall. This problem makes it impossible from *ex-post* empirical data to show that it was the factor price which caused the bias and not some land-saving bias in the innovation possibilities. However, if a factor price is exogenous to agriculture, this problem does not arise and the test proposed can properly identify causal relationships.

The price of land can be treated as endogenous. Therefore, we will neglect the series of land biases in the test, but for the other four factors the prices are governed in the long run primarily by the nonagricultural labor market and by cost conditions in the input-supplying industries. The biases of these four factors will be compared with the prices of the factors. The logic of the test is as follows: Suppose, first, that innovation possibilities are neutral, i.e., that in the absence of factor price influences the direction of technical change would be neutral. In this case induced innovation would be supported if we found that the factors with rising prices experienced a factor-saving bias while the factors with falling prices experienced a factor-using bias.

When innovation possibilities are not neutral – and we would expect this after the results of the last section – the situation is more complicated. In such a case it is possible that a factor with a rising price might experience a factor-using bias. At most, induced innovation can partly offset the fundamental bias. Indeed, the fact of a rising factor price together with a factor-using bias *establishes that innovation possibilities cannot have been neutral.*

In such a situation we have to look at the turning points of factor prices and of biases. A sharp increase of a slowly rising price should, after some time,

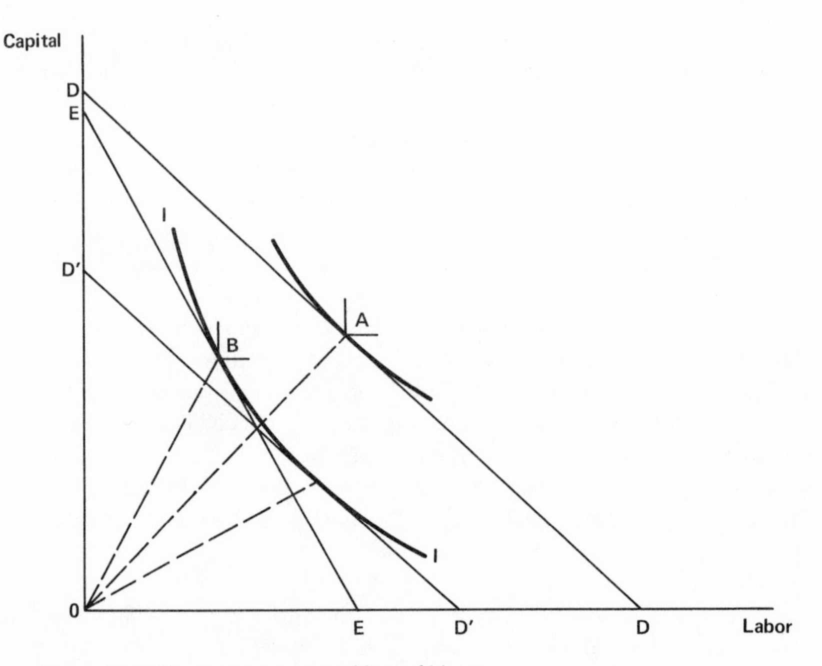

Figure 25-5. The measurement problem of biases.

be followed by a reduction in the factor-using bias to establish the presence of induced innovation. Such behavior would be consistent with a causal relationship between factor prices and biases in the direction of technical change.

The Measurement of Technical Change Biases

The basic problem of measuring biases in the Hicksian sense is most easily explained in a two-factor case.[11] The empirical difficulty of measuring biases arises from the fact that factor prices change over time, i.e., the observed changes in the factor ratio are due to two causes: (1) ordinary factor substitution along a given production function; and (2) biased technical change. Figure 25-5 illustrates this problem.

From economic data we can know the factor ratios OA and OB (for one unit of output) in two periods of time. We can also know the factor prices which prevailed during those two periods, namely, those corresponding to DD and EE. If we knew that the production function was of fixed proportions we would know that the entire change of the factor ratio between these periods was caused by technical change, and measuring biases according to (1) would be very simple. The move from A to B would have been a labor-saving technical change. But with a neoclassical production function, the observed data alone cannot tell us this. Suppose the elasticity of substitution was very large, as for the isoquant II. At equal factor prices D′D′ the capital/labor ratio would have decreased, i.e., technical change would have been labor using. The observed capital/labor ratio nevertheless decreased owing *to the rise of labor prices from* DD *to* EE. To measure biases, therefore, we need to know the parameters of the production process. In the two-factor CES case we have to know the elasticity of substitution to be able to split the observed factor ratio changes into two components, one owing to price changes and the other owing to technical change.

A similar procedure applies to the many-factor case, where we also have to know the substitution parameters of the production process to estimate biases. When many factors are present, factor share changes become a better starting point for measuring biases than factor ratio differences, and it becomes necessary to split the share changes into components owing to ordinary factor substitution and to biases, respectively. The method for this is described in Appendix 25-1.

Results

The data and results in Table 25-4, for price-corrected factor shares, and in Figure 25-6 indicate the existence of a very strong fertilizer-using bias, accompanied by a rapid decline of the price of fertilizer relative to the price of agricultural output in the United States since 1912. Over a long period of time it is safe to assume that the fertilizer price is governed by cost conditions in the fertilizer industry, i.e., that it is exogenous to agriculture. The behavior of the fertilizer bias is thus consistent with the induced innovation hypothesis.

It is also reasonable to assume that the price of labor to a great extent is governed by wage rates in the nonagricultural sector and is exogenous to agriculture. The price of labor increased throughout the period but at a much more rapid rate after about 1940. Up to 1944 technical change was labor neutral, but thereafter a strong labor-saving bias did exist.[12] This is again consistent with the induced innovation hypothesis. Between 1944 and 1968 the observed labor share dropped from almost 39.5 percent to 15.8 percent, i.e., a drop of 23.7 percent. Biased technical change alone could have explained a drop of almost 14 percent. This means that about two-thirds of the drop in the labor share must be explained by biased technical change and only one-third by simple price substitution effects.

In the long run, the price of machinery is governed primarily by cost conditions in the machinery-producing and machinery-repairing industries, as well as by fuel costs, and it is therefore also regarded as exogenous to agriculture. The overall rise of machinery prices was about the same as for labor prices, and a substantial part of it occurred before 1948. After 1948 the rate of the price rise accelerated. But despite that price rise, technical change was machinery using, not saving. Had innovation possibilities been neutral, this could not have happened. Innovation possibilities must have been machinery using regardless of the role of factor prices in determining biases. Any price-induced bias would have been machinery saving, not machinery using.

Fertilizer and labor biases are consistent with the hypothesis of neutral innovation possibilities and with the hypothesis that factor prices account for most of the biases, but the machinery bias contradicts this. It is even possible that the labor and fertilizer biases were primarily fundamental biases and that the corresponding price changes were coincidental rather than causative.

Table 25-4. Price-Corrected Shares (S_i'), Actual Factor Shares, and Factor Prices Used in Computing (S_i')

Year	Land	Labor	Machinery	Fertilizer	Other
		Price-Corrected Factor Shares a_i^{*a}			
1912 . . .	21.0	38.3	10.9	1.9	28.0
1916 . . .	21.2	36.7	11.6	1.8	28.7
1920 . . .	19.6	39.3	9.3	2.1	29.7
1924 . . .	20.0	39.7	10.3	2.2	27.8
1928 . . .	18.1	41.4	10.4	2.7	27.4
1932 . . .	18.8	40.3	14.3	2.7	24.0
1936 . . .	18.9	32.5	16.3	3.0	29.3
1940 . . .	16.8	34.3	17.6	3.9	27.5
1944 . . .	16.5	38.4	16.1	4.8	24.2
1948 . . .	17.1	37.2	13.9	5.1	26.7
1952 . . .	16.5	29.8	19.7	5.7	28.3
1956 . . .	16.3	30.6	23.1	6.5	23.4
1960 . . .	17.1	27.2	23.4	6.1	26.1
1964 . . .	17.8	25.8	22.4	6.7	27.3
1968 . . .	19.1	25.3	23.1	7.2	25.3
		Actual Factor Shares a_i			
1912 . . .	21.0	38.3	10.9	1.9	28.0
1916 . . .	21.6	36.5	11.6	1.9	28.4
1920 . . .	17.3	40.5	10.1	2.0	30.1
1924 . . .	19.7	38.5	10.3	1.7	29.7
1928 . . .	15.9	40.9	10.2	1.9	31.1
1932 . . .	18.6	37.6	12.6	1.6	29.7
1936 . . .	14.9	34.7	14.5	2.2	33.7
1940 . . .	12.0	35.3	15.1	2.3	35.2
1944 . . .	8.5	39.5	14.0	2.3	35.6
1948 . . .	9.4	37.7	12.2	2.4	38.3
1952 . . .	9.8	29.7	17.5	3.0	40.0
1956 . . .	11.5	27.4	20.1	3.3	37.8
1960 . . .	15.6	21.3	19.8	2.9	40.4
1964 . . .	17.5	18.3	18.5	3.3	42.3
1968 . . .	20.4	15.8	19.1	3.6	41.1

Table 25-4 – continued

Year	Land	Labor	Machinery	Fertilizer	Other
			Factor Prices[b]		
1912 . . .	100.0	100.0	100.0	100.0	100.0
1916 . . .	113.3	106.8	110.0	105.7	103.8
1920 . . .	79.0	104.3	81.3	85.7	105.0
1924 . . .	119.0	134.5	111.7	93.1	106.6
1928 . . .	104.8	154.1	128.5	90.0	118.9
1932 . . .	160.8	194.7	231.5	128.6	101.5
1936 . . .	69.4	113.4	189.2	99.6	110.9
1940 . . .	87.3	179.0	288.8	103.4	160.1
1944 . . .	65.2	217.2	244.2	63.0	211.7
1948 . . .	73.0	247.8	226.6	50.4	222.8
1952 . . .	91.3	274.3	301.1	53.6	214.6
1956 . . .	145.8	407.9	423.7	65.9	229.6
1960 . . .	254.1	502.7	550.3	63.0	241.5
1964 . . .	338.1	610.0	651.2	63.2	270.9
1968 . . .	481.0	766.9	735.8	58.2	280.4

Source: Hans P. Binswanger, "The Measurement of Technical Change Biases with Many Factors of Production," *American Economic Review*, 64 (December 1974), 973.

[a] Model A estimates.

[b] Relative to agricultural output prices (1912 = 100).

But the view that biases are generally caused by non-neutral innovation possibilities and not by factor prices is contradicted by the time sequence of turning points in factor prices and biases. The labor price rise strongly accelerated at the start of World War II, and it was some six to ten years later that a strong labor-saving bias emerged. Similarly, the machinery price rose rapidly between 1948 and 1952, and, again, it was about six years later that the machinery-using bias disappeared. Evenson has found a mean lag, between research initiation and benefits, of between 5½ and 8½ years in agriculture, which would support causality in the observations above.[13] It is unlikely that these changes are coincidental.

The series on land prices and biases cannot give us much more information on the induced innovation hypothesis because the land price is largely endogenous to agriculture. It is worth noting that land prices fluctuate widely, but there was apparently very little bias with respect to land.

The price of other inputs declined by about 40 percent during and after World War II but did not change much before and after. However, there was no discernible bias with respect to this conglomeration of factors.

The clearest conclusion to be drawn from the series is that fundamental biases in innovation possibilities were an important source of the machinery-using bias in United States agriculture.

There can be no question that large, sustained changes in factor prices have

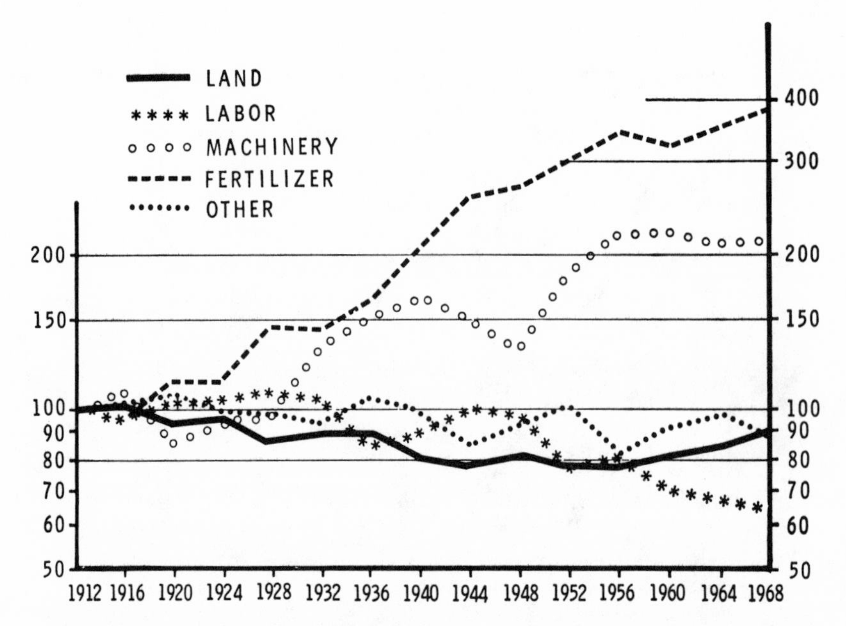

Figure 25-6. United States indices of biases in technical change, estimates of $S'_{it}/S_{i,\,1912}$ (1912 = 100).
Source: Hans P. Binswanger, "The Measurement of Technical Change Biases with Many Factors of Production," *American Economic Review*, 64:6 (December 1974), 972.

had a strong impact on the direction of technical change. In comparison with the prices of the four other inputs, the fertilizer price fell throughout most of the period. This fall was accompanied by a strong fertilizer-using bias. Similarly, about eight years after labor prices began to rise rapidly, a very strong labor-saving bias emerged. In about the same length of time, after an acceleration in the price of machinery, the machinery-using bias which had been caused by biased innovation possibilities disappeared.

Conclusions

The conscious effort of researchers and research administrators to choose projects that will maximize the returns from the research investment in their own economic environment puts the theory of induced innovation on fairly firm theoretical foundations. It has been shown that the research resource allocation mechanism used by researchers does not have to be complicated to make the direction of technical change responsive to factors and goods prices.

Once this is recognized, the empirical question remains whether innovation possibilities are flexible enough to make available to different countries paths of efficient technical change with factor ratios which can accommodate widely different factor endowments. The comparison of the development paths of six agriculturally advanced countries has shown that technology is indeed flexible enough to accommodate widely different factor ratios consistent with factor scarcities.

The comparison of factor prices with measured biases in the United States from 1912 to 1961 further confirms that fundamental biases in innovation possibilities play at least as important a role in determining the biases of technical change over time within a country as do economic factors. This seems to be particularly the case for mechanical technology, with technical change possibilities strongly biased in favor of more intensive use of machines. On the other hand, the United States series on factor prices and biases shows that factor prices do indeed either generate biases in technical change or dampen exogenous biases to make the technical change more consistent with economic factors than would be the case in the absence of induced innovation. This is fairly strong support for the empirical relevance of induced innovation.

APPENDIX

Appendix 25-1. The Methodology of the Many-Factor Test of Induced Innovation

In the two-factor case biases can be measured with the following measurement equation:

$$Q' = (k^* - \ell^*) + \sigma(r^* - w^*) \tag{1}$$

where k^* and l^* are observed proportional rates of change of K and L and r^* and w^* are rates of change of the capital rental rate and wage rate, and $\sigma(r^* - w^*)$ is the component of the capital/labor ratio change which is due to ordinary factor substitution.[14]

The measurement of biases thus consists of two steps: (A) Measure the elasticity of substitution in an independent sample. (B) Apply equation (1) to time-series data to measure the biases.

In the many-factor cases the procedure is similar to this two-step procedure. However, the biases must now be defined in terms of factor shares:

$$B_i = \frac{dS_i'}{dt} \times \frac{1}{S_i} \begin{matrix} < \\ = \\ > \end{matrix} 0 \rightarrow \begin{matrix} \text{(i-saving)} \\ \text{(neutral)} \\ \text{(i-using)} \end{matrix} \tag{2}$$

where dS_i' is the factor share change which would have occurred in the absence of factor price changes, while S_i is the actual factor share in total costs of production.

The CES production function usually used in the two-factor case is inappropriate in the many-factor case, where it implies the constraint that all partial elasticities of substitution are identical between all pairs of factors. This was considered to be too restrictive.

The basic idea outlined here was applied to the translog cost function of Christensen, Jorgensen, and Lau. The details of this procedure are fairly complicated.[15] The approach leads to a basic estimation equation of the biases as follows:

$$dS_i' = dS_i - \sum_j \gamma_{ij} \, d\ln W_j \tag{3}$$

where dS_i' are the share changes which would have occurred in the absence of factor price changes, dS_i are the actual share changes, γ_{ij} are the substitution parameters of the production process, and $d\ln W_j$ are the proportional changes of factor prices. The method thus consists of subtracting from the observed share changes that part which was caused by changing factor prices.

The resulting dS_i' can be substituted into the discrete equivalent of equation (3) to compute rates of biases. Here, however, we compute series S_i', which shows how the shares would have developed after the initial year (1912) in the absence of factor price changes:

$$S_i' = S_{i,1912} + \sum_{i=0}^{t} \Delta S_{it}'. \tag{4}$$

These series are presented in the section on U.S. vs. other countries of Table 25-3. Series of standardized values, i.e., $R_{it} = S_{it}'/S_{i,\,1912}$ are presented in Figure 25-3, which is in semilogarithmic scale. Hence, the slope of the lines indicates biases according to equation (2), while the position of the line shows the cumulative bias since 1912.

NOTES

1. Yujiro Hayami and Vernon W. Ruttan, *Agricultural Development: An International Perspective* (Baltimore: Johns Hopkins University Press, 1971).

2. Hans P. Binswanger and Vernon W. Ruttan, eds., *Induced Innovation: Technology, Institutions and Development* (Baltimore: The Johns Hopkins University Press, in press).

3. Jacob Schmookler, *Invention and Economic Growth* (Cambridge, Mass.: Harvard University Press, 1966); Zvi Griliches, "Hybrid Corn: An Explanation of the Economics

of Technical Change," *Econometrica*, 25 (October 1957), 501-522; R. E. Lucas, "Tests of a Capital Theoretical Model of Technological Change," *Review of Economic Studies*, 34 (January 1967), 175-190; Uri Ben-Zion, "Aggregate Demand and the Rate of Technical Change," in Binswanger and Ruttan, *Induced Innovation*.

4. For an example of the effect of tractor subsidies on factor ratios, regional comparative advantage, and income distribution, see John H. Sanders, "Biased Choice of Technology in Brazil," in Binswanger and Ruttan, *Induced Innovation*.

5. Hans P. Binswanger, "A Microeconomic Approach to Induced Innovation," *Economic Journal*, 84 (December 1974), 940-958.

6. For example, suppose the research administrator has the choice between two research projects, A and B. A only reduces capital or land requirements from K_0 to K_1, while B only reduces labor requirements from L_0 to L_1. In a one-period model the benefits from A and B, respectively, are

$$P_A = (K_0 - K_1)R$$

$$P_B = (L_0 - L_1)W$$

where R and W are the capital rental rate and the wage rate. If the projects have equal benefits, i.e., if

$$(K_0 - K_1)R = (L_0 - L_1)W$$

or

$$\frac{K_0 - K_1}{L_0 - L_1} = \frac{W}{R},$$

the decision-maker will have no basis for choice. If, however, he lives in an economy where the wage/rental ratio exceeds the critical ratio, he will choose Project A; he will choose Project B if the wage/rental ratio falls short of the critical ratio.

7. Hayami and Ruttan, *Agricultural Development*.

8. Simulations in a more complex model were performed to assess the possible size of such biases with various assumptions about the size of a full set of cross elasticities of factor demand. It turned out that the effects of changes in prices other than of land and labor generally had very little impact on the land/labor ratio and that consequently biases could be expected to be small.

9. Hans P. Binswanger, "A Cost Function Approach to the Measurement of Factor Demand Elasticities and Elasticities of Substitution," *American Journal of Agricultural Economics*, 56 (May 1974), 377-386. Yair Mundlak, "Elasticities of Substitution and the Theory of Derived Demand," *Review of Economic Studies*, 35 (April 1968), 225-236.

10. The conversion formula is

$$\sigma_{ij} = \frac{(\eta_{ii} - \eta_{ji}) \text{ dlog } P_i + (\eta_{ij} - \eta_{jj}) \text{ dlog } P_j}{\text{dlog } P_i - \text{dlog } P_j}$$

where η_{ij} are partial factor demand elasticities and dlog P_i are proportional changes in factor prices. For given factor demand elasticities the size of σ_{ij} depends on the extent to which the factor price ratio change comes from a change in P_i or a change in P_j. Table 2 of Binswanger, "A Cost Function Approach" gives the following values for the factor demand elasticities of A and L: $\eta_{AA} = -.3356$; $\eta_{LA} = .0308$; $\eta_{AL} = .0613$; $\eta_{LL} = -.9109$. These values imply the following values for σ_{AL}*

Price ratio change owing to	$\sigma_{AL}{}^*$
Change in P_A only	.37
Change in P_L only	.96
Equiproportional change in P_A and P_L or P_K and P_L	.67

In the international comparisons the precise source of the change in the factor price ratio is unknown. Equiproportionality of changes in the two prices involved will be assumed.

11. We define the Hicksian bias B′ as the change in factor ratios which would have occurred had factor prices stayed stable:

$$B' = \frac{d(K/L)}{dt} \times \frac{t}{K/L} \Bigg|_{\text{Factor prices}} \begin{matrix} > \text{(labor saving)} \\ = \text{(neutral)} \\ < \text{(labor using)} \end{matrix}$$

where K is capital and L is labor.

12. Lianos also found a labor-saving bias in agriculture. See Theodore P. Lianos, "The Relative Share of Labor in United States Agriculture 1949-1968," *American Journal of Agricultural Economics*, 53 (August 1971), 411-422.

13. Robert E. Evenson, "The Contribution of Agricultural Research to Agricultural Production," Ph.D. dissertation, University of Chicago, University Microfilms, Ann Arbor, Michigan, 1968.

14. Proof of equation (1): Ryuzo Sato, "The Estimation of Biased Technical Progress and the Production Function," *International Economic Review* (June 1970), 179-207, shows that rates of augmentation can be measured as follows:

$$a^* = \frac{\sigma r^* - y^* + k^*}{1 - \sigma}; \quad b^* = \frac{\sigma w^* - y^* + l^*}{1 - \sigma}. \tag{a}$$

Robert Solow proved much earlier that the bias is related to factor augmentation rates as follows:

$$A' = (1 - \sigma)(a^* - b^*). \tag{b}$$

Substituting (a) into (b) leads to equation (1).

15. Hans P. Binswanger, "The Measurement of Technical Change Biases with Many Factors of Production," *American Economic Review*, 64 (December 1974), 964-976.

26

Inducement of Technological and Institutional Innovations: An Interpretative Framework

Alain de Janvry

The economic structure of a social system can, in its essence, be characterized by two features: the level of development of its capacity to produce; and the way in which people relate to one another in their respective functions in the production process.

The capacity to produce is determined by the quantitative and qualitative stock of productive resources. Recent empirical studies have shown that the qualitative attribute of resources is indeed a fundamental source of output and productivity growth. For physical capital, this attribute takes the form of technological change; for human capital, it takes the form of increased labor skills. Since it affects both output and productivity, the rate and bias in the growth of the qualitative attribute of resources are major determinants of output growth and of changes in income levels and in the distribution of income. As a result, whoever controls the production of new technologies and skills largely determines the pace and nature of economic development. Ultimately, then, it is the social structure which conditions the rate and bias of technological and skill innovations, and these innovations are, in turn, powerful determinants of change in the social structure.

The earlier chapters in this section have focused primarily on the environmental, technological, and economic considerations that affect the direction of technical change. In this chapter institutional considerations will be examined as well. The ways in which people relate to each other in their respective functions in the production process are translated into a set of institutions that characterize and establish guidelines for these relationships. Most important are those institutions which specify the rights of ownership and which define the functions and the scope of action of the state. These institutions condition the rate and bias in technological and skill innovations and the impact of such innovations on growth and income. And, reciprocally, technological and skill innovations are powerful inducers of institutional innovations. (Throughout this chapter, for the sake of simplicity, the term *technological innovation* will be used to refer to both technological and skill innovations.)

Technological and institutional innovations are thus so closely interrelated that any change in one ultimately presses for change in the other. Which one first induces change – a question that has been raised repeatedly in the recent literature on agricultural technology – is a chicken-and-egg-type question to which only history can provide an answer. In some instances, output stagnation induces institutional change to restore the dynamics of accumulation. Here institutional change precedes technological change. Examples are the urban-induced land reform programs that followed the Punta del Este Conference in Latin America.

In other instances, the dynamics of accumulation, fueled by technological innovations, creates sharp social antagonisms that lead to institutional innovations. Examples are the peasant-led land reforms in Mexico and Bolivia and, to some extent, the initial impact of the green revolution. Here technological change precedes institutional change.

Having recognized both the endogenous nature of technological and institutional innovations within a social system and their reciprocal relationship, we now turn to the specification of a simple but, we believe, useful model of the generation of technological and institutional innovations.[1] The technological innovations considered here are those which are generated by the public sector through fundamental and applied research. In agriculture, the public sector has a major responsibility in technological innovation. As a consequence, technological and institutional innovations assume the nature of public goods, and the following model is one of supply and demand for public goods in general. Since a social system can be characterized largely by its technological and institutional structures, however, a model of supply and demand for technological and institutional change is actually an approximation of a general model of the dynamics of change in social systems.

A Dialectical Model of Technological and Institutional Innovations

After having definitely shelved the teachings of historical materialism, neoclassical economists have recently begun to theorize on the very question that is central to historical materialism: the dialectical relationship between productive forces and social relations of production and the way in which this dialectic explains the evolution from one mode of production to another; or, in other words, the relationship between technological and institutional innovations and the process through which these innovations occur.

Using the neoclassical apparatus, Hayami and Ruttan have developed a path-breaking theory of the inducement of technological and institutional innovations. Their model is linear, and its starting points are the stock of scientific knowledge – conceptualized as defining the Innovation Possibility Frontier (IPF) in the factor space – and the relative factor scarcities. Through the market mechanism, changes in relative factor scarcities determine changes in relative factor prices; these changes lead to a search within the IPF for new cost-minimizing technologies which will, of course, tend to be biased toward saving the most expensive factor. Institutional changes may, in turn, be induced by technological change in order "to enable both individuals and society to take fuller advantage of new technical opportunities under favorable market conditions."[2] The causal sequence is thus from technological to institutional change.

While useful, this model does not fully explain the process of technological and institutional change or stagnation. By its nature, a linear model cannot describe a reciprocal relationship, such as we have suggested exists between technological and institutional change. Moreover, the model in question does not uncover the dynamics of the interrelationships between supply of and demand for innovations, nor can it articulate the parameters that affect both supply and demand.

The alternative model depicted in Figure 26-1 conceptualizes the generation of technological and institutional innovations by the public sector as a dynamic process of circular and cumulative causation. In this process both the socioeconomic structure (the infrastructure) and the politico-bureaucratic structure (the superstructure) are given explicit roles. The central node of the model is a "payoff matrix" which identifies the net economic gains and losses that are expected to result from the provision of a given set of public goods (technological and institutional innovations) to a given set of interest groups in society. Some of the alternatives considered within this matrix are technological choices among commodities (e.g., food versus cash crops, cereals versus livestock); among regions (e.g., irrigated versus dry land, fertile versus marginal lands); and among technological biases (land-saving biochemicals versus labor-saving mechanical devices, modern versus intermediate mechanical techniques, etc.). Interest groups include, in particular, commercial farmers, traditional landed elites, subsistence farmers, landless agricultural workers, industrial employers, urban workers at different income levels, exporters, and government. Each social group expects to derive a specific income gain or loss

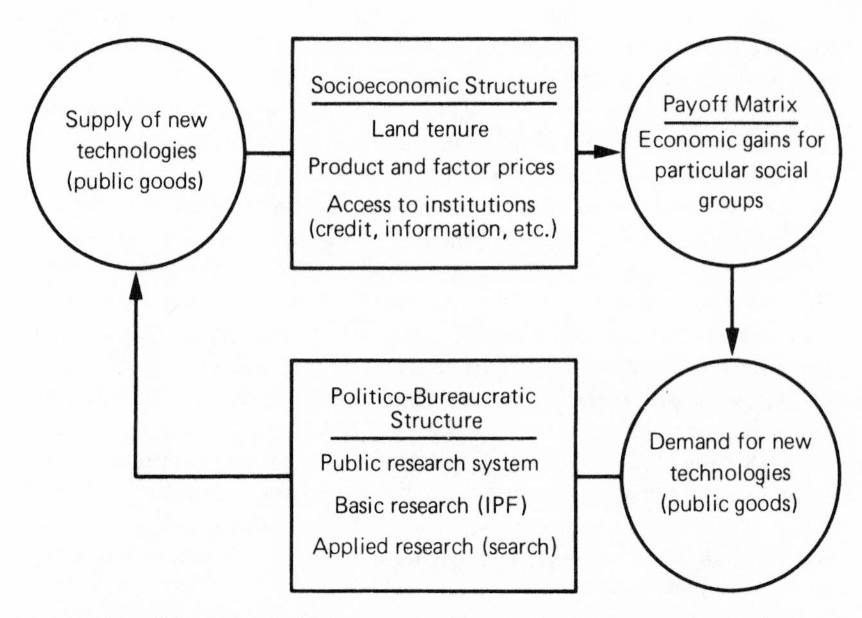

Figure 26-1. The inducement and diffusion of technological innovations (public goods).

from each particular public good. These incomes are the "payoffs" that constitute the entries in the public goods-social groups matrix. In this model the economic base of society determines its evolution.

The supply and demand mechanism for public goods is centered on the payoff matrix and conditioned by the socioeconomic structure on the one hand and by the politico-administrative structure on the other. Specifically, the demand for new public goods originates from the matrix of expected payoffs: each social group will pressure the politico-administrative structure for public goods to be (or not to be) generated depending upon the corresponding payoffs it expects to derive. Here the relative power of a social group over the politico-administrative structure determines whether the group's demands for a supply of a particular public good will be satisfied. In the case of technology, pressure on the politico-administrative structure induces allocations of funds and of human capital to research institutions and to particular lines of research. The level of scientific knowledge and the quantity of physical and human capital allocated to basic research will determine the position of the IPF. The amount of such capital allocated to applied research will determine the intensity of search for new cost-minimizing activities within the IPF. And the *organization* of national research systems will condition the way in which they will respond, in terms of both intensity and bias, to the set of demands made upon them.

The resulting supply of public goods creates, through the socioeconomic structure, specific payoffs for each social group. For agricultural technology these payoffs are determined by (1) the physical impact of the innovation in terms of yield effect and/or resource-saving and resource-substitution effects; (2) the diffusion effect, conditioned largely by the nature of the technological innovation and the particular positions of different social groups in the socioeconomic structure (the land tenure system and the degree of access each group has to institutions of credit, information, marketing, etc., are determining factors); and (3) the terms-of-trade effect, which determines the economic value of the physical and diffusion effects. All three effects – physical, diffusion, and price – are, of course, interdependent and together determine the entries of the payoff matrix as products of the interaction between supply of specific new technologies and the socioeconomic structure.

New expected payoffs induce further demands for new public goods. In activating this process, the organization of the research system is, again, of major importance because it influences the formation of such new expectations through the dialectical interactions it maintains with social groups.

Productivity of a national research system should thus be measured by the entries of the payoff matrix relative to costs incurred by the research system. If other costs are incurred in creating the observed payoff effects – for example, in the diffusion of new technology or the implementation of institutional changes – care must be taken to charge these costs against the resulting payoffs also.

The payoff matrix thus reveals the material base of the dynamics – or lack of dynamics – of the inducement and diffusion of technological innovations. It also indicates that, unless one makes severe value judgments and aggregates all payoffs into one net effect, care must be taken to identify the specific social effects of the gains and losses from technology in order to estimate the economic and social significance of such technology. Even if the aggregate payoff is used as a criterion of net social gain from technology, care must still be taken to identify possible negative payoffs (such as loss of employment and income for some social groups) in order to write them off from the total payoff and to determine needed economic compensations. Indeed, the commonly used net social gain measure is but a global approximation of the payoff matrix which largely hides the specific socioeconomic effects of technology as well as the reasons why technological innovations and change did or did not occur.

This model of the inducement of technological and institutional innovations is thus useful at the conceptual level. It permits the classification of a wealth of factors known to affect the process of innovation: the material incentives to innovate, the socioeconomic structure, the politico-bureaucratic structure, etc. It also helps to locate the blockage factors to innovation: on the demand or on the supply side; on the effective translation of demand into supply or in the materialization of supply into actual payoffs.

But this model's capacity extends beyond conceptualization. By quantifying the payoff matrix, it can actually become a powerful instrument for policy analysis and decision-making. To determine the material significance for society of alternative technological and institutional options is, however, a difficult task and one of central concern to economists and other social scientists. Although we do not pretend to resolve this question here, we can provide some guidelines on how quantification of the payoff matrix may proceed.

Estimation of the Payoff Matrix for Public Goods

The payoff vector for a public good is obtained as the change in net per capita income for each social class between the pre-public good and the post-public good situation. If the only impact on incomes between two time periods is due to the public good, *ex-post* observation of the changes in per capita incomes provides the payoff vector. But other changes have usually also occurred during the time interval; moreover, we may want to predict, *ex ante*, the expected payoff from alternative courses of action. In both instances, estimation of the payoff vector will require construction of a model that can explain the process by which personal incomes are generated and account for the changes in personal incomes brought about by introduction of the public good.

To devise such a model we propose to proceed in two stages: (1) constructing national accounts that make explicit the flows of income and expenditures for different social groups; and (2) simulating – through partial modeling – the impact that the public good has on these flows and hence ultimately on individual incomes.

Constructing Socioeconomic National Accounts

Unfortunately, the standard national account systems and input-output tables do not provide information on income and expenditure flow for social groups, being disaggregated only by types of economic activities. What is needed instead is a system of socioeconomic national accounts where a disaggregation by social class is nested within the standard disaggregation by economic sector.

To provide an example, I have constructed such an accounting system for

Chile using data compiled by Echeverria.[3] Included are four economic sectors (agricultural, nonagricultural, foreign, and financial), two types of land tenure (commercial farms, classified by the Inter-American Committee for Agricultural Development (CIDA), as large and medium multifamily farms; and peasant farms, categorized by CIDA as family and subfamily farms);[4] and three social classes within each type of tenure (the owners, the semiproletarians who receive part of their wage as the usufruct of a plot of land, and the proletarians whose whole wage is in cash). It is unfortunate that, for lack of data, the government sector must be omitted here, since its inclusion would have made it possible to account for income transfers through taxes and subsidies.

In Table 26-1 the rows correspond to income and the columns to expenditures. For the producing sectors the columns give a breakdown of the value of goods in "constant" capital (C = cost of intermediate products and depreciation of the stock of capital), "variable" capital (V = cost of the labor force), and "surplus value" (S = interest, dividends, rents, and profits). Semiproletarians are remunerated partly in land rights (VL), partly in consumption goods (VK), and the remainder in cash (VC). The land rights generate income for the semiproletarians and enter as costs for the owners valued at the opportunity cost of land for them. The payments in kind enter as income and cost for the owners and as income and expenditure for the semiproletarians. Among the columns that show the entrepreneurial activities of the owners and semiproletarians, those to the left (C, VC, VK, VL) account for production costs and those to the right account for consumption and savings, that is, for the use made of net income. The surplus value is used, in part, for consumption purposes (SC) and, in part, for savings (SS).

For each social group gross income is the sum over columns. The net income of owners and semiproletarians is the difference between gross income and the sum over rows of production costs. As Table 26-1 shows, the net income per commercial farm owner is twenty-three times higher than that of the peasant owners, fifty-two times higher than that of their semiproletarian workers, and eighty-four times higher than that of proletarians.

Simulating the Effect of Public Goods

What now has to be simulated is the impact that the introduction of a specific public good can have on the per capita net income vector. To do so, we must represent the effect of this public good on the flows of funds in the type of socioeconomic accounts referred to above.

Consider, for example, the contrast Hayami and Ruttan have drawn between two broad alternative technological options: mechanical technology (which is labor saving and yield neutral) and biochemical technology (which is land saving and, hence, yield increasing).[5] The market and distributional effects for

Table 26-1. Socioeconomic National Accounts for Chile in 1962

	Commercial Farms (1,000 escudos)											Peasant Farms (1,000 escudos)										
	Owners						Semiproletarians				Prole-	Owners						Semiproletarians				Prole-
Category	C	VC	VK	VL	SC	SS	C	VK	SC	SS	tarians	C	VC	VK	VL	SC	SS	C	VK	SC	SS	tarians
Commercial farms																						
Owners.	36.1		12.2	32.5	14.7		13.1		20.7		4.5	9.3				39.2		1.1		1.8		1.1
Semiproletarians		36.6			27.9		13.1	12.2	82.6		1.4											0.4
Proletarians		11.3																				
Peasant farms																						
Owners.					4.4						1.4	25.0		1.1	2.8	11.7						0.4
Semiproletarians					2.4						0.1		3.2					1.1	1.1	7.2		
Proletarians													2.8									
Nonagricultural sector																						
Owners.	123.0				131.5		34.3		59.4		3.9	25.0				31.2		3.0		5.2		1.0
Proletarians																						
Foreign sector																						
Imports.	21.3				9.9																	
Financial sector																						
Savings						108.2				11.1							3.6				0.9	
Total expenditure.	180.4	47.9	12.2	32.5	190.8	108.2	60.5	12.2	162.7	11.1	11.3	59.3	6.0	1.1	2.8	82.1	3.6	5.2	1.1	14.2	0.9	2.9

Table 26-1 – continued

Category	Nonagricultural Sector (1,000 escudos)					Foreign Sector (exports, 1,000 escudos)	Financial Sector (investment, 1,000 escudos)	Gross Income (1,000 escudos)	Net Income (1,000 escudos)	Active Population (1,000 persons)	Net Income per Active Person (escudos)
	Owners				Proletarians						
	C	V	SC	SA							
Commercial farms											
Owners	59.3		78.8		202.7	26.9	24.1	578.1	305.1	51.6	5,913.0
Semiproletarians					61.3		11.1	246.6	186.1	162.3	114.6
Proletarians								11.3	11.3	158.6	71.0
Peasant farms											
Owners	14.5		23.6		60.6		3.6	149.1	79.9	310.5	257.0
Semiproletarians					5.3		1.0	21.4	16.2	13.3	1,218.0
Proletarians								2.8	2.8	41.3	68
Nonagricultural sector											
Owners	2,538.8		2,793.9		1,837.6	637.9	598.7	8,824.4	3,243.0	402.6	8,055.0
Proletarians		2,403.8						2,403.8	2,403.8	1,761.7	1,365.0
Foreign sector											
Imports	565.0		54.0		236.3		251.2	1,137.7			
Financial sector											
Savings				765.8				889.6			
Total expenditure	3,177.6	2,403.8	2,950.3	765.8	2,403.8	664.8	889.7	14,264.8		2,901.9	

both types of technology in closed and open economies are depicted in Figure 26-2 (for further elaboration see chapter 6). The case of the open economy is characteristic also of an economy where price-support programs exist. In Figures 26-2.3 and 26-2.4 the supply function becomes horizontal at the minimum average variable cost level below which firms would discontinue production.

Mechanical innovations (Figures 26-2.3 and 2.4). Under the assumptions of our model, mechanization is purely cost reducing, and the resulting net social gains (NSG) accrue wholly to landowners in both closed (Figure 26-2.3) and open (Figure 26-2.4) economies, since welfare gains are capitalized in land values.[6] In this case the technological "treadmill" that coerces nonadopters into change materializes through rising land values. Adopters of mechanical innovations bid up the price of land, internalizing the welfare gains from mechanization in land values, and thus raise both the opportunity cost of holding land for owner-operators and the rent for tenants. The rise in opportunity costs forces all those that have not mechanized to do so or abandon agriculture.

Biochemical innovations (Figures 26-2.1 and 2.2). In a closed economy the net social gains from biochemicals that result from lower food prices (Figure 26-2.1) will accrue, in the form of cheaper wage goods, to consumers or to industrial employers according to the relative social power of the two groups. They accrue to landowners in an open or farm price-supported economy (Figure 26-2.2). In the open-market case, welfare gains to the whole economy arise only from increased exportable surpluses and the effect that higher foreign exchange earnings have on economic growth through the import multiplier. In the price-support case, welfare gains to landowners result from increased government costs of protectionism and thus imply an income transfer from taxpayers to landowners.

The dynamic diffusion mechanism of new biochemical techniques in the closed free-market economy (Figure 26-2.1) occurs through the Cochrane and Owen "Mill-Marshallian" treadmill of falling product prices.[7] By contrast, the diffusion mechanism occurs through the land-market treadmill in the open and in the farm price-supported economies (Figure 26-2.2). The well-known observation that, in the United States, "the net result of technological advance concurrently with public farm programs, such as we have had in the past 30 years, is mainly a continuous rise in farm real estate values"[8] indicates that this second mechanism may have been the effective one.

Conflicts of Interest and Solutions

Definite conflicts of interest exist among social groups in the use of mechanical versus biochemical techniques in the closed economy. Because no

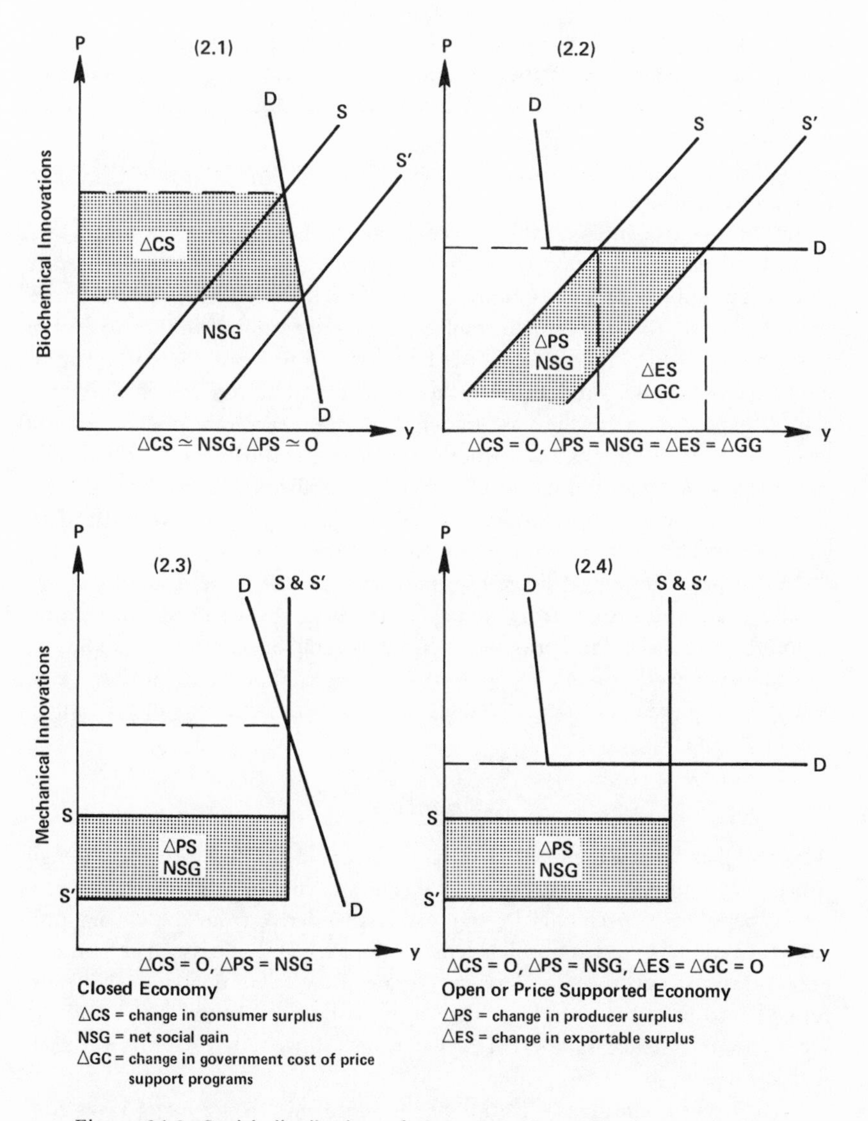

Figure 26-2. Social distribution of the economic surplus from agricultural technology.

solution exists in the game between employers-consumers and landlords over the choice between mechanical versus biological technologies, resolution of this contradiction will hinge upon the relative social power of landlords, industrial employers, and consumers. It is only when the industrial entrepreneurs' class is also the land-owning class, as Stavenhagen claims to be the case in most Latin American countries,[9] and when workers' (consumers') power is weak, that a single social class can capture the benefits both of biochemicals, through lower wage-goods and increased surplus values in industry, and of machinery, through increased land-rent earnings.

In the open and price-supported economies, by contrast, solutions exist that are of opposite nature.

In the open economy, while mechanical technology benefits only landlords, biochemicals increase the welfare both of (1) landlords, through higher land rents, and of (2) the economy at large (or at least of the industrial entrepreneurs' class), through growth resulting from relaxation of foreign exchange bottlenecks. Hence, unless landlords have specific reasons for preferring to derive gains through mechanical rather than biological technologies (which can be expected to be the case under the *latifundia* tenure system because mechanization reinforces the owner's power over farm workers, while biochemicals would weaken it and eventually jeopardize the survival of the extant social order), class and sectoral coalitions of interests would lead to choosing a technological path dominated by biochemicals.

In the price-supported economy, biochemicals still benefit landlords but require income transfers from taxpayers to them. By contrast, mechanical technologies benefit landlords while being neutral on consumer, employer, and taxpayer welfares. As a consequence, class coalitions, when they exist, will tend to pursue mechanical rather than biochemical means of technological development.

Summary

A model has been developed that specifies the dialectical process through which technological and institutional changes occur. In this model, the economic payoff that each social group expects to derive from the change provides the rational basis from which demands for such change arise. The politico-bureaucratic structure of society determines how these demands are transformed into an actual supply of new technologies and institutions. And the socioeconomic structure translates that supply into actual payoffs for each social group.

Identification and quantification of the payoff matrix are major tasks confronting social scientists because they give the key to the *ex-post* analysis of the evolution of social systems as well as to the *ex-ante* evaluation of projects of technological and institutional change. An attempt was made to indicate how such a matrix could be obtained by constructing socioeconomic accounts and identifying in those accounts the impact on per capita incomes of technological and institutional changes.

NOTES

1. If external influences are important – which is generally the case in the less developed countries – the social system in question exceeds national boundaries.

2. Yujiro Hayami and Vernon W. Ruttan, *Agricultural Development: An International Perspective* (Baltimore: Johns Hopkins University Press, 1971), pp. 59-60.

3. R. Echeverria, *Política de precios y redistribución de ingresos agrícolas* (Santiago, Chile: Instituto de Capacitación e Investigación en Reforma Agraria, 1972).

4. S. Barraclough, ed., *Agrarian Structure in Latin America* (Toronto: Lexington Books, 1973).

5. In accordance with this contrast, an aggregate two-stage production function for the agricultural sector can be specified where (1) there are four inputs – labor, land, labor-saving capital (machinery), and land-saving capital (biochemicals) denoted by L, T, K_L, and K_T, respectively; (2) L and K_L are highly substitutable and so are T and K_T; and (3) substitution possibilities between K_T or T and L or K_L are low. Consequently, with an inelastic aggregate land supply and a constant level of land-saving capital items, K_L can be only mildly output increasing since the output effect is blocked by the fixed land and land-saving inputs. Under these circumstances, mechanization is essentially labor displacing. Also, with an inelastic land supply and a constant level of labor-saving capital items, K_T is highly output increasing as long as the elasticity of supply of labor is high.

The aggregate production function for agriculture can be written then as a two-stage function:

$$Y = TF\ [f_T(T^*, K_T^*), f_L(L^*, K_L^*)]$$

where T is an index of neutral technological change and the asterisk denotes inputs measured in effective units that are related to the nominal input units through a factor-augmenting coefficient of technological change. The subfunction, f_T, is an index of "land" inputs, while f_L is an index of "power" inputs. The distributive effects of technology can be analyzed under the assumptions that land is in completely inelastic aggregate supply in the short run and that the internal market demand for agricultural products has a price elasticity of less than one. Also, to contrast the different impacts of machinery and of biochemicals, technological changes and shifts in factor supplies of these two types of capital goods can be analyzed separately, if the level of use of the other capital good is always held constant.

6. The inelasticity of supply in these two figures, which reflects the model's assumptions, implies that the marginal productivity of labor in agriculture is zero when land and land-saving capital are in fixed supply.

7. W. W. Cochrane, *Farm Prices: Myth and Reality* (Minneapolis: University of Minnesota Press, 1958); W. Owen, "The Double Developmental Squeeze on Agriculture," *American Economic Review*, 56:1 (May 1966), 43-70.

8. W. B. Black, "Discussion: Income Effects of Innovations – The Case of Labor in Agriculture," *Journal of Farm Economics*, 48:2 (May 1966), 338.

9. R. Stavenhagen, "Seven Fallacies about Latin America," *Latin America Reform or Revolution*, eds. J. Petras and M. Zeitlin (Greenwich, Conn.: Fawcett, 1968), pp. 13-31.

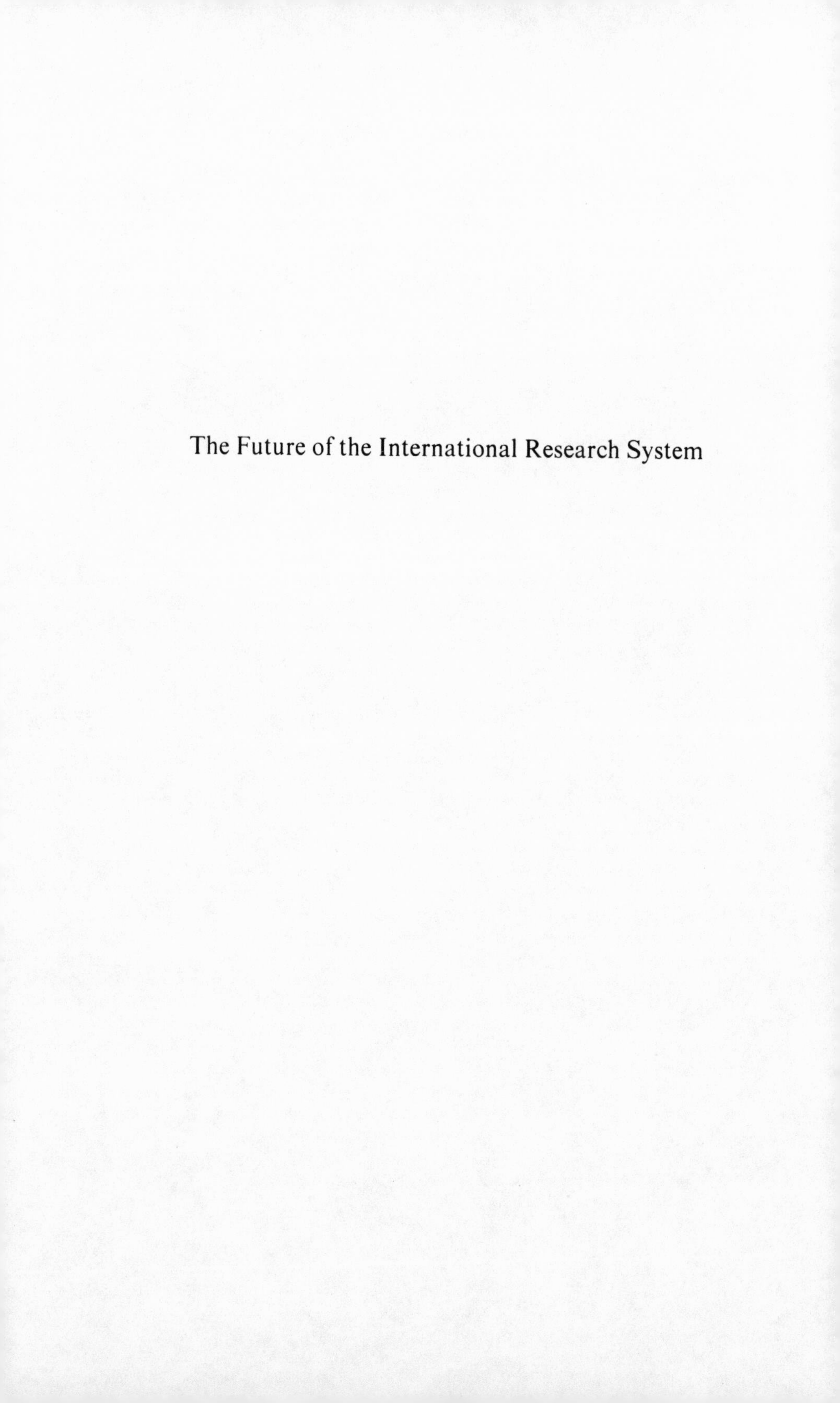

The Future of the International Research System

27

Unresolved Issues in the Evolution of the International System

A. T. Mosher

One of the purposes of the seminar which led to this volume was to discuss the relevance of a number of recent theoretical, historical, and empirical studies for the international institutes supported by members of the Consultative Group for International Agricultural Research (CGIAR). These studies lie in the fields of technology transfer, measuring economic returns to agricultural research, techniques of resource allocation within research organizations, and the dynamics of forces that influence the direction of investments in agricultural research.

This chapter attempts to draw the strands together and to articulate a few of the issues posed for the international institutes by a joint consideration of these varied types of research. Its method is, first, to present a way of delineating the interrelationships among the several components of this volume; second, to sketch a composite picture of the international institutes and their evolving programs, as these emerge from the papers and discussions; and third, to outline a few of the issues that impress me as being of considerable importance.

Interrelationships among Seminar Components

One way to summarize the interrelationships among different components of the seminar is to visualize the central area of concern of the seminar – the

A. Analytical techniques

Domain A is composed of **analytical techniques** and aids to decision making, primarily those which focus on technology development and transfer and on allocation of resources among different projects of agricultural research.

B. Agricultural development activities

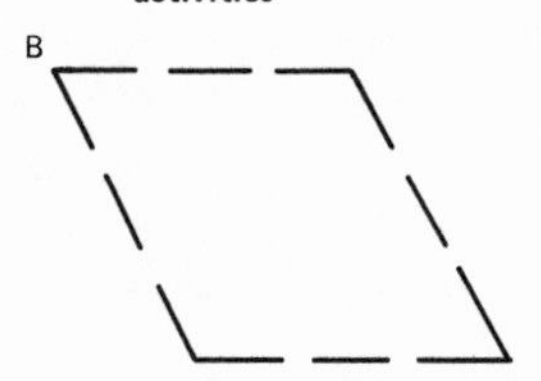

Domain B is composed of the many types of **activities** required to achieve **agricultural development.** It includes technology development and transfer, but it also includes technician training, input manufacture and distribution, farm credit, extension, marketing, many types of national policies, and the organization and administration of all of these.

C. Types of agencies

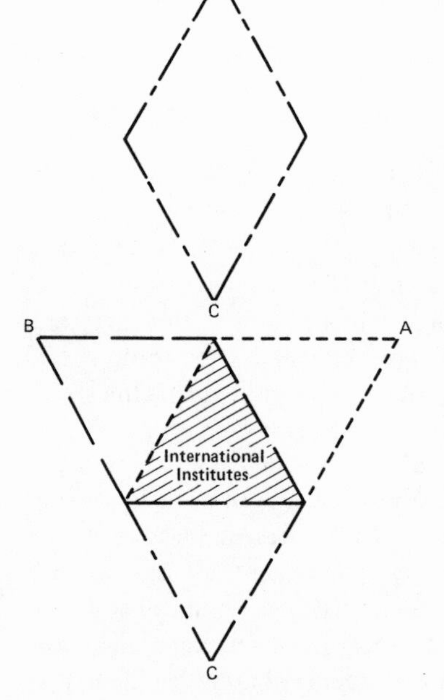

Domain C is made up of all of the various types of **agencies,** national and international, public and private, engaging in agricultural research and in technology development and transfer.

Superimposing these three domains gives us the common ground—the overlap—that represents the **activities** and **decision making concerns of the international institutes.** It is this common ground that constituted the main concern of the Airlie House seminar. A major issue with respect to the role of the international institutes concerns which activities should be inside, and which outside this overlap area, particularly with respect to Domains B and C.

Figure 27-1. The role of the international agricultural research institutes in agricultural development.

activities of the international institutes – as being represented by the common ground, or overlap, among three different domains, each of which might be represented by a diamond-shaped area, as depicted in Figure 27-1. Several major issues for the international institutes emerge from a consideration of the "domain analysis" presented in this figure.

The major question that arises with respect to domain B – agricultural development activities – is how broad the activities of the international institutes should be. Should they be limited to research and, even within that, to certain types of research? Should their programs embrace technician training? If so, of what types and in what proportion to research activities? Should their activities be extended to include still other aspects of national production programs, in addition to research and technician training?

The major question with respect to domain C – types of agencies – has to do with what the international institutes should do, what should be done through other international agencies, what should be done by national agricultural programs, what should be done by private sector research, and what formal and informal linkages should be developed among all of these.

The international institutes must consider all three domains simultaneously in order to discover which of the many types of decisions that each institute must make can benefit from the analytical and allocation techniques discussed in domain A and which must continue to be based on other criteria and utilize other techniques.

Finally, in using this "domain analysis" it must be borne in mind that the situation with respect to each domain is highly dynamic. The techniques discussed in this volume with respect to domain A are new and incomplete, and they are growing and changing every day. The programs of the international institutes differ, and each is constantly undergoing evolutionary change. New international agencies are entering domain C, and the programs of each of these keep changing. Moreover, some national programs are steadily advancing in sophistication and maturity.

A Composite Picture of the International Institutes

The papers in this volume are rich in detailing the growth of the international institutes. Together they yield a composite picture of some of the stresses which rapid growth entails. In the following paragraphs a few elements of this picture are outlined.

As the financial support for the institutes through the CGIAR has grown, the established institutes have begun to think bigger. They have launched additional activities, both at their home bases and in their respective outreach programs. They have proposed ever larger current budgets. Consequently, they face the prospect of having to compete among themselves for resources. The boards of trustees of individual institutes have, by their charters, as much fiscal and program autonomy as ever. A new element has been added, however, in the deliberations on priorities that take place in meetings of the Technical Advisory Committee (TAC) and of the CGIAR. Budget figures approved by the CGIAR are not binding, either on its members or on the institutes. Individual institutes may seek additional funding for particular activities from individual members of the CGIAR or from outside sources. The proliferation of institutes with overlapping sources of support undoubtedly has complicated the budgeting and financing processes.

As newer institutes have often adopted somewhat different objectives – though still centered largely on biological agricultural research – and as some of these institutes have assumed global and others only regional responsibilities for particular crops, significant actual or potential conflicts over the geographic scope of responsibility of each individual institute have developed.

Success in varietal experimentation has been accompanied by some disappointments. Either varieties did not spread as rapidly as had been hoped or very few farmers adopted a whole package of practices and consequently fell far below the biological potential of the varieties. As a result, the International Rice Research Institute (IRRI) and the International Maize and Wheat Improvement Center (CIMMYT) began to be more concerned about the quality of national production programs. They and the TAC seem to be ambivalent about this. In some statements, the three say that the institutes cannot become responsible for national programs and their successes or failures; in others, the institutes defend and expand their outreach activities to those programs.

A different development that tends to push institutes into a closer relationship with national research programs is the increasing emphasis on farming systems. It may be good research strategy to begin by concentrating on one or a few crops, considered separately, but sooner or later one must become concerned about how they fit into farming systems. When that time comes – as it did several years ago for IRRI and CIMMYT – the respective roles of each institute and their criteria for assessing their accomplishments need to be reconsidered.

As they had some initial successes in crop technology, the institutes generated high credibility even in agricultural fields significantly removed from their primary competence. This has led to the problem Hanson mentions of the danger that members of the institutes' staffs may pontificate beyond their competence. It has led also to an increasing number of projects of bilateral assistance being channeled through one or another international institute for administrative and technical backstopping.

Meanwhile, elsewhere on the international scene, new agencies such as the Southeast Asia Research Council in Agriculture (SEARCA) are arising with ambitions to share the field, even in agricultural research. Should they be encouraged? SEARCA is an example of a type of agency which, unlike the international institutes, is partly financed by developing countries themselves. Like the international institutes it now seeks additional financing from international and bilateral aid agencies and from foundations. Other similar agencies are likely to develop. Must the situation become increasingly competitive, or are there ways in which these various efforts can become truly complementary?

Finally, several private corporations, selling farm inputs, have gone international in their operations. They inevitably become involved in technology transfer. Recognizing the necessity of modifying their products to fit different resource endowments, they have become involved also in technology development. By the nature of their operations they are international agencies.

Issues for the International Institutes

Quite a number of issues for the international institutes emerge from the composite picture of them and their programs sketched briefly above. Six such issues which come rapidly to mind are the following:

1. How much responsibility should the international institutes assume with respect to strengthening national research systems?

2. How should relationships between two institutes which are engaged in outreach programs regarding the same crop be organized?

3. Should the institutes undertake to develop administrative talent for national research programs? For national production programs?

4. How should institutes handle the imputing to them of expertise outside their competence?

5. What should the relationship be between the institutes and organizations like SEARCA?

6. Is there a basis for a natural division of labor between governmental and private sector research? If so, would the same division hold for technology development and transfer?

Rather than attempt to catalog all such questions, however, I wish to single out four particular issues that appear to me to be crucial.

1. *Which is more central to the successes to date of the international institutes, especially those of CIMMYT and IRRI: Is it the fact that they have concentrated largely on research and training or is it their unique type of organization and method of support?*

I doubt this question can be answered definitively, but concentrating our attention on it may shed some light on what role the institutes should be asked to play in the future.

Surely, in the beginning, the concern of those who founded IRRI and CIMMYT was to accomplish high-quality research on selected food crops. The form of organization developed for the institutes was an instrumental means to that end.

Now, however, with several similar institutes well established, with their own boards of trustees, international staffing and salary patterns, and institutionalized pipelines to the financial resources of members of the CGIAR, the institutes, because of the form of operation that was created for them, have become an institutional resource that might be useful for activities other than research and related training.

At the same time, there are reasons to believe that the marriage of concentration on research and the international institute format is particularly harmonious and productive. Adequately funding and organizing national research programs in developing countries has been and remains a difficult task. The payoff is usually too long range for politicians subject to frequent reelection to be very interested in such a task. The urge to keep salaries for various governmental and university positions relatively uniform (but lower than those for similarly trained persons in the private sector or in major administrative posts) has prevented recruiting and retaining competent staff. The tradition of frequent transfers of governmental personnel in and out of research is another restraint.

Observing that essential activities of agricultural development other than research need similar release from the exigencies of national administration, there is a tendency to conclude that they also should become activities of the present, or new, international institutes. But would that conclusion be sound? Biological and engineering research, once established behind stout fences and in laboratories with adequate equipment, can go its merry way, almost regardless of the political weather around it. Economic and social studies can venture into the hinterland to collect data provided they are circumspect about topics and diplomatic in their approaches. But the closer one comes to the heart of the matter – widespread agricultural development – by engaging a number of essential issues other than research, such as price policies, land reform, tractorization, and credit policy and practice, the more stormy the political weather becomes.

Does the international institute format have certain advantages for these other agricultural development activities? Or are other types of organization equally appropriate? Or is the international institute format a positive liability here?

2. *Is achieving a minimum critical mass of research talent and other re-*

sources an important issue for the international institutes from now on, or should we be paying more attention to achieving an optimum size of staff for each institute?

Crawford poses this question in chapter 11, when he speaks of achieving a "critical mass . . . of very high skills in research" as a major element in the CGIAR decision to establish and expand support of the international institutes. However, when one looks at their recent histories one wonders how stable in size such a critical mass within the international institutes really is. CIMMYT and IRRI were much smaller when they made their major breakthroughs than they are now. They keep growing. Is the question really one of minimum critical mass, or should we now be thinking more about an optimum size of individual organization: not too small and not too big? not too specialized and not too general?

This question becomes increasingly important as one notes the rise of organizations like SEARCA and the universally recognized need to strengthen national research capacity. The international institutes have their established pipeline for resources through the CGIAR. Other international and national agencies do not have that resource but are expected to go directly to individual funding agencies. Despite frequent statements by CGIAR, TAC, and the international institutes themselves about the importance of these other types of research agencies, the very existence of the international institutes with their built-in organizational advantages would appear to give them a funding advantage over the others.

Each institute will grow as much as it is allowed to. Should there be no restraints? Are there no inherent diseconomies of scale for them – not just when considered alone but taking into account the opportunity costs consisting of the limitations that their growth imposes on financial support to other international and national research and action programs?

3. *What are the valid criteria for judging the success or failure of the research activities of the international institutes?*

It has been stated repeatedly that increasing national crop yields is the criterion by which one must judge the success or failure of each international institute. At the same time, however, several statements have been made about concentrating on increasing the productivity of farming systems rather than just the production per season of particular crops.

There are at least two shortcomings in the use of average national crop yields as a criterion in judging the performance of the research programs of international institutes.

One is that it almost forces each institute into programs related to other essential elements of agricultural development. If experimental results at an institute show a yield of X bushels per acre achieved by a given variety with a given combination of associated cultural practices, but the national average yield refuses to go above X/4 and the institute is to be judged by the national average yield, what is it to do? The obvious answer is to broaden its program to include price policies, credit administration, the extension service, managing irrigation systems, or some other policy or program affecting average yields. But to assume one or more of these responsibilities takes the institute outside its relative immunity to political winds and places it more at the mercy of national factors beyond its control. And it may not be able to be effective in one or more of these other fields even apart from the political weather.

The other shortcoming in using national average yields as the criterion for success or failure of the research program of international institutes is that it ignores the greater importance of the productivity of entire farming systems. In considering the latter it is the *value added per acre per year*, by whatever the optimum combination of crops and cultural practices may be, that is important rather than either national aggregate production or national average yield per acre per season of any particular crop. An increase in such value added per acre per year, and even in the total availability of human food produced domestically, may be achieved in ways which include an absolute decline in national aggregate production of one or more crops with respect to which an international institute has a mandate. It may also involve being satisfied with lower crop yields per season whenever it is possible, by assuming that cost, to increase the productivity of an entire farming system.

In other words, a drop either in aggregate national production or in the average crop yield per season of a particular crop may be consistent with an international institute's having done its research job well with respect to that crop.

4. *What should be the future role and responsibilities of the international institutes?*

It seems quite clear that, in the beginning, IRRI and CIMMYT were intended to be almost exclusively research institutes. Their training programs were designed to train research workers only, and their outreach programs were, primarily at least, to provide testing facilities in various countries. Over the years, however, their programs have grown not only in size but in types of activities. Statements of their role and responsibilities have become more and more varied. Such a trend is not necessarily undesirable. It may well be that there should be variety among them, and even competition.

Nevertheless, it seems to me that more attention needs to be paid to the question of what the role and responsibilities of the institutes, individually and (perhaps not so certainly) collectively, ought to be in the future, taking into account the changes, from year to year, in the circumstances under which they operate.

At least three possible future roles for the international institutes can be discerned.

Option 1. Clearly, one alternative would be for some or all of the international institutes to retain or regain the character of the exclusively research institute. This would be consistent with the initial judgment that a basic need for agricultural development is improved technology and that the facilities of an international institute, unhampered by national administrative obstacles or by other responsibilities, are essential to efficient pursuit of that goal. If this were to be done, would it not be necessary to change the criteria by which the institutes are judged? No longer would what happens to average yields or aggregate production in individual countries be considered to be the responsibility of the institutes, but only the potential performance (taking costs and returns into account) of the technologies they come up with, for different agroclimatic zones. Outreach programs would remain but only in cooperation with national *research* programs, both to strengthen them and to test technologies under a wide variety of local settings. Training might remain, but only training of research workers (and perhaps of research administrators). This option would rule out the administration by institutes of projects supported by bilateral assistance agencies to strengthen any other than national *research* systems.

Option 2. A second possibility would be to accept an evolving nature for each international institute as a welcome and fruitful development but only after a substantial breakthrough in production technology has been achieved for at least one major crop with which it deals. Choosing this option would fit my interpretation of what Hanson says, i.e., that this is already happening to CIMMYT. Is it happening to IRRI as well? Research would continue to be a central activity of each institute, but to it would be added outreach programs directed to not only national research programs but national production programs as well. Instead of being a research institute, each would become a *research-based agricultural development institute*.

Under this option, a new and different kind of allocative problem, already present for CIMMYT and IRRI, would become increasingly acute. Of the resources available to an international institute, how much should it spend on research and how much on some of its other activities? Am I right that in the past, as between research and training, such decisions have been based (and perhaps rightly so) on judgments on how much training and of what types could effectively be carried on that would complement and strengthen, or at least not unduly hamper, a research program of given size? Certainly these decisions have not been based on a comprehensive inventory, much less on quantitative estimates of relative rates of return in comparison with similarly estimated rates of return on research projects. How should they be made?

If this second option were chosen, what actually happens to national productivity in terms of value added per acre per year might remain as one measure of the institutes' performance, but not what happens to national average crop yields or aggregate physical production of particular crops. And additional evaluative criteria of various types would become important.

Option 3. A third possible role for the international institutes would be to become, more and more, *service centers to national programs*. This would entail less emphasis on the research they do themselves and more on helping national systems of research become more productive. There would be more emphasis on training (of present and other types), perhaps including training in the management of various types of agri-support activities, including but not limited to research programs. The institutes would continue to conduct symposia and to issue publications. They would provide sabbatical opportunities for research workers in a setting where sophisticated equipment and stimulating interchange with other professionals are available. Much, perhaps most, agricultural research (biological engineering at least) could probably be done with much less elaborate facilities than the institutes possess, but scattered around the world there need to be a number of places where scientists from various countries which do not have within their borders advanced agricultural research facilities can work together with other scientists with similar or complementary interests.

The present international institutes are not the only candidates to help fill this role, but they have the advantages (1) of having concentrated on crops that are important in the tropics; (2) of being strategically located in various tropical and semitropical agroclimatic zones; and (3) of having non-national forms of administration and fields of interest.

Evenson's discussion in chapter 8 is relevant to decisions with respect to the future role of the institutes. It seems to me that he is saying:

1. Investments in high-capacity conceptual-scientific skills, as represented in the early stages of IRRI and CIMMYT operations, are more productive than any other single type of investment in agricultural development activities.

2. There is constant pressure to divert organizations having those skills into less productive (but still essential) activities requiring only technical-scientific and/or technical-engineering skills.

3. With the emergence of the international institutes which can afford to be more resistant to the temptations to slip into less remunerative types of research the trend has been for these institutes to assume the role of "main" research stations of which national research programs serve as "branches."

If that is a correct interpretation, one would expect Evenson to recommend Option 1, i.e., to pull the institutes back to their initial concentration on research and away from involvement with national production programs. If that were done, then two related questions would become pertinent.

First, who is to assume support for the other types of research and for developing other agri-support services on which the translation of the research results of international institutes into higher value added per acre per year in individual countries depends? Can that question be answered in terms of comparative rates of return for individual activities? Or would it be necessary to resort to criteria of sequences and complementarities among activities regardless of individual rates of return? And are the activities requiring other than "conceptual-scientific" skills to be dependent on individual country financing, or should a portion of international financing be devoted to them?

Second, present judgments about IRRI and CIMMYT rest on past achievements. What do they now give promise of doing as an encore? Here Evenson turns somewhat pessimistic. He sees them as probably continuing to make contributions but of a much lower order of magnitude in the field of varietal development. If this is so, may there not be a time in the life of each institute when it should move to Option 2, into more general programs as research-based institutes of agricultural development? Or should the institutes skip that option and move directly to Option 3? In either case, one result would be to face quite a different set of allocative problems and a greatly increased need for effective cooperation with the activities of other international and national agencies.

I suggest the hypothesis that, for the foreseeable future, the unique strength of the international institutes will be a combination of the way in which they are organized, the manner in which they are operated, and the sources of their financial support. That strength they could retain no matter what the nature of their activities might be. The crucial question, then, would be: *within the broad spectrum of essential agricultural development activities, what should institutes of this particular kind be doing?*

If the decision is that they should concentrate solely on certain types of *research*, then the techniques being discussed for choosing among research projects are central to the efficient operation of the institutes. If, however, the decision is that, in addition to research, they should engage in selected *other types of activity*, including cooperation with national production programs, then techniques for choosing among research projects become relatively less important (though still important in an absolute sense). In the latter case, we would face the problem of searching for other decision-making techniques with respect to the broader spectrum of activities for which the international institutes have, or might be given, responsibility.

28

Uneven Prospects for Gains from Agricultural Research Related to Economic Policy

Theodore W. Schultz

The new international network of agricultural research activities which has been developed during the last three decades serves many countries. Although it is international in scope, it was not launched by governments. It began as a venture of a rare breed of research entrepreneurs who knew from experience the requirements of the agricultural sciences. At the outset it was financed by private foundations. It now consists of a wide array of interrelated research enterprises located in various low-income countries throughout the world. It is clear that the function of this network of research activities is to employ the knowledge and talents of agricultural scientists as a means of increasing the productivity of agriculture in low-income countries.

This international approach to agricultural research is now well established. It is robust. It is capable of dealing with changing circumstances. It is successful in taking advantage of new opportunities. There are many low-income countries that could have benefited from this research but have failed to do so. Others have benefited somewhat but much less than they could have. Among these are countries that have the natural endowment and the potential economic capacity to increase greatly their agricultural production. We have not, in my view, given adequate attention to the factors that account for these failures. It is my contention that, unless these factors are altered for the better, the uneven prospects for gains from agricultural research that are related to the economic policy of the respective countries will continue to thwart the potential success of this international approach to agricultural research.

I shall begin with a comment on aspects of the antiscience "movements" with special reference to the now popular misconceptions and criticisms of agricultural scientists that prevail despite the recent acute shortages of food in parts of the world. I shall then concentrate my remarks on economic issues pertaining to the future value of agricultural research, with special references to economic policies that account for the uneven prospects already referred to.

The Utility of Science Is Being Debased

The role of agricultural research in maintaining our faith in the utility and legitimacy of the sciences is stated imaginatively by Andre and Jean Mayer in a recent essay:

> Few scientists think of agriculture as the chief, or the model science. Many, indeed, do not consider it a science at all. Yet it was the first science – the mother of all sciences; it remains the science which makes human life possible; and it may well be that, before the century is over, the success or failure of Science as a whole will be judged by the success or failure of agriculture.[1]

If agriculture is the mother of sciences, motherhood is being treated rather shabbily these days. Be it the current food deficit in India or the approaching doomsday, there is a lot of rhetoric proclaiming that agricultural scientists are to blame. The effects of science on agriculture are deemed to be bad; it follows, of course, that agricultural scientists are responsible for the inordinate appetite of modern agriculture for energy and for the chemicals that pollute our soil and contaminate our food supply. It is also being said that the scientists are making agricultural production more vulnerable to changes in weather and that they cause much unemployment. In the United States during the sixties, agricultural scientists were blamed for the then mounting agricultural surpluses. Some critics even proclaim that the green revolution only confounds the excess population growth, and worst of all it is popular to say that it is unjust to poor people. Wade has recently tried to present a balanced view, but he leaves his readers with a very dismal picture of the green revolution.[2]

Agricultural scientists have good reasons for feeling ambivalent about the treatment they receive. Although our major private foundations have been generous and successful in building international agricultural research centers, United States aid in support of agricultural scientists over the years, beginning with Point Four, has been much more uneven and undependable than the

Canadian performance through the Canadian International Development Agency (CIDA) and the International Development Research Centre (IDRC). Our prestigious National Academy of Sciences, in its foreign activities department, has failed to understand the role and requirements of agricultural scientists and, in its domestic committee, has tended to look upon agricultural scientists as a lowly breed who are in general out of touch with the real physical and life sciences. The environmental movement has added to the scientists' woes. And agricultural scientists also are stung by the statements of those economists who claim that agricultural scientists are contributing to the inequality in the personal distribution of income inasmuch as they have failed to develop types of wheat, rice, and other crops that would solve the equity problems in low-income countries. I shall return to this issue.

The view that I shall take rests on two propositions: advances in the agricultural sciences are one of the necessary conditions for agriculture to succeed as the primary supplier of food; and optimum economic incentives for agriculture, i.e., for farmers, are one of the necessary conditions in determining the full possibilities for useful advances in the agricultural sciences. I shall contend that we have not given sufficient attention to the second of these two propositions. In support of my view, I shall comment briefly on parts of the papers presented at the Airlie House conference and then turn to a consideration of an economic approach that is more general than has been our wont in analyzing the real economic value of agricultural research.

The Conference Papers

In line with the plan of the Airlie House conference, the papers that were at hand before we met fell into three groups: the organization, funding, and the decision-making process of the international agricultural research activities; the economic studies of the production effects of this research during the recent past in particular countries, including estimates of costs and returns; and a set of economic models that may prove useful in undertaking future research. Judging from the quality of discussion and its fruitfulness, we should give a high mark to the plan and the papers.

Organization, Funding, and Decision-Making

The papers devoted to these topics give a rich account of what has been learned from experience. We see this result clearly as an international enterprise from Crawford's advantageous position (see chapters 11 and 29). Wortman stresses that the critical decisions come down to "considered judgments" (see chapter 14); I strongly concur. These decisions entail risk and uncertainty and other considerations that are beyond precise measurement. Each of the papers in this set adds appreciably to our insights on how this research complex is operated, how it is evaluated, and how it is adjusted to changing perceived opportunities. Three things, however, are missing.

First, no doubt out of modesty, the authors are reluctant to discuss the very important function of research entrepreneurship in the success of this enterprise. I would like to know what are the essential qualifications of research entrepreneurs. It is more than a talent for management and administration. It is undoubtedly a rare talent. There is in this area a lot of experience to draw upon. The generation of research entrepreneurs who successfully launched the new institutes will soon be replaced by a new generation. Are they being recruited, acquiring experience, and being tested? I see no evidence that this is occurring. Our discussion convinced me that the functions of the research entrepreneur are not adequately understood and that the value of their contributions to the success of this research complex is vastly underrated.

Second, there is no systematic treatment of introducing and taking advantage of *market competition* in the organization of this activity. On the contrary, the appeal by some is to ever more complex systems of communication to provide information and to integrate the decision-making process. The implied assumption of this approach is that the market is at no point as efficient in providing such information as is an extensive nonmarket communication system. This is, in my view, a false assumption. There are many points at which market competition is far more efficient in this respect. We see it in the production and distribution of high-yielding food and feed grains. Private firms, subject to competition, have demonstrated their comparative advantage. Sehgal's analysis in chapter 19 bears on this point. Once we begin to look for ways of taking advantage of market competition in shaping the organization of this research, we will find that it can serve us efficiently at many points. One further remark regarding competition: I have on other occasions expressed concern about the increasing centralization in the allocation of resources to agricultural research. It implies less competition, it assumes general indivisibility among research activities, and it constrains the research choices of individual agricultural scientists. I confess I am biased in favor of competition. What would the considered judgments of those who have inside experience tell us about the role and limits of competition in this area?

The third missing element pertains to the inherent difference in research possibilities among various crops in both the short and the long run. A good deal must be known from experience and from the state of our scientific knowledge. There are no doubt many variables that must be reckoned with in arriving at considered judgments. What are these variables and to what extent is each subject to uncertainty? What weight do research entrepreneurs give to the anticipated research effort that will be required, the time it may take, the yield effects, and the risk and uncertainty that are involved? Economists are woefully uninformed with respect to the research alternatives that face agricultural scientists and research entrepreneurs.

Production Effects of Research

To see the second set of papers as I do, the choice of language and concepts is important. In the allocation of resources, agricultural research is fundamentally an investment in future returns. It obviously is not a leisure activity to please and to enhance the welfare of agricultural scientists. The concept of research capital formation is useful in guiding our economic thinking. It leads us to look at the different forms of such capital, e.g., banks of genetic materials, laboratory facilities, experimental fields, and various mixes of scientific skills. I am attracted to Wortman's definition of "scholarly capital" (see chapter 14); it is succinct, cogent, and useful. Each form of research capital has its own rate of depreciation and, when the genetic yield reaches its maximum, all further research would consist of maintenance research. In the production of research capital, the concept of scale and the optimum combination of research inputs serve to guide us in distinguishing between less and more efficient organizations. The concept of a general economic equilibrium is always, as it should be, in the back of our minds, although the observable behavior of people, including research entrepreneurs, reveals their perceptions, decisions, and actions to regain equilibrium. It is my contention that we grossly neglect the extension of theory to analyze these equilibrating processes. Factor prices obviously matter, but so do farm product prices. Although farm product prices are in serious disarray throughout the world, the papers in this group are virtually silent on the implications of these price distortions for the optimum contribution of agricultural research.

The empirical studies that trace the diffusion of and that estimate the research costs and returns from the new high-yielding grain varieties in selected low-income countries represent a major contribution. They tell a consistent story, i.e., the realized rate of return relative to the costs of the research is much higher than the rate of return from most alternative investment opportunities in these countries. The results from Dalrymple's approach (see chapter 7) are for all practical purposes the same as those from Evenson's production function approach (chapter 9). Although data limitations abound and specification biases are ever present, the results are so robust that they must be taken seriously. The endeavor by Evenson to distinguish between applied and scientific research in this context is sufficiently important and promising that it should be placed high on our economic research agenda. With respect to developments in Colombia, the study by Hertford and his associates (chapter 4) adds two new insights: as a consequence of the strong adverse effects of PL 480 wheat on the farm price of wheat in Colombia, the payoff from the wheat research in Colombia has been very low;[3] and in the case of cotton, yields nearly quadrupled mainly since 1958 in response to the favorable price effects resulting from the exchange rate reforms in that year, and Colombia became an exporter of substantial amounts of cotton. But in this particular success story, "no significant, positive benefits were derived from the Colombian cotton research program." I venture that a careful look at the evidence pertaining to Mexico's success in cotton would reveal a similar story. Analytically, what is noteworthy in Hertford et al.'s study is that they took account of both wheat and cotton farm prices. The study is not limited to changes in factor prices.

The two papers on Brazil come closest to dealing with the dynamics of economic development, including the reduction in barriers to the export of farm products and with the resulting favorable farm price effects. The implications for agricultural research are strongly positive. According to Pastore and Alves (chapter 18), one of the main purposes of the recent large increases in federal support for agricultural research is "to gain sizable slices of the international market." The approach of de Castro and Schuh (chapter 24) is useful in interpreting the Brazilian experience.

Economic Models

There are a few papers that consist of models with no empirical analysis to show that they may be useful. I take a jaundiced view of such papers. It became evident in the conference discussion, however, that the authors had applied them. I hold the view that no paper should be published unless the author has used his model in empirical analysis. Our economic literature is plagued with an abundance of unused (useless?) models.

I am puzzled by the fact that these three groups of papers are mute on the economic policies of those low-income countries that thwart the gains to be had from agricultural research. I know that international research entrepreneurs are reluctant to criticize the policies of governments openly. It is obviously a very sensitive issue. But the issue must be faced if the production of food is to be increased, as it can be, to serve the needs of the people in these countries. The fact that the papers by economists say so little bearing on this issue is odd indeed.

Economic Policy Matters

We have not allowed ourselves to see the effects that governmental policies have on the economic value of agricultural research. If we did, as we should have, we would find that in many low-income countries these effects are highly adverse to the attainment of the optimum benefits from both international and national agricultural research enterprises. We find it convenient not to see these consequences because the economic policy of any government, other than our own, is a very sensitive issue. We play it safe by proclaiming that it is improper for us to criticize such policies. This uncritical view of what sovereign countries do leads us into a serious inconsistency. Our endeavor to help solve the food problem in these countries tends to be, under such circumstances, inconsistent with their sovereign behavior. In my view, a large part of the poor performance of agriculture in many low-income countries is a consequence of bad economic policies. In this context, it is a mistake on our part not to take a carefully considered, critical view of such policies, granted that in doing so we would enter upon some very sensitive issues pertaining to sovereignty.

I shall restrict my remaining remarks to three such issues: the gains that consumers derive from agricultural research; the effects of such research on the personal distribution of income; and the adverse effects of farm product price distortions on the optimum contribution of agricultural research.

Consumer Gains from Agricultural Research

We are remiss in not showing the extent to which consumers gain from agricultural research. Merely to say that there may be a consumer surplus is not sufficient. To contend, as we do, that farmers as producers are the real beneficiaries of agricultural research is far from true. Farm families as consumers may gain and, under special circumstances, would derive the full gain. Suppose that the costs of producing rice were reduced significantly by means of new variety, and suppose further that the small rice farmers who adopt this variety are wholly self-sufficient in what they produce and consume. In this special case, all the benefits would accrue to the farm family as consumer gains. The real income of the family would be increased. The transfer of these gains from a reduction in real costs is much more complex in a market economy under competition. The event that leads to a reduction in costs in the first instance creates a disequilibrium. Farmers make adjustments to regain equilibrium. The rate at which and the extent to which the gains are transferred to consumers depend on the elasticity of demand and the shift in demand during the full adjustment period. If all the product were consumed within the country, as a first approximation, the gains would be transferred to domestic consumers (including, of course, farm families). If all the product were exported, it would be the consumers in the importing countries who would gain and the extent to which they did would depend upon the effects that the additional exports of the country have on the world price. Although there are various possibilities, general economic theory tells us that the gains in agricultural productivity derived from research are, in general, transferred over time through competition to consumers. Farmers cannot hold on to these gains for long, although those who are among the first to reduce their costs profit for a time and until competition has had its say. Farmers, of course, benefit in their capacity as consumers.

We know from Peterson's research that the marked decline in the price of meat from poultry was in considerable part a consequence of agricultural research.[4] Between 1910-14 and the 1971 crop year, "the real price of food grains declined by 37 percent."[5] I wish we knew how much of this consumer gain was derived from agricultural research. Before World War I, the price of a ton of wheat was substantially higher than that of rice. Since the mid fifties, however, the price of wheat has in general been about one-half that of rice. My guess is that agricultural research over this period was much more effective in reducing the real cost of producing wheat in the world than it was in the case of rice. How much is to be credited to research and how the many equilibrating processes in agriculture have performed is still to be determined. There is, however, the reduction in the costs of producing poultry, wheat, and many other farm products. This being true, the possibilities of reducing these costs further in the future should enter as a basic consideration in making economic policy.

Effects of Research on Personal Income Distribution

Although we know all too little about the factors that alter the personal distribution of income, we nevertheless know enough about the income distribution effects that are associated with the history of agricultural modernization to correct some of the widely held misconceptions about this issue. I shall restrict my remarks to consumers, landlords, and farmers.

The primary research effects on income inequality occur as a consequence of the gains in agricultural productivity that are transferred to *consumers*. Any reduction in the real costs of producing farm products benefits the consumers. The real income of low-income families is increased thereby, and relatively more than the real income of high-income families. We know that a sales tax on food is regressive. By the same economic logic, the research effects here under consideration are progressive in the way the benefits are distributed among poor and rich families. Lower farm-food costs, therefore, are important in reducing the inequality in personal income. There is much evidence which shows that the primary accumulative effect of agricultural modernization, including agricultural research, has not been unjust to poor people; on the contrary, it improves their lot more than it has that of the rich. We owe our agricultural scientists a great deal in this connection.

Another income distribution effect occurs as a consequence of changes in factor shares. It is widely observed that the share of income of farm *landlords* declines as the modernization of agriculture proceeds. It is my contention that agricultural research accounts for a considerable part of this decline in the economic importance of farm land as a factor in production, measured in terms of the fraction of the income that accrues as land rent. We fail to see this process because our analytical approaches rest on Hicks's Law. No one has been more cogent than Pen in showing the limitations of Hicks's Law in analyzing the historical decline in farm land rents as a share of income.[6] Some of the credit for the fading away of this landlord class belongs to our agricultural scientists, because they have come up with substitutes for farm land and because they have increased the human capital that is required.

The effects of agricultural research on the distribution of personal income among *farmers* is an issue that is dominated by confusion. Farm families may gain as consumers, and in this context, as Hayami and Herdt have shown, small rice farmers tend to benefit more than the large rice farmers.[7] To the extent that the research provides a substitute for farm land and to the extent that farmers are owners of farm land, the rent that accrues to them will tend to decline over time. As modern farming becomes more complex and as the adoption of new research results becomes more important in farming, the human capital of farmers becomes increasingly important. Under such dynamic conditions, the better educated farmers, in terms of their acquired human capital, win out under competition.[8]

Contrary to much that has been said about the bad effects of the high-yielding wheat varieties on the distribution of personal income in Indian agriculture, there are now two studies in depth that show that in fact the income inequality has been somewhat reduced.[9] These effects could not have been predicted; we had to wait for the evidence to find out. Under other economic conditions, the results could be otherwise. Important as it is that economic policies not bypass and not discriminate against small farmers, agricultural scientists who are endeavoring to develop more "efficient" plants (and animals, too) in terms of their genetic capacities and chemists who are engaged in developing cheaper and better chemicals should *not* be placed under the constraint that the fruits of their research be applicable only to small farms.

Effects of Price Distortions on the Contribution of Agricultural Research

Last and most important on my agenda is the pricing problem. In my view, the distortions in farm product prices greatly reduce the potential economic value of agricultural research. But we have not come to grips with this problem, except for (a) the distortion in wheat prices in Colombia and its adverse effects on wheat research and (b) the awareness of the importance of export prices in Brazil. The countries of Western Europe and Japan overprice farm products, and as a consequence the economic value of research is thereby overvalued. The high internal price of rice in Japan and of wheat in France is a misleading indicator of the real value that agricultural research adds to income. The success of research in Japan must be discounted substantially for this reason. The effects of overpricing in Western Europe are much more serious in the sense that they reduce the farm price incentives in some of the low-income countries.

Our primary concern, however, is to improve the organization of and to increase the investment in agricultural research as a means of increasing the supply of food in low-income countries. But the persistent, serious, underpricing of food and feed grains in most of these countries greatly reduces the potential possibilities of achieving this objective. None of the recent rhetoric on shortages of food grains has dealt with this basic issue, namely, that the cheap food policies of many of these countries not only hampers, but keeps agriculture far below its optimum production. In my view, *it is the task of outside economists to provide the analysis that will determine the extent to which low-income countries are themselves responsible for their shortages of food.* It is this implied interaction between agricultural research and production that we did not face throughout the Airlie House conference. To take the internal prices of these countries as given is to dodge our analytical responsibility.

Hard as it is to ascertain the real farm product prices in these countries, the prices that I have at hand tend to support the following economic inferences.[10]

First, the state monopoly marketing boards throughout West and East Africa have been and are pricing their best farm crops to death. By "best," I mean the crops in which these countries have a real comparative advantage. If these marketing boards had been established for the sole purpose of reducing the economic incentives to produce these crops, it would be difficult to see how they could have accomplished that purpose more successfully. To give an example, while I was in Senegal recently, I observed that the farm price for groundnuts was decidedly less than one-half of the real export price. The marketing board was collecting revenue and the farmers were being discouraged further by the marked rise in the price of fertilizer. The implications for research on these key crops in these parts of Africa are obvious.

Second, the economic potential of the Argentine as a major supplier of wheat and corn has been greatly impaired by the underpricing of these crops within that country. The farm price incentives have long been one-third and more below the real value of these crops. Is it any wonder that yields have benefited so little from wheat and corn research?

Third, in the production of rice, I see Thailand and Burma as having a very large unrealized economic potential. But it is not forthcoming. The economy of Burma is in serious disarray. In Thailand, the politically convenient export tax on rice has long been the means to keep the price that rice producers receive a third to two-fifths below its real economic value. The new high-yielding dwarf rices and their costly complement, i.e., investments to control the water, continue to remain uneconomic. In contrast, the pricing, the use made of research, and the impressive gains in the production of corn in Thailand reflect a story of real economic success.

Finally, another view of farm price distortions is evident when one looks at the prevailing internal relative prices. I take as my standard that the real economic value of rice is about twice that of wheat per ton and that of corn somewhat less than for wheat. In India, the farm harvest price of rice has tended to be somewhat below that of wheat. The implication seems obvious: if the farm price of rice were twice as high as that of wheat, the incentive to modernize rice production would turn sharply in favor of rice. To return to Thailand, the wholesale price of rice in Bangkok has been virtually the same as that for corn. In seven of the years between 1960 and 1973, the price of corn was actually higher. Yet the real economic value of rice is more than twice that of corn; whereas the price of corn is in line with its real value, rice, the key crop, is vastly underpriced.

Summary

We have learned a lot about the organization of agricultural research. It is true that the basic decisions come down to "considered judgments." It is also true that research entrepreneurs and competition are important, but they were omitted in the several useful papers on organization. We now have robust evidence that the payoff on investment in agricultural research has been very high. But there is more to be said on the interactions between agricultural research and the economy. The gains from research over time benefit primarily consumers. As consumers benefit from lower costs of farm foods, low-income families gain relatively more than high-income families, and to this extent the inequality in income is reduced. The economic importance of the landowner class declines in part as a consequence of the contributions of agricultural scientists. Within agriculture, the income distribution effects in low-income countries are not necessarily adverse to small farmers. From a policy point of view, however, what is important is that there are appropriate economic policies for reducing the inequality among farm families; it is a mistake to burden agricultural scientists with this particular inequality problem for the reasons I have presented. The neglect of the farm product pricing problem between and within countries is, in my view, the most serious omission. The existing distortions in these prices greatly reduces the potential economic value of agricultural research. I realize that it is a very sensitive problem, in all probability too sensitive an issue for those who are *inside* to enter upon. Nevertheless, solutions must be found if in fact many of the low-income countries are to succeed in producing enough food to satisfy their demand.

NOTES

1. Andre and Jean Mayer, "Agriculture: The Island Empire," in *Science and Its Public: The Changing Relationship*, summer 1974 issue of *Daedalus*, pp. 83-95. See also the excellent essay by Edward Shils, "Faith, Utility and Legitimacy of Science," in the same issue of *Daedalus*.

2. Nicholas Wade, "Green Revolution (I)" and "Green Revolution (II)," *Science*, 186 (December 20, 1974), 1093-1096, and (December 27, 1974), 1186-1192.

3. See also the study by Leonard Dudley and Roger J. Sandilands, "The Side Effects of Foreign Aid: The Case of Public Law 480 Wheat in Colombia," *Economic Development and Cultural Change*, 23 (January 1975), 325-336.

4. Willis Peterson, "Returns to Poultry Research in the United States," Ph.D. dissertation (Chicago: University of Chicago, 1966).

5. D. Gale Johnson, "World Food Problems," Agricultural Economics Research Paper no. 74:8, Department of Economics, University of Chicago, July 10, 1974.

6. Jan Pen, *Income Distribution: Facts, Theories, Policies*, trans. Trevor S. Preston (New York: Praeger, 1971). See especially pp. 208-214.

7. Yujiro Hayami and Robert W. Herdt, "The Impact of Technological Change in Subsistence Agriculture on Income Distribution" (Los Baños: International Rice Research Institute, 1974).

8. I deal with this issue in considerable detail in "The Value of the Ability to Deal with Disequilibria," Human Capital Paper, no. 74:1, revised, University of Chicago, December 20, 1974.

9. K. Satyanarayana and M. A. Muralidharan, "Impact of New Farm Technology on the Economy of Cultivator Households in Punjab," New Delhi, Indian Agricultural Research Institute, 1974; Katar Singh, "The Impact of New Agricultural Technology on Farm Income Distribution in the Aligarh District of Uttar Pradesh," *Indian Journal of Agricultural Economics*, 28:2 (April-June 1973).

10. Christine Collins of Economic Research Service, USDA, has been most generous of her time in helping me with data on prices. It would require a major paper to interpret fully the prices that I have from her along with those I have had at hand. She is, of course, not responsible for what I say in the paragraphs that follow.

29

The Future of the International System: A View from the Inside[1]

J. G. Crawford

I felt cheered by the importance so clearly attached by all speakers at the Airlie House conference to the growth of international agricultural research. True, at times I felt that some speakers too easily saw the problem of deciding the correct allocation of resources to research as a matter of mathematical and economic formulas which could determine the probable costs/benefits relationships before the event. There is no doubt that research has paid off and handsomely, but I doubt if the founders of the International Center for the Improvement of Maize and Wheat (CIMMYT) and the International Rice Research Institute (IRRI) could have made any very convincing calculations of "payoff" before the event. The truth is that those who invest in research – especially in international research of the kind we discussed at the conference – have little more than past experience and informed judgment to guide them.

This is not to disparage analysis of the economic and social implications of projected research if success be assumed. If judgment is the basis of the allocation decisions to be made, it will be the better for good analytical work, of which I fear there is not yet enough in the system. As Schultz has said (see chapter 28), the decisions being made in the international system of which the Technical Advisory Committee (TAC) is a part "entail risk and uncertainty and other considerations that are beyond precise measurement." Accordingly, "considered judgment" is still the basis of decision-making. Nevertheless, the more informed the consideration the better – especially as we approach a limit to finance available for further investments in new centers or for new lines of research to be organized in other ways than in new centers. In such conditions we are bound to try to project and compare the significance of the results of more than one way of spending limited resources.

Certainly I have to claim that TAC's advice to the Consultative Group for International Agricultural Research (CGIAR) has been well considered but, alas, it has not always been possible to offer the refinement of analysis which might have been possible under conditions of less pressure to act fast. TAC was four years old in June 1975, and, as I have indicated in chapter 11, much has happened. In monetary terms the expansion in CGIAR funding from, say, $10 million in 1971 to about $45 million in 1975 is evidence of rapid action if nothing else. We cannot guarantee that all the new ventures (or expansions of old ones) will succeed as hoped. Nevertheless, our judgments have been made in relation to priorities consciously adopted by TAC with CGIAR endorsement. These priorities need review from time to time, for they will play an increasingly important role if, as must be expected, the rising curve of funds available for expansion of effort begins to flatten out.

In developing my view of the future I will address myself briefly to six headings: "A Unique System"; "The Approach So Far Developed"; "What Do We Expect of the Centers?" "Are Centers the Only Approach?" "The Link with National Research"; and "Evaluating International Research."

A Unique System

My background paper (chapter 11) does, I hope, bring out the uniqueness of the system now developed. In a paragraph, it can be said that those who came together in the CGIAR believed strongly that international agricultural research could contribute to the expansion of agricultural output in developing countries. This expansion was (and is) seen as necessary for higher food production and for stimulating general economic development and employment via higher farm incomes. In this belief in research the members were encouraged by the results already apparent in respect of IRRI and CIMMYT. It was especially felt that, without decrying the importance of national research (of which more later), more extensive *international* research could lead to the development of new technologies without which farm incomes could rise but little if at all, especially in those nations of the world where population was already pressing heavily on limited land resources.

The members (governmental and nongovernmental alike) were limited in the belief that a concentration, in international centers, of very high skills in research could be *one* way of achieving widely applicable results. It was not thought to be the only way—an important fact I discuss later in the fourth section of this chapter. While the early attention of the CGIAR and of TAC (as its adviser) was in fact given to the existing and new centers, TAC, which had ample scope for initiative, was encouraged to think in different terms if it wished.

The Approach So Far Developed

The thinking of TAC on priorities is substantially reflected in my background paper (chapter 11). Food was given absolute priority in the sense that TAC expressed opposition to suggestions that it make recommendations in respect of nonfood crops. Within food, crop priority was given to food grains. It is fair to say that our concern was production, and we did not in our early work debate extensively the question of maximizing production versus optimizing farm income distribution, whatever that may mean. We did, however, keep before us the need to strive for technologies which were applicable to small-scale farming for the simple reason that many of the world's underfed people are poor, so-called subsistence farmers. Moreover, many of the farms supplying the market are small and lack adequate institutional support for their development. As our work evolved the recognition of these facts became stronger, and it has been reflected in our growing concern for a more effective farm application of research, whether international or national in character.

A few further related facts about our past work need to be mentioned to make easier any statement about the future. Let me outline these as follows:

1. Our past approach was to promote far more work through crops – rice, wheat, maize, potatoes, sorghum, millet, and certain legumes – and through animals for food. Our choice of products related to their essentiality to the mass of people in the very broad regions of developing countries which we chose to recognize.

2. Early on we were faced with strong pressure for separate research on factors of production: water use and management, fertilizers, pesticides, biological insect control, and so on. We preferred that these matters be dealt with in relation to the crop or animal under research at a center – to ensure if at all possible that an integrated package of technology would emerge from the work of the center. Some significant departures from this approach are under examination now, and the future may well see specific programs in fertilizers (see item 4 below), entomological work, and even in water management.

3. In fact the first apparent break from a strict crop or product-oriented approach came with the recognition by TAC of the importance of farming *systems* work. Nevertheless, this was bound to happen if our objective was to develop integrated and more productive technologies. A farmer's response to

the adoption of a new technology for a given crop is often influenced by how this fits into his overall system. Rotational livestock, pasture, and cropping systems and multiple cropping practices and possibilities (especially under irrigation) are reminders that the final objective is to increase output per unit area per unit of time. This is especially so when land, water, and inputs such as fertilizer are scarce. There is an increasing recognition of systems work in all centers. This does not mean that their research on the central crop or animal components of the system has been assigned lower priority. It does necessarily mean that staffs and budgets are increased – a matter that will call for care and attention by TAC and CGIAR.

4. The second break from the first approach may emerge from the study we have lately begun on fertilizers. TAC is now examining what research should be encouraged in (a) developing new chemical fertilizers for tropical conditions; (b) making more effective use of existing fertilizers, including organic manures; and (c) widening the scope for biological fixation of nitrogen. This last illustrates an important future role TAC may assume (see later discussion).

5. We have repeatedly looked at the water question, but it has proved difficult to find an effective way to advance research of a kind likely to be widely applicable. Nevertheless we may be able, building on work done in Israel and some other countries, to develop proposals especially relating to the conservation of water usage in arid land farming systems. Meanwhile it is important as a factor in existing centers and figures prominently, for example, in the work of the International Crops Research Institute for the Semi-Arid Tropics (ICRISAT) on dryland farming systems.

The five points above, so summarily presented, do represent significant shifts without abandonment of the basic role of crop improvement in terms of yield and quantity. Indeed, the shifts occur because of our central emphasis on securing a technology applicable in the farming areas of those developing countries heavily dependent on their agricultural systems for food and economic growth.

Whether our judgment is sound will ultimately have to be assessed post hoc. It is in this connection that I think both Wortman and Ulbricht are right. Our judgments are "considered" as Wortman puts it, if only because, as I understand Ulbricht, we see no alternative in the form of usable *ex-ante* calculations of cost/benefit ratios. Such calculations can be made but only on the basis of assumed results, obtained in a *given* time and applied with an assumed geographical coverage and assumed results in the form of yields per hectare. We do not make such calculations! Nevertheless, events and experience are forcing us to encourage more analytical work, more national institutional building, and more resources for investment in the extension of the use of research if the benefits, which can be confidently expected in the research centers, are to be reaped in the farmers' fields.

What Do We Expect of the Centers?

This question is raised sharply by the preceding paragraph. The answer, in my view, is very much in terms of Mosher's second option. This option has three components:

a. At the center, core research on one or more crops. (TAC prefers *few* (two or three) to too many.)

b. Training component at the center (see later remarks on national links).

c. Research relations with other research bodies — national, international, and regional systems. (Here there are limits in the managerial capacity at the center and the financial and skills capacity of the cooperators to support their end of agreed programs.)

From these three components we expect (1) that technologies will evolve to serve the farming systems in the areas being served by the research in question and capable of being tested in the field in many countries (and not merely at the research center); and (2) that technology will be related to the circumstances of the countries in need (i.e., to the structure of small farms, drylands or wet lands, etc.). We do *not* expect every particular outcome (e.g., a particular high-yielding variety) to be applicable to all situations. Hence our repeated stress on the links with national systems and the need to build up their capacity to undertake adaptive (as well as original) research.

Since we expect the centers to produce usable technology packages (subject to locational modifications), should they carry the *responsibility* for the adoption in each country for which they show promise? The answer is *no*. Centers can be criticized if their research results have requirements which are beyond any reasonable expectation of being adopted by small and medium-sized farmers, but if the final constraints on adoption lie only within the capacity of the national governments (with or without technical assistance from external sources) then the centers cannot carry the responsibility for any final failure. For, indeed, that failure may result simply from the inability or unwillingness of national governments to supply the inputs, price incentives, and credit facilities necessary if the farmers are to adopt new technologies which promise higher yields but call also for higher investment.

This view does not downgrade the contribution the centers can and do make through components (b) and (c) of Mosher's second option. Currently, it is clear that many smaller countries — in economic terms, weak in their agricultural productivity — are seeking more direct help from centers, even to the extent of seeking staff from the centers virtually to develop and manage their research cum extension effort. Fears of food shortages ventilated at the World Food Conference have no doubt stimulated these requests, especially from countries once self-sufficient in food but now facing rising import bills. I have in mind, for example, the clear interests many countries have in seeking direct help from CIMMYT in developing their maize and wheat production programs. This help appears to relate to national extension as much if not more than it does to national research for, in some cases, effective research exists alongside farms continuing low-yield traditional farming systems.

This need for stronger national systems of research and extension stands between the centers and the full application of their work. Until "the international system" can generate more national recognition of agricultural priorities within each country and increased international support for developing more efficient national systems, much of the dividend of international research will be lost. I will return to this matter later, but before doing so I wish to dispose of the notion (implicit in much of the conference discussion) that "international" research necessarily implies new centers whenever a felt need for additional research arises.

Are Centers the Only Approach?

I will not spend much time on this matter, but I believe my examples have important implications for the future. TAC believes that it is not always necessary or desirable to think in terms of new centers. I take three examples:

1. There is need for raising the productivity and extending the use of soya bean in developing countries. TAC believes the already outstanding work of the University of Illinois in this field should be supported to this end. TAC's "INTSOY" proposal would virtually make the university an international center behaving and governing in the manner of IRRI, CIMMYT, and the like. The proposal is still under examination, but the political difficulties (both university and international in character) of using a center in a developed country have yet to be overcome. I believe ways will be found and, once found, will open the way to a more effective use of established institutions in developed countries in the interests of developing countries.

2. The second example also relates to developed countries, but not exclusively so. Biological fixation of nitrogen is associated with the legumes. There are hopes that a capacity on the part of cereals might be developed, a long shot which is no longer regarded as hopeless by workers in countries like Britain, the United States, Australia, and India. The case for establishing a major new center for this work in a developing country is weak: the risks are too great at this stage. Moreover, it is unnecessary. TAC believes that basic work of this kind can rightly be stimulated by "the system" by supporting

periodic workshops to renew progress by the pioneers. TAC believes it would be right to support research if these workshops reveal prospects for breakthroughs which may later prove sufficiently promising to justify risk capital investment. That point may not be immediate, but, when it comes, support from an international system could be critical.

3. My third example is not dramatic, nor so far has it been remarkable for early success. The West African Rice Development Association (WARDA) is an effort to provide regional cooperation in developing rice varieties appropriate to the environments of member countries in West Africa. Such efforts need strong backing from the centers (IRRI and the International Institute of Tropical Agriculture (IITA), in this case) if they are to succeed. Again, in Latin America an effort to improve legume yields may be made by a regional cooperative program similarly coordinated and backstopped by the International Center for Tropical Agriculture (CIAT). These are early days for such experiments, but TAC is not and ought not be afraid to think in terms other than of international centers in the grand manner of their progenitors.

The Link with National Research

Again I return to the problem of the capacity of national systems to absorb and use the results of international research.[2] More and more is being said about the problem of linkage between the work of the international and national systems. From the beginning, the "Bellagio" meetings (which predated the establishment of the CGIAR and have continued since), TAC, and the CGIAR have strongly recognized the linkage in certain forms. Increasingly these forms are seen to be not enough.

There is no lack of support for units at the international research centers for training young research workers – a training which includes the development of understanding about technological packages required by high-yielding varieties or new multiple cropping systems. Again, through the work of both the scientists and the economists there is a good deal being learned about constraints on new technologies to be found in the field.

The collaborative programs between centers and national systems do produce two-way communications, or feedbacks, and we do wrong to underestimate their value. Yet many scientists feel frustrated when national farming systems do not reflect more quickly the new technologies they (the scientists) consider – and usually with good reason – to be available. Why?

The truth is, of course, that research can produce a useful technological package but that this requires (a) adaptation to often widely varying local conditions within national boundaries; (b) national policies of extension, supply of inputs, price policies, credit systems, and the like designed to enable wide adaptation within national units; and (c) a recognition of (a) and (b) by the national policy makers and administrators, if progress in overcoming severe economic constraints is to be made in accordance with what is often an urgent national need for faster agricultural growth.

It is asking the centers too much if we expect them to accept responsibility for (a), (b), and (c). They can and do assist significantly in (a) through their relations with national research systems; they may be drawn willy-nilly into some aspects of (b);[3] but they can hardly be expected to help much with (c). Who can? The TAC's answer so far has been the U.N. systems and especially the World Bank, FAO, and UNDP, together with the larger national aid bodies. In all these international agencies things are now moving faster, and practical policy is evolving. Nevertheless, I suspect that some new and faster techniques may need to be developed if the very large number of weak national agricultural systems are to be strengthened rapidly enough to profit from the work of international centers as well as from the stronger national systems available in some developing countries and many developed countries. A "Bellagio" conference held in Montreal in June 1975 recorded progress. TAC, in whose terms of reference the subject is included, is striving to develop workable ideas for consideration by the CGIAR. It is bound to do so in the effort to maximize the dividends expected to flow from the research of the many centers now supported by the CGIAR on TAC advice.

Evaluating International Research

This brings me to the final portion of my remarks, which relates to the evaluation of the work of the centers. I will report first a major decision of the CGIAR in this respect and then comment on the wider tasks of assessing the impact of the work of the centers as measured by the results in producing areas of the world.

For some time the various centers have had both inhouse and external reviews, especially in preparation of their developing programs. The CGIAR, with the concurrence of the centers, has asked TAC to undertake quinquennial reviews of the entire work of each center. The reviews are not seen as an accounting exercise nor as a means of grading individual scientists. Rather they are an assessment of programs under way, of the relative emphasis given to various objectives, and of their prospect for success, together with some attempt to weigh the pros and cons of possible new directions of effort.

The review teams will comprise both TAC members and outside invitees, selected after consultation with the center in question. The terms of reference will include not only the core program but also center-national linkages and intercenter linkages in the work of major "worldwide" programs.

As Wortman would agree, a final test is the impact on actual national programs – itself rather difficult for a quinquennial review mission except perhaps for those countries in which linked programs operate. For this reason it becomes important for TAC to develop or encourage others to make more systematic assessments of the impact of international research and linked national research on national production programs.

Dalrymple, Evenson, and others are doing valuable ex post facto studies of the kind needed. Nevertheless, some continuous monitoring system needs working out – either by TAC or, better, by TAC with the help of individuals and of bodies like FAO, the World Bank, and the new organization shortly to be established, the International Food Policy Research Institute (IFPRI). Some system of monitoring for results would also bring to light many problems and constraints capable of resolution by national and international effort.

In preparing (by experience) methodology for our quinquennial reviews and our monitoring of results "in the field," we may well meet some of the critical questions raised in the papers at the Airlie House conference.

I finish where I began. From the inside I see a unique system which has produced much and promises more. None of us inside is smug about it. This conference has sharply, fairly, and helpfully reminded us all of the questions that have not only to be asked but also to be answered. In many cases we know the problem but not yet the answers; but if I and my colleagues in TAC seek answers somewhat pragmatically I hope this will be understood. The many blueprints available in the papers in this volume will suggest ideas and approaches. In the end, however, judgments will have to be made by TAC and the CGIAR, not to mention the review missions in the field.

NOTES

1. These remarks do not represent a paper prepared before the Airlie House conference but rather one listener's reaction to the various issues presented by speakers in the main conference sessions. Reference to my paper provided as background material for the conference (chapter 11) will be a help in understanding some of the points made here. The word *inside* in the title for these remarks really refers to my position as chairman of the Technical Advisory Committee (TAC) which is, as the background paper explains, the principal adviser to the Consultative Group on International Agricultural Research (CGIAR) on the expansion of international agricultural research activities.

2. In the pressure of limited time to speak on many of the issues raised in the main sessions one inevitably telescopes words. Thus it needs always to be remembered that important parts of the results of international centers often derive from the two-way linkages in research which mark the relationship between the established centers and cooperating national centers.

3. Since these words were spoken, it has been clearer to me that many small nations are looking to the centers for help – especially in national demonstration programs. This is a difficult matter which needs early resolution in TAC and the CGIAR.

Indexes

Name Index

Note: a page number followed by "f" refers to a figure appearing on that page, one followed by "t" to a table, and one followed by "n" to a note.

Subject Index

Note: a page number followed by "f" refers to a figure appearing on that page, one followed by "t" to a table, and one followed by "n" to a note.